T0178497

Lecture Notes in Artificial Intelligence   **13633**

Subseries of Lecture Notes in Computer Science

Series Editors

Randy Goebel
University of Alberta, Edmonton, Canada

Wolfgang Wahlster
DFKI, Berlin, Germany

Zhi-Hua Zhou
Nanjing University, Nanjing, China

Founding Editor

Jörg Siekmann
DFKI and Saarland University, Saarbrücken, Germany

More information about this subseries at https://link.springer.com/bookseries/1244

JingTao Yao · Hamido Fujita · Xiaodong Yue ·
Duoqian Miao · Jerzy Grzymala-Busse ·
Fanzhang Li (Eds.)

Rough Sets

International Joint Conference, IJCRS 2022
Suzhou, China, November 11–14, 2022
Proceedings

 Springer

Editors
JingTao Yao
University of Regina
Regina, SK, Canada

Hamido Fujita
Iwate Prefectural University
Takizawa, Iwate, Japan

Xiaodong Yue
Shanghai University
Shanghai, China

Duoqian Miao 🆔
Tongji University
Shanghai, China

Jerzy Grzymala-Busse
University of Kansas
Lawrence, KS, USA

Fanzhang Li
Soochow University
Suzhou, Jiangsu, China

ISSN 0302-9743   ISSN 1611-3349 (electronic)
Lecture Notes in Artificial Intelligence
ISBN 978-3-031-21243-7   ISBN 978-3-031-21244-4 (eBook)
https://doi.org/10.1007/978-3-031-21244-4

LNCS Sublibrary: SL7 – Artificial Intelligence

This Springer imprint is published by the registered company Springer Nature Switzerland AG
The registered company address is: Gewerbestrasse 11, 6330 Cham, Switzerland

Preface

This volume contains the papers selected for presentation at IJCRS 2022, the 2022 International Joint Conference on Rough Sets, held on November 11–14, 2022, Suzhou, China.

Conferences in the IJCRS series are held annually and comprise four main tracks relating the topic rough sets to other topical paradigms: rough sets and data analysis covered by the RSCTC conference series from 1998, rough sets and granular computing covered by the RSFDGrC conference series since 1999, rough sets and knowledge technology covered by the RSKT conference series since 2006, and rough sets and intelligent systems covered by the RSEISP conference series since 2007.

Owing to the gradual emergence of hybrid paradigms involving rough sets, it was deemed necessary to organize Joint Rough Set Symposiums, first in Toronto, Canada, in 2007, followed by Symposiums in Chengdu, China in 2012, Halifax, Canada, 2013, Granada and Madrid, Spain, 2014, Tianjin, China, 2015, where the acronym IJCRS was proposed, continuing with the IJCRS 2016 conference in Santiago de Chile, IJCRS 2017 in Olsztyn, Poland, IJCRS 2018 in Quy Nhon, Vietnam, IJCRS 2019 in Debrecen, Hungary, IJCRS 2020 in La Habana, Cuba (held online), and IJCRS 2021 in Bratislava, Slovakia (hybrid).

Following the success of the previous conferences, IJCRS 2022 continued the tradition of a very rigorous reviewing process. The 28 papers included in this proceedings were selected from 42 submissions. Every submission, including invited keynote papers, was reviewed by at least two PC members and domain experts. Additional expert reviews were sought when necessary. On average, each submission received 3 reviews. Some papers received five reviews. As a result, only top-quality papers were chosen for presentation at the conference. Final camera-ready submissions were further reviewed by PC and conference chairs. Some authors were requested to make additional revisions. We would like to thank all the authors for contributing their papers. Without their contribution, this conference would not have been possible.

The IJCRS 2022 program was further enriched by five Keynote Speeches presented by former presidents of the International Rough Set Society. We are grateful to our keynote speakers, Davide Ciucci, Andrzej Skowron, Dominik Slezak, Roman Slowinski, Guoyin Wang, and Yiyu Yao for their visionary talks on rough sets, granular computing and three-way decisions.

The IJCRS 2022 program also included two workshops: Conceptual Knowledge Discovery and Machine Learning Based on Three-way Decisions and Granular Computing, and Uncertainty in Three-way Decisions, Granular Computing, and Data Science. We thank the workshop organizers Jinhai Li, Dun Liu, Georg Peters, Jianjun Qi, Xianyong Zhang, Huilai Zhi, and Jie Zhou for their contribution.

IJCRS 2022 would not have been successful without the support of many people and organizations. We wish to thank the members of the Steering Committee for their invaluable suggestions and support throughout the organization process. We are indebted to the PC members and external reviewers for their effort and engagement in providing a rich and rigorous scientific program.

We greatly appreciate the co-operation, support, and sponsorship of various institutions, companies, and organizations, including Suzhou University, China, the University of Regina, Canada, and the International Rough Set Society. We are also grateful to Springer for the sponsorship of the Best Student Paper Awards.

We acknowledge the use of the EasyChair conference system for paper submission and review. We are grateful to Springer for their support and co-operation publishing the proceedings as a volume of LNCS/LNAI.

September 2022

JingTao Yao
Hamido Fujita
Xiaodong Yue
Duoqian Miao
Jerzy Grzymala-Busse
Fanzhang Li

Organization

Steering Committee

Andrzej Skowron (Co-chair)	Polish Academy of Sciences, Poland
Guoyin Wang (Co-chair)	Chongqing University of Posts and Telecommunications, China
Davide Ciucci	Università di Milano-Bicocca, Italy
Jiye Liang	Shanxi University, China
Roman Slowinski	Poznan University of Technology, Poland
Yiyu Yao	University of Regina, Canada
Wei-Zhi Wu	Zhejiang Ocean University, China
Tianrui Li	Southwest Jiaotong University, China
Yang Gao	Nanjing University, China

Conference Chairs

Duoqian Miao	Tongji University, China
Jerzy Grzymala-Busse	University of Kansas, USA
Fanzhang Li	Soochow University, China
Mingrong Shen	Soochow University, China

Program Committee Chairs

JingTao Yao	University of Regina, Canada
Hamido Fujita	Iwate Prefectural University, Japan
Xiaodong Yue	Shanghai University, China

Special Event Chairs

Jusheng Mi	Hebei Normal University, China
Łukasz Sosnowski	Polish Academy of Sciences, Poland

Workshop Chairs

Deyu Li	Shanxi University, China
Pawan Lingras	St. Mary's University, Canada

Special Session Chairs

Qinghua Hu   Tianjin University, China
Hong Yu   Chongqing University of Posts and
                              Telecommunications, China

PhD School Chairs

Davide Ciucci   Università di Milano-Bicocca, Italy
Yuhua Qian   Shanxi University, China

Best Paper Committee

Davide Ciucci   Università di Milano-Bicocca, Italy
Tianrui Li   Southwest Jiaotong University, China
Pawan Lingras   St. Mary's University, Canada

Local Organization Chairs

Xizhao Luo   Soochow University, China
Xiaoyan Xia   Soochow University, China
Hongyun Zhang   Tongji University, China

Publication Chairs

Nouman Azam   National University of Computer and Emerging
                              Sciences, Pakistan
Bing Zhou   Chongqing University of Posts and
                              Telecommunications, China

Publicity Chairs

Qinghua Zhang   Chongqing University of Posts and
                              Telecommunications, China
Yan Zhang   Cal State University, San Bernardino, USA
Zhihua Wie   Tongji University, China

Web Chair

Juncheng Jia   Soochow University, China

Program Committee

Qiusheng An	Shanxi Normal University, China
Piotr Artiemjew	University of Warmia and Mazury, Poland
Nouman Azam	National University of Computer and Emerging Sciences, Pakistan
Joaquín Borrego-Díaz	University of Seville, Spain
Andrea Campagner	Università degli Studi di Milano-Bicocca, Italy
Yuming Chen	Xiamen University of Technology, China
Hongmei Chen	Southwest Jiaotong University, China
Zehua Chen	Taiyuan University of Technology, China
Davide Ciucci	Università di Milano-Bicocca, Italy
Chris Cornelis	Ghent University, Belgium
Jianhua Dai	Hunan Normal University, China
Tingquan Deng	Harbin Engineering University, China
Thierry Denoeux	Université de Technologie de Compiègne, France
Deyu Li	Shanxi University, China
Shifei Ding	China University of Mining and Technology, China
Weiping Ding	Nantong University, China
Soma Dutta	Vistula University, Poland
Hamido Fujita	Iwate Prefectural University, Japan
Can Gao	Shenzhen University, China
Yang Gao	Nanjing University, China
Salvatore Greco	University of Portsmouth, UK
Jerzy Grzymala-Busse	University of Kansas, USA
Shen-Ming Gu	Zhejiang Ocean University, China
Christopher Henry	University of Winnipeg, Canada
Mengjun Hu	University of Regina, Canada
Qinghua Hu	Tianjin University, China
Xuegang Hu	Hefei University of Technology, China
Bing Huang	Nanjing Audit University, China
Amir Hussain	Edinburgh Napier University, UK
Masahiro Inuiguchi	Osaka University, Japan
Ryszard Janicki	McMaster University, Canada
Andrzej Janusz	University of Warsaw, Poland
Richard Jensen	Aberystwyth University, UK
Xiuyi Jia	Nanjing University of Science and Technology, China
Chunmao Jiang	Harbin Normal University, China
Bin Jie	Hebei Normal University, China
Guangming Lang	Changsha University of Science and Technology, China

Fanchang Li	Soochow University, China
Huaxiong Li	Nanjing University, China
Jinhai Li	Kunming University of Science and Technology, China
Jinjin Li	Minnan Normal University, China
Kewen Li	China University of Petroleum, China
Lei-Jun Li	Hebei Normal University, China
Tianrui Li	Southwest Jiaotong University, China
Tong-Jun Li	Zhejiang Ocean University, China
Yuefeng Li	Queensland University of Technology, Australia
Jiuzhen Liang	Changzhou University, China
Shujiao Liao	Minnan Normal University, China
Guoping Lin	Minnan Normal University, China
Yaojin Lin	Hefei University of Technology, China
Pawan Lingras	St. Mary's University, Canada
Baoxiang Liu	North China University of Science and Technology, China
Caihui Liu	Gannan Normal University, China
Dun Liu	Southwest Jiaotong University, China
Guilong Liu	Beijing Language and Culture University, China
Wenqi Liu	Kunming University of Science and Technology, China
Jesús Medina	University of Cádiz, Spain
Jusheng Mi	Hebei Normal University, China
Duoqian Miao	Tongji University, China
Marcin Michalak	Silesian University of Technology, Poland
Fan Min	Southwest Petroleum University, China
Hung Son Nguyen	University of Warsaw, Poland
Witold Pedrycz	University of Alberta, Canada
Shenglei Pei	Qinghai Nationalities University, China
Georg Peters	Munich University of Applied Sciences, Germany
James Peters	University of Manitoba, Canada
Lech Polkowski	Polish-Japanese Institute of Information Technology, Poland
Jianjun Qi	Xidian University, China
Jin Qian	East China Jiaotong University, China
Yuhua Qian	Shanxi University, China
Taorong Qiu	Nanchang University, China
Sheela Ramanna	University of Winnipeg, Canada
Marek Reformat	University of Alberta, Canada
Sergio Ribeiro	Pontificia Universidade Catolica do Parana, Brazil
Hiroshi Sakai	Kyushu Institute of Technology, Japan

Lin Shang	Nanjing University, China
Ming-Wen Shao	Chinese University of Petroleum, China
Yanhong She	Xian Shiyou University, China
Andrzej Skowron	Polish Academy of Sciences, Poland
Dominik Slezak	University of Warsaw, Poland
Roman Slowinski	Poznan University of Technology, Poland
Jingjing Song	Macau University of Science and Technology, China
Lukasz Sosnowski	Polish Academy of Sciences, Poland
Jaroslaw Stepaniuk	Bialystok University of Technology, Poland
Lin Sun	Henan Normal University, China
Zbigniew Suraj	University of Rzeszów, Poland
Marcin Szczuka	University of Warsaw, Poland
Anhui Tan	Zhejiang Ocean University, China
Shusaku Tsumoto	Shimane University, Japan
Renxia Wan	North Minzu University, China
Baoli Wang	Yuncheng University, China
Changzhong Wang	Bohai University, China
Guoyin Wang	Chongqing University of Posts and Telecommunications, China
Lai Wei	Shanghai Maritime University, China
Ling Wei	Northwest University, China
Wei Wei	Shanxi University, China
Zhihua Wei	Tongji University, China
Marcin Wolski	Maria Curie-Sklodowska University, Poland
Wei-Zhi Wu	Zhejiang Ocean University, China
Shuyin Xia	Chongqing University of Posts and Telecommunications, China
Bin Xie	Hebei Normal University, China
Jun Xie	Taiyuan University of Technology, China
Jianfeng Xu	Nanchang University, China
Jiucheng Xu	Henan Normal University, China
Weihua Xu	Southwest University, China
Zhan-Ao Xue	Henan Normal University, China
Lin Xun	Henan Normal University, China
Hailong Yang	Shaanxi Normal University, China
Jilin Yang	Sichuan Normal University, China
Tian Yang	Hunan Normal University, China
Xibei Yang	Jiangsu University of Science and Technology, China
Xin Yang	Southwestern University of Finance and Economics, China

President Forum Talk Abstracts

We present abstracts of keynote talks in this section. This year, 2022, is the 40th anniversary of rough set theory, we invited former presidents of the International Rough Set Society (https://www.roughsets.org/) to deliver keynote talks to share with us their views of rough set theory and insights of its future development. There were six past presidents, Andrzej Skowron (Poland), Roman Słowiński (Poland), Dominik Ślęzak (Poland), Guoyin Wang (China), Yiyu Yao (Canada), Davide Ciucci (Italy), deceived wonderful talk at IJCRS 2022.

Rough Sets and Fuzzy Sets in Interactive Granular Computing

Andrzej Skowron

Systems Research Institute, Polish Academy of Sciences, Newelska 6,
01-447 Warsaw and Multidisciplinary Research Center, UKSW,
Marii Konopnickiej 1, 05-092 Dziekanow Lesny

Abstract. This paper is an attempt to present the ground that a change is needed in the way of viewing mathematical tools, such as fuzzy sets, rough sets, in the context of classifying and approximating concepts pertaining to the real physical complex phenomenon. The paper argues in favour of developing models going beyond the pure mathematical manifold. The main idea is not to develop a theory only based on gathered data, rather to incorporate the methods of perception and real physical interactions through which the data is obtained. In this regard, a primary proposal has been put forward to model fuzzy sets and rough sets in the framework of Interactive Granular Computing (IGrC).

Dominance-Based Rough Set Approach as a Breakthrough in Reasoning About Ordinal Data

Roman Słowiński

Institute of Computing Science, Poznan University of Technology, 60-965 Poznan, and Systems Research Institute, Polish Academy of Sciences, 01-447 Warsaw, Poland

Abstract. Shortly after the first publication of the Zdzisław Pawlak on Rough Sets, attempts were made to adapt this concept to ordinal data [1]. Ordinal data characterize decision situations where potential actions (objects) are described by attributes with ordinal scales, and decision classes are ordered from the best to the worst. The major difference between nominal and ordinal data is the type of possible inconsistency: in nominal data it means violation of indiscernibility (objects a and b being indiscernible by nominal condition attributes have been assigned to different decision classes), while in ordinal data inconsistency means violation of dominance (object a having at least as good evaluations as object b on all ordinal condition attributes has been assigned to a worse class than b). The rough set concept has been adapted to ordinal data in the methodology called Dominance-based Rough Set Approach (DRSA) [2]. It substitutes the approximation of sets by granules of indiscernible or similar objects in the condition attribute space with the approximation of upward and downward unions of ordered sets by dominance cones in the ordinal condition attribute space. DRSA made a particular breakthrough in preference learning for decision aiding [3]. The preference model is induced from dominance-based rough approximations of ordered decision classes or gradual relations in terms of certain and possible "if ..., then ..." decision rules. This methodology was first adapted to multiple criteria ordinal classification, choice, and ranking. Then, to decision under uncertainty and time preference, case-based reasoning, and interactive multiobjective evolutionary optimization. It gained importance in preference modeling for transparency and explainability of decision rules, as well as for their capacity of handling interacting attributes. It appears, moreover, that the assumption admitted by DRSA about the ordinal character of evaluations on condition and decision attributes is not a limiting factor in knowledge discovery from non-ordinal data, because the presence or the absence of a property can be represented in ordinal terms. Precisely, if two properties are related, the presence, rather than the absence of one property should make more (or less) probable the presence of the other property. This is even more apparent when the presence or the absence of a property is graded or fuzzy. This observation led to a straightforward hybridization of DRSA with fuzzy sets [4]. Since the presence of properties, possibly fuzzy, is the base of information granulation, DRSA can also be seen as a general framework for granular computing [5].

References

1. Pawlak, Z., Słowiński, R.: Decision analysis using rough sets. Int. Trans. Oper. Res. **1**(1), 107–114 (1994)
2. Greco, S., Matarazzo, B., Słowiński, R.: Rough sets theory for multicriteria decision analysis. Eur. J. Oper. Res. **129**, 1–47 (2001)
3. Słowiński, R., Greco, S., Matarazzo, B.: Rough set methodology for decision aiding. In: Kacprzyk, J., Pedrycz, W. (eds.) Springer Handbook of Computational Intelligence. Springer Handbooks, pp. 349–370. Springer, Berlin (2015). https://doi.org/10.1007/978-3-662-43505-2_22
4. Palangetic, M., Cornelis, Ch., Greco, S., Słowiński, R.: Fuzzy extensions of the dominance-based rough set approach. Int. J. Approximate Reasoning **129**, 1–19 (2021)
5. Palangetic, M., Cornelis, Ch., Greco, S., Słowiński, R.: Granular representation of OWA-based Fuzzy rough sets. Fuzzy Sets Syst. (2021). https://doi.org/10.1016/j.fss.2021.04.018

Rough Sets in Industry

Dominik Slezak

Institute of Informatics, University of Warsaw, Warsaw, Poland
and QED Software, Warsaw, Poland

Abstract. We discuss two real-world use cases of deployment of the paradigms of rough sets in commercial software solutions. The rst use case refers to rough set approximations and their utilization in the internals of an analytical database engine. The second use case refers to rough-set-based decision reducts and their utilization in a software system which is aimed at monitoring and diagnosing the machine learning models. We claim that in both of those cases, the next step toward a wider industry applicability should correspond to the means for handling complex, unstructured and multimodal data sources.

MGCC: Multi-Granularity Cognitive Computing

Guoyin Wang

Chongqing Key Laboratory of Computational Intelligence,
Chongqing University of Posts and Telecommunications,
Chongqing 400065, P. R. China

Abstract. Cognitive computing aims to develop a coherent, unified, universal mechanism with inspiration of mind's capabilities. Granular human thinking is a kind of cognition mechanism for human problem solving. Multi-Granularity cognitive computing (MGCC) is introduced to integrate the information transformation mechanism of traditional intelligent information processing systems and the multi-granularity cognitive law of human brain in this paper. The data-driven granular cognitive computing model (DGCC) developed in 2017 is a typical theoretical model for implementing MGCC. MGCC is a valuable model for developing highly intelligent systems consistent with human cognition. The theoretical research issues and some applications about MGCC are introduced.

Three-Way Decision, Three-World Conception, and Explainable AI

Yiyu Yao

Department of Computer Science, University of Regina,
Regina, SK S4S 0A2, Canada

Abstract. Three-way decision is about thinking, problem-solving, and computing in threes or through triads. By dividing a whole into three parts, by focusing on only three things, or by considering three basic ingredients, we may build a theory, a model, or a method that is simple-to-understand, easy-to-remember, and practical-to-use. This philosophy and practice of triadic thinking appear everywhere. In particular, there are a number of three-world or tri-world models in different fields and disciplines, where a complex system, a complicated issue, or an intricate concept is explained and understood in terms of three interrelated worlds, with each world enclosing a group of elements or representing a particular view. The main objective of this paper is to review and reinterpret various three-world conceptions through the lens of three-way decision. Three-world conceptions offer more insights into three-way decision with new viewpoints, methods, and modes. They can be used to construct easy-to-understand explanations in explainable artificial intelligence (XAI).

Orthopartitions in Knowledge Representation and Machine Learning

Davide Ciucci

Universita degli Studi di Milano-Bicocca, DISCo,
Viale Sarca 336, 20126, Milan, Italy

Abstract. Orthopartitions are partitions with uncertainty. We survey their use in knowledge representation (KR) and machine learning (ML). In particular, in KR their connection with possibility theory, intuitionistic fuzzy sets and credal partitions is discussed. As far as ML is concerned, their use in soft clustering evaluation and to define generalized decision trees are recalled. The (open) problem of relating an orthopartition to a partial equivalence relation is also.

Contents

Invited Papers, IRSS President Forum

Orthopartitions in Knowledge Representation and Machine Learning

Davide Ciucci$^{(\boxtimes)}$ ⓘ, Stefania Boffa ⓘ, and Andrea Campagner ⓘ

Università degli Studi di Milano-Bicocca, DISCo, Viale Sarca 336, 20126 Milano, Italy
davide.ciucci@unimib.it

Abstract. Orthopartitions are partitions with uncertainty. We survey their use in knowledge representation (KR) and machine learning (ML). In particular, in KR their connection with possibility theory, intuitionistic fuzzy sets and credal partitions is discussed. As far as ML is concerned, their use in soft clustering evaluation and to define generalized decision trees are recalled. The (open) problem of relating an orthopartition to a partial equivalence relation is also sketched.

Keywords: Credal partition · Decision tree · Rough sets · Intuitionistic fuzzy sets · Orthopairs · Partial labels · Partition · Possibility theory · Soft clustering

1 Introduction

Let us suppose to have a set of objects that we want to group according to some criteria. Due to intrinsic ambiguity or lack of knowledge, it may happen that we are not able to precisely define the groups. *Not precisely* may be understood in different ways: we are undecided if an object belongs to a certain group; an object can belong to more than one group; an object can belong to one or more groups with a degree of possibility or probability, etc. The problem to describe this situation and how to manage the obtained groups is interesting per se from a theoretical standpoint and it has been widely studied in the context of clustering in machine learning.

In the last years, to cope with this issue, we introduced and studied the notion of orthopartition, i.e., a partition with uncertainty, and subsequently its fuzzy version. An orthopartition is a collection of orthopairs, that is pairs of disjoint sets (A, B), such that $A \cap B = \emptyset$, representing the equivalence classes (with uncertainty) of a partition.

In the present work, we recall what is an orthopartition and survey the main results obtained up to now. We also put forward a new problem and give the first comments on it in Sect. 2.3. We will follow two directions: knowledge representation in Sect. 2 and machine learning in Sect. 3.

At first, the basic definitions of orthopartition and its uncertainty measures are reported. Then, in Sect. 2.1 the link with possibility theory is given. In

J. Yao et al. (Eds.): IJCRS 2022, LNAI 13633, pp. 3–18, 2022.
https://doi.org/10.1007/978-3-031-21244-4_1

Sect. 2.2 the fuzzy version of orthopartitions is provided, with some operations and order relations on it. Then, the link between fuzzy orthopartitions and credal partitions is studied. In Sect. 2.3, we sketch the problem of finding a partial equivalence relation corresponding to an orthopartition and vice versa.

Then, we turn our attention to the role of orthopartitions in Machine Learning. In Sect. 3.1, we explain how they can be used in Soft Clustering, in particular in rough and three-way clustering, to provide an external clustering evaluation. In Sect. 3.2, orthopartitions are applied to the problem of learning from partial labels (with Decision Trees) and to generalize Decision Trees with the ability to abstain from a precise label. We finally conclude with some on-going works and future perspectives.

2   Knowledge Representation

As said in the introduction, an orthopair O is a pair of disjoint sets (A, B) on a universe U. It has been introduced firstly in [17] as an abstraction of a rough sets and then studied in [1,12,18,19]. The two sets (A, B) represent a partially known set X with A representing the elements that surely belong to X and B those that surely do not belong to X. Of course, A, B tri-partition the universe and by $Bnd = U \setminus (A \cup B)$ we mean the objects on which we are uncertain (clearly, Bnd is taken from the *boundary* of a rough set). Often, an orthopair is denoted as (P, N) where P stands for *positive* and N for *negative*. We also notice that an orthopair is formally equivalent to an interval set, though their interpretation may be different [18,27].

If (P, N) is interpreted as an equivalence class of a partition with uncertainty, then a collection of orthopairs can be understood as a partition with uncertainty.

Definition 1. *An* orthopartition *is a set* $\mathcal{O} = \{O_1, ..., O_n\}$ *of orthopairs such that the following axioms hold:*

(Ax O1) $\forall O_i, O_j \in \mathcal{O}, O_i, O_j$ *are disjoint, that is* $P_i \cap P_j$, $P_i \cap Bnd_j$ *and* $P_j \cap Bnd_i$ *are empty*
(Ax O2) $\bigcup_i (P_i \cup Bnd_i) = U$; *(coverage requirement)*
(Ax O3) $\forall x \in U$ *if* $(x \in Bnd_i)$ *then* $\exists j \neq i$ *such that* $(x \in Bnd_j)$ *(an object cannot belong to only 1 boundary)*

Example 1. Let U be the set of the first 10 integers, i.e. $U = \{1, 2, \ldots, 10\}$. Then, the collection $\{O_1, O_2, O_3\}$ where: $O_1 = (\{1, 2\}, \{9, 10\})$, $O_2 = (\{9\}, \{1, 2\})$, $O_3 = (\emptyset, \{1, 2, 9\})$ is an orthopartition of U. On the other hand, the set $\{O_1, O_2\}$ is not an orthopartition since it does not satisfy axiom O_3: 10 belongs to only one boundary.

An orthopartition represents an underlying partition, which is not precisely known. Hence, several standard partitions can be consistent with a given partition, i.e., the orthopartition could collapse in one of its consistent partitions if full knowledge would be available.

Definition 2. *A set S is* consistent *with an orthopair $O = (A, B)$ if for all $x \in A$ then $x \in S$ and for all $x \in B$ then $x \notin S$.*

A partition π is consistent *(resp, weakly consistent) with an orthopartition $\mathcal{O}$ iff $\forall O_i \in \mathcal{O}, \exists!$ (resp., $\exists! \vee \not\exists$) $S_i \in \pi$ s.t. S is consistent with O_i.*

Example 2. Let $\mathcal{O} = \{O_1, O_2, O_3\}$ the orthopartition given in Example 1. Then, two partitions consistent with $\mathcal{O}$ are: $\Pi_1 = \{\{1, 2, 3\}, \{7, 8, 9, 10\}, \{4, 5, 6\}\}$ and $\Pi_2 = \{\{1, 2\}, \{9\}, \{3, 4, 5, 6, 7, 8, 10\}\}$.

Let us denote with $\Pi_\mathcal{O}$ the collection of all partitions consistent with $\mathcal{O}$. We notice that the difference between consistency and weak consistency, lies in the fact that the former requires that each orthopair represents an equivalence class, whereas in the case of weak consistency there can be more orthopairs than equivalence classes.

In order to measure the uncertainty intrinsic to an orthopartition the notions of Ellerman and Shannon entropy can be generalized to the present setting.

Definition 3. *Given an orthopartition $\mathcal{O}$ and let $\Pi_\mathcal{O}$ be the set of all the partitions consistent with $\mathcal{O}$, we define the* Ellerman (or logical) average entropy *as*

$$H_E(\mathcal{O}) = \frac{1}{|\Pi_\mathcal{O}|} \sum_{\pi \in \Pi_\mathcal{O}} h_E(\pi) \tag{1}$$

and the Shannon average entropy *as*

$$H_S(\mathcal{O}) = \frac{1}{|\Pi_\mathcal{O}|} \sum_{\pi \in \Pi_\mathcal{O}} h_S(\pi) \tag{2}$$

with $h_E(\pi)$ and $h_S(\pi)$, respectively, the Ellerman and Shannon entropy of the standard partition π.

An algorithm to compute H_E with polynomial complexity can be found in [14]. On the other hand no efficient algorithm for H_S has been defined.

Moreover, let us notice that the average entropy can be bounded by the lower and upper entropies as follows

$$H_{i_*} = min\{h_i(\pi) | \pi \in \Pi_\mathcal{O}\} \tag{3a}$$
$$H_i^* = max\{h_i(\pi) | \pi \in \Pi_\mathcal{O}\} \tag{3b}$$

for $i \in \{E, S\}$. The definition of average, lower and upper entropy can then be used as a basis to define the mutual information between orthopartitions, by means of the following definition:

$$m_i(\mathcal{O}_1, \mathcal{O}_2) = \hat{H}_i(\mathcal{O}_1) + \hat{H}_i(\mathcal{O}_2) - \hat{H}_i(\mathcal{O}_1 \wedge \mathcal{O}_2) \tag{4}$$

where $\hat{H}_i = \frac{H_{i_*} + H_i^*}{2}$ and $\mathcal{O}_1 \wedge \mathcal{O}_2$ is the *meet* orthopartition, defined by $\mathcal{O}_1 \wedge \mathcal{O}_2 = \{(A_{i1} \cap A_{i2}, B_{i1} \cup B_{i2}) | O_{i1} \in \mathcal{O}_1 \text{ and } O_{j2} \in \mathcal{O}_2\}$. Intuitively, in analogy with the definition of mutual information in information theory, the mutual information between orthopartitions represents a measure of similarity between the informational content of two orthopartitions. As such, it has maximum value when $\mathcal{O}_1 = \mathcal{O}_2$ and both are standard partitions.

2.1   Possibility Theory

Orthopairs can have different interpretations and one is to look them as *partial models*, that is, partial assignments of truth values [5]. More precisely, let $\mathcal{V}$ be a set of propositional variables, an orthopair (A, B) on $\mathcal{V}$ corresponds to assigning *true* to the variables of A and *false* to those of B, while the truth value of all remaining variables is unknown.

Thus a correspondence between orthopairs and Boolean possibility distributions [20] can be put forward. Indeed, let Π be the collection of all Boolean possibility distributions whose domain is made of all evaluation functions on $\mathcal{V}$, i.e. $\Pi = \{\pi \mid \pi : \{0,1\}^{\mathcal{V}} \rightarrow \{0,1\}\}$. Then, each orthopair on $\mathcal{V}$ generates a Boolean possibility distribution of Π, but not all Boolean possibility distributions of Π can be obtained starting from an orthopair. Then, each collection of orthopairs on $\mathcal{V}$ generates a distribution of Π, and vice versa each distribution of Π is associated to at least a set of orthopairs.

This link has been explored in [9], where it has been shown that orthopartitions on $\mathcal{V}$ can be identified with a special class of Π. Furthermore, a necessary and sufficient condition for a distribution of Π to be generated by an orthopartition on $\mathcal{V}$ has been given. We can consequently classify possibility distributions of Π on the basis of their relations with orthopairs and orthopartitions as depicted in Fig. 1.

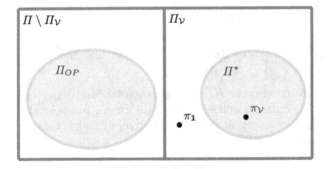

Fig. 1. Possibility distributions classification determined by their correspondence with orthopairs and orthopartitions.

Indeed, the possibility distributions of $\Pi_{\mathcal{V}}$ are those that can be represented by an individual orthopair on $\mathcal{V}$. Then, the possibility distributions in $\Pi^* \subseteq \Pi_{\mathcal{V}}$ have the additional feature that cannot be generated by sets made of two or more orthopairs. On the other hand, all possibility distributions of Π corresponding to orthopartitions form a set $\Pi_{OP} \cup \{\pi_{\mathcal{V}}, \pi_1\}$, with $\Pi_{OP} \subset \Pi \setminus \Pi_{\mathcal{V}}$, $\pi_{\mathcal{V}}$ is the distribution generated by $(\mathcal{V}, \emptyset)$, and π_1 is the distribution generated by $(\emptyset, \emptyset)$. We thus notice that there exist possibility distributions that cannot be represented by an orthopartition.

2.2 Fuzzy Orthopartitions

Fuzzy orthopartitions are introduced in [6,8] as collections of intuitionistic fuzzy sets satisfying a list of axioms to model partitions including both uncertainty and vagueness.

Let us recall that an intuitionistic fuzzy set (IFS) A of a universe U [2] is a pair of functions $\mu_A : U \to [0,1]$ and $\nu_A : U \to [0,1]$ such that $\mu_A(u) + \nu_A(u) \leq 1$ for each $u \in U$. The two functions μ_A and ν_A are respectively called the *membership* and *non-membership* functions of A. Moreover, the *hesitation margin* of A $h_A : U \to [0,1]$ is given by $h_A(u) = 1 - (\mu_A(u) + \nu_A(u))$ and it expresses the uncertainty contained in A.

Definition 4. *A* fuzzy orthopartition *is a set* $\mathcal{O} = \{(\mu_1, \nu_1), \dots, (\mu_n, \nu_n)\}$ *of intuitionistic fuzzy sets of U such that the following axioms hold:*

(Ax OR1) $\sum_{i=1}^{n} \mu_i(u) \leq 1$ *for each $u \in U$;*
(Ax OR2) $\mu_i(u) + h_j(u) \leq 1$ *for each $i \neq j$ and $u \in U$;*
(Ax OR3) $\sum_{i=1}^{n} \mu_i(u) + h_i(u) \geq 1$ *for each $u \in U$;*
(Ax OR4) *for each $i \in \{1, \dots, n\}$ and $u \in U$ with $h_i(u) > 0$, there exists $j \in \{1, \dots, n\} \setminus \{i\}$ such that $h_j(u) > 0$.*

The first two axioms capture the idea that intuitionistic fuzzy sets in $\mathcal{O}$ must represent mutually disjoint blocks of U, and *(Ax OR3)* corresponds to a covering condition. Lastly, axiom *(Ax OR4)* allows fuzzy orthopartitions to be extensions of crisp orthopartitions given by Definition 1.

Indeed, fuzzy orthopartitions are generalizations of both orthopartitions based on classical sets and Ruspini (fuzzy) partitions defined as follows.

Definition 5. *[24] A* Ruspini partition π *of U is a collection of functions* $\pi_1, \dots, \pi_n : U \to [0,1]$ *such that $\pi_1(u) + \dots + \pi_n(u) = 1$ for each $u \in U$.*

Theorem 1. *[6] Let $\mathcal{O} = \{(\mu_1, \nu_1), \dots, (\mu_n, \nu_n)\}$ be a fuzzy orthopartition of U, then the following statements are true:*

(a) if $\mu_1, \dots, \mu_n, \nu_1, \dots, \nu_n$ are Boolean functions, then $\mathcal{O}$ is an orthopartition;
(b) if $h_1, \dots, h_n$ are null functions, then $\mathcal{O}$ is a Ruspini partition.

Logical Entropy of Fuzzy Orthopartitions. The definition of lower and upper entropy of fuzzy orthopartitions is based on the concepts of *consistency* and *logical entropy* of Ruspini partitions.

Definition 6. *Let $\mathcal{O} = \{(\mu_1, \nu_1), \dots, (\mu_n, \nu_n)\}$ be a fuzzy orthopartition and let $\pi = \{\pi_1, \dots, \pi_n\}$ be a Ruspini partition of U. Then, π is* fuzzy consistent *with $\mathcal{O}$ if and only if $\mu_i(u) \leq \pi_i(u) \leq \mu_i(u) + h_i(u)$ for each $u \in U$ and $i \in \{1, \dots, n\}$.*

Definition 6 arises by viewing fuzzy orthopartitions as Ruspini partitions with uncertainty, that is a fuzzy orthopartition $\mathcal{O}$ must became a Ruspini partition (consistent with $\mathcal{O}$) once the uncertainty is solved. In the sequel, we simply say "consistent" instead of "fuzzy consistent" and we use the symbol $\Pi_{\mathcal{O}}$ to denote the set of all Ruspini partitions consistent with $\mathcal{O}$.

Let $u, u' \in U$, we call $dit\, \pi(u, u') = \max\{|\pi_i(u) - \pi_i(u')|$ such that $i \in \{1, \ldots, n\}\}$ the *degree of distinction* of (u, u'). It is interpreted as the capacity of π to distinguish u and u' by means of its fuzzy sets $\pi_1, \ldots, \pi_n$.

Definition 7. *The* logical entropy *of a Ruspini partition π is defined as*

$$\mathcal{H}(\pi) = \frac{\sum_{(u,u') \in U \times U} dit\, \pi(u, u')}{|U \times U|}. \tag{5}$$

It is easy to understand that $\mathcal{H}(\pi)$ measures how much π is able to distinguish the elements of U by means of its fuzzy sets. We now use $\mathcal{H}$ to bound the value of a fuzzy orthopartition entropy.

Definition 8. *Let $\mathcal{O}$ be a fuzzy orthopartition of U, the* lower *and* upper entropy *are respectively given by $\mathcal{H}_*(\mathcal{O}) = \min\{\mathcal{H}(\pi) \mid \pi \in \Pi_{\mathcal{O}}\}$ and $\mathcal{H}^*(\mathcal{O}) = \max\{\mathcal{H}(\pi) \mid \pi \in \Pi_{\mathcal{O}}\}$.*

Since $\mathcal{H}_*(\mathcal{O}) \leq \mathcal{H}^*(\mathcal{O})$, the interval $\mathcal{I}_{\mathcal{O}} = [\mathcal{H}_*(\mathcal{O}), \mathcal{H}^*(\mathcal{O})]$ can be seen as an entropy measure too. Moreover, it holds that $\mathcal{I}_{\mathcal{O}} \subseteq [0, 1 - \frac{|\Delta_U|}{|U \times U|}]$, where $\Delta_U = \{(u, u) \mid u \in U\}$. By comparing the lower and upper entropy of fuzzy orthopartitions with other entropy measures, the following results are obtained:

Theorem 2. *[8] Let $\mathcal{O}$ be a fuzzy orthopartition of U, then the following statements are true:*

(a) if $\mathcal{O}$ is a Ruspini partition then $\mathcal{H}_*(\mathcal{O}) = \mathcal{H}^*(\mathcal{O})$ and they can be computed by (5);

(b) if $\mathcal{O}$ is an orthopartition then $[H_*(\mathcal{O}), H^*(\mathcal{O})] \subseteq [\mathcal{H}_*(\mathcal{O}), \mathcal{H}_*(\mathcal{O})]$, where $H_*(\mathcal{O})$ and $H^*(\mathcal{O})$ are respectively the lower and upper entropies of $\mathcal{O}$ already defined in equations (3);

(c) if O is a partition then $\mathcal{H}_*(\mathcal{O}) = \mathcal{H}^*(\mathcal{O})$ and they coincide with the logical entropy defined by Ellerman in [22].

The upper and lower entropy of $\mathcal{O}$ can be computed by solving an optimization problem: $\mathcal{H}^*(\mathcal{O})$ and $\mathcal{H}_*(\mathcal{O})$ correspond to the maximum and minimum points of a non-linear function subject to a list of constraints. The latter can be converted into a liner programming problem, which is solved by using one of the standard techniques in linear programming as the Simplex method. Additionally, the Ruspini partition π_{med} such that $\mathcal{H}(\pi_{med})$ is the arithmetic mean of $\mathcal{H}_*(\mathcal{O})$ and $\mathcal{H}^*(\mathcal{O})$, can be found by determining the solutions of a non-linear system. More details about such procedures are found in [6,8].

Operations and Orderings on Fuzzy Orthopartitions. Operations and orderings on fuzzy orthopartitions are based on the following operators and relations on intuitionistic fuzzy sets.

Definition 9. *[3, 10] Let (μ_i, ν_i) and (μ_j, ν_j) be intuitionistic fuzzy sets of U, then*

(a) $(\mu_i, \nu_i) \cap (\mu_j, \nu_j) = (\min\{\mu_i, \mu_j\}, \max\{\nu_i, \nu_j\})$;

(b) $(\mu_i, \nu_i) * (\mu_j, \nu_j) = (\frac{\mu_i + \mu_j}{2}, \frac{\nu_i + \nu_j}{2})$;

(c) $(\mu_i, \nu_i) \preceq (\mu_j, \nu_j)$ *if and only if* $\mu_i(u) \leq \mu_j(u)$ *and* $\nu_i(u) \geq \nu_j(u)$ *for each* $u \in U$;

(d) $(\mu_i, \nu_i) \preceq^* (\mu_j, \nu_j)$ *if and only if* $\mu_i(u) \leq \mu_j(u)$ *and* $\nu_i(u) \leq \nu_j(u)$ *for each* $u \in U$.

In the sequel, given $n \in \mathbb{N}$, we use the symbol OR to denote the collection of all sequences of n IFSs forming an orthopartition of U. Furthermore, we consider $\mathcal{O}_1 = ((\mu_1^1, \nu_1^1), \ldots, (\mu_n^1, \nu_n^1))$ and $\mathcal{O}_2 = ((\mu_1^2, \nu_1^2), \ldots, (\mu_n^2, \nu_n^2))$ in OR.

Definition 10. *Let $\mathcal{O}_1, \mathcal{O}_2 \in OR$, we put*

(a) $\mathcal{O}_1 \cap_{OR} \mathcal{O}_2 = ((\mu_1^1, \nu_1^1) \cap (\mu_1^2, \nu_1^2), \ldots, (\mu_n^1, \nu_n^1) \cap (\mu_n^2, \nu_n^2))$;

(b) $\mathcal{O}_1 *_{OR} \mathcal{O}_2 = ((\mu_1^1, \nu_1^1) * (\mu_1^2, \nu_1^2), \ldots, (\mu_n^1, \nu_n^1) * (\mu_n^2, \nu_n^2))$;

(c) $\mathcal{O}_1 \preceq_{OR} \mathcal{O}_2$ *if and only if* $(\mu_i^1, \nu_i^1) \preceq (\mu_i^2, \nu_i^2)$ *for each* $i \in \{1, \ldots, n\}$;

(d) $\mathcal{O}_1 \preceq_{OR}^* \mathcal{O}_2$ *if and only if* $(\mu_i^1, \nu_i^1) \preceq^* (\mu_i^2, \nu_i^2)$ *for each* $i \in \{1, \ldots, n\}$.

It can be proved that

- OR_α^* is closed under $\cap_{OR}$, where $OR_\alpha^* = \{\mathcal{O} \in OR \mid \nu(u) < \alpha$ and $h(u) > 0$ for each $(\mu, \nu) \in O$ and $u \in U\}$;
- OR is closed under $*_{OR}$.

Moreover, $\cap_{OR}$ and $*_{OR}$ are both commutative and idempotent, $\cap_{OR}$ is associative, and $*_{OR}$ is distributive w.r.t. $\cap_{OR}$. If we confine to OR_α^* with $\alpha \leq 1 - \frac{1}{n}$, then $\preceq_{OR}$ is the ordering associated with $\cap_{OR}$. Indeed, let $\mathcal{O}_1, \mathcal{O}_2 \in OR$, $\mathcal{O}_1 \cap_{OR} \mathcal{O}_2 = \mathcal{O}_1$ if and only if $\mathcal{O}_1 \preceq \mathcal{O}_2$. Both $\preceq_{OR}$ and $\preceq_{OR}^*$ are ordering relations, namely they are reflexive, anti-symmetric, and transitive.

The next theorem connects the lower and upper entropy with the operations and relations of the previous definition.

Theorem 3. *[8] The following statements are true:*

- If $\mathcal{O}_1 \preceq_{OR} \mathcal{O}_2$ then $\mathcal{I}_{\mathcal{O}_1 \cap_{OR} \mathcal{O}_2} \subseteq \mathcal{I}_{\mathcal{O}_1}$;
- If $\mathcal{O}_1 \preceq_{OR}^* \mathcal{O}_2$ then $\mathcal{I}_{\mathcal{O}_1 *_{OR} \mathcal{O}_2} \subseteq \mathcal{I}_{\mathcal{O}_1}$.

Let us point out that $\preceq_{OR}^*$ is an extension of the relation $\preceq^*$ on OR, where $(\mu_i, \nu_i) \preceq^* (\mu_j, \nu_j)$ means that (μ_i, ν_i) *is less fuzzy than* (μ_j, ν_j), i.e. (μ_j, ν_j) is closer than (μ_i, ν_i) to a fuzzy set because it contains less uncertainty. The same meaning can be extended to $\preceq_{OR}^*$: let $\mathcal{O}_1, \mathcal{O}_2 \in OR$, $\mathcal{O}_1 \preceq_{OR}^* \mathcal{O}_2$ states that $\mathcal{O}_2$ is closer than $\mathcal{O}_1$ to a Ruspini partition.

If $\mathcal{O}_1 \preceq_{OR}^* \mathcal{O}_2$, then we will say that $\mathcal{O}_2$ is a *refinement* of $\mathcal{O}_1$. Indeed, in a dynamic situation, we can imagine $\mathcal{O}_2$ as an evolution of $\mathcal{O}_1$ once more information is known about the elements of the universe U.

Let $I_{OR} = \{\mathcal{I}_{\mathcal{O}} \mid \mathcal{O} \in OR\}$, we consider an ordering relation on I_{OR}:

$$\mathcal{I}_{\mathcal{O}_1} \leq_{I_{OR}} \mathcal{I}_{\mathcal{O}_2} \text{ if and only if } \mathcal{H}_*(\mathcal{O}_1) \leq \mathcal{H}_*(\mathcal{O}_2) \text{ and } \mathcal{H}^*(\mathcal{O}_2) \leq \mathcal{H}^*(\mathcal{O}_1). \quad (6)$$

In simple words, $\mathcal{I}_{\mathcal{O}_1} \leq_{I_{OR}} \mathcal{I}_{\mathcal{O}_2}$ means that $\mathcal{I}_{\mathcal{O}_2}$ is a closed subinterval of $\mathcal{I}_{\mathcal{O}_2}$. Therefore, it is easy to understand that the entropy associated with $\mathcal{O}_2$ is less than the one associated to $\mathcal{O}_1$. The next theorem shows that the lower and upper entropies (in the form of closed subintervals of [0,1] equipped by $\leq_{I_{OR}}$) are monotonic with respect to $\preceq_{OR}^*$.

Theorem 4. *[8] Let $\mathcal{O}_1, \mathcal{O}_2 \in OR$. If $\mathcal{O}_1 \preceq_{OR}^* \mathcal{O}_2$ then $\mathcal{I}_{\mathcal{O}_1} \leq_{I_{OR}} \mathcal{I}_{\mathcal{O}_2}$.*

Let us focus on the meaning of Theorem 4. If $\mathcal{O}_2$ is a refinement of $\mathcal{O}_1$ (i.e., the uncertainty contained in $\mathcal{O}_2$ is less than that relative to $\mathcal{O}_1$), then $\mathcal{O}_1$ and $\mathcal{O}_2$ could correspond to the same Ruspini partition π once the uncertainty is eliminated from $\mathcal{O}_1$ and $\mathcal{O}_2$ (i.e., for each $u \in U$, the degrees of uncertainty $h_1^i(u), \ldots, h_n^i(u)$ with $i \in \{1, 2\}$ are distributed among the blocks of $\mathcal{O}_i$ in order to obtain π). Then, we have more detailed information about the entropy of π by taking into account $\mathcal{I}_{\mathcal{O}_2}$ instead of $\mathcal{I}_{\mathcal{O}_1}$ by considering that $\mathcal{I}_{\mathcal{O}_2} \subseteq \mathcal{I}_{\mathcal{O}_1}$.

Fuzzy Orthopartitions and Credal Partitions. Credal partitions are relevant structures in evidential clustering used to represent partitions in cases of partial knowledge concerning the membership of elements to classes [21]. Assuming that $C = \{C_1, \ldots, C_n\}$ is a standard partition of a universe $U = \{u_1, \ldots, u_l\}$, a credal partition is mathematically defined as a collection $m = \{m_1, \ldots, m_l\}$ of basic belief assignments. By a basic belief assignment (bba), we mean a function $m_i : 2^C \to [0, 1]$ verifying the condition $\sum_{A \subseteq C} m_i(A) = 1$. We also assume here that m_i is normalized, namely $m_i(\emptyset) = 0$. Let $A \subseteq C$, the *mass of belief* $m_i(A)$ quantifies the evidence supporting the claim "u_i belongs to a block of A" [25,26]. Credal partitions subsume the concept of fuzzy probabilistic partitions, which are composed of all Bayesian bbas, i.e., bbas assigning a non-zero degree only to the singletons of 2^U [4].

A correspondence between credal partitions and fuzzy orthopartitions is provided in [7], where to a fuzzy orthopartition we attach the following semantics. Let (μ_i, ν_i) be an IFS on U, $\mu_i(u)$ and $\nu_i(u)$ are respectively the degrees of belief that "u belongs to the class i" and "u does not belong to the class i". Moreover, according to this interpretation, $pl_i(u) = 1 - \nu_i(u)$ is the degree of *plausibility* that "u belongs to the class i". Therefore, fuzzy orthopartitions like credal partitions, are understood as extensions of fuzzy probabilistic partitions.

Let us focus on ORP and $\mathcal{M}$ being the sets of all fuzzy orthopartitions and credal partitions related to a partition $C = \{C_1, \ldots, C_n\}$ of a universe $U = \{u_1, \ldots, u_l\}$. In the next definition, a class of credal partitions is associated to a given fuzzy orthopartition.

Definition 11. *Let $\mathcal{O} \in ORP$. Then, we put $\mathcal{F}(\mathcal{O}) = \{m \in \mathcal{M} \mid m_j(\{C_i\}) = \mu_i(u_j) \text{ and } \sum_{\{A \mid C_i \in A\}} m_j(A) = pl_j(u_i), \forall i \in \{1, \ldots, n\} \text{ and } j \in \{1, \ldots, l\}\}$.*

Vice versa, a fuzzy orthopartition is associated to each credal partition as follows.

Definition 12. *Let* $m \in \mathcal{M}$. *Then, we consider* $\mathcal{O}_m = \{(\mu_1, \nu_1), \ldots, (\mu_n, \nu_n)\}$ *such that for each* $i \in \{1, \ldots, l\}$ *and* $j \in \{1, \ldots, n\}$,

$$\mu_j(u_i) = m_i(C_j) \quad and \quad \nu_j(u_i) = 1 - \sum_{\{A \mid C_j \in A\}} m_i(A).$$

(Of course, we have, $pl_j(u_i) = \sum_{\{A \mid C_j \in A\}} m_i(A)$).

It can be proved that

(a) for each $\mathcal{O} \in ORP$, $\mathcal{F}(\mathcal{O})$ can be made of 0, 1, or infinite credal partitions;
(b) for each $m \in \mathcal{M}$, $\mathcal{O}_m$ is a fuzzy orthopartition;
(c) for each $\mathcal{O} \in ORP$ and $m \in \mathcal{F}(\mathcal{O})$, $\mathcal{O}_m = \mathcal{O}$.

According to the previous points, we finally conclude that fuzzy orthopartitions can be considered more general than credal partitions that are made of normalized bbas.

2.3   Partial Equivalence Relations

It is well-known that any partition is in bijection with an equivalence relation. Thus, we can naturally ask if given an orthopartition, a corresponding relation can be defined. The intuition is that to an orthopartition there corresponds a *partial* equivalence relation, i.e., an equivalence relation of which we know that some elements are equivalent, some are not equivalent and some other connections are possible but unknown. This kind of partial equivalence relation can be expressed by means of two relations: an equivalence R_E and a similarity R_S one, which represent the necessary and the possible relationships occurring among objects. In other words, we can model a partial equivalence relation with an orthopair of relations (R_E, R_N) on the cartesian product of the set of objects $X \times X$, such that:

- R_E is an equivalence relation (symmetric, reflexive, transitive);
- $Bnd_R = R_S$ is a similarity relation (symmetric, reflexive);
- there holds a kind of transitivity property: if $(x_i, x_j) \in R_S$ and $(x_j, x_k) \in R_E$ then $(x_i, x_k) \in R_S$.

Example 3. Let $U = \{1, 2, \ldots, 10\}$ and $\mathcal{O} = \{O_1, O_2, O_3\}$ an orthopartition of U where: $O_1 = (\{1, 2\}, \{9, 10\})$, $O_2 = (\{9\}, \{1, 2\})$ and $O_3 = (\emptyset, \{1, 2, 9\})$. Then, the partial equivalence relation defined by $\mathcal{O}$ is

$$R_E = \{(x_i, x_i), (x_1, x_2), (x_2, x_1)\}$$
$$R_S = R_E \cup P \cup P^S$$

where
$$P = \{(x_1, x_3), (x_1, x_4), \ldots, (x_1, x_8), (x_2, x_3), \quad (x_2, x_4), \ldots, (x_2, x_8), (x_9, x_3),$$

$(x_9, x_4), \ldots, (x_9, x_8), (x_{10}, x_3), (x_{10}, x_4), \ldots, (x_{10}, x_8)\}$ and P^S is made of the symmetric pairs in P. We notice that supposing that the uncertainty in R_S will be solved, not all pairs in R_S could be part of the final equivalence relation. For instance, given that $(x_1, x_2) \in R_E$ we could not have both (x_1, x_3) and (x_3, x_9), since (x_1, x_9) is not listed as a possible relationship in R_S.

It can be proved that given an orthopartition, there exists a partial equivalence relation representing it. The converse, i.e., whether given a pair of equivalence-similarity relations there exists an orthopartition O such that the partitions obtained from R_E, R_S are consistent (or weakly consistent) with O is still an open problem. In any case, the obtained orthopartition would not be unique.

Example 4. Let us consider the following table: Assuming the standard indis-

	a	b	c
x_1	1	1	0
x_2	1	*	0
x_3	0	2	1

cernibility relation in rough set theory, from the table we get the pair $R_E = \{(x_i, x_i)\}$, $R_S = R_E \cup \{(x_1, x_2), (x_2, x_1)\}$. Once the missing value in the table will be known, one of the two partitions could be defined: $\Pi_1 = \{\{x_1, x_2\}, \{x_3\}\}$ and $\Pi_2 = \{\{x_1\}, \{x_2\}, \{x_3\}\}$.

In this case, there does not exist an orthopartition O such that Π_1 and Π_2 are consistent with O and no other partition is consistent with O. On the other hand both partitions are weakly consistent with $O_1 = \{(\{x_3\}, \{x_1, x_2\}), (\{x_1\}, \{x_3\}), (\emptyset, \{x_1, x_3\})\}$ and $O_2 = \{(\{x_3\}, \{x_2, x_1\}), (\{x_2\}, \{x_3\}), (\emptyset, \{x_2, x_3\})\}$.

3 Machine Learning

In this section, we explain how orthopartitions can be used in machine learning in two directions: to define evaluation measures for soft clustering and to generalize the standard decision-tree approach for partial label settings and with the possibility to abstain from taking a decision.

3.1 Rough and Three-Way Clustering

As previously mentioned, an orthopartition represents a partition which is only partially known, due to uncertainty or insufficient information. The problem of how to represent uncertain assignments of objects to sets has also been studied from an application-oriented perspective in the setting of soft clustering. This latter refers to clustering techniques (i.e., machine learning algorithms that aim to group objects that can be considered *similar*, in some sense) in which the assignment of objects to clusters is allowed to be uncertain or partial. Among such soft clustering methods, the uncertainty representation framework of orthopartitions has relevant connections with rough and three-way clustering.

Considering rough clustering, as initially proposed in [23], a single cluster C_i is defined as an interval set $C_i = (\underline{C_i}, \overline{C_i})$, where $\underline{C_i}$ is the cluster's lower bound and $\overline{C_i}$ is its upper bound. The set $C = \{(\underline{C_1}, \overline{C_1}), (\underline{C_2}, \overline{C_2}), \ldots, (\underline{C_n}, \overline{C_n})\}$ is then a rough clustering if and only if it satisfies the following properties:

(Ax RC1) $|\{C_i \mid x \in \underline{C_i}\}| \leq 1, \forall x \in U$: an object x can be assigned to at most a cluster's lower bound,

(Ax RC2) $x \in \underline{C_i} \implies x \in \overline{C_i}, \forall x \in U$: if an item x is in a cluster's lower bound, it is assigned to its upper bound as well,

(Ax RC3) $\{C_i \mid x \in \underline{C_i}\} = \emptyset \iff |\{C_i \mid x \in \overline{C_i}\}| \geq 2, \forall x \in U$: an object x not assigned to any cluster's lower bounds must be contained in at least two clusters' upper bound.

In the case of three-way clustering [28], by contrast, a different set of defining axioms is adopted. Indeed, even though clusters are represented as interval sets, a collection $T = \{T_1, \ldots, T_n\}$ is said to be a three-way clustering if it satisfies the following properties:

(Ax TWC1) $\forall T_i, \underline{T_i} \neq \emptyset$, i.e., the lower bounds of the clusters are required to contain at least an object;

(Ax TWC2) $\bigcup_i \overline{T_i} = U$ (coverage requirement);

(Ax TWC3) $\forall T_i, T_j$ it holds that $\underline{T_i} \cap \underline{T_j} = \emptyset$.

It is easy to observe that the axioms for rough clustering and three-way clustering are not equivalent: a cluster lower bound is allowed to be empty in a rough clustering, while an object is allowed to belong to only one boundary in three-way clustering. Nonetheless, both structures can be unified through orthopartions [14]. Clearly, any rough clustering $C = \{(\underline{C_i}, \overline{C_i})\}_i$ can be equivalently represented as an orthopartition $\mathcal{O}(C) = \{(\underline{C_i}, \overline{C_i}^c)\}_{C_i \in C}$: it is not hard to see that Axs RC1-3 corresponds to Axs O1-3. Similarly, three-way clustering $T = \{(\underline{T_i}, \overline{T_i})\}_i$ can be represented as an orthopartition $\mathcal{O}(T) = \{(\underline{T_i}, \overline{T_i}^c)\}_{T_i \in T} \cup \{O_\eta\}$, where O_η is a noise cluster (P, N) defined as $P_{O_\eta} = \emptyset$, $Bnd_{O_\eta} = \{x \in U : \exists! T_i \in T \text{ s.t. } x \notin \underline{T_i} \wedge x \in \overline{T_i}\}$ and $N_{O_\eta} = (P_{O_\eta} \cup Bnd_{O_\eta})^c$. Intuitively, O_η contains in its boundary all objects that can be considered as outliers for the clusters in T, i.e., all objects which belong to only one upper bound.

More recently, the correspondence between rough and three-way clustering proved by means of orthopartitions has been applied also to the study of more general forms of soft clustering, including fuzzy clustering (where assignment of objects to clusters is represented in terms of fuzzy partitions) and evidential clustering (where assignment of objects to clusters is modeled by means of belief functions). In particular, in [15] an algorithm to convert rough or three-way clusterings into evidential clustering and vice-versa was proposed based on the representation of the former two classes of structures in terms of orthopartitions, while in [16] it was shown that any soft clustering can be represented in terms of a probability distribution over orthopartitions.

The correspondence between orthopartitions and different forms of soft clustering, aside from its theoretical significance, has been applied to the evaluation of clustering results: indeed, the entropy and mutual information metrics defined in Sect. 2 have been applied to compare the results of a rough or three-way clustering algorithm with a given ground truth cluster assignment [14]; while the correspondence between soft clusterings and distributions of orthopartitions has been applied to design generalized metrics to evaluate the results of soft clustering algorithms [16].

3.2 Decision Trees

The theory of orthopartitions has also been applied in the setting of Machine Learning to the design of Decision Tree algorithms, focusing in particular either on weakly-supervised or three-way Decision Trees [11,13].

In the case of a three-way Decision Tree, the aim is to construct a Machine Learning model that is able to partially abstain on the more uncertain instances. Let $\mathcal{DT} = \langle S, A, d \rangle$ be a decision table with $S = \{x_1, ..., x_{|S|}\}$ a set of objects; $A = \{a_1, ..., a_m\}$ a set of attributes (features) and d a decision that assigns a label to each object. Let the possible values for a given feature a be v_i^a and $S_i^a = \{x \in S | v_a(x) = v_i^a\}$ be the set of instances that have value v_i^a for feature a. If a is a continuous attribute, then, given a threshold value v_i^a, one can consider:

$$v_a(x_k) = \begin{cases} 1 & v_a(x_k) \geq v_i^a \\ 0 & \text{otherwise} \end{cases}. \tag{7}$$

The optimal classification C_i^a for S_i^a is the classification (i.e., the set of labels) obtained by solving locally on the tree nodes an optimisation procedure that assigns to each node the corresponding set of labels associated with the minimal risk. That is, if $Pr(y|S_i^a) = \frac{|\{x_k \in S_i^a : d(x_k) = y\}|}{|S_i^a|}$, then:

$$C_i^a = \arg\min_{Z \subseteq Y} \alpha(Z) \cdot \sum_{y \in Z} Pr(y|S_i^a) + \sum_{y \notin Z} Pr(y|S_i^a) \cdot \frac{\sum_{y' \in Z} \epsilon_{y'y}}{|Z|} \tag{8}$$

where $\alpha(Z)$ is the cost of abstention associated with set of labels Z, whereas $\epsilon_{y'y}$ is the cost of predicting y' when the true label is instead y. Note, that whenever $|C_i^a| > 1$, then the three-way Decision Tree does not predict a single label but rather a set, associated with an epistemic semantics: the correct class is in C_i^a, but it is not known which it is. Since this classification determines an orthopartition $\mathcal{O}_a$, the *mutual information* of $\mathcal{O}_a$ w.r.t. S can be computed as described previously in Sect. 2 and then we select the feature a^* which results in the maximum mutual information value, subsequently recurring on the subsets of S determined by feature a^* until a termination criterion is met. In [11] it is shown that ensembles of such three-way Decision Trees can significantly out-perform both standard supervised machine learning algorithm as well as other state-of-the-art three-way classifiers, both in terms of accuracy as well as coverage (i.e., the number of points that are assigned to single classes rather than to sets).

In the case of weakly-supervised learning, orthopartitions are used to design Decision Tree algorithms that are able to learn from partially labelled data. In a standard Decision Tree learning algorithm, one considers, at each internal node N_i and for each attribute a, a possible split point p_a (the technique to determine the split point depends on the specific adopted learning algorithm) and evaluate that split by computing its induced mutual information:

$$
\begin{aligned}
L(p_a, N_i) &= P(v_a \geq p_a | N_i) \cdot \sum_{y \in Y} P(y | v_a \geq p_a, N_i) \cdot log_2 P(y | v_a \geq p_a, N_i) \\
&+ P(v_a < p_a | N_i) \cdot \sum_{y \in Y} P(y | v_a < p_a, N_i) \cdot log_2 P(y | v_a < p_a, N_i) \quad (9) \\
&= P(v_a \geq p_a | N_i) \cdot H(S | v_a \geq p_a, N_i) \\
&+ P(v_a < p_a | N_i) \cdot H(S | v_a < p_a, N_i)
\end{aligned}
$$

Then, an attribute a and the corresponding split point p_a is selected, by identifying the minimal value of $L(p_a, N_i)$. In a leaf node N_l, a decision label $d(N_l)$ is selected, according to the Bayes optimal decision rule:

$$
d(N_l) = argmax_{y \in Y} P(y | N_l) \quad (10)
$$

These formulas are not directly applicable in the weakly supervised setting, since the labeling of the instances does not form a partition but, more generally, an orthopartition: indeed, instances whose class is only partially known can be assigned to the boundaries of the corresponding orthopairs (then, each orthopair corresponds to a class). This means, however, that the definitions of entropy given for orthopartitions in Sect. 2 can be used for training the weakly supervised Decision Tree. We can, thus, redefine the value of $L(P_a, N_i)$ by using the mutual information for orthopartitions, comparing attributes and split points in the same way as for classical Decision Trees. In regard to inference, i.e., obtaining predictions for new instances, it can be observed that each leaf N_l defines a *basic belief assignment* (bba) m:

$$
m_{N_l}(Z \subseteq Y) = \frac{|\{x \in S | t(x) = Z \wedge x \in N_l\}|}{|N_l|} \quad (11)
$$

from which a *pignistic probability* distribution can be computed:

$$
p_{bet}(y) = \sum_{y \in Z} \frac{m(Z)}{|Z|} = \frac{1}{|N_l|} \cdot (|P_y| + \sum_{x \in Bnd_y} \frac{1}{Bnd(x)}) \quad (12)
$$

Thus, predictions can be made based on the obtained probability distribution over the class labels, and the same probability distributions can also be combined, e.g. when training an ensemble model such as a Random Forest. In particular, in [11], it was shown that such ensembles of weakly-supervised Decision Trees can have significantly better performance than other state-of-the-art weakly supervised machine learning methods, with a reduced computational cost: indeed, the cost to train such Decision Trees is asymptotically the same as that of training standard Decision Trees.

4 Conclusion

We reported the basic definitions and the results obtain in the study of partitions with uncertainty. Of course, there are still some open issues. The on-going works include, at first, the development of the relation-orthopartition link just sketched in Sect. 2.3. Then, we are extending the correspondance between credal partitions and fuzzy orthopartitions to the case of non-normalized bbas. In the case of machine learning, we are completing a comparison of rough and three-way clustering algorithms using the measures given in Sect. 3.1.

Possible future works include: a comparison of soft clustering algorithms by means of their representations as probability distributions [15]; the investigation of the possibility to apply the fuzzy orthopartitions and their entropy to soft clustering; the comparison and/or integration with other uncertainty representation theories. Finally, it could be interesting to construct orthopartitions as approximations of fuzzy partitions following the approach provided in [29], where special partitions, made by rough sets and called *three-way fuzzy partitions*, are generated from a fuzzy partition using the notion of shadowed sets. In general, we believe that orthopartitions and their generalized versions could represent a useful tool to deal with uncertainty and vagueness whenever the available knowledge is expected to be organized as a partition.

References

1. Aguzzoli, S., Boffa, S., Ciucci, D., Gerla, B.: Finite IUML-algebras, finite forests and orthopairs. Fundam. Informaticae **163**(2), 139–163 (2018)
2. Atanassov, K.T.: Review and new results on intuitionistic fuzzy sets. Math. Found. Artif. Intell. Semin. Sofia (1988). Preprint IM-MFAIS1-88. Reprinted: Int. J. Bioautom. **20**(S1), S7–S16 (2016)
3. Atanassov, K.: Intuitionistic fuzzy sets. In: Intuitionistic fuzzy sets, pp. 1–137. Springer, Berlin (1999). https://doi.org/10.1007/978-3-7908-1870-3
4. Bezdek, J.C., Keller, J., Krisnapuram, R., Pal, N.R.: Fuzzy Models and Algorithms for Pattern Recognition and Image Processing. THFSS, vol. 4. Springer, Boston (1999). https://doi.org/10.1007/b106267
5. Blamey, S.: Partial logic. In: Gabbay, D., Guenthner, F. (eds.) Handbook of Philosophical Logic. Synthese Library, vol. 166, pp. 1–70. Springer, Dordrecht (1986)
6. Boffa, S., Ciucci, D.: Fuzzy orthopartitions and their logical entropy. In: Ciaramella, A., Mencar, C., Montes, S., Rovetta, S. (eds.) Proceedings of WILF 2021. CEUR Workshop Proceedings, vol. 3074. CEUR-WS.org (2021)
7. Boffa, S., Ciucci, D.: A correspondence between credal partitions and fuzzy orthopartitions. In: Le Hégarat-Mascle, S., Bloch, I., Aldea, E. (eds.) Belief Functions: Theory and Applications. BELIEF 2022. Lecture Notes in Computer Science, vol. 13506, pp 251–260. Springer, Cham (2022). https://doi.org/10.1007/978-3-031-17801-6_24
8. Boffa, S., Ciucci, D.: Logical entropy and aggregation of fuzzy orthopartitions. Fuzzy Sets Syst. (2022). https://doi.org/10.1016/j.fss.2022.07.014
9. Boffa, S., Ciucci, D.: Orthopartitions and possibility distributions. Fuzzy Sets Syst. (2022). https://doi.org/10.1016/j.fss.2022.04.022

10. Burillo, P., Bustince, H.: Estructuras algebraicas en conjuntos ifs. In: II Congresso Nacional de Logica y Tecnologia Fuzzy, Boadilla del monte, Madrid, Spain, pp. 135–147 (1992)
11. Campagner, A., Cabitza, F., Ciucci, D.: The three-way-in and three-way-out framework to treat and exploit ambiguity in data. Int. J. Approximate Reasoning **119**, 292–312 (2020)
12. Campagner, A., Ciucci, D.: Measuring uncertainty in orthopairs. In: Antonucci, A., Cholvy, L., Papini, O. (eds.) ECSQARU 2017. LNCS (LNAI), vol. 10369, pp. 423–432. Springer, Cham (2017). https://doi.org/10.1007/978-3-319-61581-3_38
13. Campagner, A., Ciucci, D.: Three-way and semi-supervised decision tree learning based on orthopartitions. In: Medina, J., et al. (eds.) IPMU 2018. CCIS, vol. 854, pp. 748–759. Springer, Cham (2018). https://doi.org/10.1007/978-3-319-91476-3_61
14. Campagner, A., Ciucci, D.: Orthopartitions and soft clustering: soft mutual information measures for clustering validation. Knowl.-Based Syst. **180**, 51–61 (2019)
15. Campagner, A., Ciucci, D., Denœux, T.: Belief functions and rough sets: Survey and new insights. Int. J. Approximate Reasoning **143**, 192–215 (2022)
16. Campagner, A., Ciucci, D., Denœux, T.: A distributional approach for soft clustering comparison and evaluation. In: Le Hégarat-Mascle, S., Bloch, I., Aldea, E. (eds.) Belief Functions: Theory and Applications. BELIEF 2022. Lecture Notes in Computer Science, vol. 13506. Springer, Cham (2022). https://doi.org/10.1007/978-3-031-17801-6_1
17. Ciucci, D.: Orthopairs: a simple and widely used way to model uncertainty. Fundam. Inform. **108**(3–4), 287–304 (2011)
18. Ciucci, D.: Orthopairs and granular computing. Granular. Computing **1**, 159–170 (2016)
19. Ciucci, D., Dubois, D., Lawry, J.: Borderline vs. unknown: comparing three-valued representations of imperfect information. Int. J. Approx. Reason. **55**(9), 1866–1889 (2014)
20. Ciucci, D., Dubois, D., Lawry, J.: Borderline vs. unknown: comparing three-valued representations of imperfect information. Int. J. Approximate Reasoning **55**(9), 1866–1889 (2014)
21. Denœux, T., Masson, M.H.: Evclus: evidential clustering of proximity data. IEEE Trans. Syst. Man Cybern. Part B (Cybernetics) **34**(1), 95–109 (2004)
22. Ellerman, D.: An introduction to logical entropy and its relation to shannon entropy (2013)
23. Lingras, P., West, C.: Interval set clustering of web users with rough k-means. J. Intell. Inf. Syst. **23**(1), 5–16 (2004)
24. Ruspini, E.H.: A new approach to clustering. Inf. Control **15**(1), 22–32 (1969)
25. Shafer, G.: A Mathematical Theory of Evidence. Princeton University Press, Princeton (1976)
26. Smets, P., Kennes, R.: The transferable belief model. Artif. intell. **66**(2), 191–234 (1994)
27. Yao, Y.: Interval sets and interval-set algebras. In: Baciu, G., Wang, Y., Yao, Y., Kinsner, W., Chan, K., Zadeh, L.A. (eds.) Proceedings of the 8th IEEE International Conference on Cognitive Informatics, ICCI 2009, 15–17 June 2009, Hong Kong, China, pp. 307–314. IEEE Computer Society (2009)

28. Yu, H.: A framework of three-way cluster analysis. In: Polkowski, L., et al. (eds.) IJCRS 2017. LNCS (LNAI), vol. 10314, pp. 300–312. Springer, Cham (2017). https://doi.org/10.1007/978-3-319-60840-2_22
29. Zhao, X.R., Yao, Y.: Three-way fuzzy partitions defined by shadowed sets. Inf. Sci. **497**, 23–37 (2019)

Rough Sets and Fuzzy Sets in Interactive Granular Computing

Andrzej Skowron[1,2]([✉]) [iD] and Soma Dutta[3] [iD]

[1] Systems Research Institute, Polish Academy of Sciences, Newelska 6,
Warsaw 01-447, Poland
`skowron@mimuw.edu.pl`
[2] Multidisciplinary Research Center, UKSW, Marii Konopnickiej 1, 05-092
Dziekanów Leśny, Poland
[3] University of Warmia and Mazury in Olsztyn, Słoneczna 54, 10–710 Olsztyn, Poland
`soma.dutta@matman.uwm.edu.pl`

Abstract. This paper is an attempt to present some grounds that a change is needed in the way of viewing mathematical tools, such as fuzzy sets, rough sets, in the context of classifying and approximating concepts pertaining to the real physical complex phenomenon. The paper argues in favour of developing models going beyond the pure mathematical manifold. The main idea is not to develop a theory only based on gathered data, rather to incorporate the methods of perception and real physical interactions through which the data is obtained. In this regard, a primary proposal has been put forward to model fuzzy sets and rough sets in the framework of Interactive Granular Computing (IGrC).

Keywords: Interactions · (Interactive) Granular computing · Perception · Complex granule (c-granule) · Informational granule (ic-granule) · Grounding problem · Control of c-granule · Information system · Decision system · Rough sets · Fuzzy sets

1 Introduction

Existing approaches to soft computing, such as rough sets, fuzzy sets, and other tools used in machine learning lack in considering the processes of perception and interaction with the physical world while modeling a (vague) concept. There are two prevalent traditions of mathematical modelling. One, which is purely mathematical, considers that the sets are given. For example, in rough set approach, the starting point is the universe of objects and an indiscernibility or a similarity relation; further developments for approximating concepts are done on this basis. In the context of fuzzy sets, the universe of objects and some basic fuzzy membership functions, relative to some concepts, are assumed a priori; based on those fuzzy membership functions other complex concepts are induced.

In the second tradition of modelling, may be called constructive, it is assumed that objects are partially perceived by means of some features or attributes, and

J. Yao et al. (Eds.): IJCRS 2022, LNAI 13633, pp. 19–29, 2022.
https://doi.org/10.1007/978-3-031-21244-4_2

only a partial information about these objects in the form of vectors of values of attributes is available. On this basis the indiscernibility or similarity relations are defined and further developments are carried out (assuming often that the information about the sets of objects is also partial). This approach is often followed in rough sets. In the context of fuzzy sets also often this approach is used, i.e., the fuzzy membership functions are constructed based on a set of attributes and then they are ascribed over the universe of objects.

Both of these traditional modelling do not take into account how the process of perceiving the values of attributes is realised, where and how to access the concerned objects in the physical space, and why those attributes are selected. Hence, clearly the perception and action are out of the scope of such practices of modelling. However, this is crucial for many tasks, especially, when the subject of analysis is a complex phenomena in the real physical world. Consequently, characterization of the state of the complex physical phenomena by a fixed set, or a priori set of attributes becomes irrelevant. From a similar concern, the researchers in [11] proposed to extend Turing test by embedding into it the challenges related to action and perception. So, for an intelligent agent it is important that the model should incorporate the information such as, how a function representing a particular vague concept is learned from the uses of the community, which parameters are to be considered as crucial in (approximately) defining a vague concept, how the values of these parameters are observed or measured etc; otherwise a non-human system cannot derive the relevant information about the so far unseen cases.

Hence, we need an extension of the existing approaches where apart from the information about a physical object, a specification of how the information label of a physical object is physically linked to the actual object also can be incorporated.

Till now, in different works (see, *e.g.*, [4,5,9,14,15]), we tried to introduce what do we mean by Interactive Granular Computing (IGrC) and how it is different from other existing theories from the perspective of modeling computations in intelligent systems dealing with complex phenomena.

Research on IGrC aims at developing a foundation for modeling computations in the intelligent systems dealing with complex phenomena. Reasoning performed on such computations should guide the intelligent systems toward achieving their goals. Many evidences in favour of the need of such foundations can be found in different domains such as multi-agent systems, robotics, machine learning, cyber physical systems, Internet of things and different branches of computational intelligence. However, what we need is a common foundation for the computing model. In particular, the issue of combining together different research directions such as reasoning, action and perception is pointed out in discussion on the necessity of the modification of the Turing [8,11]. In IGrC it is emphasized that the relevant computing model should be based on combination of abstract and physical objects, and this is the reason that for modeling such systems we should go beyond the purely mathematical objects[1]. The aim of the current paper is to outline the consequences of developing perceptual approaches to rough sets and fuzzy sets based on IGrC. The need for such approaches is motivated by the practical applications. As an example one can consider the problem of estimating the risk of traffic on a particular crossroads, in a given town; formulation of such a problem depends on the perception of the current situation in the physical world, rather than a fixed mathematical model of it.

This paper is organized as follows. In Sect. 2 we outline the role of the control of a c-granule, which is the basic building block of IGrC model. In Sect. 3 we discuss the approach to rough sets as well as to fuzzy sets in the framework of perception based approximation of concepts. In Conclusion we point out the necessity of developing new methods of reasoning in the further developments of rough sets as well as fuzzy sets.

2   A Brief Description of the Basic Building Blocks of IGrC

In Interactive Granular Computing (IGrC) (see papers on interactive computations on https://dblp.uni-trier.de/pers/hd/s/Skowron:Andrzej) the necessity of introducing complex granules (c-granules, for short) is recognised.

The computations in the IGrC model are realized on the interactive complex granules and the progress of the computation process is based on the consequences of the interactions occurring in the physical world. Hence, the computational models in IGrC cannot be constructed solely in an abstract mathematical space. The proposed model of computation based on complex granules seems to be of fundamental importance for developing intelligent systems dealing with complex phenomena, in particular in the areas such as Data Science, Internet of Things, Wisdom Web of Things, Cyber Physical Systems, Complex Adaptive Systems, Natural Computing, Software Engineering, and applications based on Blockchain Technology, etc.

In the proposed approach, we assume that physical objects exist in the physical space and are embedded into its parts. The physical objects are interacting in the physical space, and thus some collections of physical objects may create dynamical systems in the physical space. So, it is important to explain how properties of these objects and interactions among them can be perceived by the c-granules. In our attempt, to design c-granules with the ability of perceiving physical objects and their interactions, this is realised by the control mechanism of a c-granule. The control of a c-granule works based on the informational complex granules (ic-granules) lying within its scope and a special kind of reasoning mechanism over them, may be called judgment.

Informational complex granules (ic-granules) are constructed over two basic ingredients: abstract and physical; we may count these two ingredients respec-

---

[1] For more details the readers are referred to papers on IGrC on https://dblp.org/pid/s/AndrzejSkowron.html.

tively as informational and physical objects. The abstract ingredients of an ic-granule contains families of formal specifications of spatio-temporal windows labelled by the information specific for a given c-granule or a family of c-granules, and this information is expressed in a formal or natural language. The information layer of an ic-granule may contain formulas and their (expected or real) degrees of satisfiability at a given moment of local time of the control of a c-granule related to some collection of physical objects, as well as the formal specifications of the spatio-temporal windows indicating the location and (perception) time of those physical objects. It can be a Boolean atomic formula of the form $a =_t v$ specifying that the value of the attribute a is v at a given moment of local time t, or can be a more compound expression, encoded in an information system containing results of measurements (over time) in the form of vectors of values for attributes of the objects related to the given formal specifications of the spatio-temporal windows, present in the informational layer of the ic-granule.

The physical layer of any ic-granule is basically a c-granule and is divided into three parts, namely soft_suit, link_suit and hard_suit. Each of these parts is a collection of physical objects.

The behavior of the control of a given c-granule can be divided into cycles. Each cycle of the control of a given c-granule starts from a current configuration (i.e., a family) of the ic-granules lying within its scope. This configuration contains a distinguished ic-granule with information representing the perception of the current situation. Each cycle may execute several steps such as, modification, deletion, suspension of ic-granules or generation of some new ic-granules from the current configuration. It should be noted that a special kind of ic-granule, called implementation ic-granule, is used for generation of new ic-granules from their formal specifications. Once a new configuration of ic-granules is created the control measures the features of the new physical objects in the scope of the newly developed ic-granules and/or matches or aggregates the information with that of the previous ic-granules using its judgment mechanism. The cycle ends when the control gathers perception, to a satisfactory degree, about the current configuration, and becomes able to take a relevant decision with respect to the goal of the computation process.

Formal specification of many complex tasks or formal specification of the needs of the c-granule may be thought of as a complex game consisting of a family of complex vague concepts, labelled by the actions or plans (represented by the relevant formal specifications in the information layers of the respective ic-granules) that to be performed when the concepts are satisfied to a satisfactory degree. These complex vague concepts may describe, e.g., invariants which should be preserved to a satisfactory degree, conditions representing degrees of risk of disaster in the environment perceived by the system, safety properties of trajectory of granular computations, conditions representing the quality of the current path from the point of view of carrying out computation toward the target goals, or risk concerning a possibility that the current needs are no longer achievable etc. It should be noted that these complex vague concepts (usually described in a fragment of a natural language) should be learned from the data and domain knowledge with the use of physical laws [2]. During perception of the current situation in the physical world the control uses its judgment tools over information gathered from information layers of dynamically changing (by control and the environment) configurations of the ic-granules. Moreover, the concepts as well as their labels, involved in a complex game, evolve with time. Hence, the control should have some adaptive strategies allowing relevant modification of the complex game.

We would like to emphasize the role of reasoning, called judgment, for the control of a c-granule. The further development of judgment methods will play the principal role in further development of intelligent systems [6]. These reasoning methods are far beyond the existing deductive methods, which are already well developed in mathematical logic or even inductive reasoning that are widely used in Machine Learning. The required reasoning mechanism should take into account the experience as well as explanation of the behaviour of the (intelligent) agent. In order to develop such a reasoning mechanism a new computing model, which can make it possible to perform reasoning from sensory measurement to perception (*i.e.*, understand the perceived situation), is required. We propose to base on such a model in the context of interactive complex granules (c-granules), where the informational parts associated to them are grounded in the physical world.

3   Rough Sets and Fuzzy Sets in IGrC

Modelling of rough sets [12] and fuzzy sets [16] in the IGrC framework requires some substantial changes in the fuzzy and rough set approaches. For example, a fuzzy set is defined by a membership function f from a given set X of objects into the interval $[0, 1]$ of reals. In this way, fuzzy sets are completely embedded in the "mathematical manifold" [3]. As the fuzzy concepts are embedded in the real world, in the IGrC framework, their semantics should be perceived by the control of the c-granules[2] . Thus, based on the perception of the real situations (objects), by the control of a c-granule, the estimation of the membership functions (and their values) should be done. One should be aware that physical situations (objects) cannot be omitted in such perception based modeling of fuzzy sets. Hence, instead of pure mathematical objects such as functions some non-pure mathematical constructs should be also used while modeling fuzzy sets from the perspective of c-granules. Partial information about the perceived situations is used by the control of a c-granule as input for the estimation of the mathematical membership functions, and these functions can only be treated as temporary models. These estimated functions might be adapted by the relevant strategies based on the changes in the perceived data recorded by the c-granule at further points of time. In the process of constructing an estimation for the membership function the control of a c-granule strives to better understand the

---

[2] The discussed approach is consistent with the idea of perception presented in the book [10] by Alva Nöe.

perceived situations, taking into account different constraints e.g., time or other resources.

The control of a c-granule is also responsible for providing the specifications for conducting the perception process in the environment in order to obtain the relevant data for estimating the functions. On the basis of perceived data by the c-granule estimations of the membership functions are constructed and adapted according to the observed changes in the perceived data over the period of time.

Thus, one can notice that the perception process of the current situation happens over a period of time in which the control of the c-granule collects the necessary information in order to select a proper estimation of the membership function. This includes (i) reasoning properly leading the control to focus on the measurements and/or exploration of the relevant fragments of the physical space (ii) generation of the relevant configurations of physical objects and initiating interactions among them (iii) reasoning about the properties of the physical objects which are not directly measurable with the use of the physical laws and/or domain knowledge, (iv) providing the right dialogue strategies with users and/or domain experts, etc.

Analogous comments can be made about rough sets. An attempt to link rough sets and IGrC for developing the perceptual rough set approach, requires introducing changes in definitions of the basic concepts of the existing rough set approach. In particular, this concerns the definition of attributes in information systems (decision systems) (see Fig. 1 and Fig. 2).

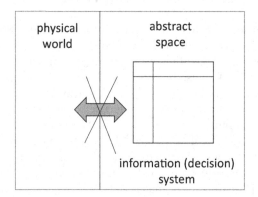

Fig. 1. In the existing approach to rough sets interactions with the physical world are eliminated. Attributes are mathematical functions.

Figures 1 and 2 illustrate the differences in construction of information (decision) systems in the existing approach to rough sets and the approach to rough sets in IGrC.

In the context of IGrC, the attributes are grounded in the physical world and their values are perceived using constructs which are not purely mathematical functions as it is in the case of existing rough set approach. These constructs are

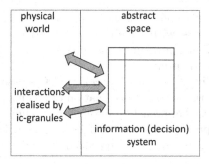

Fig. 2. Rough sets in IGrC: (i) perceiving values of attributes are based on interactions with the physical world, (ii) attributes are not pure mathematical functions; they are realised by ic-granules.

basically the ic-granules. They behave like physical pointers, linking the abstract objects (i.e., information) and the physical objects from the real-physical world (see also grounding problem [7]). The ic-granules make it possible to perceive the relevant fragments of the current situation in the physical world. Pointing out the relevant fragment is done using the formal specifications of the spatio-temporal windows available in the information layers of those ic-granules.

In this regard, let us note a few important features of the architecture of the control of the c-granule, which would help to learn a mathematical function based on the perception of the concerned situation.

– The c-granule contains an implementation module which enables its control to generate some relevant configurations of interacting physical objects based on the specification available in the language of the actuators.
– The control has the ability to perceive the properties of the generated configurations of the physical objects, their interactions, and reason about the perceived properties.
– The generated configurations of physical objects should be robust (to a highest possible degree) with respect to unexpected interactions with the environment.

The control of the c-granule makes decisions on the basis of acquired information about the currently perceived situation, and different (combined) reasoning methods. Some of them help to directly acquire new information about the currently perceived situation by activating sensors or actuators in the relevant parts of the physical space; some other provide strategies for extracting the relevant facts from the (domain) knowledge bases helping to better understand the situation. Some other methods can use physical laws making it possible to infer properties of the physical objects which are not directly accessible or measurable. The control of the c-granule performs such reasoning about the perceived situation aiming to select the most relevant decisions in the currently perceived situation. These selections are done by triggering concepts responsible for activating these decisions. Hence, one can see that, in particular, the aim of such reasoning performed by the control of the c-granule is to decide to which of these triggering concepts belongs, with certainty, to the currently perceived situation, *i.e.*, to which of the lower approximations of these triggering concepts belongs to the currently perceived situation.

Hence, in the intelligent systems based on IGrC, reasoning about computing the membership to the lower approximation of a concept can be much more compound than so far used approaches in rough sets. The issues related to developing the rough set approach in IGrC will be discussed in more detail in our future paper concerning the rough set approach in IGrC.

4   Conclusions

We have discussed some aspects of perceptual rough sets and fuzzy sets approaches in the framework of IGrC. From this discussion follow important consequences related to (inducing) construction of models of approximations in the rough set approach as well as construction of the fuzzy membership functions in the fuzzy sets, especially when we deal with complex phenomena.

It is worthwhile to cite here the opinion of Frederick Brooks [1]:

> *Mathematics and the physical sciences made great strides for three centuries by constructing simplified models of complex phenomena, deriving, properties from the models, and verifying those properties experimentally. This worked because the complexities ignored in the models were not the essential properties of the phenomena. It does not work when the complexities are the essence.*

The above quotation applies to the decision support systems and intelligent systems dealing with complex phenomena. It states that the traditional modeling used so far is not satisfactory. The existing modeling bases on the assumption that humans can create fixed mathematical models for describing the phenomena, that are perceived or observed by them. However, in the case of complex phenomena such decision making tool does not fit well. In traditional modeling of the fuzzy sets, the fuzzy membership functions are used, while in rough sets the information (decision) systems are used as purely mathematical objects on which the next considerations are carried out. In the proposed approach this is not the case. The considered decision support systems or intelligent systems, dealing with complex phenomena, are continuously linked to the physical space of objects (including humans) and interact with them. The perception process [13], controlled by the control of the c-granule, aims at perceiving the relevant data and accordingly adaptating the currently used models. This requires to develop the control mechanism on so called ic-granules creating links between the abstract and the physical objects.

So, according to the proposed approach to rough sets and fuzzy sets it is necessary to develop a new reasoning method from the perspective of interactive granular computation which can derive conclusions about the induced models and adapt new features based on dynamically changing nature of the task environment. Information systems or decision systems or fixed membership functions are considered in the context of the control of a c-granule and they are dynamically changing in time. One can get analogy of the considered systems to the laboratories in which continuously some new facts are perceived about the current situation in the physical space, and thus the abstract space and the physical space are continuously linked to make it possible to initiate, transmit and perceive interactions between physical objects.

Fig. 3. The illustrative process of information flow in a laboratory in the framework of IGrC (u_1, u_2 – users)

Let us consider a simple illustrative example related to sending messages between c-granules. Figure 3 presents a scheme representing functionality of a medical laboratory and this is stored in the information layer of the c-granule u_1. This information is used by the control of u_1 when the control decides to send a message to the patient regarding performing test t_1; the patient is represented in the figure by c-granule u_2. The localization of u_2 is already known to the control of u_1 and is represented by the relevant spatio-temporal window. In our example, this specification can be considered as the cellular phone number, say no, of u_2. Now, the first task of the control of u_1 is to encode the abstract description of message concerning the test t_1 and the number no of the cellular phone into the physical representation as the relevant physical state of a buffer in the next c-granule 'sender' (see Fig. 3). This is realized by the control of u_1 after it sends a message concerning t and no to the implementational module of u_1. This implementational module realizes this encoding by a special ic-granule linking a part of informational layer consisting of t and no with the buffer in 'sender' creating the hard_suit of this ic-granule. After this, control of u_1 may initiate process of sending t to the owner of the cellular phone with the number no by sending again to its implementational module a message 'send now'. This message is encoded by implementational module in the physical state of the c-granule sender what is recognized by c-granule 'network' as a command to transmit signal with encoded information through its cellular network. Control of the c-granule 'network' is responsible for selecting a proper root in the cellular network through which the signal with the encoded message will be transferred. Finally, this signal reaches to the c-granule 'receiver' linked to the c-granule u_2. In the buffer of c-granule 'receiver' the incoming signal changes the physical state of a buffer to the state corresponding to the message t and c-granule u_2 stores its abstract representation in the relevant information system. It should be noted that it is assumed that the control of the c-granule u_1 has in its informational layer all necessary information concerning c-granules presented in the figure. It is also worthwhile mentioning that the expected results by the control of u_1

concerning transferring a message t to u_2 may be disturbed due to unexpected interactions of the considered c-granules with their physical environment.

One can easily see that the reasoning methods used so far, *e.g.*, in rough sets based on partial inclusions of sets leading to conclusions related to approximations are only some simple examples in the very wide variety of reasoning methods necessary to develop systems working with approximations of concepts used by the decision support systems or intelligent systems dealing with complex phenomena (see Fig. 4).

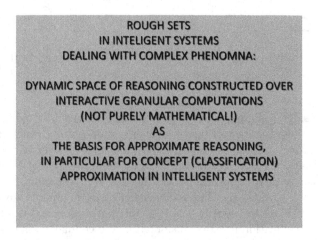

Fig. 4. Rough sets in decision support systems or intelligent systems dealing with complex phenomena.

Acknowledgments. For more details on Interactive Granular Computing the readers are referred to EU ACADEMY 2020 ANNUAL REPORT's from 2019 to 2021, papers listed at https://dblp.org/pid/s/AndrzejSkowron.html as well as recent invited talks at conferences (*e.g.*, the Fourth International Conference on Intelligence Science (ICIS2020) (https://www.icis2020nitdgp.com), 13th International Conference on Computational Collective Intelligence (ICCI 2021) 29 September - 1 October 2021, Rhodes, Greece, the International Conference on Rough set theory and Mathematical Applications (ICRSMA2021), India, 8–9 January 2021, and Modelling Rough and Fuzzy Sets in Interactive Granular Computing (IGrC), invited talk at International e-Conference on Rough Sets - Generalizations and Applications, November 11- 13, 2021, Organized by Fuzzy and Rough Sets Association, Agartala, Tripura, India or keynote at Information Processing and Management of Uncertainty in Knowledge-Based Systems (IPMU 2022) July 11–15, 2022, Milan, Italy).

References

1. Brooks, F.P.: The Mythical Man-Month: Essays on Software Engineering. Addison-Wesley, Boston (1975). (extended Anniversary Edition in 1995)

2. Deutsch, D., Ekert, A., Lupacchini, R.: Machines, logic and quantum physics. Neural Comput. **6**, 265–283 (2000). https://doi.org/10.2307/421056
3. Dourish, P.: Where the Action Is. The Foundations of Embodied Interaction. The MIT Press, Cambridge (2004)
4. Dutta, S., Skowron, A.: Toward a computing model dealing with complex phenomena: interactive granular computing. In: Nguyen, N.T., Iliadis, L., Maglogiannis, I., Trawiński, B. (eds.) ICCCI 2021. LNCS (LNAI), vol. 12876, pp. 199–214. Springer, Cham (2021). https://doi.org/10.1007/978-3-030-88081-1_15
5. Dutta, S., Skowron, A.: Interactive granular computing model for intelligent systems. In: Shi, Z., Chakraborty, M., Kar, S. (eds.) ICIS 2021. IAICT, vol. 623, pp. 37–48. Springer, Cham (2021). https://doi.org/10.1007/978-3-030-74826-5_4
6. Gerrish, S.: How Smart Machines Think. MIT Press, Cambridge (2018)
7. Harnad, S.: The symbol grounding problem. Physica D **42**, 335–346 (1990). https://doi.org/10.1016/0167-2789(90)90087-6
8. Hodges, A.: Alan Turing, logical and physical. In: Cooper, S.B., Löwe, B., Sorbi, A. (eds.) New Computational Paradigms. Changing Conceptions of What is Computable, pp. 3–15. Springer, New York (2008). https://doi.org/10.1007/978-0-387-68546-5_1
9. Jankowski, A.: Interactive Granular Computations in Networks and Systems Engineering: A Practical Perspective. Lecture Notes in Networks and Systems. Springer, Heidelberg (2017). https://doi.org/10.1007/978-3-319-57627-5
10. Nöe, A.: Action in Perception. MIT Press, Cambridge (2004)
11. Ortiz, C.L., Jr.: Why we need a physically embodied Turing test and what it might look like. AI Mag. **37**, 55–62 (2016). https://doi.org/10.1609/aimag.v37i1.2645
12. Pawlak, Z.: Rough sets. Int. J. Comput. Inf. Scie. **11**, 341–356 (1982). https://doi.org/10.1007/BF01001956
13. Pylyshyn, Z.W.: Computation and Cognition: Toward a Foundation for Cognitive Science. The MIT Press, Cambridge (1884)
14. Skowron, A., Jankowski, A., Dutta, S.: Interactive granular computing. Granular Comput. **1**(2), 95–113 (2016). https://doi.org/10.1007/s41066-015-0002-1
15. Skowron, A., Jankowski, A.: Rough sets and interactive granular computing. Fundamenta Informaticae **147**, 371–385 (2016). https://doi.org/10.3233/FI-2016-1413
16. Zadeh, L.A.: Fuzzy sets. Inf. Control **8**(3), 338–353 (1965). https://doi.org/10.1016/S0019-9958(65)90241-X

MGCC: Multi-Granularity Cognitive Computing

Guoyin Wang[(⊠)] [iD]

Chongqing Key Laboratory of Computational Intelligence, Chongqing University of Posts and
Telecommunications, Chongqing 400065, People's Republic of China
wanggy@cqupt.edu.cn

Abstract. Cognitive computing aims to develop a coherent, unified, universal
mechanism with inspiration of mind's capabilities. Granular human thinking is
a kind of cognition mechanism for human problem solving. Multi-Granularity
cognitive computing (MGCC) is introduced to integrate the information transformation mechanism of traditional intelligent information processing systems
and the multi-granularity cognitive law of human brain in this paper. The data-
driven granular cognitive computing model (DGCC) developed in 2017 is a typical
theoretical model for implementing MGCC. MGCC is a valuable model for developing highly intelligent systems consistent with human cognition. The theoretical
research issues and some applications about MGCC are introduced.

Keywords: Granular computing · Cognitive computing · Data-driven granular
cognitive computing · Multi-granularity cognitive computing

1 Introduction

We are already living in a big data and intelligence era now. Generally speaking, there
are two ways for dealing with big data. A human could analyze the data and get a result
himself/herself, and also use a computer to deal with the data using some computing
models and algorithms. Thus, there might be a big problem. Could we guarantee that the
result generated by the computer is the same as the result get by the human? Is there any
difference or contradiction between these two results? That is, is there any difference or
contradiction between the intelligent computing and brain cognition? There are a lot of
real life examples showing that they are different and contradict each other. Anti-face
recognition is a typical case with such difference [1]. A deep neural network (DNN) could
not recognize a noised face which can be easily recognized by human beings. It shows
that the recognition mechanism of DNNs is inconsistent with that of human beings. It is

This work is supported in part by the National Key Research and Development Program of
China under grant 2021YFF0704100, National Natural Science Foundation of China under Grant
61936001, Natural Science Foundation of Chongqing under Grants cstc2019jcyj-cxttX0002 and
cstc2021ycjh-bgzxm0013, Project of Chongqing Municipal Education Commission under Grant
HZ2021008.

also found that ImageNet trained CNNs are strongly biased towards recognizing textures rather than shapes, which is in stark contrast to human behavioural evidence and reveals fundamentally different classification strategies [2]. This contradiction problem should be an important theoretical problem to be addressed in cognitive computing study.

Cognitive computing aims to develop a coherent, unified, universal mechanism with inspiration of mind's capabilities. It is mind inspired computing with the goal of developing more accurate models to simulate the human brain/mind senses, reasons, and responds to stimulus [3]. Granular human thinking is a kind of cognition mechanism for human problem solving [4]. Multi-granularity computing (MGrC) is a model for studying and implementing the granular human thinking. It is regarded as an umbrella term to cover theories, methodologies, techniques, and tools that make use of granules in complex problem solving [5–10]. In this paper, the MGCC model is introduced, which integrates the information transformation mechanism of traditional intelligent information processing systems and the multi-granularity cognitive law of human brain.

The DGCC model developed in 2017 [11] is a typical concrete theoretical model for implementing MGCC. In recent years, a lot of theoretical researches and real life applications based on DGCC are conducted [12]. These achievements show that MGCC is a valuable model for developing highly intelligent systems consistent with human cognition. In this paper, some related recent research achievements about MGCC are introduced.

2 Multi-Granularity Cognitive Computing

2.1 Contradiction Between Intelligent Computing and Brain Cognition

The information transformation and processing in traditional intelligent systems are always from finer granularity layers to coarser granularity layers. For example, in data mining (machine learning, or knowledge discovery), the information transformation is unidirectionally from data to knowledge. In image recognition processes, low level features are extracted from pixels at first, while high level features are generated later. However, it is found that there is a global precedence (GP) law in human cognition process [13, 14]. People always recognize the large characters in the global level at first and then the small characters in the local level as shown in Fig. 1 [11, 13–15].

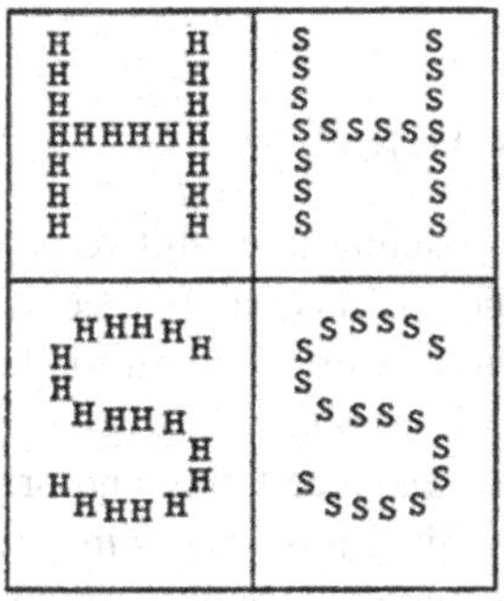

Fig. 1. Global precedence [13–15]

Thus, we can find there is contradiction between the information transformation mechanism "from finer granularity to coarser granularity" of traditional intelligent systems and the "global precedence" cognitive law of human brain. They should be integrated together in order to resolve this contradiction. In traditional intelligent systems, such as machine learning systems, data mining systems, et al., data space and knowledge space are expressed separately. This leads to the independence of data and knowledge. The mapping and reasoning from data to knowledge could not be established. The separate expression of data space and knowledge space is a big problem for the integration of the information transformation mechanism of traditional intelligent systems and the cognitive law of human brain.

2.2 Data-Driven Granular Cognitive Computing

Wang proposed the DGCC model in 2017 [11]. In DGCC, data and knowledge are expressed together in a multi-granularity knowledge expressing space, where, data is the knowledge represented in the lowest granularity layer while knowledge is the data represented in high granularity layers. The following nine theoretical issues to be studied for implementing a DGCC model were discussed in detail in [11].

1) Multiple granularity representation of data, information and knowledge.
2) Integration of the human cognition of "from coarser to finer" and the information processing of "from finer to coarser".
3) Transformation of the uncertainty of big data in a multiple granularity space.
4) Multiple granularity joint computing model and problem solving mechanism.
5) Dynamical evolution mechanism in a multiple granularity knowledge space.
6) Effective progressive variable granularity computing method.
7) Intelligent computation forwarding.
8) Distributed multiple granularity machine learning method.
9) Multiple granularity mechanism of associative memory with forgetting.

Since data and knowledge are integrated and expressed together in a multi-granularity knowledge expressing space, the information transformation mechanism of "from finer granularity to coarser granularity" and the "global precedence" cognitive law could be studied using its two transformations of "bottom-up" and "top-down". This is the key idea of DGCC.

2.3 The Formation Process of MGCC

The formation process of multi-granularity cognitive computing is shown in Fig. 2. It originated from the set theory and the uncertainty theory. The set theory established by Cantor in the 19th century is the basis of modern mathematics [16]. Frege proposed the vague uncertainty problem of set boundary region in 1904 [17]. Zadeh used the membership function to describe this uncertainty and proposed the fuzzy set theory in 1965 [18]. In 1982, Pawlak described this uncertainty with two certain sets of upper approximation and lower approximation, proposed the rough set theory and established the concept of knowledge granularity [19]. Li synthesized vague uncertainty and random uncertainty, proposed the cloud model in 1995 [20–22]. He established the qualitative and quantitative transformation of an uncertain concept. Zhang studied the variable granularity solving of complex problems and proposed the quotient space theory in 1990 [4]. Zadeh firstly proposed and discussed information granulation in 1979, which is the origin of granular computing [23, 24]. Fuzzy set, rough set, cloud model and quotient space constitute the theoretical basis of granular computing research. Chen studied the basic expression problem of human perception and proposed the "global precedence" topological perception theory in 1982 [13, 14]. Summarizing the researches of granular computing for decades, based on the integration of the "global precedence" cognitive law and multi-granularity computing mechanism, Wang proposed the DGCC model [11], which is a concrete implementation case of MGCC, and explained its three major scientific problems including nine scientific topics, that is, unified expression of data and knowledge, collaborative problem solving, and integration of cognition and computation.

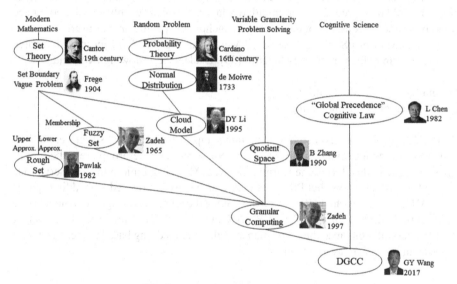

Fig. 2. Formation process of MGCC

There are already some theoretical research achievements about MGCC in recent years, such as knowledge distance measure [25, 26], multi-granularity knowledge space description [27, 28], granularity optimization and selection [29, 30], multi-granularity knowledge space generation [28, 31, 32], integration of cognition and computation [33], multi-granularity explanation for deep neural network [34], et al. MGCC has also been applied in many real life fields such as information security [35], production safety [36], multimedia processing [37], industry control [38], health and hygiene [39], decision making and management [40], et al. Some brief introductions of recent research achievements about MGCC are available in [12]. In this paper, several typical application research examples of MGCC are introduced and analyzed.

3 Multi-Granularity Social Network Alignment

Aligning users across networks is a basic task in many cyberspace security applications. It aims to identify person identities across networks. It is important for link prediction, user recommendation, information dissemination and users' behavior analysis, et al. In [35], a multi-granularity social network alignment model is developed. Pseudo anchors are introduced to implement variable granularity problem solving.

Graph representation learning based models could preserve structural proximity as well as content similarities. They have shown superior performance for this task. Unfortunately, the objective of structural proximity preserving usually results in "overly-close" embedding for the nodes in a dense neighborhood structure. This makes it hard to differentiate them from each other in the embedding space, and thus hard to align users across social networks. Generally speaking, from the perspective of coarse granularity, users in the embedding space can be separated based on their communities. However, this strategy still results in indistinguishable specific users in the same community. This problem could be solved in a finer granularity layer. A global "evenly distributed" space could be learned through implanting pseudo anchors to specific anchor users.

A framework of PSeudo anchor based Meta Learning (PSML) is proposed to solve this problem in a finer granularity layer in [35]. It takes the following two strategies:

1) Implanting pseudo anchors.
2) Meta-learning for fine tuning.

Pseudo anchors are expected to have more impact to their local structure, but less impact on the nodes topologically far away. Nodes in its close group are pulled away from the other nodes during the learning process. Macro and micro layer observations on real life datasets show that PSML can learn an evenly distributed embedding space.

PSML is also integrated into several the state of the art network alignment models like IONE, ABNE, SNNA, DALUAP, DEEPLINK and MGCN. Experiment results show that it can successfully enhance the performance of these embedding based models. Detailed experiment results are available in [35].

4 Multi-Granularity Sketch-Based Face Image Retrieval

SketcH-based face image retrieval is a difficult cross-modal image retrieval problem. In traditional sketch-based face image retrieval systems, completed face sketches are used for retrieval. However, it requires strong drawing skills and is much time cost for drawing a complete face sketch. In some cases, it might be impossible for a sketch artist to draw a complete high quality face sketch since he/she has only very limited memory of the real face. In addition, there is no interaction between the drawing process and retrieval process. The retrieval process is a black box for sketch artists. It is difficult for a sketch artist to draw a complete high quality sketch for face image retrieval. It leads to the low efficiency in traditional sketch-based face image retrieval practices.

As we know, a sketch artist draws a face sketch from coarser granularity to finer granularity. A face sketch is always drawn stroke by stroke. It inspires us whether it is possible to perform the retrieval in the drawing process dynamically, from coarser granularity to finer granularity?

The retrieval result of a partial face sketch could be feedback to a sketch artist in real time. The feedback information of the retrieval system would be very helpful for inspiring he/him to draw the face sketch continually. It would speed up the drawing process, and improve the drawing quality. In this way, a sketch artist could interact with the retrieval system. There could be interaction between the drawing process and retrieval process as shown in Fig. 3. Thus, the target could be generated with a partial sketch only. Sketch artist may not need to draw a complete high quality sketch for retrieval at all.

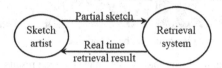

Fig. 3. Interaction between drawing process and retrieving process

A multi-granularity dynamical sketch-based face image retrieval system is developed in [37]. A multi-granularity sketch dataset as shown in Fig. 4 is generated based on FS2K [41] as a training dataset. It has 2006 face photos and each face photo has a drawing episode of a sequence of 70 incomplete sketches (140420 sketches in total). In addition, 50 art-majored college students are invited to submit sketch drawing episodes for 100 face photos as shown in Fig. 5.

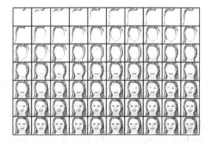
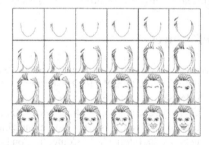

Fig. 4. Multi-granularity sketch dataset   **Fig. 5.** Hand drawn sketch drawing episode

Some sketch-based face image retrieval results are shown in Fig. 6. In this figure, the face pictures in each line are the retrieval result of the sketch in the most left column. The faces in red box are the targets. The experiment results show that with the help of the feedback and interaction of retrieval process, a target face could be generated quickly using a sketch with a few strokes. It is proved that the interaction of the drawing process and retrieving process is very important and helpful. Multi-granularity dynamical sketch-based face image retrieval could be used to improve traditional sketch-based face image retrieval.

a) With feedback and interaction

b) Without feedback and interaction

Fig. 6. Sketch-based face image retrieval results

5 Conclusions

In this paper, MGCC is introduced to integrate the "from fine to coarse" multi-granularity information transformation mechanism of traditional intelligent information processing systems and the "from coarse to fine" multi-granularity cognitive law of human brain. It solved their contradiction problem. It is a success implementation of cognitive computing. With the development of cognitive science, more concrete cognitive computing models will be further proposed in the future. It would also push the development of artificial intelligence study in a great degree.

References

1. Bose, A.J., Aarabi, P.: Adversarial attacks on face detectors using neural net based con-strained optimization. In: 2018 IEEE 20th International Workshop on Multimedia Signal Processing (MMSP), pp. 1–6. IEEE (2018)
2. Geirhos, R., Rubisch, P., Michaelis, C., Bethge, M., Wichmann, F.A., Brendel, W.: ImageNet-trained CNNs are biased towards texture; increasing shape bias improves accuracy and robustness. In: 7th International Conference on Learning Representations (ICLR). OpenReview.net (2019)
3. Modha, D.S., Ananthanarayanan, R., Esser, S.K., Ndirango, A., Sherbondy, A.J., Singh, R.: Cognitive computing. Commun. ACM **54**(8), 62–71 (2011)
4. Zhang, L., Zhang, B.: Quotient Space Based Problem Solving: A Theoretical Foundation of Granular Computing. Elsevier Science, Amsterdam (2014)
5. Wang, G., Yang, J., Xu, J.: Granular computing: from granularity optimization to multi-granularity joint problem solving. Granul. Comput. **2**(3), 105–120 (2016). https://doi.org/10.1007/s41066-016-0032-3
6. Xu, J., Wang, G.Y., Yu, H.: Review of big data processing based on granular computing. Chin. J. Comput. **38**(8), 1497–1517 (2015)
7. Lin, T.Y.: Granular computing. In: Wang, G., Liu, Q., Yao, Y., Skowron, A. (eds.) International Workshop on Rough Sets, Fuzzy Sets, Data Mining, and Granular-Soft Computing, vol. 2639, pp. 16–24. Springer, Heidelberg (2003). https://doi.org/10.1007/3-540-39205-X_3

8. Yao, J.T., Vasilakos, A.V., Pedrycz, W.: Granular computing: perspectives and challenges. IEEE Trans. Cybernet. **43**(6), 1977–1989 (2013)
9. Yao, Y.: Perspectives of granular computing. In: 2005 IEEE International Conference on Granular Computing, vol. 1, pp. 85–90. IEEE (2005)
10. Yao, Y.: Artificial intelligence perspectives on granular computing. In: Pedrycz, W., Chen, SM. (eds.) Granular Computing and Intelligent Systems, vol 13, pp. 17–34. Springer, Heidelberg (2011). https://doi.org/10.1007/978-3-642-19820-5_2
11. Wang, G.: DGCC: data-driven granular cognitive computing. Granul. Comput. **2**(4), 343–355 (2017). https://doi.org/10.1007/s41066-017-0048-3
12. Wang, G.Y., Fu, S., Yang, J., Guo, Y.K.: A review of research on multi-granularity cognition based intelligent computing. Chin. J. Comput. **45**(6), 1161–1175 (2022)
13. Han, S., Chen, L.: The relationship between global properties and local properties-global precedence. Adv. Psychol. Sci. **4**(1), 36–41 (1996)
14. Chen, L.: Topological structure in visual perception. Science **218**(4573), 699–700 (1982)
15. Navon, D.: Forest before trees: the precedence of global features in visual perception. Cogn. Psychol. **9**(3), 353–383 (1977)
16. Edwards, H.M.: Kronecker's views on the foundations of mathematics. In: Proceedings of the Symposium on the History of Modern Mathematics, pp. 65–77. Elsevier (1989)
17. Grattan-Guinness, I.: The Search for Mathematical Roots, 1870–1940: Logics, Set Theories and the Foundations of Mathematics from Cantor through Russell to Godel. Princeton University Press, Princeton (2000)
18. Zadeh, L.: Fuzzy sets. Inf. Control **8**(3), 338–353 (1965)
19. Pawlak, Z.: Rough sets. Int. J. Comput. Inform. Sci. **11**(5), 341–356 (1982)
20. Li, D.Y., Meng, H.J., Shi, X.M.: Membership clouds and membership cloud generators. J. Comput. Res. Dev. **32**(6), 15–20 (1995)
21. Li, D.Y., Du, Y.: Artificial Intelligence with Uncertainty, 1st edn. Chapman and Hall/CRC, London (2007)
22. Wang, G., Xu, C., Li, D.: Generic normal cloud model. Inf. Sci. **280**, 1–15 (2014)
23. Zadeh, L.A.: Fuzzy sets and information granularity. Adv. Fuzzy Set Theory Appl. **11**, 3–18 (1979)
24. Zadeh, L.A.: Toward a theory of fuzzy information granulation and its centrality in human reasoning and fuzzy logic. Fuzzy Sets Syst. **90**(2), 111–127 (1997)
25. Xia, D., Wang, G., Yang, J., Zhang, Q., Li, S.: Local knowledge distance for rough approximation measure in multi-granularity spaces. Inf. Sci. **605**, 413–432 (2022)
26. Yang, J., Wang, G., Zhang, Q., Wang, H.: Knowledge distance measure for the multigranularity rough approximations of a fuzzy concept. IEEE Trans. Fuzzy Syst. **28**(4), 706–717 (2020)
27. Duan, J., Wang, G., Hu, X.: Equidistant k-layer multi-granularity knowledge space. Knowl.-Based Syst. **234**, 107596 (2021)
28. Xu, J., Wang, G., Li, T., Pedrycz, W.: Local-density-based optimal granulation and manifold information granule description. IEEE Trans. Cybernet. **48**(10), 2795–2808 (2018)
29. Li, S., Yang, J., Wang, G., Zhang, Q., Hu, J.: Granularity selection for hierarchical classification based on uncertainty measure. IEEE Trans. Fuzzy Syst. (2022). https://doi.org/10.1109/TFUZZ.2022.3161747
30. Zhang, Q., Cheng, Y., Zhao, F., Wang, G., Xia, S.: Optimal scale combination selection integrating three-way decision with Hasse diagram. IEEE Trans. Neural Netw. Learn. Syst. (2021). https://doi.org/10.1109/TNNLS.2021.3054063
31. Xia, S., Zhang, Z., Li, W., Wang, G., Giem, E., Chen, Z.: GBNRS: a novel rough set algorithm for fast adaptive attribute reduction in classification. IEEE Trans. Knowl. Data Eng. **34**(3), 1231–1242 (2022)
32. Xia, S., et al.: Ball k-means: fast adaptive clustering with no bounds. IEEE Trans. Pattern Anal. Mach. Intell. **44**(1), 87–99 (2022)

33. Xu, C., Wang, G.: Bidirectional cognitive computing model for uncertain concepts. Cogn. Comput. **11**(5), 613–629 (2019)
34. Bao, H., Wang, G., Li, S., Liu, Q.: Multi-granularity visual explanations for CNN. Knowl.-Based Syst. (2022). https://doi.org/10.1016/j.knosys.2022.109474
35. Yan, Z., et al.: Towards improving embedding based models of social network alignment via pseudo anchors. IEEE Trans. Knowl. Data Eng. (2021). https://doi.org/10.1109/TKDE.2021.3127585
36. Wang, G., Dai, J., Li, H.: Production safety management and decision-making based on multi-granularity cognitive computing. Bull. Natl. Nat. Sci. Found. China **35**(5), 752–758 (2021)
37. Dai, D., Tang, X., Xia, S., Liu, Y., Wang, G., Chen, Z.: Multi-granularity association learning framework for on-the-fly fine-grained sketch-based image retrieval. CoRR abs/2201.05007 (2022)
38. Yu, H., Yang, Q., Wang, G., Xie, Y.: A novel discriminative dictionary pair learning constrained by ordinal locality for mixed frequency data classification. IEEE Trans. Knowl. Data Eng. (2020). https://doi.org/10.1109/TKDE.2020.3046114
39. Xiao, B., et al.: PAM-DenseNet: a deep convolutional neural network for computer-aided covid-19 diagnosis. IEEE Trans. Cybernet. (2021). https://doi.org/10.1109/TCYB.2020.3042837
40. Yu, H., He, D.N., Wang, G.Y., Li, J., Xie, Y.F.: Big data for intelligent decision making. Acta Automatica Sinica **46**(5), 878–896 (2020)
41. Fan, D.P., Huang, Z., Zheng, P., Liu, H., Qin, X., Van Gool, L.: Facial-sketch synthesis: a new challenge. Mach. Intell. Res. **19**(4), 257–287 (2022)

Three-way Decision, Three-World Conception, and Explainable AI

Yiyu Yao$^{(\boxtimes)}$

Department of Computer Science, University of Regina, Regina, SK, Canada S4S 0A2
Yiyu.Yao@uregina.ca

Abstract. Three-way decision is about thinking, problem-solving, and computing in threes or through triads. By dividing a whole into three parts, by focusing on only three things, or by considering three basic ingredients, we may build a theory, a model, or a method that is simple-to-understand, easy-to-remember, and practical-to-use. This philosophy and practice of triadic thinking appears everywhere. In particular, there are a number of three-world or tri-world models in different fields and disciplines, where a complex system, a complicated issue, or an intricate concept is explained and understood in terms of three interrelated worlds, with each world enclosing a group of elements or representing a particular view. The main objective of this paper is to review and re-interpret various three-world conceptions through the lens of three-way decision. Three-world conceptions offer more insights into three-way decision with new viewpoints, methods, and modes. They can be used to construct easy-to-understand explanations in explainable artificial intelligence (XAI).

Keywords: Three-way decision · Three-world conception · Three-world model · Thinking in threes · Trilevel thinking · SMV space · Explainable AI

1 Introduction

With the ever-increasing power, functionality, and applications of intelligent machines and systems, the issue of the explainability takes center stage. The recent research trend in explainable artificial intelligence (XAI) suggests that a machine must effectively explain its internal processes and decisions, in order to gain human understanding, trust, and acceptance [2,12]. As a prerequisite for producing effective explanations, it is necessary to study human ways to

Y. Yao : I would like to express my thanks to Professor Duoqian Miao and Professor JingTao Yao for organizing the IRSS President's forum and for encouraging me to write this paper. I am grateful to the reviewers for their encouraging and constructive comments. This work was partially supported by a Discovery Grant from NSERC, Canada.

J. Yao et al. (Eds.): IJCRS 2022, LNAI 13633, pp. 39–53, 2022.
https://doi.org/10.1007/978-3-031-21244-4_4

perceive, think, and act. With an understanding of human ways to think, understand, and act, a machine may explain its processes and decisions by building a model aligned with human mental models. Driven by such motivation, this paper explores particular mental models, namely, three-way decision as thinking in threes, three-world conception as thinking through three worlds, and the relationships between the two, as well as their applications in XAI.

There are two related types of issues around the notion of an explanation.[1] One type concerns the meaning, functionality, and properties of the explanation, as well as various formal models of explanation. In the context of XAI, an intelligent machine explains its working processes and results for the purpose of facilitating human understanding and building human trust. In a wide context of scientific enquiry and discovery, one of the goals and tasks of science is to explain the world, i.e., to seek "mathematically formulated and experimentally validated impersonal principles that explain a wide variety of phenomena" [36]. The other type focuses on the communication of an explanation, involving the structures and the construction process of the explanation. To some degree, an appropriate structure plays a crucial role in constructing an easy-to-represent, easy-to-communicate, and easy-to-understand explanation. The focus of this paper is on the latter type of issues. By applying the principles of three-way decision, I discuss ways to construct and communicate explanations with triadic structures.

The rest of the paper is organized around three objectives. Section 2 provides an overview of a theory of three-way decision with the objective to establish a basis for this study. The objective of Sect. 3 is to introduce, in light of three-way decision, a framework for studying three-world conceptions, that is, thinking through three worlds. In particular, I examine three -world models. The objective of Sect. 4 is to outline a possible application of three-way decision and three-world conception in constructing human-friendly explanations in data science, human-machine co-intelligence, and explainable artificial intelligence (XAI).

2   An Overview of Three-Way Decision

In 2009, I introduced the concept of three-way decision (3WD) [39] to provide a semantically sound interpretation of the three types of decision rule (i.e., acceptance, rejection, and undecided) derived through Pawlak rough sets [18,19] and probabilistic rough sets [40]. Further studies have shown that three-way decision is a much richer concept, with wide-ranging applications. Since 2012, I have been refining a new theory of three-way decision, consisting of thinking, problem-solving, and computing in threes [41–43,45,46]. Three-way decision has fostered

---

[1] The two types presented here are related to the distinction, suggested by Achinstein [1], of an "explaining act" and an explanation as a "product" of an explaining act. Ruben [23] made a similar distinction through "process and product." The first type is more about an explanation itself. The second type relies on an understanding of an "explaining act" that includes both the formulation and the communication of an explanation.

new research areas, such as three-way classification, three-way clustering, three-way data analytics, three-way formal concept analysis, three-way approximations of fuzzy sets, three-way conflict analysis, three-way recommendation systems, three-way granular computing, and many others. The field has grown substantially since its inception, with researchers from around the world contributing to a significant number of papers, edited books, journal special issues, workshops, and special sessions on three-way decision. For the current state of research and development of the art, science, and practice of three-way decision, a reader may consult the reports by Yang and Li [37], Wei et al. [35], and Yao [38] based on networks analysis and bibliometrics analysis.

Thinking in threes (i.e., triads consisting of three things) or triadic thinking is perhaps one of the most common mental models, metaphors, and structures, such as a tripartite scheme, a three-part theory, a three-element structure, a three-pillar framework, a three-word slogan, a three-character story, a three-generation classification, a three-level architecture, a three-version design of a product, a third grey option in addition to commonly used dichotomies (e.g., Yes and No, black and white, good and bad, positive and negative), a third middle point through the balancing and synthesis of the two opposites, and many more [3, 4, 15, 22, 33, 43, 46]. We humans and particularly scientists have an intriguing preference for a ternary patterned theory, model, or explanation of reality [20]. As an illustration, we may give three examples of thinking in threes. The first example is building a model of explanation for explainable artificial intelligence (XAI) based on the What-Why-How triad[2]: What are the results? Why are the results meaningful? How are the results derived? The second example is the MIT Sloan Management Review's short podcast, Three Big Points[3], in which each episode presents a mold-breaking idea in ten minutes with three useful takeaways. The third example is the effective use of threes in writing a great paper[4]: the three C's of paper structure consisting of the Context for introduction, the Content for results, and the Conclusion for discussion; the ABC (Accurate, Brief, and Clear) of straightforward writing; the DEF (Declarative, Engaging, and Focused) for choosing a title. In particular, advice on straightforward writing is summarized in three sentences: "Never choose a long word when a short one will do. Use simple language to communicate your results. Always

---

[2] This example will be further examined in the later part of the paper. For an actual application, we may point at the earlier expert system MYCIN that uses the What-Why-How triad, in which an explanation subsystem focuses mainly on Why and How questions to justify the decision of the system or to educate the user [32]. The triad is equally useful for enhancing human intelligence and guiding human behavior [46]. For example, the Golden Circle leadership model, introduced by Sinek [29], is based on the Why-How-What triad, which advises that every organization and everyone of us should know the three most important things: why we do (i.e., purpose and goals), how we do, and what we do. The same Why-How-What triad was used by Clear [6] in his three-level model of behavior change, focusing on what we believes, what we do, and what we get.

[3] https://sloanreview.mit.edu/audio-series/three-big-points/, accessed May 20, 2022.

[4] https://www.nature.com/articles/d41586-019-01362-9, accessed May 20, 2022.

aim to distill your message down into the simplest sentence possible." We can find many examples that explore the power triads for crafting great, powerful, and memorable speeches [10].

These examples show that we do commonly build an argument, a model, or a theory by thinking in threes. To provide further supporting evidence, it may be more constructive by giving three good reasons why we humans think in threes. The first explanation is the cognitive basis. It has long been recognized that we humans can only hold up a few things in the short-term working memory [7,16]. While there does not exist a general agreement on the exact number, which may range from two to nine, three seems to be a pivoting one. Another related result is our subitizing ability to tell immediately, without counting, the number of items presented to us when the number of items is small, typically fewer than six [14]. This may explain why the very first three Roman numbers are written as one, two, and three vertical lines, respectively, the very first three Chinese numbers are written as one, two, and three horizontal lines, respectively, and the pattern breaks at and after the fourth number. The third result is our natural ability to form patterns in order to make sense of the reality and our experiences. Three seems to be the minimum number of things required to form a meaningful and useful pattern. Drawing from these results of human cognition, thinking in threes comes naturally and may be an innate capacity.

The second explanation is the evolutionary basis. From an evolutionary point of view, we are better at older skills than at newer skills. Counting a few things and thinking about a small number of things, as evidenced by the 'one, two, three, four, many' and 'one, two, many' types of numerical systems [8], may be older skills in the process of human evolution. We, in fact, learned counting and thinking in small numbers at a younger age. Thus, we excel at skills of thinking in small numbers. It may be argued that thinking in threes is one of the products of evolution or early childhood learning.

The third explanation is the cultural basis. The number three plays an essential role across many cultures [9,25]. The number three typically represents completeness, harmony, and perfection, as expressed by the following quotations [25]:

- All good things come in threes. (Folk saying)
- A threefold cord is not quickly broken. (Bible)
- All was divided into three. (Homer)
- A whole is that which has a beginning, middle and end. (Aristotle)
- The Triad is the form of the completion of all things. (Nichomachus of Gerasa)
- Three is the formula of all creation. (Honoré de Balzac)
- The One engenders the Two, the Two engenders the Three and the Three engenders all things. (Tao Te Ch'ing)

Using a triad of three things for perceiving, understanding, interpreting, and representing the reality seems to be a universal practice across different cultures. Triads are perhaps one of the most used structures when crafting a story, a speech, a theory, or a worldview. For example, Schneider [25] stated, "Whenever there are three, as the three knights, three musketeers, three wise men, or three wishes, there is *throughness*, rebirth, transformation, and success." To a large extent, our cultural immersion experience further re-enforces an inclination and a preference towards thinking in threes.

Given the omnipresence of triadic thinking on the one hand and a lack of a formal theory on the other, a theory of three-way decision has been proposed and received much attention in recent years [41–43, 46]. The theory is about a systematic study of thinking, problem-solving, and computing in threes. By attaching specific interpretations and meanings to various triads, we can obtain different models and modes of three-way decision. In the rest of this paper, I interpret a triad in terms of three worlds, which gives rise to thinking through three worlds.

3   Thinking Through Three Worlds

This section examines three triadic structures, namely, a Venn diagram of three sets, a triangle, and a concentric tricircle, for thinking through three worlds.

3.1   The Concept of Worlds

The concept of "the world" is perhaps one of the most commonly used notions or metaphors for us to describe, view, and understand the reality and our relationships to the reality. The word "world," particularly, 'the world,' is used in various contexts with multiple meanings [34]. According to Webel [34], "the world" is "a linguistic and historical construction" and "an abstraction, a concept, or idea." It is how the "meaning-creating organisms frame the boundaries of their being-in-this-world." The view of "world as idea" [26, 34] provides a starting point for exploring how we use the concept of worlds to understand the reality and to guide our conducts, namely, how to observe the world, how to make sense of the world, and how to change the world.

We may categorize and characterize things into different worlds in many ways, for example, from a temporal, spatial, functional, positional, or contextual consideration. We typically divide various aspects of the reality, for example, a group of geographical regions, a timeline of developments, a discourse of discussion, a family of human activities, etc., into a number of different and interrelated worlds. By restricting to a particular world, we limit our investigation within that world in the context of other worlds. Conceptually, we can talk about the inside, the outside, and the boundary of a world, which offers three interpretations and understandings of the same world. By considering different worlds, we can make comparisons, study their interconnections and influences, and shift our attention by switching between different worlds. While a single world presents a local view, multiple worlds give rise to a global view.

Our extensive living experiences on the planet earth as "the world," our relentless search for a better world, and our constant cultivation of a superior inner world all suggest the value of "world as idea." Conceptualizing the reality in terms of different worlds leads to both intuitive and in-depth understandings. By combining the principles of three-way decision as thinking in threes and the view of "world as idea," we immediately arrive at a paradigm of thinking through three worlds. There are abundant examples of three-world thinking. In the contexts of information processing, knowledge management, problem solving, and human experience, for example, we have:

- The three-world theory of the reality and knowledge by Popper [21], consisting of World 1 of physical objects, World 2 of mental activities, and World 3 of human-created things.
- The theory of three worlds of mathematics by Tall [31], consisting of conceptual embodiment, operational symbolism, and axiomatic formalism.
- The classification of three worlds of knowledge by Mouton [17], consisting of the worlds of everyday life (lay knowledge), science (scientific knowledge), and metascientific reflection (metascience).
- The theory of triadic game design by Harteveld [13] through balancing the three worlds of reality, meaning, and play.
- The theory of collective human experience by Shaw [28] in terms of the three worlds of commonsense, religion, and science.

Other examples of three-world thinking in more general contexts include various triads, such as the material-intellectual-spiritual three worlds, the three worlds above-below-upon the earth (i.e., heaven, hell, and earth), the three worlds of yours-mine-theirs, etc.

It becomes evident that three-world thinking, with an understanding of "world as idea," offers a new direction for expanding the study of three-way decision as triadic thinking. In the rest of this section, I examine three particular models by organizing and arranging the three worlds in three different ways.

3.2   A Venn Diagram Model of Three Worlds

One methodology of the three-world view and analysis is to divide the discourse of discussion into three possibly overlapping and relatively independent worlds. There may exist multiple ways to construct three worlds. Any particular three-world configuration is only one of the many possible simplifications or representations of the reality. In general, the division between the three worlds is not a clear cut and some issues may appear in two or all three worlds. The Venn diagram in Fig. 1(a) depicts such a set-theoretic view of three-world thinking. Each world represents a particular view and focuses on some particular aspects. While a set covers issues in a world, the complement of the set covers issues not in the world. An intersection of two or three worlds represents their joint issues. With three worlds, the eight disjoint and possibly non-empty regions are, in terms of set intersection, $A \cap B \cap C$, $A \cap B \cap \bar{C}$, $A \cap \bar{B} \cap C$, $A \cap \bar{B} \cap \bar{C}$, $\bar{A} \cap B \cap C$, $\bar{A} \cap B \cap \bar{C}$, $\bar{A} \cap \bar{B} \cap C$, $\bar{A} \cap \bar{B} \cap \bar{C}$, where $\bar{A}$ denotes the set complement of A. In so doing, we can systematically investigate issues in the eight regions.

Alternatively, we may consider only regions constructed by using set intersection, representing issues in the overlapping regions of different worlds. In this

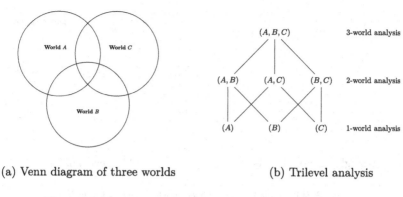

(a) Venn diagram of three worlds   (b) Trilevel analysis

Fig. 1. Thinking through three worlds with a Venn diagram

way, a three-world method offers a trilevel seven-element analysis in Fig. 1(b), where the comma corresponds to set intersection. The result is, in fact, a set-theoretic model of three-way decision [47]. The bottom level of 1-world analysis focuses on each world independently, the middle level of 2-world comparative analysis shifts attention to issues brought by interactions of two worlds, and the top level 3-world integrative analysis looks into more complicated interactions of three worlds. To have a holistic view, it is necessary to have investigations at the three levels, both individually and jointly.

Tall's [31] three-world model of mathematical thinking may be interpreted based on the Venn diagram of three worlds. While each individual world focuses on a particular type of mathematical methods and skills, a join of two worlds shifts the focus to the integration and combination of the respective methods and skills. Mouton's [17] classification of three worlds of knowledge and Shaw's [28] three-world theory of collective human experience may be similarly explained based on the Venn diagram of three worlds.

3.3   A Triangle Model of Three Worlds

For studying relationships, influences, and transformations of different worlds, a triangle of three worlds, given in Fig. 2(a), may be an appropriate config-uration [46]. In the triangle, each world is linked with the other two worlds. Links between two worlds may have many different interpretations, for example, dependency, transformation, support, and others. In this way, a triangle may, in fact, offer various models. Figure 2(b) describes a model of trilevel analysis based on a triangle configuration of three worlds, where ⇝ denotes support or transformation. We examine individual worlds at the bottom level, relationships between two different worlds at the middle level, and relationships among three worlds at the top level.

Popper's [21] three-world model of human knowing and knowledge is typically interpreted as a triangle. World 1 of physical objects exists first. Through World 2 of mental activities and processes, humans observe and make sense of World 1.

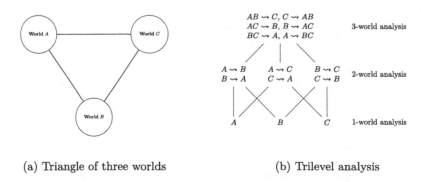

(a) Triangle of three worlds   (b) Trilevel analysis

Fig. 2. Thinking through three worlds with a triangle

The results are human-created things that exist in World 3 as abstract ideas and/or in World 1 as physical objects. The things in World 3, created by World 2, may be used to change World 1. Humans are constantly searching for a better world by exploring the three worlds and their relationships [21].

Gu and Zhu [11] proposed a tripartite WSR (wuli-shili-renli) model as a basis of a systems methodology of management. The W (wuli) is about regularities in objective existence, the S (shili) is about ways of seeing and doing, and the R (renli) is about patterns underlying human relations. It is possible to interpret the WSR model based on a triangle of three worlds: W represents the natural world (domains of natural sciences), R represents the human world (human and human society, domains of psychology, social sciences, humanities, etc.), and S represents the applied world (pragmatic problem-solving, human conduct, domains of management science, engineering, operational research, etc.). Theories and knowledge discovered in both W and R worlds are used to guide human conduct in S world, which may change both W and R worlds. To be a better problem-solver, one must integrate the three worlds.

Stern [30] suggested a triadic conception of the reality, in which the reality is conceived and represented as "unified and wholistic as well as differentiated" three worlds: physical world of matter/energy, theoretical world of meaning, and phenomenological world of experience. Furthermore, Stern gave a simplified diagram by enclosing the triangle of the three worlds in a circle representing the unity and wholeness.

One can easily observe both similarities and differences of these three three-world models. Although the contents of the three models are useful and important by themselves, what most interests us is the common triadic structure. On the one hand, the three models have their respective different divisions, understandings, and representations of the reality. On the other hand, they agree upon a three-world triadic structure. It is their agreement on the use of a triangle of three worlds that supports and applies the principles of three-way decision as thinking in threes.

3.4 A Concentric Tricircle Model of Three Worlds

In some situations, we may build three worlds sequentially such that one is on top of another. There are at least two possible ways to depict such a structure [46]. The concentric tricircle of Fig. 3(a) gives us a sense of an inner-outer relationship, or a core-shell relationship, among the three worlds. Typically, an inner world determines an outer world, and the core is more important and serves as a foundation for constructing the outer ones.

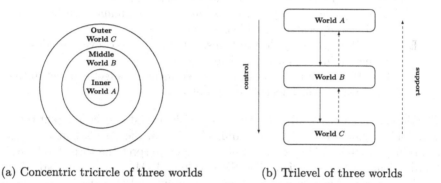

(a) Concentric tricircle of three worlds   (b) Trilevel of three worlds

Fig. 3. Thinking through three worlds with a concentric tricircle or a trilevel

The inner-outer layered interpretation of a concentric tricircle makes it a commonly used architecture for explanation. For example, in understanding a computer system, the inner kernel represents machine hardware, the middle layer represents system software, and outer layer represents application software. In the Golden Circle leadership model by Sinek [29], the three circles are labeled, respectively, by WHY, HOW, and WHAT. By moving inside-out, a successful leader starts with WHY (i.e., purpose and goals) and moves towards WHAT. Similarly, in the model of behavior change by Clear [6], the three circles correspond to Identify, Processes, and Outcome. We build habits by moving inside-out in the identity-directed way. More examples of three-world thinking based on a concentric tricircle can be found in another paper [46].

Figure 3(b) of three levels gives us a sense of a top-down or a bottom-up relationship among the three worlds. Typically, a world at a higher level controls its lower level and, at the same time, is supported by its lower level. The earlier discussions have shown that three-level models arise naturally in the Venn diagram model and the triangle model of three-world thinking. Trilevel thinking is an important mode of three-way decision. Many examples of trilevel thinking can be found in another paper [44].

4 Three-world Thinking for Building Explanations

The triadic structures of three worlds offer architecture and a scheme for us to make sense of the reality and ourselves. Depending on different contexts and applications, we may have different interpretations of a triad of three worlds. In this section, I discuss the notion of an SMV (Symbols-Meaning-Value) space [48, 49] as a concrete interpretation of the three-way conception for the purpose of building and communicating explanations.

Weaver [27] insightfully divided communication problems into three categories, which is quoted here:

> Relative to the broad subject of communication, there seem to be problems at three levels. Thus it seems reasonable to ask, serially:
> LEVEL A. How accurately can the symbols of communication be transmitted? (The technical problem.)
> LEVEL B. How precisely do the transmitted symbols convey the desired meaning? (The semantic problem.)
> LEVEL C. How effectively does the received meaning affect conduct in the desired way? (The effectiveness problem.)

The three levels focus on different types of problems and answer different types of questions, from easier ones to more difficult ones. In the case of human communication through speaking and writing, we may interpret the three levels by the Words-Meaning-Impact triad. The SMV (Symbols-Meaning-Value) space generalizes Weaver's ideas to a much broader context and provides a structure for trilevel or triadic thinking in many other fields. Considering any theory or model, the SMV space suggests that we need to explain the theory at three levels: the content of the theory, the meaning of the theory, and the utility of the theory[5].

In an attempt to build a conceptual model for explaining data science, I explored a close connection between the SMV space and the widely used DKW (Data-Knowledge-Wisdom) hierarchy [48]. In terms of the three-world thinking, World S is about data (i.e., raw symbols), World M is about knowledge (i.e., meaning of data), and World V is about wisdom (i.e., value from wise use of knowledge). The three-level structure reflects the dependency and transformation between data, knowledge, and wisdom. A conceptual model of data science needs to consider the issues in the three worlds of the data, the knowledge hidden in the data, and the value of the knowledge, as well as the issues arisen from the interactions of the three worlds. Broadly speaking, three goals of data

---

[5] As an example, we may take a look at the many different interpretations and explanations of a Chinese classic, "I Ching" (The Book of Changes). "I Ching" has shaped every aspects of Chinese ways of seeing, knowing, and living (for example, culture, art, politics, science, etc.) throughout the Chinese history. Many scholars have interpreted and explained, and are continually searching for new interpretations and explanations, this classic text from many different angles. The notion of SMV space may shed a new light by organizing some of the existing interpretations and explanations at the three levels: (1) images and numbers at the S (Symbols) level, (2) meaning and principles at the M (Meaning) level, and (3) living and practice, according to its meaning and principles, at the V (Value) level. Although this organization may not be hundred percent appropriate or accurate, it does provide a good enough approximation in terms of the text itself, the meaning of the text, and the value of the text.

science are (a) to make data a kind of resources through data collection, storage, retrieval, etc., (b) to make data meaningful through data analysis and knowledge discovery, and (c) to make data valuable through practical application of the knowledge in data for making wise decisions and taking the right actions.

In another attempt to build a conceptual model for explaining human-machine co-intelligence, I viewed the SMV space as an architectural system or a metaphorical structure used by an intelligent being to understand and organize itself, its environments, and relationships with others [49]. Human-machine co-intelligence emerges from human-machine symbiosis in the SMV space. There are three fundamental principles of human-machine co-intelligence. The principle of unified oneness: Human-machine co-intelligence is the third intelligence that is based on human intelligence and machine intelligence on the one hand and is above both on the other hand. Human-machine co-intelligence is not possessed by either humans or machines, but through their seamless unification and integration. The principle of division of labor: Human-machine co-intelligence combines the computational power of machines and the cognitive power of humans through proper division of labor. Moving from the World S, to the World M, and to the World V, humans are doing more work and the machines are doing less. The principle of coevolution: Humans and machines mutually adapt to each other, learn from each other, and work with each other as equal partners. Human-machine co-intelligence exploits a mutualism symbiosis in which both humans and machines benefit and, at the same time, avoids a parasitism symbiosis in which one hurts the other. In this respect, in addition to their own SMV spaces, humans and machines share a common SMV space[6]. The notion of SMV space is a structure and a starting point for explaining human-machine co-intelligence.

I now turn my attention to the possibility of applying the SMV space to explainable AI. I have the view that the concept of SMV space suggests a plausible trilevel scheme for constructing an easy-to-understand explanation in explainable artificial intelligence. The SMV triad leads to a trilevel results-meaning-value (RMV) framework of explanation. Like data, the results from a system may be considered as the raw materials that need, or can be used to construct, an explanation. An intelligent system explains its results, outcome, or output (e.g., recommendations, actions, behaviors, etc.), the meaning of the results, and the value of the results at three separate levels. Moreover, at each level, it is possible to apply the ideas of the Venn diagram or the triangle configurations of three worlds to focus on three related questions characterized by the What-Why-How triad. Table 1 summarizes the main features of this 3×3

---

[6] A few important issues regarding AI and human-machine relations are relevant to the discussion here, such as alignment and control. Christian [5] argued that artificial intelligence systems, in particular machine learning, need to be aligned with human values. Russell [24] pointed out that advances in AI may pose a potential risk to the human race by out of control superhuman AI. Future AI research must ensure that machines remain beneficial to humans and we humans must retain "absolute power over machines that are more powerful than us." By living together in the three worlds of SMV, namely, symbols/data, meaning/knowledge, value/wisdom, humans and machines may coexist in harmony.

architecture of explanations. A 'What' question is about the existence, a 'Why' question is about the reasons/motivations, and a 'How' question is about the processes/applications. By focusing on three fundamental questions of What, Why, and How at each of the three levels, an explanation follows a clearly defined logic, is easy-to-understand, and covers three important aspects.

Table 1. 3 × 3 architecture of explanations

SMV	Explanation level	Questions
Value	Value	**What** is the value of the results?
		Why are the results valuable?
		How to use the results?
Meaning	Meaning	**What** is the meaning of the results?
		Why are the results meaningful?
		How to interpret the results?
Symbols	Results	**What** are the results?
		Why are certain input/conditions required?
		How does the system derive the results?

A trilevel explanation with three basic questions at each level reflects the principles of triadic thinking. Generally speaking, at a given time, it is possible to focus on the discussion at each level without much interference from the other two levels. In other words, we may need to consider only three questions at a particular level, instead of nine questions at all three levels simultaneously. The labels of the three levels and the three questions at each level in Table 1 may be interpreted more liberally. Depending on different applications, it is possible to use other labels and to ask other types of questions. Nevertheless, the essential components and the structure of the 3 × 3 architecture remain unchanged. The 3 × 3 architecture provides a very general framework. In some situations, it may be only necessary to consider some of the nine issues when constructing an explanation. This is particularly true if the results from a system are simple and/or self-explanatory.

5   Conclusion

Three-way decision and three-world conception mutually support each other. On the one hand, three-world models enrich the studies of three-way decision by offering new views, models, and methods. On the other hand, the fundamental philosophy and principles of three-way decision may find new applications in three-world models. In this paper, I explored in brief the connections of three-way decision and three-world conceptions. Thinking through three worlds offers the necessary simplicity and flexibility for building a theory, a model, an argument, etc. In particular, I examined three models of three-world thinking based on, respectively, a Venn diagram, a triangle, and a concentric tricircle (or a trilevel) organization of three worlds.

I motivated this study by stating that three-way decision is a human-inspired theory. Since humans frequently and naturally think in threes, theories, models, or methods are easy to grasp and understand if they are constructed based on a tripartite architecture. Therefore, explanations from any intelligent systems may be built in a human-friendly way by following a tripartite scheme. It may be fruitful to apply the principles and ideas of three-way decision and three-world thinking to address the issues of the quality and effectiveness of explanations in explainable artificial intelligence (XAI). In this paper, I only presented a proposal for an important research direction, which may be called "three-way decision for explainable AI." Although I gave an outline of a trilevel framework for building explanations based on the notion of an SMV (Symbols-Meaning-Value) space, many fundamental questions remain unanswered. Based on the discussion in the paper, we can explore the new territory of three-way decision and three-world thinking for XAI.

References

1. Achinstein, P.: The Nature of Explanation. Oxford University Press, New York (1983)
2. Adadi, A., Berrada, M.: Peeking inside the black-box: a survey on explainable artificial intelligence (XAI). IEEE Access **6**, 52138–52160 (2018)
3. Assagioli, R.: The Balancing and Synthesis of the Opposites. Psychosynthesis Research Foundation, New York (1972)
4. Boer, C.: Thinking in threes: how we human love patterns. Kindle Edition (2014)
5. Christian, B.: The Alignment Problem: Machine Learning and Human Values. W.W. Norton & Company, New York (2020)
6. Clear, J.: Atomic Habits: An Easy & Proven Way to Build Good Habits & Break Bad Ones. Avery, New York (2018)
7. Cowan, N.: The magical number 4 in short-term memory: a reconsideration of mental storage capacity. Behav. Brain Sci. **24**, 87–185 (2000)
8. Deakin, M.A.B.: The Name of the Number. ACER Press, Camberwell, Victoria (2007)
9. Dundes, A.: The number three in American culture. In: Dundes, A. (ed.) Every Man His Way: Readings in Cultural Anthropology, pp. 401–424. Prentice-Hall, Englewood Cliffs (1968)
10. Gallo, C.: Talk Like TED: The 9 Public-Speaking Secrets of the World's Top Minds. Sy. Martin's Press, New York (2014)
11. Gu, J.F., Zhu, Z.C.: Knowing Wuli, sensing Shili, caring for Renli: methodology of the WSR approach. Syst. Pract. Action Res. **13**, 11–20 (2000)
12. Gunning, D., Stefik, M., Choi, J., Miller, T., Stumpf, S., Yang, G.Z.: XAI - Explainable artificial intelligence. Sci. Robot. **4**, 7120 (2019)
13. Harteveld, C.: Triadic Game Design: Balancing Reality. Meaning and Play. Springer-Verlag, London (2011). https://doi.org/10.1007/978-1-84996-157-8
14. Kaufman, E.L., Lord, M.W., Reese, T.W., Volkmann, J.: The discrimination of visual number. Am. J. Psychol. **62**, 498–525 (1949)

15. Logan, D., King, J., Fischer-Wright, H.: Tribal Leadership: Leveraging Natural Groups to Build a Thriving Organization. Harper Business, New York (2011)
16. Miller, G.A.: The magical number seven, plus or minus two: some limits on our capacity for processing information. Psychol. Rev. **63**, 81–97 (1956)
17. Mouton, J.: Understanding Social Research. Van Schaik Publishers, Hatfield, Pretoria (1996)
18. Pawlak, Z.: Rough sets. Int. J. Comput. Inf. Sci. **11**, 341–356 (1982)
19. Pawlak, Z.: Rough Sets. Theoretical Aspects of Reasoning About Data. Kluwer Academic Publishers, Dordrecht (1991)
20. Pogliani, L., Klein, D.J., Balaban, A.T.: Does science also prefer a ternary pattern? Int. J. Math. Educ. Sci. Technol. **37**, 379–399 (2006)
21. Popper, K..: In Search of a Better World: Lectures and Essays from Thirty Years. Routledge, New York (1994)
22. Radej, B., Golobič, M.: Complex Society: In the Middle of a Middle World. Vernon Press, Wilmington, Delaware (2021)
23. Ruben, D.H.: Explaining Explanation. Routledge, New York (1990)
24. Russell, S.J.: Human Compatible: Artificial Intelligence and the Problem of Control. Viking, New York (2019)
25. Schneider, M.S.: A Beginner's Guide to Constructing the Universe: The Mathematical Archetypes of Nature, Art, and Science. Harper, New York (1994)
26. Schopenhauer, A.: The World as Will and Idea, Vol. I, II, III, 7th Edition. Kegan Paul, Trench, Trübner & Co., London (1909)
27. Shannon, C.E., Weaver, W.: The Mathematical Theory of Communication. The University of Illinois Press, Urbana (1949)
28. Shaw, V.N.: Three Worlds of Collective Human Experience: Individual Life, Social Change, and Human Evolution. LNCS (LNAI), Springer, Cham (2019). https://doi.org/10.1007/978-3-319-98195-6_16
29. Sinek, S.: Start With Why: How Great Leaders Inspire Everyone to Take Action. Portfolio/Penguin, New York (2009)
30. Stern, H.W.. In support of a triadic conception of reality. https://harrisstern.medium.com/in-support-of-a-triadic-conception-of-reality-38a784229e9d Accessed 9 June 2022
31. Tall, D.: How Humans Learn to Think Mathematically: Exploring the Three Worlds of Mathematics. Cambridge University Press, New York (2013)
32. Van Melle, W.: MYCIN: a knowledge-based consultation program for infectious disease diagnosis. Int. J. Man-Mach. Stud. **10**, 313–322 (1978)
33. Watson, P.: Ideas: A History, from Fire to Freud. Weidenfeld & Nicolson, London (2005)
34. Webel, C.P.: The World as Idea: A Conceptual History. Routledge, New York (2022)
35. Wei, W.J., Miao, D.Q., Li, Y.X.: A bibliometric profile of research on rough sets. In: IJCRS 2019, LNCS, vol. 11499, pp. 534–548 (2019). https://doi.org/10.1007/978-3-030-22815-6_41
36. Weinberg, S.: To Explain the World: The Discovery of Modern Science. Harper Perennial, New York (2015)
37. Yang, B., Li, J.: Complex network analysis of three-way decision researches. Int. J. Mach. Learn. Cybern. **11**(5), 973–987 (2020). https://doi.org/10.1007/s13042-020-01082-x
38. Yao, J.T.: The impact of rough set conferences. In: IJCRS 2019, LNCS, vol. 11499, pp. 383–394 (2019). https://doi.org/10.1007/978-3-030-22815-6_30

39. Yao, Y.Y.: Three-way decision: an interpretation of rules in rough set theory. In: RSKT 2009, LNCS, vol. 5589, pp. 642–649 (2009). https://doi.org/10.1007/978-3-642-02962-2_81

40. Yao, Y.Y.: Three-way decisions with probabilistic rough sets. Inf. Sci. **180**, 341–353 (2010)

41. Yao, Y.Y.: An outline of a theory of three-way decisions. In: RSCTC 2012, LNCS, vol. 7413, pp. 1–17 (2012). https://doi.org/10.1007/978-3-642-32115-3_1

42. Yao, Y.Y.: Three-way decisions and cognitive computing. Cogn. Comput. **8**, 543–554 (2016)

43. Yao, Y.Y.: Three-way decision and granular computing. Int. J. Approximate Reasoning **103**, 107–123 (2018)

44. Yao, Y.Y.: Tri-level thinking: models of three-way decision. Int. J. Mach. Learn. Cybern. **11**, 947–959 (2020)

45. Yao, Y.Y.: Three-way granular computing, rough sets, and formal concept analysis. Int. J. Approximate Reasoning **116**, 106–125 (2020)

46. Yao, Y.: The geometry of three-way decision. Appl. Intell. **51**(9), 6298–6325 (2021). https://doi.org/10.1007/s10489-020-02142-z

47. Yao, Y.Y.: Set-theoretic models of three-way decision. Granular Comput. **6**, 133–148 (2021)

48. Yao, Y.Y.: Symbols-Meaning-Value (SMV) space as a basis for a conceptual model of data science. Int. J. Approximate Reasoning **144**, 113–128 (2022)

49. Yao, Y.Y.: Human-machine co-intelligence through symbiosis in the SMV space. Appl. Intell. 1–21 (2022). https://doi.org/10.1007/s10489-022-03574-5

Rough Set Theory and Applications

Scikit-Weak: A Python Library for Weakly Supervised Machine Learning

Andrea Campagner[1(✉)], Julian Lienen[2], Eyke Hüllermeier[3],
and Davide Ciucci[1]

[1] Dipartimento di Informatica, Sistemistica e Comunicazione, University of
Milano–Bicocca, Viale Sarca 336/14, 20126 Milano, Italy
`a.campagner@campus.unimib.it`
[2] Department of Computer Science, Paderborn University, Warburger Str. 100,
33098 Paderborn, Germany
[3] Institute of Informatics, University of Munich (LMU), Munich Center for Machine
Learning (MCML), Akademiestr. 7, 80799 Munich, Germany

Abstract. In this article we introduce and describe SCIKIT-WEAK, a
Python library inspired by SCIKIT-LEARN and developed to provide an
easy-to-use framework for dealing with weakly supervised and imprecise
data learning problems, which, despite their importance in real-world
settings, cannot be easily managed by existing libraries. We provide a
rationale for the development of such a library, then we discuss its design
and the currently implemented methods and classes, which encompass
several state-of-the-art algorithms.

Keywords: Weakly supervised learning · Imprecise data · Rough
sets · Generalized risk minimization · Imprecisiation

1   Introduction

In the recent years, applications of machine learning (ML) have spread into
both research and industry. Arguably, one of the major driving forces behind
this growth has been the wide availability of a multitude of publicly available
ML libraries, chiefly among them the Python ML eco-system [1,9,21,22], centred
around the SCIKIT-LEARN library[1] [23]. While such libraries offer a wide array
of methods that can be applied to various ML tasks, including supervised, semi-
supervised and fully unsupervised learning. By providing high-level APIs not
requiring deeper knowledge, they drastically improved the accessibility.

However, not all ML tasks fit neatly into the above mentioned categories.
In particular, weakly supervised learning [29] refers to machine learning tasks
situated in the spectrum between supervised and unsupervised learning [24],
encompassing various tasks such as multiple-instance learning [30], learning from
aggregate data [8] and learning from imprecise data [15]. In this latter case, in

---

[1] https://scikit-learn.org.

J. Yao et al. (Eds.): IJCRS 2022, LNAI 13633, pp. 57–70, 2022.
https://doi.org/10.1007/978-3-031-21244-4_5

particular, the data and annotations can be imprecise or partial: Some examples include semi-supervised learning as mentioned above, but also more general tasks such as soft labels learning [10,11,25], in which partial labels are represented through belief functions; learning from fuzzy labels [12,15], in which partial labels are represented through possibility distributions, and superset learning [4,16,20], in which partial labels are represented by exclusive sets of alternatives.

Despite the importance and practical relevance of weakly supervised learning in a variety of settings, including learning from anonymized data [26], learning from multi-rater data [8] and self-regularized learning [19], out-of-the-box libraries and frameworks to deal with such tasks are still missing and no libraries currently exist to easily manage this type of data in Python. In this article we introduce SCIKIT-WEAK, the first, to the authors' knowledge, Python library, inspired by and compatible with SCIKIT-LEARN, that provides easy-to-use methods and classes for dealing with weakly supervised learning problems. More in particular, the current version of the library focuses on the implementation of algorithms to deal with imprecise data learning problems. We provide a rationale for the development of such a library, followed by a discussion of its design and the currently implemented methods and classes, which encompass several state-of-the-art algorithms. Furthermore we briefly show the use of SCIKIT-WEAK, highlighting its interoperability with SCIKIT-LEARN, through a purposely simple but illustrative code example.

2   Background and Design Philosophy

In this section, we provide a basic background on weakly supervised learning, and specifically so to learning from imprecise data, describe the general design philosophy of SCIKIT-WEAK and illustrate an exemplary application of the library through a simple code example.

2.1   Background

In the supervised learning setting, a problem instance is defined by an instance space X and a target space Y, along with a probability distribution $\mathcal{D}$ over $X \times Y$. A finite sample of data $S = \{(x_1, y_1), \ldots, (x_n, y_n)\}$, called *training set*, is assumed to be sampled from $\mathcal{D}$ and to be available for learning. In rough set terminology we can describe S by means of a *decision table*[2], that is a triple (U, Att, Y), where $U \subseteq X$ is a finite set of instances in the instance space X, Att is a set of features with each feature $f : X \to V_f$, and t is a target feature with $t : X \to Y$, where Y denotes the target space. We note that while the definition of t may suggest that the association between instances and target labels is deterministic (hence, a mapping), this is not necessarily the case as the

---

[2] Compared to the usual definition of a training set considered in the ML literature the definition of a decision table in rough set theory distinguishes instances in U from their representation in terms of features.

dependency between X and Y is probabilistic and described by the unknown data generating distribution $\mathcal{D}$.

By contrast, in weakly supervised learning, and more specifically in learning from imprecise labels, the target feature is not assumed to be precisely known, but is instead only given in an imprecise form. In general, instead of the true target t, one can only observe the values of d, that is, a function $d : X \to D(Y)$, where $D(Y)$ is a set of *structures* over Y. As before, more in general, we may assume that instances are sampled from a distribution $\tilde{\mathcal{D}}$ defined over $X \times D(Y)$. As described in the introduction, weakly supervised learning aims at modeling learning problems in which knowledge about the supervision in a learning problem is not precisely or completely specified, but is only given in terms of imprecise beliefs or knowledge. Then, different tasks are defined based on the considered type of structures, for example:

- When $D(Y) = Y \cup \{\bot\}$, that is, each instance x is associated with either a label $y \in Y$ or no label at all ($\bot$), then the corresponding learning problem is called *semi-supervised learning*;
- When $D(Y) = 2^Y$, that is, each instance x is associated with a set of possible labels $\tilde{y} \subset Y$, then the corresponding learning problem is called *superset learning* or *partial-label learning*;
- When $D(Y) = [0,1]^Y$, that is, each instance x is associated with a possibility distribution $\pi_x : Y \to [0,1]$ over Y, then the corresponding learning problem is called *learning from fuzzy labels*;
- When $D(Y) = 2^{\mathbb{P}(Y)}$, that is, each instance x is associated with a set of probability distributions $\mathcal{Q}_x \subseteq \mathbb{P}(Y)$ over Y (that is, a *credal set*), then the corresponding leaning problem is called *credal learning*.

Thus, a weakly supervised problem instance is defined by a *weakly supervised training set* $W = \{(x_1, d_1), \ldots, (x_n, d_n)\}$ and the corresponding weakly supervised decision table $W = (U, Att, d)$, where, as above, $d : X \to D(Y)$. Given a weakly supervised decision table W, an *instantiation* of W is a standard decision table $I = (U, Att, \tilde{t})$, that is *compatible* with W (denoted $I \sim W$). For example:

- If $D(Y) = Y \cup \{\bot\}$, then $I \sim W$ iff $\forall x \in U, d(x) \neq \bot \implies \tilde{t}(x) = d(x)$ and $d(x) \neq \bot \implies \tilde{t}(x) \in Y$;
- If $D(Y) = 2^Y$, then $I \sim W$ iff $\forall x \in U, \tilde{t}(x) \in d(x)$;
- If $D(Y) = [0,1]^Y$, then $I \sim W$ iff $\forall x \in U, \pi_x(\tilde{t}(x)) > 0$.;
- If $D(Y) = 2^{\mathbb{P}(Y)}$, then $I \sim W$ iff $\forall x \in U, \exists p \in \mathcal{Q}_x$ s.t. $p(\tilde{t}) > 0$.

Notably, while we gave a binary definition of compatibility, a *graded* notion of compatibility can be defined for the learning from fuzzy labels and credal learning settings. Focusing on the first case for simplicity, for example, given two instantiations I_1, I_2 compatible with W, one could say that I_1 has stronger compatibility than I_2 when $\forall x \in U, \pi_x(\tilde{t}_1(x)) \geq \pi_x(\tilde{t}_2(x))$. See also [6] for possible alternative definitions of graded compatibility.

2.2   Design Philosophy

SCIKIT-WEAK is an open-source library, freely available via GitHub[3] and PyPi[4], that has been designed with two main aims:

- To provide a variety of easy-to-use tools and functionalities to enable data analysis grounding on weakly supervised data;
- To be inter-operable with SCIKIT-LEARN main functionalities and API.

To address the first aim, SCIKIT-WEAK is implemented through a module hierarchy that offers a variety of classes and functions to meet the main needs of a machine learning pipeline: data representation (through the `data_representation` module); pre-processing (through the `utilities` and `feature_selection` modules) and learning (through the `classification` module). Section 3 gives a comprehensive overview over each module.

To address the second aim, SCIKIT-WEAK conforms to the API of SCIKIT-LEARN. For example, classes in SCIKIT-WEAK's `feature_selection` module inherit from `sklearn.base.TransformerMixin` and thus exhibit the usual `fit`, `transform`, `fit_transform` interface. Thus, SCIKIT-WEAK classes can be used anywhere, and in the same way, a corresponding SCIKIT-LEARN class would be used, e.g., inside a `Pipeline`, enabling greater modularity and inter-operability.

Aside from SCIKIT-LEARN compatibility, to further facilitate use, SCIKIT-WEAK documentation, generated using SPHINX[5], is freely available online[6] and the library ships with an integrated suite of unit tests to ensure its correct functionality.

2.3   Code Example

To demonstrate the ease-of-use and the interoperability of SCIKIT-WEAK with SCIKIT-LEARN, consider the following example. First, starting from a standard supervised learning problem, weak supervision is generated (lines 11–18) by applying `DiscreteEstimatorSmoother`: this employs an underlying base classifier (in the example, a `KNeighborsClassifier`) to generate fuzzy labels. Then, a weakly supervised kNN model is instantiated (line 21; cf. Section 3.4) and a 5-fold cross validation is computed using the SCIKIT-LEARN implementation (lines 24 – 30), in order to fit and evaluate the weakly supervised model: this step, in particular, shows the interoperability between SCIKIT-WEAK and SCIKIT-LEARN base functionalities.

```
1 from scikit_weak.data_representation import
     DiscreteFuzzyLabel
2 from scikit_weak.classification import
     WeaklySupervisedKNeighborsClassifier
```

---

[3] https://github.com/AndreaCampagner/scikit-weak.
[4] https://pypi.org/project/scikit-weak/.
[5] https://sphinx-doc.org/.
[6] https://scikit-weak.readthedocs.io.

```
 3
 4 from sklearn.datasets import load_iris
 5 from sklearn.neighbors import KNeighborsClassifier
 6 from sklearn.model_selection import cross_val_score
 7
 8 import numpy as np
 9
10 # Construct exemplary weak supervision
11 X, y = load_iris(return_X_y=True)
12 smooth = DiscreteEstimatorSmoother(KNeighborsClassifier(
       n_neighbors=10), type="fuzzy")
13 y_fuzzy = smooth.fit_transform(X, y)
14
15 # Instantiate weakly-supervised KNN classifier
16 clf = WeaklySupervisedKNeighborsClassifier(k=5)
17
18 # Accuracy metric
19 def accuracy(estimator, X, y_soft):
20     y_pred = estimator.predict(X)
21     y_true = np.array([np.argmax(y.to_probs()) for y in
       y_soft])
22     return np.mean(y_true == y_pred)
23
24 # Perform 5-fold cross-validation
25 cv_scores = cross_val_score(clf, X, y_soft, cv=5, scoring=
       accuracy)
```

3   Contents and Documentation

In this section, we describe the main sub-modules and classes implemented in the SCIKIT-WEAK library.

3.1   Data Representation

SCIKIT-WEAK offers a flexible set of object classes representing weak target information [13,15], which can be found in the corresponding data_representation module and is depicted in Fig. 1.

The basic representation is given by the abstract class GenericWeakLabel that defines a standard interface that should be implemented by every concrete class of weak targets, such as the ability to randomly sample an element through the sample_value method. SCIKIT-WEAK primarily distinguishes between continuous and discrete weak labels, which are described in the following.

Continuous Weak Labels. Continuous weak labels are represented as instances of the abstract class ContinuousWeakLabel, whose main concrete sub-class is IntervalLabel. An object of this kind represents an interval-valued target

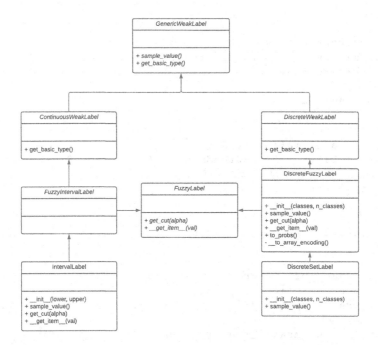

Fig. 1. The class hierarchy of data representation formats included in the module `data_representation`.

specified by its lower and upper bounds l and u, e.g., as often observed in weakly supervised regression problems. Without any further specification, each element within $[l, u]$ is considered to be equally plausible. Moreover, this class features to sample an element uniformly within this interval.

Discrete Weak Labels. Discrete weak labels can be represented as instances of the abstract class `DiscreteWeakLabel`, whose main concrete sub-classes are `DiscreteFuzzyLabel` and `DiscreteSetLabel`. As discrete target representation, objects of the former class maintain possibilities $\pi_x(y) \in [0, 1]$ over elements Y, e.g., classes as typically considered in classification problems. These possibilities represent upper probabilities of the true underlying probability distribution over Y. `DiscreteFuzzyLabel` supports a sampling mechanism to draw labels according to the possibilities. Moreover, discrete fuzzy labels can represent agnostic label information, i.e., assigning full possibility $\pi_x = (1, \ldots, 1)$ to any value in Y without further distinction. Semi-supervised learning is a typical setting where such data occurs, as parts of the data are completely unlabeled and target information is agnostic. To simplify the management of the type of data that occur frequently in superset and partial-label learning, namely, that a set of elements in Y have full plausibility, while all other elements are totally implausible, SCIKIT-WEAK also implements the `DiscreteSetLabel` class.

3.2   Utilities

The utilities module collects general utility functions that can be used for pre-processing, optimization or data checking and analysis. In particular, the module contains the smoothers sub-module, that encompasses several methods to transform supervised datasets into weakly supervised datasets; as well as the losses sub-module, that contains some commonly used loss functions for model evaluation and optimization-based learning.

DiscreteEstimatorSmoother. DiscreteEstimatorSmoother is a class to transform a supervised learning problem into a weakly supervised one, which uses an underlying classifier for imprecisiation. The need for this class stems from the fact that most existing benchmark datasets are precise, and hence cannot be used to test weakly supervised learning algorithms. Thus, DiscreteEstimatorSmoother allows to convert a standard supervised benchmark into a weakly supervised one. It supports transformation of standard labels to either DiscreteSetLabel or DiscreteFuzzyLabel. In the case of transformation to DiscreteFuzzyLabel objects, the underlying classifier given as input is trained on the supervised data given as input, and the output confidence scores are then normalized and used as values for the corresponding DiscreteFuzzyLabel. In the case of transformation to DiscreteSetLabel instances, only labels whose normalized confidence scores are greater than a parameterized threshold ϵ are considered as output.

DiscreteRandomSmoother. Related to the previous method, instances of the class DiscreteRandomSmoother realize the transformation from supervised to weakly supervised problems based on random sampling. Therefore, the class supports transformation of standard labels to either DiscreteSetLabel or DiscreteFuzzyLabel. To this end, discrete random smoother offers two sampling strategies: either according to the random set model, or according to the random membership model. In the random set model, labels in the corresponding DiscreteSetLabel are sampled at random, according either to probability p_incl (for the correct label) or p_err (for the incorrect labels). Formally, given instance (x, y) and the corresponding set-valued label S, it holds that $P(y \in S) = $ p_incl and $\forall y' \neq y, P(y' \in S) = $ p_err. In the random membership model, possibility degrees for the labels are sampled uniformly from the set of possible possibility degrees given as input in parameter prob_ranges.

3.3   Feature Selection

SCIKIT-WEAK offers a selection of methods to control model complexity and data dimensionality through the feature_selection module, which comprises different classes to perform weakly supervised feature selection and dimensionality reduction. In particular, the current version of the library implements two rough set-based feature selection algorithms (namely, classes RoughSetSelector and GeneticRoughSetSelector) and a dimensionality reduction algorithm (DELIN).

RoughSetSelector. `RoughSetSelector` performs weakly supervised feature selection using rough set-based reduct search [5,7]. The class supports datasets whose weakly supervised labels are either instances of the class `DiscreteSet` `Label` or `DiscreteFuzzyLabel`, and offers several choices in regard to the search strategy (brute-force or greedy search), the class of reducts to search for (super-set reducts, C-reducts, λ-reducts), and the rough set model to be used (k-neighborhood or radius neighborhood rough sets). When the weakly supervised labels are instances of `DiscreteSetLabel`, both brute-force and greedy search aim to find minimal superset reducts. A superset reduct is a reduct for an instantiation of the weakly supervised dataset given as input. The brute-force search strategy examines all subsets of features $R \subseteq Att$ exhaustively to check whether they are superset reducts. The algorithm is guaranteed to return all the minimal-size superset reducts, but, however, the computational complexity is exponential $(O(|X| \cdot 2^{|Att|}))$. By contrast, the greedy search strategy starts with the full set of features Att and iteratively removes one feature as long as the remaining set of feature is a superset reduct. The algorithm is not guaranteed to return a minimal-size superset reduct, but global search is supported via random restarts. The complexity of greedy search is $O(|X| \cdot |Att|^2)$. When the weakly supervised labels are instances of `DiscreteFuzzyLabel`, brute-force and greedy search aim to find either C- or λ-reducts. A C-reduct $R \subseteq Att$ is a superset reduct for an instantiation I_R for which $\nexists R' \subseteq Att$ superset reduct for an instantiation $I_{R'}$ such that both $|R'| \leq |R|$ and $\min_{x \in S} \pi_x(\tilde{t}_{I_R}(x)) \leq \min_{x \in S} \pi_x(\tilde{t}_{I_{R'}}(x))$. A λ-reduct $R \subseteq Att$ is a superset reduct for an instantiation I_R that minimizes $(1 - \lambda)(\min_{x \in S} \pi_x(\tilde{t}_{I_R}(x))) - \lambda \frac{|R|}{|Att|}$ among all superset reducts. Both brute-force and greedy search perform feature selection by searching for superset reducts on the α-cuts of the fuzzy-labeled dataset given as input, and then selecting among the retrieved reducts those that satisfy the constraints of being either a C-reduct or a λ-reduct. Thus, the complexity of brute-force search is $O(|X|^2 \cdot 2^{|Att|})$ while the complexity of greedy search is $O(|X|^2 \cdot |Att|^2)$.

GeneticRoughSetSelector. The class `GeneticRoughSetSelector` offers functionality to perform weakly supervised selection by reduct search using genetic algorithms [6]. The class supports datasets whose weakly supervised labels are instances of `DiscreteFuzzyLabel`. `GeneticRoughSetSelector` aims to find either C-reducts, D-reducts or λ-reducts for the weakly supervised dataset given as input, supporting every type of weakly supervised label. A D-reduct $R \subseteq Att$ is a superset reduct for an instantiation I_R for which $\nexists R' \subseteq Att$ superset reduct for an instantiation $I_{R'}$ s.t. both $|R'| \leq |R|$ and $\exists x \in S, \pi_x(\tilde{t}_{I_R}(x)) < \pi_x$ $(\tilde{t}_{I_{R'}}(x))$. The genetic algorithm-based search is guided by one of three possible fitness functions, corresponding to the above mentioned reduct classes:

$$Fitness_C = \langle r, p \rangle, \tag{1}$$

$$Fitness_\lambda = (1 - \lambda)p - \lambda \frac{r}{|Att|}, \tag{2}$$

$$Fitness_D = \langle r, s \rangle, \tag{3}$$

where $p = \min_{x \in S} \pi_x(\tilde{t}_I(x))$, $r = \begin{cases} |A| & F \text{ is a super-reduct} \\ \infty & \text{otherwise} \end{cases}$, and $s \in [0,1]^{|U|}$ is
a vector s.t. $s_x = \pi_x(\tilde{t}_I(x))$. Note, in particular, that only $Fitness_\lambda$ is single-valued, while the other two fitness functions are multi-valued. Consequently, for these latter two fitness functions, the implementation employs a multi-objective optimization algorithm. Irrespective of the fitness function adopted, the computational complexity of GeneticRoughSetSelector is $O(|X| \cdot |Att|)$. With regard to selection and cross-over, GeneticRoughSetSelector employs non-dominated tournament selection and single-point cross-over, respectively. For mutation, candidate reducts are mutated by random addition or deletion of features according to a Bernoulli distribution. By contrast, instantiations are mutated according to a two-step procedure. First, for each instance x, a binary value is randomly sampled from a Bernoulli distribution, then, if the above mentioned value was equal to 1, a new target label is sampled using the method sample_value of the corresponding GenericWeakLabel instance.

DELIN. DELIN is a weakly supervised dimensionality reduction algorithm, based on the combination of linear discriminant analysis and weakly supervised k-NN [2,27,28]. The class supports datasets whose weakly supervised labels are instances either of the class DiscreteSetLabel or DiscreteFuzzyLabel. DELIN requires one to determine a-priori the number of dimensions to be selected via the parameter n. Intuitively, the algorithm works in iterations, each of which consists of two steps: first, WeaklySupervisedKNeighborsClassifier is applied to the data, then linear discriminant analysis is applied to the original data w.r.t. the confidence scores given as output of the first step. Compared to the algorithm originally proposed in [27,28], the DELIN class has two main modifications: first, it supports not only DiscreteSetLabel but also DiscreteFuzzyLabel instances; second, singular value decomposition is used in the computation of linear discriminant analysis to avoid stability issues. The computational complexity of DELIN is $O(|X| \cdot |Att|^2)$.

3.4 Classification

Aside from the pre-processing and dimensionality reduction methods described in the previous sections, SCIKIT-WEAK also offers a wide selection of weakly supervised classification algorithms contained in the classification module.

WeaklySupervisedKNeighborsClassifier. As one of two neighborhood-based methods, `WeaklySupervisedKNeighborsClassifier` is a simple generalization of k-nearest neighbors classification to the setting of weakly supervised data [3,17], and is compatible with every instance of `DiscreteWeakLabel`. The number of neighbors can be controlled through parameter `k`, while the class supports any metric callable (through the `metric` parameter, default is the Euclidean metric). For efficiency reasons, SCIKIT-LEARN's `NearestNeighbors` is used to speed-up neighbors search: the computational complexity is $\Omega(|X| \cdot log|X|)$, with an additional complexity of $\Omega(log|X|), O(|X|)$ at inference time.

WeaklySupervisedRadiusClassifier. `WeaklySupervisedRadiusClassifier` is yet another simple generalization of radius-based neighbors classification to weakly supervised data [17], and is compatible with every instance of `DiscreteWeakLabel`. The radius within which to search for neighbor instances can be controlled through the `radius` parameter, while the class supports any metric callable (through the `metric` parameter, default is the Euclidean metric). Similarly as for class `WeaklySupervisedKNeighborsClassifier`, `NearestNeighbors` is used to speed-up neighbors search: the computational complexity is $\Omega(|X| \cdot log|X|)$, with an additional complexity of $\Omega(log|X|), O(|X|)$ at inference time.

GRMLinearClassifier. `GRMLinearClassifier` is an optimization-based classification method that attempts to directly minimize the generalized risk for a linear model [15]. Currently, it supports instances of `DiscreteFuzzyLabel` and implements two different linear classification algorithms, namely, logistic regression (by setting `loss` parameter to *"logistic"*) or linear SVM (by setting `loss` parameter to *"hinge"*). More in detail, given loss function l, `GRMLinearClassifier` attempts to solve the following optimization problem:

$$\underset{W}{\operatorname{argmin}} \frac{1}{|X|} \sum_{(x,\pi) \in S} l_F(\pi, W \cdot x)$$

where $l_F : [0,1]^Y \times \mathbb{R}^Y \rightarrow \mathbb{R}$ is the generalized risk [15], defined as

$$l_F(\pi, W \cdot x) = \int_0^1 \min_{y \in \pi^\alpha} l(y, W \cdot x) d\alpha. \tag{4}$$

Optimization is implemented by means of gradient descent, relying on TENSORFLOW[7] for efficient computation. In particular, the class supports every TENSORFLOW optimizer (through the `optimizer` parameter, default is stochastic gradient descent *"sgd"*). In general, the optimization problem described above is non-convex, thus convergence to a global optimum is not guaranteed and no convergence checking is implemented. Training is performed for a fixed number of iterations (set through parameter `max_epochs`), therefore complexity is on the order of $O(|X| \cdot |Att|)$. To avoid overfitting, `GRMLinearClassifier` supports weight regularization, set through the `regularizer` parameter.

---

[7] https://tensorflow.org.

RRLClassifier. RRLClassifier is an efficient ensemble-based method for weakly supervised classification based on a generalization of tree ensemble-based learning [8]. RRLClassifier trains an ensemble of standard supervised classifiers (by default, SCIKIT-LEARN's ExtraTreeClassifier [14], but the type of classifier can be set through parameter estimator) by drawing random samples from the weakly supervised data given as input. For each instance label Y in the training set, and each classifier h_i to be ensembled, a sample label is obtained by calling y.sample_value(). Thus, RRLClassifier supports every instance of GenericWeakLabel. Optionally, bootstrapping (as in random forests) can be applied (through parameter resample, by default set to False) to ensure increased diversity among the classifiers in the ensemble. The computational complexity of RRLClassifier is $O(k \cdot |Att||X| \cdot log|X|)$, where k is the number of classifiers to be ensembled (set through parameter n_estimators).

LabelRelaxationNNClassifier. As one example of a credal learning classifier, LabelRelaxationNNClassifier provides an implementation of the label relaxation loss [19] to train probabilistic neural network classifiers $H : X \rightarrow \mathbb{P}(Y)$ with $\mathbb{P}(Y)$ denoting the space of probability distributions over Y. Commonly, training of such models H involves a gradient-descent based optimization of a probabilistic loss $l : \mathbb{P}(Y) \times \mathbb{P}(Y) \rightarrow \mathbb{R}_+$, where degenerate probability distributions p_y with $p_y(y) = 1$ and $p_y(\cdot) = 0$ otherwise are considered as surrogate targets for an observed class labels $y \in Y$, typically resulting into overconfident models. To achieve better calibrated models by a more faithful target modeling, label relaxation replaces the degenerate distribution p_y assigned to an instance x by a credal set $\mathcal{Q}_{\pi_x}$ in accordance with a possibility distribution π_x that assigns a fixed possibility $\pi_x(y') = \alpha \in [0, 1]$ to the labels $y' \neq y$ and $\pi_x(y) = 1$. This credal set $\mathcal{Q}_{\pi_x}$ is then used as target within a generalized loss formulation adopting Eq. (4) to train models, which is implemented in the class LabelRelaxationLoss. LabelRelaxationNNClassifier allows one to specify the imprecisiation parameter α (parameter lr_alpha), as well as hyperparameters related to stochastic gradient descent (SGD) optimization. Moreover, the base network to be trained can be specified by its hidden layer depth and widths. The computational complexity depends on the parameterization of the SGD procedure, resulting in a complexity similar to GRMLinearClassifier. As before, we use TENSORFLOW as optimization framework.

CSSLClassifier. Another credal learning method is provided in the class CSSL-Classifier, which implements so-called credal self-supervised learning (CSSL) [18] to induce probabilistic classifiers in a semi-supervised learning scenario. To this end, CSSL maintains credal sets $\mathcal{Q}_{\pi_x}$ as used in LabelRelaxationLoss expressing the model's belief about the true target for previously unlabeled instances, proceeding from agnostic credal sets of the form $\mathcal{Q}_{\pi_x} = \mathbb{P}(Y)$ with $\pi_x(y') = 1 \, \forall y' \in Y$. These credal sets successively shrink with increased training progress and thus higher model confidence by reducing the degree of imprecisiation in π_x. CSSLClassifier allows one to specify a base model (parameter estimator, e.g., an instance of `LabelRelaxationNNClassifier`), the number of iterations (`n_iterations`), a class prior distribution used within the credal set construction (`p_data`) and the buffer size of the model prediction history also employed in the credal set construction (`p_hist_buffer_size`). In each iteration, the base model is retrained on the complete data and the credal sets are adjusted according to the updated model.

4   Conclusion

In this article, we introduced SCIKIT-WEAK, a Python library for weakly supervised learning and data analysis, currently focusing on the handling of learning from imprecise data problems. To the authors knowledge, SCIKIT-WEAK is the first library providing such functionality in Python, and thus we believe it could advance the applicability of the Python data science ecosystem to non-standard and weakly supervised learning problems. We described the fundamental design concepts underlying the library and documented the main implemented functionalities and classes. We also illustrated the use of the library by means of a simple example. The SCIKIT-WEAK is an open source project and we hope that additional contributors can help maintain the library as well as implement new functionalities: indeed, being freely and openly available on GitHub, and being implemented completely in Python, we believe developers could easily extend and add new functionalities to the existing library. In particular, we envision the following next steps for the development of the library:

- To extend the suite of implemented weakly supervised data representation, so as to encompass additional and more general learning settings such as those mentioned in the introduction;
- To provide more efficient and robust implementations of the currently implemented classes, e.g., by off-loading time-sensitive routines to low-level or device code, or by implementing more extensive type checking and tests;
- To enrich the library with sample weakly supervised datasets that can be used for prototyping, testing as well as benchmarking purposes.

References

1. Abadi, M., et al.: TensorFlow: large-scale machine learning on heterogeneous systems (2015). https://www.tensorflow.org/, software available from tensorflow.org
2. Bao, W.X., Hang, J.Y., Zhang, M.L.: Partial label dimensionality reduction via confidence-based dependence maximization. In: Proceedings of the 27th ACM SIGKDD Conference on Knowledge Discovery & Data Mining, pp. 46–54 (2021)
3. Bezdek, J.C., Chuah, S.K., Leep, D.: Generalized k-nearest neighbor rules. Fuzzy Sets Syst. **18**(3), 237–256 (1986)
4. Cabannes, V., Bach, F., Rudi, A.: Disambiguation of weak supervision with exponential convergence rates. arXiv preprint arXiv:2102.02789 (2021)

5. Campagner, A., Ciucci, D.: Feature selection and disambiguation in learning from fuzzy labels using rough sets. In: Ramanna, S., Cornelis, C., Ciucci, D. (eds.) IJCRS 2021. LNCS (LNAI), vol. 12872, pp. 164–179. Springer, Cham (2021). https://doi.org/10.1007/978-3-030-87334-9_14

6. Campagner, A., Ciucci, D.: Rough-set based genetic algorithms for weakly supervised feature selection. In: et al. International Conference on Information Processing and Management of Uncertainty in Knowledge-Based Systems, vol. 1602, pp. 761–773. Springer, Cham (2022). DOIurlhttps://doi.org/10.1007/978-3-031-08974-9_60

7. Campagner, A., Ciucci, D., Hüllermeier, E.: Rough set-based feature selection for weakly labeled data. Int. J. Approximate Reasoning **136**, 150–167 (2021)

8. Campagner, A., Ciucci, D., Svensson, C.M., Figge, M.T., Cabitza, F.: Ground truthing from multi-rater labeling with three-way decision and possibility theory. Inf. Sci. **545**, 771–790 (2021)

9. Chollet, F., et al.: Keras. https://keras.io (2015)

10. Côme, E., Oukhellou, L., Denoeux, T., Aknin, P.: Learning from partially supervised data using mixture models and belief functions. Pattern Recogn. **42**(3), 334–348 (2009)

11. Denoeux, T.: Maximum likelihood estimation from uncertain data in the belief function framework. IEEE Trans. Knowl. Data Eng. **25**(1), 119–130 (2011)

12. Denœux, T., Zouhal, L.M.: Handling possibilistic labels in pattern classification using evidential reasoning. Fuzzy Sets Syst. **122**(3), 409–424 (2001)

13. Destercke, S.: Uncertain data in learning: challenges and opportunities. Conformal and Probabilistic Prediction with Applications, pp. 322–332 (2022)

14. Geurts, P., Ernst, D., Wehenkel, L.: Extremely randomized trees. Mach. Learn. **63**(1), 3–42 (2006)

15. Hüllermeier, E.: Learning from imprecise and fuzzy observations: Data disambiguation through generalized loss minimization. Int. J. Approximate Reasoning **55**(7), 1519–1534 (2014)

16. Hüllermeier, E., Beringer, J.: Learning from ambiguously labeled examples. Intell. Data Anal. **10**(5), 419–439 (2006)

17. Kuncheva, L.: Fuzzy Classifier Design, vol. 49. Springer, Heidelberg (2000)

18. Lienen, J., Hüllermeier, E.: Credal self-supervised learning. In: Advances in Neural Information Processing Systems, vol. 34: Annual Conference on Neural Information Processing Systems 2021, NeurIPS 2021, 6–14 December 2021, virtual, pp. 14370–14382 (2021)

19. Lienen, J., Hüllermeier, E.: From label smoothing to label relaxation. In: Proceedings of the 35th AAAI Conference on Artificial Intelligence, virtual, 2–9 February (2021)

20. Liu, L., Dietterich, T.G.: A conditional multinomial mixture model for superset label learning. In: Advances in Neural Information Processing Systems, pp. 548–556 (2012)

21. Löning, M., Bagnall, A., Ganesh, S., Kazakov, V., Lines, J., Király, F.J.: sktime: a unified interface for machine learning with time series. arXiv preprint arXiv:1909.07872 (2019)

22. McKinney, W., et al.: pandas: a foundational python library for data analysis and statistics. Python High Perform. Sci. Comput. **14**(9), 1–9 (2011)

23. Pedregosa, F., et al.: Scikit-learn: machine learning in python. J. Mach. Learn. Res. **12**, 2825–2830 (2011)

24. Poyiadzi, R., Bacaicoa-Barber, D., Cid-Sueiro, J., Perello-Nieto, M., Flach, P., Santos-Rodriguez, R.: The weak supervision landscape. In: 2022 IEEE International Conference on Pervasive Computing and Communications Workshops and other Affiliated Events (PerCom Workshops), pp. 218–223. IEEE (2022)

25. Quost, B., Denoeux, T., Li, S.: Parametric classification with soft labels using the evidential EM algorithm: linear discriminant analysis versus logistic regression. Adv. Data Anal. Classif. **11**(4), 659–690 (2017)

26. Sakai, H., Liu, C., Nakata, M., Tsumoto, S.: A proposal of a privacy-preserving questionnaire by non-deterministic information and its analysis. In: 2016 IEEE International Conference on Big Data (Big Data), pp. 1956–1965. IEEE (2016)

27. Wu, J.H., Zhang, M.L.: Disambiguation enabled linear discriminant analysis for partial label dimensionality reduction. In: Proceedings of the 25th ACM SIGKDD International Conference on Knowledge Discovery & Data Mining, pp. 416–424 (2019)

28. Zhang, M.L., Wu, J.H., Bao, W.X.: Disambiguation enabled linear discriminant analysis for partial label dimensionality reduction. ACM Trans. Knowl. Discov. Data (TKDD) **16**(4), 1–18 (2022)

29. Zhou, Z.H.: A brief introduction to weakly supervised learning. Natl. Sci. Rev. **5**(1), 44–53 (2018)

30. Zhou, Z.H., Sun, Y.Y., Li, Y.F.: Multi-instance learning by treating instances as non-IID samples. In: Proceedings of the 26th Annual International Conference on Machine Learning, pp. 1249–1256 (2009)

Applying Rough Set Theory for Digital Forensics Evidence Analysis

Khushi Gupta, Razaq Jinad, and Zhou Bing$^{(\boxtimes)}$

Sam Houston State University, Huntsville, TX, USA
{kxg095,raj032,bxz003}@shsu.edu

Abstract. With the growth of digital crime and the pressing need for strategies to counteract these forms of criminal activities, there is an increased awareness of the importance of digital forensics. However, due to the poor quality or the availability of incomplete information, the evidence gathered from a crime scene may not always be optimal in practical situations. Digital evidence can be present in different kinds of devices and in many different forms, much of which is found in an imprecise format making it very difficult to be analyzed. We propose the use of Rough Set theory for the classification of digital evidence. Rough Set theory is a computational model which is an effective tool for analyzing uncertainty and incomplete information. In this paper, we apply a Rough Set model to two digital forensics datasets proving Rough Set to be a valid tool that can be used for digital forensics investigations. We applied two algorithms for feature selection namely, Recursive feature elimination and Fuzzy Rough feature selection. Additionally, various algorithms such as Support Vector Machine (SVM), Naïve Bayes, Decision Tree (J48), Logistic Regression, and Rough Set theory were used for classification. Rough Set when used for both feature extraction and classification gives higher accuracy compared to other algorithms.

Keywords: Rough set · Computational forensics · Digital forensics

1 Introduction

According to NIST [24], "Digital forensics is the field of forensic science that is concerned with retrieving, storing and analyzing electronic data that can be useful in criminal investigations. This includes information from computers, hard drives, mobile phones, and other data storage devices". Digital forensics being in its early stages as a discipline has increasingly received immense attention in the past few years. It plays a significant role in society when it comes to justice, security, and privacy [23]. Digital evidence can now be found on various kinds of devices and has become even more valuable during investigations. With the increased importance and relevance also comes new challenges presented by the field of digital forensics. One of the major challenges includes the volume of digital evidence that may be collected during an investigation [6]. This evidence can

© The Author(s), under exclusive license to Springer Nature Switzerland AG 2022
J. Yao et al. (Eds.): IJCRS 2022, LNAI 13633, pp. 71–84, 2022.
https://doi.org/10.1007/978-3-031-21244-4_6

sometimes be imperfect due to measurement error, or it can be partially missing [13] making it difficult to analyze it and thus get hold of a suspect. It is part of criminalistic sciences which deals with digital evidence recovery and exploitation in the solution of criminal cases through the application of scientific principles. As part of evidence analysis, imprecise evidence must be used to elicit hypotheses concerning events, actions, and facts to obtain evidence to present in court. It also involves examining fragmented incomplete knowledge and reconstructing and aggregating complex scenarios involving time, uncertainty, causality, and alternative possibilities [1]. Although in recent years, significant progress has been made in computer forensics, limitations still exist which makes it difficult to meet the objective requirements of forensic technology. Forensic tools need to move in the direction of being intelligent and automated. This can be done through the usage of new intelligent information processing techniques such as Rough Set theory in the data analysis phase [29].

Soft computing is a combination of methods that complement each other when dealing with ambiguous real-life decision systems. Rough Set theory (RST) is one of the soft computing models that deals with incomplete knowledge and thus provides a mechanism for concept approximation. In this paper, we use two datasets, the SpamBase dataset which is an imprecise and incomplete forensics-related dataset, and the Pen-Based dataset which comparatively does not have a lot of errors. We then use Rough Set theory with other feature reduction and classification algorithms to compare and identify which combination of algorithms gives the best results.

The main contributions of this paper are summarized as follows:

- Use Rough Set theory for feature selection and classification of a digital forensic evidence dataset, proving its usage in the field.
- Using different datasets with missing features to conduct experiments to illustrate the performance of Rough Set.
- Conducting experiments to compare the effectiveness of Rough Set over other algorithms for classifying an imprecise dataset.

The rest of the paper is organized as follows. Section 2 reviews some of the preliminary concepts. Section 3 gives a detailed overview of the existing work. The experimental study and discussion of results are presented in Sect. 4. Lastly, the conclusion and future work are presented in Sect. 5.

2   Preliminaries

In this section, we recall some basic notions related to some of the concepts used in this paper.

2.1   Computational Models

Computational models are mathematical models that are simulated using computation to study complex systems [10]. These models are widely being used in diverse fields such as computer science, chemistry, physics, etc. Computational models are increasingly used to help us make decisions wisely by translating observations into future events [10]. The outcome is used to understand, manage, and predict the mechanisms of complex processes and systems. As the need to increase the understanding of real-world phenomena grows rapidly, computer-based simulations and modeling tools are increasingly being accepted as viable means to study such problems.

Soft Computing. Soft computing provides a blend of data processing mechanisms that deal with vague or imprecise knowledge with the help of approximate models and gives solutions to complex real-life problems. It is tolerant of imprecision, uncertainty, partial truth, and approximations. Soft computing is based on techniques such as Rough Sets, fuzzy logic, genetic algorithms, artificial neural networks, machine learning, and expert systems [16]. Soft computing plays a crucial role in learning complex data structures and patterns and classifying them to make intelligent decisions [19]. Soft computing has been widely used in various applications, such as machine vision, pattern detection, data segmentation, data mining, adaptive control, biometrics, and information assurance.

Rough Set. Proposed by Professor Pawlak in 1982, the Rough Set theory is an important mathematical tool to deal with imprecise, inconsistent, incomplete information and knowledge [31]. It handles data reduction and generates precise decision rules that help in extracting correct patterns from the data [8]. Additionally, no additional information and parameters are required to analyze the data and it is structured as a decision table [21]. Since its development, Rough Set theory has been able to formulate computationally efficient and mathematically sound techniques for addressing issues such as pattern discovery, decision rule formulation, data reduction, principal component analysis, and inference interpretation based on available data [5].

Because of the development of computer science and technology, especially the development of computer networks, a large amount of information is provided for people every second of the day. With the growing amount of information, the requirement for information analysis tools is also becoming higher and higher, and people hope to automatically acquire the potential knowledge from the data. Especially in the past years, knowledge discovery (rule extraction, data mining, machine learning, etc.) has attracted much attention in the field of artificial intelligence. Rough Set theory has had a significant impact in the data analysis field and thus has attracted a lot of attention from researchers from around the globe [5].

Rough Set theory uses two precise boundary lines to describe imprecise concepts. Therefore, in a sense, the Rough Set theory is a certain mathematical tool to solve uncertain problems. Because of its novel and unique method and easy operation, Rough Set theory has become an important information processing tool in the field of intelligent information processing [6] and it has been widely used in machine learning, knowledge discovery, data mining, decision support, and analysis, etc. The primary benefit of Rough Set theory in data analysis is that it does not require any preliminary or supplementary data information, unlike probability distributions required in statistics, basic probability assignments required in evidence theory, a membership grade, or the value of possibility required in fuzzy set theory [27].

Rough Set for Feature Selection. Feature selection is used to select features that are most predictive of a given outcome [12]. Rough Set was introduced for data analysis in pattern recognition, data mining, and machine learning [21]. It is integral and powerful for discovering relations between a class and its attributes in a dataset. Rough Set provides a powerful mechanism for reducing attributes without altering the basic properties of the system. This process is referred to as attribute reduction or feature selection [25]. This process thus, retains the minimal set of attributes that preserves the information of interest. The main advantage of Rough Set attribute reduction (RSAR) is that no additional parameters are required other than the supplied data. In Rough Set theory, an information table is defined as a tuple $T = (U, A)$ where U and A are two finite, non-empty sets, U the universe of primitive objects, and A the set of attributes. Each attribute or feature $a \in A$ is associated with a set V_a of its value called the domain of A. We may partition the attribute set A into two subsets C and D, called condition and decision attributes, respectively. Let $P \subset A$ be a subset of attributes. The indiscernibility relation, denoted by *IND(P)*, is an equivalence relation defined in [30]:

$$(P) = (x, y) \in U \times U : \forall \, a\epsilon \, P, a(x) = a(y) \tag{1}$$

where $a(x)$ denotes the value of feature a of object x. If $(x, y) \in IND(P)$, x and y are said to be indiscernible with respect to P. The family of all equivalence classes of *IND(P)* (Partition of U determined by P) is denoted by *U/IND(P)*. Each element in *U/IND(P)* is a set of indiscernible objects with respect to P. Equivalence classes *U/IND(C)* and *U/IND(D)* are called condition and decision classes. For any concept $X \subseteq U$ and attribute subset $R \subseteq A$, X could be approximated by the R-lower approximation and R- upper approximation using the knowledge of R [30]. The lower approximation of X is the set of objects of U that are surely in X, defined as [30]:

$$R_* (X) = \cup \; E \in U/IND(R) : E \subseteq X \tag{2}$$

The upper approximation of X is the set of objects of U that are possibly in X, defined as [30]:

$$R^* (X) = \cup E \in U/IND(R)E \cap X \neq \phi \tag{3}$$

The boundary region is defined as: [30]

$$BND_R(X) = R^*(X) - R_*(X) \tag{4}$$

If the boundary region is empty, that is, $R_* (X) = R^* (X)$ concept X is said to be R-definable. Otherwise, X is a Rough Set with respect to R. The positive region of decision classes $U/IND(D)$ with respect to condition attributes C is denoted by: [30]

$$POS_c (D) = \cup R_* (X) \tag{5}$$

It is a set of objects of U that can be classified with certainty to classes $U/IND(D)$ employing attributes of C. A subset $R \subseteq C$ is said to be a D-reduct of C if

$$POS_R (D) = POS_C (D) \tag{6}$$

$POS_R (D) = POS_C (D)$ condition is satisfied and there is no $R' \subset R$ such that [30]

$$POS'_R(D) = POS_C(D) \tag{7}$$

Hence, a reduct is the minimal set of attributes preserving the positive region. There may exist many reducts in an information table.

Rough Set theory has increasingly been used as a feature selection step to reduce the number of features in a dataset for a more effective classification process. The feature selection process is done by using the reduct attribute of Rough Set. Reducts in Rough Sets are the minimal features in a dataset or database that can sufficiently characterize the dataset or database.

Rough Set for feature selection is achieved by an accompanying search method. Search methods are algorithms or techniques that are used to test different combinations of features or attributes in the dataset. The attribute reduct (Rough Set) is then used to select the best combination. In this project, Particle Swarm Optimization is used in conjunction with Rough Set as the search algorithm.

Particle Swarm Optimization. Particle swarm optimization (PSO) algorithm was proposed by Eberhart and Kennedy in 1995 [28]. It is a stochastic optimization technique based on swarm intelligence. PSO algorithm simulates animals' social behavior, including insects, herds, birds, and fishes [4]. PSO intends to find the optimal solution in a high-dimensional solution space. It achieves this by maximizing or minimizing a function to find the optimum solution. A function can have different local maximums and minimums with one global maximum or minimum. PSO is a heuristic algorithm as it can sacrifice accuracy and completeness for speed.

Rough Set for Classification. Rough Set can be used on classification problems and datasets. It looks at the structural interactions within imprecise and noisy data. It can only work on discrete data. Rough Set theory works with the establishment of equivalence classes within the given training data [14]. All the data tuples forming an equivalence class are indiscernible, that is, the samples are identical with respect to the attributes describing the data.

Computational Forensics. Computational Forensics is an emerging research domain that is a hypothesis-driven investigation of specific scientific problems using computers with the primary goal to get an in-depth understanding of the forensic discipline [31]. It involves applying computer science, applied mathematics, and statistics techniques (modeling and computer simulation) to study and solve forensics problems. A systematic methodology for computer forensics must have a comprehensive research, development, and investigation process which focuses on the needs of the forensic problem. The potential of computational forensics can have a great impact on forensics. Some of the most promising contributions made by computational forensics include:

- Increased efficiency and reporting on investigation results and deductions.
- Perform the often-time-consuming testing using systematic testing foundations which can also be tested on a large scale of data.
- Help synthesize unequally distributed datasets and noisy data and simulate meaningful influences.

3   Related Work

Computational forensics is an emerging research domain with various computational methods being used to solve various kinds of problems. Rough Set is one of the computational methods that are excellent with incomplete and imprecise data and it's previously been used to solve various kinds of forensic problems.

Singh et al. [26] use Rough Set to process the physiological and facial characteristics of a criminal thus, helping in their identification. The behavioral attributes used are a person's gait patterns and the way of speaking, his/her physic can be represented in the form of age, gender, and height while facial features used include face category, face tone, eyebrows type, eye shape, nose shape, and lip size. This information provided by the eyewitness is usually vague and imprecise and thus this information is processed by a Rough Set.

Another application of a Rough Set includes online signature detection [7]. The authors used global features extracted as time functions of various dynamic properties of signatures to identify a signature. A database of 2160 signatures from 108 subjects was built. Thirty-one features were identified and extracted from each signature. The Rough Set approach was used to reduce the features to a set of nine features that captured the essential characteristics required to accurately identify a signature. Rough Set demonstrated a 100% correct classification rate proving it to be sustainable and effective for online signature identification.

The field of steganography (an art of secret transmission by embedding data, not multimedia) has also benefitted from Rough Set theory. Lang et al. [18] talks about a novel steganalysis approach based on Rough Set theory. It considers characteristics of both embedded messages and digital images and gives decisions from huge data set of steganographic signatures. Through the experiments, the method is proved to be effective and applicable. The steganalysis method based on Rough Set theory can be used to mine the hidden rules in a huge data set which is applicable for the implementation and scaling the experiment.

Another study [20] used a decision Rough Set -positive region reduction for steganalysis feature selection. A Rough Set model is used to reduce the dimension and improve the efficiency of the steganalysis algorithm. Rich model steganalysis features usually result in a large computation cost but using the proposed method can significantly reduce the feature dimension and maintain detection accuracy.

Rough Sets have even found their roots in the application of identifying insider threats. This paper [32] monitors users' abnormal behavior which is applied to insider threat identification to build a user's behavior attribute information database based on weights changeable feedback tree augmented Bayes network. This data can however be massive, and this experiment uses dimensionality reduction based on Rough Sets to establish the process information model of the user's behavior attribute.

Additionally, Rough Sets have even been used for dimensionality minimization to enable pattern classifiers to be effective. The main limitation of Rough Set-based classification/selection is that all data should be discrete and thus real-valued and noisy data cannot be used. This paper [17] investigates two approaches based on Rough Set extensions, namely fuzzy-rough and tolerance Rough Sets, that address these problems and retains dataset semantics. The methods are compared experimentally and utilized for the task of forensic glass fragment identification. Other studies that use Rough Sets for feature and dimensionality minimization include [9]. This study uses the set of permissions required by any android app during installation time as the feature set which are used in permission-based detection of android malware. This feature set is then reduced to minimize computational overhead by choosing an optimal and meaningful set of features/attributes. A selection technique based on Rough Set and improvised particle swarm optimization (PSO) algorithm is thus proposed for this study.

Lastly, Rough Sets have been used in social network analysis of law information privacy protection [26]. The paper uses the hierarchical structure of data as domain knowledge for the Rough Set theory. The Rough Set modeling for complex hierarchical data is studied for hierarchical data of the decision table. The theoretical research results are applied to hierarchical decision rule mining and k-anonymous privacy protection data mining research, which enriches the connotation of Rough Set theory and has important theoretical and practical significance for further promoting the application of this theory.

To the best of our knowledge, very few studies have tackled the use of Rough Sets to make predictions on imperfect data to solve forensic problems. Our study tackles the use of soft computing using Rough Set theory to make predictions on digital forensic problems in a prevalent situation of imprecise data.

4   Experimental Study

In this section, we compare the proposed Rough Set approach with other existing machine learning models.

4.1 Datasets Used

We used the Spambase [2] and the Pen-Based Recognition of Handwritten Digits dataset from the UCI machine learning repository for our experiment. The
spam dataset classifies email either as spam or non-spam (two-class values 1
and 0). The spam emails in the dataset were collected from postmasters and
individuals who have filed for spam and the non-spam emails come from filed
work and personal emails. It is a moderately big database with 4601 instances
and 57 attributes. This dataset was used because emails are one of the most
important sources of evidence in digital forensic investigations. Additionally, the
database contains some errors that can be efficiently addressed by Rough Set
theory. Some of the issues with the database include missing values, duplicate
data, class imbalance, and outliers (extreme values outside the range of what
is expected). The second dataset, the Pen-Based Recognition of Handwritten
Digits was created by collecting 250 samples from 444 writers to with the aim
to identify a person through their handwriting. It contains 10992 instances with
16 attributes. However, no significant errors were found in the dataset.

These errors in the SpamBase dataset occur in many real-world applications
thus, we found it to be the most applicable dataset to use for the experiment
and test the efficiency of Rough Set theory on digital forensics-related datasets
which usually contain all of the above issues.

4.2 Methodology

Preprocessing. Data preprocessing is the process of preparing raw data to
make it suitable for a machine learning model to analyze. This is a vital step as
real-world data contains noise, missing values, and unusable data values. In this
experiment, however, no preprocessing was done on the datasets before feature
reduction. This is because Rough Set is a competent and an efficient algorithm
for classifying inconsistent and noisy data.

Feature Selection. Feature selection is the process of selecting the most significant features from a given dataset [3]. It is an effective way to eliminate
redundant and irrelevant data. In this experiment, we used two techniques for
feature selection namely, Recursive Feature Elimination (RFE) and Fuzzy Rough
Feature Selection.

Recursive Feature elimination is a sci-kit-learn python machine learning
library that provides a feature selection method that fits a model and removes
the weakest feature(s) until the number of specified features has been reached
[11]. We specified the RFE algorithm to select fifteen features for both the Spam-
Base and PenBased datasets. Recursive feature elimination is a recursive process
as the name suggests. The estimator is first trained with the initial set of features. During the feature selection process, the importance of each feature is
obtained, and to reduce the complexity of the model, significant features are
chosen, and the least important features are removed. This process continues
until the specified number of features is reached.

Fuzzy Rough Feature Elimination is used to reduce noisy data without the need for the user to supply information. It eliminates redundant and misleading attributes in the dataset for faster and more accurate classifications [12]. The number of features using the fuzzy rough feature selection was determined by the library on its own. We remained with forty-five features on the SpamBase dataset and sixteen features on the pen-based dataset after the Fuzzy Rough feature selection process.

Classification. Classification is the process of using existing data (training data) to identify the category of new data that is fed in. For this research, we use several classification algorithms alongside a soft computing model (Rough Set theory) to compare their effectiveness to an imprecise and inaccurate forensics dataset. The following algorithms were used; Support Vector Machine (SVM), Naïve Bayes, Decision Tree (J48), Logistic Regression, and Rough Set theory. Throughout the experiment, we utilized WEKA (Waikato Environment for Knowledge Analysis) which is a machine learning software suite that includes the functionality of data processing, classification, clustering, association, regression and visualization.

4.3   Results Analysis

The results were analyzed using the following measures as shown in Figs. 1 and 2: Precision, Recall, and F-Measure. Precision and recall are indicators of the accuracy of the machine learning model. Precision is used to measure the positive patterns that are correctly predicted from the total predicted patterns in a positive class [15]. In essence, a classifier cannot label a negative sample as positive. Recall is used to measure the fraction of positive patterns that are correctly classified. Thus, recall is the ability of the classifier to find all the positive samples. Lastly, F measure is a metric that represents the harmonic mean between the recall and precision values.

The results produced from the experiment are depicted in Table 1 and 2. Table 1 gives the results for the SpamBase dataset when both Fuzzy Rough feature elimination and Recursive feature elimination are combined with different classification algorithms. As depicted in Fig. 1a, Fuzzy Rough feature selection gave a higher precision value for most of the classification algorithms except SVM. Figures 2a and 2b present an overall picture of Precision, Recall, and F-Measure values when each of the feature selection algorithms were used.

Table 2 gives the results for the Pen-Based dataset. It displays the results when both the feature selection methods are utilized in combination with different classification algorithms. The results for the SVM classification algorithm are blank since the SVM algorithm could not classify some of the classes in the Pen-Based dataset hence, Weka outputs a question mark (?) for the average evaluation metric when one of the classes cannot be classified by a certain classification algorithm. The precision values gotten for both the Fuzzy rough feature selection and Recursive feature selection algorithms are very similar as depicted by Fig. 1b. Additionally, Fig. 3 presents an overall picture of precision, recall and f measure when each of the feature selection algorithms were utilized.

As seen in Table 2, when Fuzzy Rough is used as feature selection, the best performing algorithm is logistic regression instead of Rough Set as anticipated. This may be due to the fact that the Pen-Based dataset does not have significant errors in terms of inaccurate or missing values. On the contrary, when used with Recursive feature elimination, Rough Set yields the highest accuracy. The results from the experiment show that using Rough Set resulted in higher accuracy values when compared to the other machine learning models. This is true for both when using RFE or using Rough Set for feature selection as shown in Table 1 for the SpamBase dataset since it contained inaccurate and missing values. However, as shown in Table 2, Rough Set does not yield the highest percentage when used in the PenBased dataset due to the lack of significant errors, Rough Set does a better job at classifying.

Table 1. Results for the SpamBase dataset.

	Fuzzy rough set for feature selection			RFE for feature selection		
	Precision	Recall	F-Measure	Precision	Recall	F-Measure
Rough set	*0.931*	*0.930*	*0.929*	*0.866*	*0.866*	*0.866*
SVM	0.826	0.825	0.825	0.861	0.858	0.859
Naïve Bayes	0.838	0.782	0.809	0.749	0.523	0.615
Decision tree	0.921	0.920	0.920	0.866	0.863	0.864
Logistic regression	0.924	0.924	0.924	0.867	0.862	0.864

Table 2. Results for the PenBased dataset.

	Fuzzy rough set for feature selection			RFE for feature selection		
	Precision	Recall	F-Measure	Precision	Recall	F-Measure
Rough set	*0.946*	*0.944*	*0.944*	*0.925*	*0.924*	*0.924*
SVM	-	0.113	–	–	0.119	–
Naïve Bayes	0.868	0.862	0.864	0.860	0.859	0.859
Decision tree	0.737	0.731	0.733	0.750	0.739	0.744
Logistic regression	**0.952**	**0.952**	**0.952**	0.916	0.914	0.914

We faced a few challenges throughout the experiment. Some of those challenges include computational inefficiency. Despite Rough Set models being very successful in data analysis, they still encounter challenges with large datasets [22]. Rough Set method is computationally time-consuming; thus, it requires scalable implementations on large datasets.

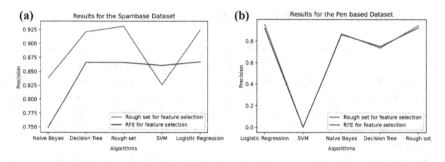

Fig. 1. Comparison of the two feature selection algorithms on the (a) SpamBase dataset (b) Pen-Based dataset

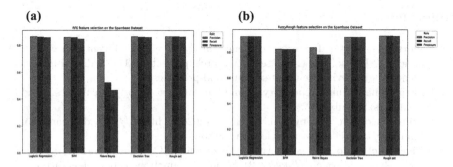

Fig. 2. Metrics of the (a) RFE feature selection and (b) Fuzzy Rough feature selection on the SpamBase dataset

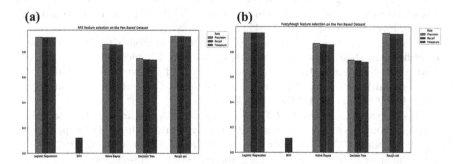

Fig. 3. Metrics of the (a) RFE feature selection and (b) Fuzzy Rough feature selection on the Pen based dataset

5   Conclusion and Future Work

In this paper, we propose using a soft computing model called Rough Set theory in comparison to other machine learning models as a valid tool to be applied for digital forensic investigations. As per our results, Rough Set outweighed all the other models during the classification of an imprecise, inconsistent, and incomplete dataset which are frequently found in the case of digital evidence collected. Therefore, Rough Set theory can successfully be applied by a forensic investigator.

In the future, we plan to expound on this research by fine-tuning some of the parameters of the Rough Set model to see whether better results can be produced despite some of the limitations mentioned in this research.

References

1. Digforasp - digital forensics: evidence analysis via intelligent systems and practices. https://www.umu.se/en/research/projects/ca17124–digital-forensics-evidence-analysis-via-intelligent-systems-and-practices/
2. UCI machine learning repository: spambase data set. https://archive.ics.uci.edu/ml/datasets/Spambase
3. Feature selection in python sklearn (2020). https://www.datacamp.com/community/tutorials/feature-selection-python
4. An introduction to particle swarm optimization (PSO) algorithm (2021). https://www.analyticsvidhya.com/blog/2021/10/an-introduction-to-particle-swarm-optimization-algorithm/
5. Abbas, Z., Burney, S.: A survey of software packages used for rough set analysis. J. Comput. Commun. **4**, 10–18 (2016). https://doi.org/10.4236/jcc.2016.49002
6. Adedayo, O.M.: Big data and digital forensics. In: 2016 IEEE International Conference on Cybercrime and Computer Forensic (ICCCF), pp. 1–7 (2016). https://doi.org/10.1109/ICCCF.2016.7740422
7. Al-Mayyan, W., Own, H.S., Zedan, H.: Rough set approach to online signature identification. Digit. Signal Process. **21**(3), 477–485. Elsevier (2011)
8. Andhalkar, S., Momin, B.F.: Rough set theory and its extended algorithms. In: 2018 Second International Conference on Intelligent Computing and Control Systems (ICICCS), pp. 1434–1438. IEEE (2018)
9. Bhattacharya, A., Goswami, R.T., Mukherjee, K.: A feature selection technique based on rough set and improvised PSO algorithm (PSORS-FS) for permission based detection of Android malwares. Int. J. Mach. Learn. Cybern. **10**(7), 1893–1907. Springer (2019). https://doi.org/10.1007/s13042-018-0838-1
10. Clader, M., et al.: Computational modelling for decision-making: where, why, what, who and how. Roy. Soc. Open Sci. **5**(6), 172096 (2018)
11. Chen, X.W., Jeong, J.C.: Enhanced recursive feature elimination. In: Sixth International Conference on Machine Learning and Applications (ICMLA 2007), pp. 429–435 (2007). https://doi.org/10.1109/ICMLA.2007.35
12. Cornelis, C., Verbiest, N., Jensen, R.: Ordered weighted average based fuzzy rough sets. In: Yu, J., Greco, S., Lingras, P., Wang, G., Skowron, A. (eds.) RSKT 2010. LNCS (LNAI), vol. 6401, pp. 78–85. Springer, Heidelberg (2010). https://doi.org/10.1007/978-3-642-16248-0_16

13. Dawid, A.P., Mortera, J.: Forensic identification with imperfect evidence. Biometrika **85**(4), 835–849 (1998). https://www.jstor.org/stable/2337487. [Oxford University Press, Biometrika Trust]
14. Han, J., Kamber, M., Pei, J.: 9 - Classification: advanced methods. In: Han, J., Kamber, M., Pei, J. (eds.) Data Mining (Third Edition), pp. 393–442. The Morgan Kaufmann Series in Data Management Systems, Morgan Kaufmann, Boston (2012). https://doi.org/10.1016/B978-0-12-381479-1.00009-5, https://www.sciencedirect.com/science/article/pii/B9780123814791000095
15. Hossin, M., Sulaiman, M.N.: A review on evaluation metrics for data classification evaluations. Int. J. Data Min. Knowl. Manage. Process **5**, 01–11 (2015). https://doi.org/10.5121/ijdkp.2015.5201
16. Ibrahim, D.: An overview of soft computing. Proc. Comput. Sci. **102**, 34–38 (2016). https://doi.org/10.1016/j.procs.2016.09.366, https://www.sciencedirect.com/science/article/pii/S1877050916325467
17. Jensen, R., Shen, Q.: Tolerance-based and fuzzy-rough feature selection. In: 2007 IEEE International Fuzzy Systems Conference, pp. 1–6. IEEE (2007)
18. Lang, R., Lu, H.: A general steganalysis method based on rough set theory. In: 2009 Asia Pacific Conference on Postgraduate Research in Microelectronics & Electronics (PrimeAsia), pp. 241–244. IEEE (2009)
19. Lian, S., Heileman, G.L., Noore, A.: Special issue on soft computing for digital information forensics. Soft. Comput. **15**(3), 413–415 (2011). https://doi.org/10.1007/s00500-009-0531-0
20. Ma, Y., Luo, X., Li, X., Bao, Z., Zhang, Y.: Selection of rich model steganalysis features based on decision rough set α -positive region reduction. IEEE Trans. Circ. Syst. Video Technol. **29**(2), 336–350, IEEE (2018)
21. Mohtashami, M., Eftekhari, M.: Using a novel merit for feature selection based on rough set theory. In: 2018 6th Iranian Joint Congress on Fuzzy and Intelligent Systems (CFIS), pp. 68–70. IEEE (2018)
22. Qian, Y., et al.: Local rough set: a solution to rough data analysis in big data. Int. J. Approximate Reasoning **97**, 38–63 (2018). https://doi.org/10.1016/j.ijar.2018.01.008
23. Quick, D., Choo, K.K.R.: Impacts of increasing volume of digital forensic data: a survey and future research challenges. Digit. Investig. **11**(4), 273–294, Elsevier (2014)
24. robin.materese@nist.gov: Digital evidence (2016). https://www.nist.gov/digital-evidence. Accessed 25 Oct 2021
25. Shen, Q., Jensen, R.: Rough sets, their extensions and applications. Int. J. Autom. Comput. **4**(3), 217–228. Springer (2007). https://doi.org/10.1007/s11633-007-0217-y
26. Singh, A.K., Baranwal, N., Nandi, G.C.: A rough set based reasoning approach for criminal identification. Int. J. Mach. Learn. Cybern. **10**(3), 413–431 (2017). https://doi.org/10.1007/s13042-017-0699-z
27. Skowron, A., Dutta, S.: Rough sets: past, present, and future. Nat. Comput. **17**(4), 855–876 (2018). https://doi.org/10.1007/s11047-018-9700-3
28. Wang, D., Tan, D., Liu, L.: Particle swarm optimization algorithm: an overview. Soft Comput. **22**(2), 387–408. Springer (2018). https://doi.org/10.1007/s00500-016-2474-6
29. Wang, Y., Lee, H.C.: Research on some relevant problems in computer forensics. In: Conference of the 2nd International Conference on Computer Science and Electronics Engineering (ICCSEE 2013), pp. 1564–1571. Atlantis Press (2013)

30. Zhang, M., Yao, J.T.: A rough sets based approach to feature selection. In: IEEE Annual Meeting of the Fuzzy Information, 2004. Processing NAFIPS 2004, vol. 1, pp. 434–439. IEEE (2004)
31. Zhang, Q., Xie, Q., Wang, G.: A survey on rough set theory and its applications. CAAI Trans. Intell. Technol. **1**(4), 323–333 (2016). https://doi.org/10.1016/j.trit.2016.11.001, https://www.sciencedirect.com/science/article/pii/S2468232216300786
32. Zhang, T., Zhao, P.: Insider threat identification system model based on rough set dimensionality reduction. In: 2010 Second World Congress on Software Engineering, vol. 2, pp. 111–114 (2010). https://doi.org/10.1109/WCSE.2010.106

Binary Boundaries and Power Set Space of Graded Rough Sets and Their Correlative ECG (Electrocardiogram) Data Analysis

Qian Wang[1,2], Xiaoxue Wang[1,2], and Xianyong Zhang[1,2,3(✉)]

[1] School of Mathematical Sciences, Sichuan Normal University, Chengdu 610066, China
[2] Institute of Intelligent Information and Quantum Information, Sichuan Normal University, Chengdu 610066, China
[3] National-Local Joint Engineering Laboratory of System Credibility Automatic Verification, Sichuan Normal University, Chengdu 610066, China
xianyongzh@sina.com.cn

Abstract. Graded rough sets (GRSs) act as a bidirectional quantitative model of three-way decision, but their approximation operators cannot preserve union, intersection and complement operations. Aiming at GRSs, this paper mines binary boundaries and constructs the power set space, so the corresponding ECG (electrocardiogram) data analysis is eventually performed. Based on union and intersection inequalities of approximation operators, four types of binary boundaries and their operators are first proposed to generate fundamental union and intersection equations, and both their quantitative semantics regarding dual membership grades and their degenerate properties on quantitative parameters are revealed. Then, union, intersection and complement operations of approximation sets are redefined by boundaries to acquire the set operation preservation of approximation operators, so the power set space of GRSs is established to induce homomorphisms regarding the classical power set space. Finally, the binary boundaries in power set space are utilized for ECG dataset analysis, and experimental results demonstrate the effectiveness of theoretical structures and in-depth properties. This study adopts double viewpoints of operator theory and set theory to enrich GRSs, its quantitative extension underlies uncertainty modeling and granular computing, while its mathematical structures facilitate data mining in terms of parameter optimization.

Keywords: Graded rough set · approximation operator · set operation · binary boundary · power set space · ECG data analysis

1 Introduction

Rough sets are a fundamental uncertainty methodology of granular computing and three-way decision, and their uncertainty and corresponding data analysis primarily come from boundaries. Rough sets exhibit two interpretations regarding operator theory and set theory [1]. The operator theory thinks that rough sets extend classical sets by adding upper and lower approximation operators to classical set operators, while the set theory

J. Yao et al. (Eds.): IJCRS 2022, LNAI 13633, pp. 85–99, 2022.
https://doi.org/10.1007/978-3-031-21244-4_7

thinks that rough sets never introduce new operators but change set operations. Nowadays, rough sets have been widely researched and applied in the uncertainty modeling, information measurement, classification learning, outlier detection, medical diagnosis, etc.

The traditional rough sets can be called Pawlak-RSs [2], and their upper and lower approximations rely on strict inclusion relationships; thus, they become a qualitative model to lack fault tolerance. For quantitative improvement, the probability rough sets [3–5] utilize the relative information of probability to become a relatively quantitative model, and they actually include multiple models related to three-way decision, such as the decision-theoretic rough sets (DTRSs) and variable precision rough sets (VPRSs). In contrast, the graded rough sets (GRSs) rely on the absolute information of grade to become an absolutely quantitative model. GRSs mainly focus on absolute cardinalities of internal and external membership degrees, and they introduce natural numbers for uncertainty modeling, so they exhibit quantitative bidirectionality and intuitiveness. GRSs gain a series of important and meaningful results. For example, Yao and Lin [6] initially propose GRSs when using modal logics to make the generalization of rough sets; Liu et al. [7] construct two-universe GRSs and relevant properties; Huang et al. [8] design intuitionistic fuzzy graded covering rough sets; Xue et al. [9] propose multi-granulation graded rough intuitionistic fuzzy sets models based on dominance relation. In particular, GRSs also make construction of information fusion by extensively combining probability rough sets or probability measures. For example, Zhang et al. [10] advocate the complete double-quantization to combine absolutely quantitative information and relatively quantitative information by comparatively analyzing VPRSs and GRSs. Fang and Hu [11] define probabilistic GRSs and double-quantitative DTRSs.

Clearly, GRSs have the theoretical significance and applied values by virtue of their quantitative characteristic, and their studies on relative quantization is worth reinforcing. For Pawlak-RSs, the approximation operators partly hold the set operations. In this regard, Zhang [12] et al. propose two kinds of boundaries and operators, they further redefine union, intersection and complement operations of approximate sets, so they finally construct the power set space. These results can be extended from Pawlak-RSs to GRSs, in view of the quantitative expansion. In other words, GRSs approximation operators do not maintain set operations at all, and thus the new definition of approximate set operations and the follow-up construction of power set space are worth exploring. For this case, this paper mainly mines binary boundaries and uses unitary boundaries, and we establish the approximate set operations and power set space of GRSs, so we deeply reveal the uncertainty semantics and operation homomorphism; at last, four binary boundary operators and corresponding structure properties of power set space are applied to ECG (electrocardiogram) data analysis, which serves as an important topic in intelligent processing and medical applications [13, 14]. In summary, this paper deepens the two interpretations of GRSs from both the operator theory and the set theory, and it quantitatively advances the qualitative results of Pawlak-RSs in [12]. In terms of data experiments, all results are effectively verified and they are helpful to intelligent detection and medical diagnosis of ECG data analysis.

2 Rough Set Approximation Operators

This section uses Refs. [2, 6, 12, 15] to review the approximate operators of Pawlak-RSs and GRSs. The discourse domain U and the equivalence relation R compose approximate space (U, R), and the knowledge granule $[x]_R$ is the equivalence classes of sample $x \in U$. Suppose two sets $X, Y \subseteq U$ and power set space $2^U = \{Z : Z \subseteq U\}$, and $\sim$ denotes the complement operation of sets.

Definition 1([2]). In Pawlak-RSs, the upper and lower approximation sets of X are

$$\overline{R}X = \{x : [x]_R \cap X \neq \varnothing\}, \ \underline{R}X = \{x : [x]_R \subseteq X\},$$

and the corresponding upper and lower approximation operators are noted as $\overline{R}, \underline{R} : 2^U \to 2^U$. The positive, negative, and boundary regions of X are

$$Pos_R(X) = \underline{R}X, Neg_R(X) = \sim \overline{R}X, Bnd_R(X) = \overline{R}X - \underline{R}X,$$

and the corresponding three-way operators are supposed to be $Pos_R, Neg_R, Bnd_R : 2^U \to 2^U$.

Proposition 1([12]). We have the following equalities by adding boundary operations. That is,

1) $\overline{R}(X \cap Y) \cup Obnd_R(X, Y) = \overline{R}X \cap \overline{R}Y$,
2) $\underline{R}(X \cup Y) = \underline{R}X \cup \underline{R}Y \cup Ibnd_R(X, Y)$.

Here,

$$Obnd_R(X, Y) = \{ x : [x]_R \cap (X \cap Y) = \varnothing, [x]_R \cap (X - Y), (Y - X) \neq \varnothing \},$$

$$Ibnd_R(X, Y) = \{ x : [x]_R \subseteq X \cup Y, [x]_R \cap (X - Y), (Y - X) \neq \varnothing \},$$

are called the outer boundary and inner boundary, and their corresponding outer and inner boundary operators are labeled by $Obnd_R, Ibnd_R : 2^U \times 2^U \to 2^U$.

Definition 2([12, 15]). In GRSs, the quantification parameter $k \in N$ is offered, and then the internal and external membership degrees of $[x]_R$ to X are

$$ig([x]_R, X) = |[x]_R \cap X|, \tag{1}$$

$$og([x]_R, X) = |[x]_R| - |[x]_R \cap X|. \tag{2}$$

The upper and lower approximation sets of X are

$$\overline{R}_k X = \{x : ig([x]_R, X) > k\}, \underline{R}_k X = \{x : og([x]_R, X) \leq k\}, \tag{3}$$

and the corresponding upper and lower approximation operators are $\overline{R}_k, \underline{R}_k : 2^U \to 2^U$. The upper and lower boundary of X are

$$\overline{bnd}_R^k(X) = \overline{R}_k X - \underline{R}_k X, \underline{bnd}_R^k(X) = \underline{R}_k X - \overline{R}_k X, \tag{4}$$

and the corresponding upper and lower boundary operators are $\overline{bnd}_R^k$, $\underline{bnd}_R^k : 2^U \to 2^U$.
Moreover, the positive, negative and boundary regions of X are

$$Pos_R^k(X) = \underline{R}_k X \cap \overline{R}_k X,$$

$$Neg_R^k(X) =\sim (\overline{R}_k X \cup \underline{R}_k X),$$

$$Bnd_R^k(X) = \sim (Pos_R^k(X) \cup Neg_R^k(X)),$$

and the corresponding operators are Pos_R^k, Neg_R^k, $Bnd_R^k : 2^U \to 2^U$.

Proposition 2([6]). Dual operators $\overline{R}_k$ and $\underline{R}_k$ have the following properties on union, intersection and complement operations.

1) $\overline{R}_k(X \cup Y) \supseteq \overline{R}_k X \cup \overline{R}_k Y$;
2) $\overline{R}_k(X \cap Y) \subseteq \overline{R}_k X \cap \overline{R}_k Y$;
3) $\overline{R}_k(\sim X) =\sim \underline{R}_k X$;
4) $\underline{R}_k(X \cup Y) \supseteq \underline{R}_k X \cup \underline{R}_k Y$;
5) $\underline{R}_k(X \cap Y) \subseteq \underline{R}_k X \cap \underline{R}_k Y$;
6) $\underline{R}_k(\sim X) =\sim \overline{R}_k X$.

Both the internal and external membership degrees are absolute measures, and they underlie the bidirectionality and intuitiveness of GRSs. Thus, GRSs have intuitive quantification semantics, and they also quantitatively expand the qualitative model of Pawlak-RSs, which corresponds to a special parameter value $k = 0$. For parameter settings, the threshold k can be determined according to the granular cardinality distribution and actual needs, or merely by expert experience. As shown by Proposition 2, the approximation operators of GRSs do not maintain set operations at all, and thus this fact leads to both the binary boundary mining and the power set space construction as follows.

3   Binary Boundaries of Graded Rough Sets

Aiming at the union and intersection inequations of GRSs approximation operators (Proposition 2), this section mines four binary boundaries (and matching operators) to construct union and intersection equations, thus acquiring improvements.

Lemma 1. We have the following calculation and determination for approximation sets, i.e.,

$$\overline{R}_k(X \cup Y) - \overline{R}_k X \cup \overline{R}_k Y = \{x : ig([x]_R, X \cup Y) > k, ig([x]_R, X) \le k, ig([x]_R, Y) \le k\}.$$

$$\overline{R}_k X \cap \overline{R}_k Y - \overline{R}_k(X \cap Y) = \{x : ig([x]_R, X \cap Y) \le k, ig([x]_R, X) > k, ig([x]_R, Y) > k\}.$$

$$\underline{R}_k(X \cup Y) - \underline{R}_k X \cup \underline{R}_k Y = \{x : og([x]_R, X \cup Y) \le k, og([x]_R, X) > k, og([x]_R, Y) > k\}.$$

$$\underline{R}_k X \cap \underline{R}_k Y - \underline{R}_k(X \cap Y) = \{x : og([x]_R, X \cap Y) > k, og([x]_R, X) \le k, og([x]_R, Y) \le k\}.$$

Definition 3. The upper-inner boundary, upper-outer boundary, lower-inner boundary, and lower-outer boundary of X and Y are respectively defined as

$$\overline{Ibnd}_R^k(X, Y) = \{x : ig([x]_R, X \cup Y) > k, ig([x]_R, X) \leq k, ig([x]_R, Y) \leq k\},$$

$$\overline{Obnd}_R^k(X, Y) = \{x : ig([x]_R, X \cap Y) \leq k, ig([x]_R, X) > k, ig([x]_R, Y) > k\},$$

$$\underline{Ibnd}_R^k(X, Y) = \{x : og([x]_R, X \cup Y) \leq k, og([x]_R, X) > k, og([x]_R, Y) > k\},$$

$$\underline{Obnd}_R^k(X, Y) = \{x : og([x]_R, X \cap Y) > k, og([x]_R, X) \leq k, og([x]_R, Y) \leq k\}.$$

Furthermore, the upper-inner boundary operators, upper-outer boundary operators, lower-inner boundary operators, and lower-outer boundary operators on $2^U \times 2^U \rightarrow 2^U$ are respectively denoted by

$$\overline{Ibnd}_R^k(X, Y), \overline{Obnd}_R^k(X, Y), \underline{Ibnd}_R^k(X, Y), \underline{Obnd}_R^k(X, Y)$$

Theorem 1. In GRSs, approximation sets and four boundaries can formulate operation equalities, that is,

1) $\overline{R}_k(X \cup Y) = \overline{R}_k X \cup \overline{R}_k Y \cup \overline{Ibnd}_R^k(X, Y);$
2) $\overline{R}_k(X \cap Y) \cup \overline{Obnd}_R^k(X, Y) = \overline{R}_k X \cap \overline{R}_k Y;$
3) $\underline{R}_k(X \cup Y) = \underline{R}_k X \cup \underline{R}_k Y \cup \underline{Ibnd}_R^k(X, Y);$
4) $\underline{R}_k(X \cap Y) \cup \underline{Obnd}_R^k(X, Y) = \underline{R}_k X \cap \underline{R}_k Y.$

For Proposition 2, Lemma 1 makes the difference comparison, and thus it reveals the difference factors that affect the equation. Thus, Definition 3 establishes four kinds of boundaries (as well as corresponding operators), while Theorem 1 modifies the union and intersection inequalities of Proposition 2 to become ideal equalities. The four kinds of boundaries (operators) act on two sets to present the duality. They are mainly related to the system of upper and lower approximations with two underlying sets, so they are partially related to the unitary boundary of a single set, such as $\overline{bnd}_R^k(X), \underline{bnd}_R^k(X)$. Univariate boundaries depict single-concept boundaries and related uncertainty; in contrast, binary boundaries adhere to the double-concept boundaries and relevant uncertainty, as shown by Fig. 1.

Figure 1 does not lose the generality. Thus, the sets X, Y are intersected, and they divide the universe U into four pieces:

$$X - Y, Y - X, X \cap Y, \sim (X \cup Y).$$

The little rectangle identifies the GRSs binary boundary (or its constituent granules), and this block usually has overlaps with the four-divided parts. However, the four boundaries have different degrees of overlap, so they adopt four distinctive perspectives to describe the binary-set system and its uncertainty. Definition 3 and Fig. 1 determine the general distribution of the four binary boundaries, so they can explain their quantitative semantics regarding the dual membership degrees. For example, the upper-inner boundary $\overline{Ibnd}_R^k(X, Y)$ is composed of the following equivalent class $[x]_R$, whose inner membership degree of $X \cup Y$ (i.e., the number of intersecting elements) is greater than threshold k, while whose inner membership degrees of X, Y (i.e., the numbers of intersecting elements) both are not greater than number k.

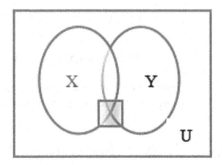

Fig. 1. Schematic diagram of binary boundary of GRSs

Lemma 2. Dual membership degrees have properties as follows:

1) $ig([x]_R, X \cup Y) = ig([x]_R, X) + ig([x]_R, Y) - ig([x]_R, X \cap Y)$,
2) $og([x]_R, X \cup Y) = og([x]_R, X) + og([x]_R, Y) - og([x]_R, X \cap Y)$.

The system of binary boundaries involves the characterization of dual membership degrees for sets $X \cup Y, X \cap Y, X, Y$, and Lemma 2 provides the usual degree relationships for the four sets.

Theorem 2. Binary boundaries (operations) have the following properties.

1) If $X \cup Y = U$, then
   $\overline{Ibnd}_R^k(X, Y) = \{x : |[x]_R| > k, ig([x]_R, X) \le k, ig([x]_R, Y) \le k\}$,
   $\underline{Ibnd}_R^k(X, Y) = \{x : og([x]_R, X) > k, og([x]_R, Y) > k\}$.
2) If $X \cap Y = \varnothing$, then
   $\overline{Obnd}_R^k(X, Y) = \{x : ig([x]_R, X) > k, ig([x]_R, Y) > k\}$,
   $\underline{Obnd}_R^k(X, Y) = \{x : |[x]_R| > k, og([x]_R, X) \le k, og([x]_R, Y) \le k\}$.
3) In general cases,
   $\overline{Ibnd}_R^k(X, \sim X) = \{x : |[x]_R| > k, ig([x]_R, X) \in [|[x]_R| - k, k]\}$,
   $\overline{Obnd}_R^k(X, \sim X) = \{x : ig([x]_R, X) \in (k, |[x]_R| - k)\}$,
   $\underline{Ibnd}_R^k(X, Y) = \{x : og([x]_R, X) \in (k, |[x]_R| - k)\}$,
   $\underline{Obnd}_R^k(X, Y) = \{x : |[x]_R| > k, og([x]_R, X) \in [|[x]_R| - k, k]\}$.

Proof. 1) If $X \cup Y = U$, then $ig([x]_R, X \cup Y) = |[x]_R|$, $og([x]_R, X \cup Y) = 0$. From Definition 3, $\overline{Ibnd}_R^k(X, Y)$ and $\underline{Ibnd}_R^k(X, Y)$ naturally have the above simplifications.

2) If $X \cap Y = \varnothing$, then $ig([x]_R, X \cup Y) = 0, og([x]_R, X \cup Y) = |[x]_R|$. From Definition 3, $\overline{Obnd}_R^k(X, Y)$ and $\underline{Obnd}_R^k(X, Y)$ similarly have their above simplifications.

3) Suppose $X \cup Y = U$ and $X \cap Y = \varnothing$, and then $Y = \sim X$. Thus from Lemma 2 and the above items 1) 2), we can get the following formulas:

$$ig([x]_R, X) + ig([x]_R, \sim X) = |[x]_R|,$$
$$og([x]_R, X) + og([x]_R, \sim X) = |[x]_R|.$$

Moreover,

$$ig([x]_R, \sim X) = |[x]_R| - ig([x]_R, X) \leq k \Leftrightarrow ig([x]_R, X) \geq |[x]_R| - k,$$

so according to the above item 1), we have

$$\overline{Ibnd}_R^k(X, \sim X) = \{x : |[x]_R| > k, ig([x]_R, X.) \in [|[x]_R| - k, k]\}.$$

Similarly, the remaining results of three binary boundaries can be derived, so we prove item 3). □

Based on Lemma 2, Theorem 2 gives the simplified results of binary boundaries, when X, Y are in three cases of overlap, exclusion, and partition respectively. The proof process correspondingly simplifies the systematic relationships of dual memberships, which are provided by Lemma 2. The overlap and exclusion cases are suitable for inner and outer boundary simplifications, respectively, while the superimposed division is suitable for the binary classification and its applications.

Theorem 3 Suppose $k = 0$, then we have

1) $\overline{Ibnd}_R^0(X, Y) = \varnothing$;
2) $\overline{Obnd}_R^0(X, Y) = Obnd_R(X, Y)$;
3) $\underline{Ibnd}_R^0(X, Y) = Ibnd_R(X, Y)$;
4) $\underline{Obnd}_R^0(X, Y) = \varnothing$.

Based on Theorem 3, the four binary boundaries of GRSs quantitatively expand the two qualitative binary boundaries of Pawlak-RSs, and the latter become a special case when $k = 0$.

In summary, binary boundaries (operators) of GRSs have the union/intersection operation modifications, dual membership degree semantics, and quantitative expansion. With regard to the operator theory, four binary operators are introduced; with regard to the set theory, four bounded sets are introduced.

At last, we offer the calculation algorithm on four binary boundaries, and Algorithm 1 is clear.

Algorithm 1 Calculation of four binary boundaries

Input: $(U, R), X, Y, k$;

Output: $\overline{Ibnd}_R^k(X,Y), \overline{Obnd}_R^k(X,Y), \underline{Ibnd}_R^k(X,Y), \underline{Obnd}_R^k(X,Y)$.

1. Calculate the equivalence class $[x]_R$.

2. Calculate the union and intersection sets $X \cup Y$ and $X \cap Y$.

3. for $x \in U$

4.   By Definition 2, compute 8 measures:

   $ig([x]_R, X)$, $ig([x]_R, Y)$, $ig([x]_R, X \cup Y), ig([x]_R, X \cap Y)$,

   $og([x]_R, X)$, $og([x]_R, Y)$, $og([x]_R, X \cup Y), og([x]_R, X \cap Y)$.

5.   Comparing the above values to threshold k, judge the belongingness of sample x to four boundaries

   $\overline{Ibnd}_R^k(X,Y), \overline{Obnd}_R^k(X,Y), \underline{Ibnd}_R^k(X,Y), \underline{Obnd}_R^k(X,Y)$.

6. end for

7. Acquire and return

   $\overline{Ibnd}_R^k(X,Y), \overline{Obnd}_R^k(X,Y), \underline{Ibnd}_R^k(X,Y), \underline{Obnd}_R^k(X,Y)$.

---

4   Power Set Space of Graded Rough Sets

Within the framework of GRSs, this section uses the binary boundaries (operators) to redefine the union and intersection operations of approximation sets, and it uses the unary boundaries (operators) to determine the complement operations of approximation sets, so it finally constructs the power set space.

Definition 4. Regarding GRSs, new union and intersection operations (noted as $\underset{*}{\cup}, \underset{*}{\cap}$) of approximation sets are defined as follows:

$$\overline{R}_k X \underset{*}{\cup} \overline{R}_k Y = \overline{R}_k X \cup \overline{R}_k Y \cup \overline{Ibnd}_R^k(X, Y); \tag{5}$$

$$\overline{R}_k X \underset{*}{\cap} \overline{R}_k Y = \overline{R}_k X \cap \overline{R}_k Y - \overline{Obnd}_R^k(X, Y); \tag{6}$$

$$\underline{R}_k X \underset{*}{\cup} \underline{R}_k Y = \underline{R}_k X \cup \underline{R}_k Y \cup \underline{Ibnd}_R^k(X, Y); \tag{7}$$

$$\underline{R}_k X \underset{*}{\cap} \underline{R}_k Y = \underline{R}_k X \cap \underline{R}_k Y - \underline{Obnd}_R^k(X, Y); \tag{8}$$

Theorem 4. Regarding GRSs, the union and intersection operations are formally preserved as follows:

1) $\overline{R}_k X \underset{*}{\cup} \overline{R}_k Y = \overline{R}_k (X \cup Y)$;

2) $\overline{R}_k X \cap \overline{R}_k Y = \overline{R}_k (X \cap Y)$;

3) $\underline{R}_k X \underset{*}{\cup} \underline{R}_k Y = \underline{R}_k (X \cup Y)$;

4) $\underline{R}_k X \underset{*}{\cap} \underline{R}_k Y = \underline{R}_k (X \cap Y)$.

By Theorem 1, Definition 4 uses four binary boundaries to adjust the classical $\cup$, $\cap$ operations, so it redefines the union and intersection operations $\underset{*}{\cup}$, $\underset{*}{\cap}$ of GRSs approximation sets. Theorem 4, proved by Theorem 1 and Definition 4, naturally obtains good properties of operation maintenance for GRSs approximate operators.

Definition 5. Regarding GRSs, the new complement operator $\underset{*}{\sim}$ for approximation is determined as follows:

$$\underset{*}{\sim} \overline{R}_k X = {\sim} \underline{R}_k X, \; \underset{*}{\sim} \underline{R}_k X = {\sim} \overline{R}_k X. \tag{9}$$

Theorem 5. Regarding GRSs, the complement operation offers

$$\underset{*}{\sim} \overline{R}_k X = \overline{R}_k ({\sim} X), \; \underset{*}{\sim} \underline{R}_k X = \underline{R}_k ({\sim} X).$$

Corollary 1. We have two equations:

1) $\underset{*}{\sim} \overline{R}_k X = ({\sim} \overline{R}_k X) \cup \overline{bnd}_R^k (X) - \underline{bnd}_R^k (X)$,

2) $\underset{*}{\sim} \underline{R}_k X = ({\sim} \underline{R}_k X) \cup \underline{bnd}_R^k (X) - \overline{bnd}_R^k (X)$.

Definition 5 redefines the complement operator $\underset{*}{\sim}$ of GRSs approximations. Based on Proposition 2, Theorem 5 naturally acquires the good result of complement operational preservation. According to Corollary 1, the $\underset{*}{\sim}$ operation mainly uses the upper and lower boundaries (operators) to adjust the original ${\sim}$ operation.

Thus far, GRSs approximation sets are endowed with new union, intersection, complement operations $\underset{*}{\cup}$, $\underset{*}{\cap}$ and $\underset{*}{\sim}$. These operations differ from classical set operations $\underset{*}{\cup}$, $\underset{*}{\cap}$ and $\underset{*}{\sim}$, but they have associated corrections and depend on set X or Y. From the perspective of the set theory [1], the new operations $\underset{*}{\cup}$, $\underset{*}{\cap}$ and $\underset{*}{\sim}$ change the classical operations $\cup$, $\cap$ and ${\sim}$, and then the GRSs approximation operators maintain the set union and intersection operations; thus, Proposition 2 is improved to Theorems 4 and 5. In this way, we can determine the power set space and relevant homomorphism for GRSs. Here, $(2^U, \cup, \cap, {\sim})$ means the classical power set space.

Definition 6. Regarding GRSs, the upper and lower power sets are.

$$\overline{R}_k 2^U = \{ \overline{R}_k X : X \subseteq U \}, \underline{R}_k 2^U = \{ \underline{R}_k X : X \subseteq U \}, \tag{10}$$

so they induce two corresponding structures of upper and lower approximate power set spaces:

$$(\overline{R}_k 2^U, \underset{*}{\cup}, \underset{*}{\cap}, \underset{*}{\sim}), (\underline{R}_k 2^U, \underset{*}{\cup}, \underset{*}{\cap}, \underset{*}{\sim}).$$

Theorem 6 $(2^U, \cup, \cap, \sim)$ is homomorphic with

$$(\overline{R}_k 2^U, \underset{*}{\cup}, \underset{*}{\cap}, \underset{*}{\sim}), (\underline{R}_k 2^U, \underset{*}{\cup}, \underset{*}{\cap}, \underset{*}{\sim})$$

where $\overline{R}_k$, $\underline{R}_k$ act as the homomorphic surjections.

Definition 6 uses the GRSs approximate power set and the $\underset{*}{\cup}$, $\underset{*}{\cap}$, $\underset{*}{\sim}$ operations to establish the GRSs power set space, and the approximation operations have the closeness and dependence. By Theorems 4 and 5, surjections $\overline{R}_k$ and $\underline{R}_k$ maintain the set operations of the two power set spaces, so the homomorphism of Theorem 6 reveals the in-depth connection between the two algebraic structures. Moreover, $\overline{R}_k$ and $\underline{R}_k$ generally cannot motivate the algebraic isomorphism between the two power set spaces.

The power set space of GRSs and its homomorphisms are primarily based on the newly defined operations $\underset{*}{\cup}$, $\underset{*}{\cap}$ and $\underset{*}{\sim}$, and the former space also quantitatively expands the power set space of qualitative Pawlak-RSs in [12]. Furthermore, the difference operations and inclusion relations of GRSs approximation sets can be corrected and perfected, and thus the operation preservation can be similarly obtained. For example, if we define $\underset{*}{-}$ and $\underset{*}{\subseteq}$ as follows:

$$\overline{R}_k X \underset{*}{-} \overline{R}_k Y = \overline{R}_k X \underset{*}{\cap} (\underset{*}{\sim} \overline{R}_k Y),$$

$$\underline{R}_k X \underset{*}{\subseteq} \underline{R}_k Y \Leftrightarrow \underline{R}_k X \underset{*}{\cup} \underline{R}_k Y = \underline{R}_k Y \text{ or}$$

$$\underline{R}_k X \underset{*}{\subseteq} \underline{R}_k Y \Leftrightarrow \underline{R}_k X \underset{*}{\cap} \underline{R}_k Y = \underline{R}_k X,$$

then we can obtain

$$\overline{R}_k (X - Y) = \overline{R}_k X \underset{*}{-} \overline{R}_k Y, X \subseteq Y \Rightarrow \underline{R}_k X \underset{*}{\subseteq} \underline{R}_k Y.$$

5 ECG (Electrocardiogram) Data Analysis Based on Four Binary Boundaries in Power Set Space of Graded Rough Sets

ECG signals are an important tool for analyzing and judging the type of heart rate aberration. At present, computer-based ECG analysis and diagnosis have entered a more mature stage. Because ECG data are related to a more complex time series, the morphological characteristics of each wave are closely related to pathology; this case causes the difficulty for computer ECG diagnostic analysis to reach the effect of expert diagnosis. Therefore, "optimizing the old methods and exploring the new methods to improve the accuracy of diagnosis" is still an important problem to be solved in ECG data analysis. In this section, the above theoretical results regarding binary boundaries and power set space of GRSs are experimentally verified by ECG data, and relevant studies are expected to provide deep thinking and valuable auxiliaries for ECG data analysis and relevant medical diagnosis.

Next, we mainly focus on calculation of four binary boundaries of ECG data. Concretely, we select 10 groups of ECG data with good waveforms from 48 MIT-BIH

ECG databases (https://archive.physionet.org/cgi-bin/atm/ATM), whose numbers are 101, 103, 105, 121, 122, 123, 214, 217, 220, 234. The 6 extracted sets of amplitude features implement the pretreatment by three-part division with equal lengths, and the relevant data with matrix scale 10×6 are given in Table 1.

Table 1. ECG amplitude information table

ID	PB-P	P-Q	Q-R	S-R	T-S	T-TB
u_1	1	3	1	1	1	1
u_2	1	3	3	3	1	1
u_3	3	3	2	2	1	1
u_4	1	3	1	1	1	1
u_5	1	2	3	1	1	1
u_6	1	3	2	3	1	1
u_7	1	3	3	2	1	1
u_8	1	1	1	3	3	3
u_9	1	3	3	3	1	1
u_{10}	1	3	3	3	1	1

At first, set up $X = \{ u_1, u_2, u_5 \}$, $Y = \{ u_4, u_5, u_9 \}$. (1) When k = 1, by Definitions 2 and 3, we can get

$$\underline{R}_1(X) = \underline{R}_1(Y) = \{ u_1, u_3, u_4, u_5, u_6, u_7, u_8 \} , \underline{R}_1(X \cup Y) = U,$$

$$\underline{R}_1(X \cap Y) = \{ u_3, u_5, u_6, u_7, u_8 \} , \bar{R}_1(X) = \bar{R}_1(Y) = \bar{R}_1(X \cap Y) = \varnothing,$$

$$\overline{Ibnd}_R^1(X, Y) = \{ u_1, u_2, u_4, u_9, u_{10} \} , \overline{Obnd}_R^1(X, Y) = \varnothing,$$

$$\underline{Ibnd}_R^1(X, Y) = \{ u_2, u_9, u_{10} \} , \underline{Obnd}_R^1(X, Y) = \{ u_1, u_4 \} .$$

This result conforms to Theorem 1. (2) When k = 0, we similarly have

$$\overline{Ibnd}_R^0(X, Y) = \underline{Obnd}_R^0(X, Y) = \varnothing, \underline{Ibnd}_R^0(X, Y) = \{ u_1, u_4 \} = Ibnd_R(X, Y)$$

$$\overline{Obnd}_R^0(X, Y) = \{ u_1, u_2, u_4, u_9, u_{10} \} = Obnd_R(X, Y).$$

This case verifies Theorem 3. (2) Now consider

$$X = \{ u_2, u_4, u_6, u_8, u_{10} \} , Y = \ \sim X = \{ u_1, u_3, u_5, u_7, u_9 \} , k = 1.$$

By Definition 3, the four boundaries become

$$\overline{Ibnd}_R^1(X, Y) = \underline{Obnd}_R^1(X, Y) = \{ u_1, u_4 \} , \underline{Ibnd}_R^1(X, Y) = \overline{Obnd}_R^1(X, Y) = \varnothing$$

According to Definition 2, the cardinality of $[u_i]_R$ and the internal and external membership degrees of X, Y offer

$$|[u_i]_R| : (2, 3, 1, 2, 1, 1, 1, 1, 3, 3),$$

$$ig([u_i]_R, X) : (1,2,0,1,0,1,0,1,2,2), \quad og([u_i]_R, X) : (1, 1, 1, 1, 1, 0, 1, 0, 1, 1),$$

$$og([u_i]_R, X) : (1,1,1,1,1,0,1,0,1,1), \quad ig([u_i]_R, Y) : (1, 1, 1, 1, 1, 0, 1, 0, 1, 1).$$

By Definition 3, the four boundaries become

$$\overline{Ibnd}_R^1(X, Y) = \underline{Obnd}_R^1(X, Y) = \{ u_1, u_4 \}, \quad \underline{Ibnd}_R^1(X, Y) = \overline{Obnd}_R^1(X, Y) = \varnothing.$$

This result justifies Theorems 2 and 3.

Then, the ECG data are analyzed by using binary boundaries based on change chains. For this purpose, we suppose $X = \{u_1, u_3, u_5, u_7, u_9\}$, and we concern subset and threshold chains:

$$Y : Y_1 = U \supset Y_2 = U - \{u_9\} \supset ... \supset Y_{10} = \{u_1\},$$
$$k : k_1 = 0 < k_2 = 1 < ... < k_{11} = 10.$$

The following is a calculation of the four binary boundary sets on the ECG data table, by using Algorithm 1, and a part of the results are shown in Table 2.

Through Table 2, $Y = Y_6$ and $k = 1$ are used as examples for ECG data analysis. At this time, the upper-inner boundary and lower-inner boundary are not empty:

$$\overline{Ibnd}_R^1(X, Y) = \underline{Ibnd}_R^1(X, Y) = \{ u_2, u_9, u_{10} \} .$$

Herein, the overlap numbers of u_2, u_9, u_{10} about $X \cup Y$ are all greater than k, while the overlap values of u_2, u_9, u_{10} about X, Y are all less than or equal to k. So it is more reasonable to use $X \cup Y$ for diagnostic analysis of objects u_2, u_9, u_{10}, and this treatment is better than considering sets X and Y separately.

The obtained results are shown in Fig. 2. There, the xoy plane represents the dual-chain net consisting of parameter k and set Y, and the z-axis represents the element number in each boundary.

According to Table 2 and Fig. 2, the case of $k = 1$ and $Y = \{Y_3, Y_4, Y_5, Y_6, Y_7, Y_8, Y_9\}$ produces the upper-inner boundary and lower-inner boundary:

$$\overline{Ibnd}_R^1(X, Y) = \underline{Ibnd}_R^1(X, Y) = \{ u_2, u_9, u_{10} \} .$$

As a result, for the diagnostic analysis of u_2, u_9, u_{10}, together taking $X \cup Y$ is better than taking X and Y separately. Similarly, the case of $k = 2$ and $Y = \{Y_3, Y_4, Y_5, Y_6, Y_7, Y_8, Y_9\}$ induce the lower-outer boundary:

$$\underline{Obnd}_R^1(X, Y) = \{ u_2, u_9, u_{10} \} .$$

Table 2. Four boundaries based on subset and threshold chains of ECG data analysis

Y	k	$\overline{Ibnd}_R^k$	$\overline{Obnd}_R^k$	$\underline{Ibnd}_R^k$	$\underline{Obnd}_R^k$
$Y_1=U$	0	$\varnothing$	$\varnothing$	$\varnothing$	$\varnothing$
	1	$\varnothing$	$\varnothing$	$\varnothing$	$\varnothing$
	...	...	...	...	...
	9	$\varnothing$	$\varnothing$	$\varnothing$	$\varnothing$
	10	$\varnothing$	$\varnothing$	$\varnothing$	$\varnothing$
$Y_2=U-\{u_9\}$	0	$\varnothing$	$\varnothing$	$\varnothing$	$\varnothing$
	1	$\varnothing$	$\varnothing$	$\varnothing$	$\varnothing$
	...	...	...	...	...
	9	$\varnothing$	$\varnothing$	$\varnothing$	$\varnothing$
	10	$\varnothing$	$\varnothing$	$\varnothing$	$\varnothing$
...	...	...	...	...	...
$Y_6=\{u_1,u_2,u_3,u_4,u_5\}$	0	$\varnothing$	$\{u_2,u_9,u_{10}\}$	$\varnothing$	$\varnothing$
	1	$\{u_2,u_9,u_{10}\}$	$\varnothing$	$\{u_2,u_9,u_{10}\}$	$\varnothing$
	...	...	...	...	...
	9	$\varnothing$	$\varnothing$	$\varnothing$	$\varnothing$
	10	$\varnothing$	$\varnothing$	$\varnothing$	$\varnothing$
...	...	...	...	...	...
$Y_{10}=\{u_1\}$	0	$\varnothing$	$\varnothing$	$\varnothing$	$\varnothing$
	1	$\varnothing$	$\varnothing$	$\varnothing$	$\varnothing$
	...	...	...	...	...
	9	$\varnothing$	$\varnothing$	$\varnothing$	$\varnothing$
	10	$\varnothing$	$\varnothing$	$\varnothing$	$\varnothing$

This result means the outer membership degrees between u_2, u_9, u_{10} and $X \cap Y$ are greater than k; from Definition 2, the non-overlapping information between u_2, u_9, u_{10} and $X \cap Y$ is greater than k. Meanwhile,

$$\overline{Obnd}_R^k(X,Y) = \varnothing,$$

so the overlapping information between u_2, u_9, u_{10} and $X \cap Y$ is less than k. Therefore, it is not appropriate to take the set $X \cap Y$ for medical diagnosis analysis.

6   Conclusion

This paper targets at the non-preservation of set operations of GRSs approximation operators, and constructs four types of binary boundaries (and matching operators) to obtain

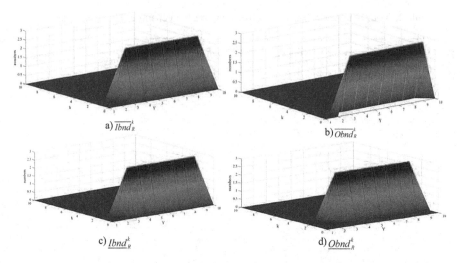

a) $\overline{Ibnd}_R^k$

b) $\overline{Obnd}_R^k$

c) $\underline{Ibnd}_R^k$

d) $\underline{Obnd}_R^k$

Fig. 2. Three-dimensional figures of four boundaries on ECG data analysis

relevant uncertainty descriptions. Then combining the two kinds of unitary boundaries (operators), new union, intersects and complement operations $\underset{*}{\cup}, \underset{*}{\cap}, \underset{*}{\sim}$ of GRSs approximations are systematically constructed to obtain a perfect property that GRSs approximation operators maintain set operations. Furthermore, the GRSs power set space is constructed, and we obtain the homomorphic characteristics between the new space and the classical power set space. Finally, ECG data analysis is made to validate four binary boundaries and their properties in power set space, and the concrete data analysis provides some beneficial guidances for ECG diagnosis.

This article deepens GRSs. With regard to the operator theory, four binary boundary operators have been added to the existing operational system of GRSs:

$$(2^U, \cup, \cap, \sim, \bar{R}_k, \underline{R}_k, \overline{bnd}_R^k, \underline{bnd}_R^k).$$

With regard to the set theory, the new operations $\underset{*}{\cup}, \underset{*}{\cap}$ and $\underset{*}{\sim}$ expand the classical union, intersection and complement operations, and they further induce the homomorphism between power set spaces. The relevant results quantitatively expand the qualitative results of Pawlak-RSs (i.e., this paper generalizes Ref. [12]), and they are worth further generalizing to probability rough sets. Four binary boundary operators are used to divide ECG data, and relevant uncertainty processing and intelligent decision should be deeply explored, so as to better advance the ECG data mining and medical diagnosis.

Acknowledgments. The authors thank the reviewers for their valuable suggestions. The work was supported by Sichuan Science and Technology Program of China (2021YJ0085), and Joint Research Project of Laurent Mathematics Center of Sichuan Normal University and National-Local Joint Engineering Laboratory of System Credibility Automatic Verification (ZD20220101).

References

1. Yao, Y.Y.: Two views of the theory of rough sets in finite universes. Int. Approx. Reason. **15**, 291–317 (1996)
2. Pawlak, Z.: Rough sets. Int. J. Comput. Inform. Sci. **11**(5), 341–356 (1982)
3. Ma, X.A.: Fuzzy entropies for class-specific and classification-based attribute reducts in three-way probabilistic rough set models. Int. J. Mach. Learn. Cybern. **12**(2), 433–457 (2020).
4. Sun, B.Z., Chen, X.T., Zhang, L.Y., Ma, W.M.: Three-way decision making approach to conflict analysis and resolution using probabilistic rough set over two universes. Inf. Sci. **507**, 809–822 (2020)
5. Maldonado, S., Peters, G., Weber, R.: Credit scoring using three-way decisions with probabilistic rough sets. Inf. Sci. **507**, 700–714 (2020)
6. Yao, Y.Y., Lin, T.Y.: Generalization of rough sets using modal logics. Intell. Autom. Soft Comput. **2**(2), 103–119 (1996)
7. Liu, C.H., Miao, D.Q., Zhang, N.: Graded rough set model based on two universes and its properties. Knowl. Based Syst. **33**, 65–72 (2012)
8. Huang, B., Guo, C.X., Li, H.X., Feng, G.F., Zhou, X.Z.: An intuitionistic fuzzy graded covering rough set. Knowl. Based Syst. **107**, 155–178 (2016)
9. Xue, Z.A., Lv, M.J., Han, D.J., Xin, X.W.: Multi-granulation graded rough intuitionistic fuzzy sets models based on dominance relation. Symmetry **10**(10), 446 (2018)
10. Zhang, X.Y., Mo, Z.W., Xiong, F., Cheng, W.: Comparative study of variable precision rough set model and graded rough set model. Int. J. Approx. Reason. **53**(1), 104–116 (2012)
11. Fang, B.W., Hu, B.Q.: Probabilistic graded rough set and double relative quantitative decision-theoretic rough set. Int. J. Approx. Reason. **74**, 1–12 (2016)
12. Zhang, X.Y., Mo, Z.W., Shu, L.: Operators of approximation and approximate power sets space. J. Electron. Sci. Technol. China **2**(2), 91–96 (2004)
13. Li, F.H., Chen, K.H., Lie, J., Zhan, Y.W., Manogaran, G.: Automatic diagnosis of cardiac arrhythmia in electrocardiograms via multigranulation computing. Appl. Soft Comput. **80**, 400–413 (2019)
14. Zhang, X.Y., Wang, X.X., Mo, Z.W., Wang, Q., Tang, X.: Three-level and three-way vagueness similarity measurements for electrocardiogram (ECG) data analysis. In: 2021 16th International Conference on Intelligent Systems and Knowledge Engineering (ISKE), p. 471–478. IEEE (2021)
15. Zhang, X.Y., Xie, S.C., Mo, Z.W.: Graded rough set. J. Sichuan Normal Univ. (Nat. Sci.) **1**, 12–16 (2010)

Neighborhood Approximate Reducts-Based Ensemble Learning Algorithm and Its Application in Software Defect Prediction

Zhiyong Yang, Junwei Du, Qiang Hu, and Feng Jiang[✉]

Qingdao University of Science and Technology, Qingdao 266100, Shandong, China
jiangfeng@qust.edu.cn

Abstract. Ensemble learning is a machine learning paradigm that integrates the results of multiple base learners according to a certain rule to obtain a better classification result. Ensemble learning has been widely used in many fields, but the existing methods still have the problems of difficult to guarantee the diversity of base learners and low prediction accuracy. In order to overcome the above problems, we considered ensemble learning from the perspective of attribute space division, defined the concept of neighborhood approximate reduction through neighborhood rough set theory, and further proposed an ensemble learning algorithm based on neighborhood approximate reduction, called ELNAR. ELNAR algorithm divides the attribute space of the data set into multiple subspaces. The basic learners trained based on the data sets corresponding to different subspaces have great differences, so as to ensure the strong generalization performance of the ensemble learner. In order to verify the effectiveness of ELNAR algorithm, we applied ELNAR algorithm to software defect prediction. Experiments on 20 NASA MDP data sets show that ELNAR algorithm can better improve the performance of software defect prediction compared with the existing ensemble learning algorithms.

Keywords: Neighborhood approximate reducts · Ensemble learning · Software defect prediction · Neighborhood rough set

1 Introduction

Ensemble learning is a learning method that uses a series of learners to learn, and uses some rules to integrate each learning result, so as to obtain a better learning effect than a single learner. The characteristics of ensemble learning determine its good generalization ability. In recent years, ensemble learning has been widely used in intrusion detection [1], text classification [2] and software defect prediction [3] and software defect prediction [4].

In ensemble learning, there are two main strategies to generate base learners: 1) method based on training sample disturbance. 2) Method based on attribute

© The Author(s), under exclusive license to Springer Nature Switzerland AG 2022
J. Yao et al. (Eds.): IJCRS 2022, LNAI 13633, pp. 100–113, 2022.
https://doi.org/10.1007/978-3-031-21244-4_8

space perturbation. For the former, the representative algorithms is Bagging [5] algorithm. These algorithms generate the base learner by resampling the training samplesm, where Bagging algorithm resamples the examples and trains the base learner in the way of equal probability. For the second method of generating basic learners, the representative algorithm is RSM [6] (random subspace method), which randomly divides the attribute subspace of the training set into different attribute subsets, and constructs different basic learners based on multiple attribute subsets.

Nowadays, many scholars have studied ensemble learning and put forward many new ensemble learning methods [7–10]. For instance, in the research of ensemble learning based on training sample disturbance, Liu et al. [7] proposed a self-paced ensemble learning method based on "classification hardness". This method considers the "classification hardness" distribution on the data set and undersamples according to the hardness distribution. García et al. [8] proposed a new dynamic ensemble learning model. This model firstly preprocesses the data and generates base learners, then associates the voting weights of samples in the current data neighborhood, and adaptively adjusts them according to the data distribution. Besides, Liu et al. [9] proposed a dual balanced ensemble learning method, called DuBE, which focuses on inter-class and intra-class imbalance. Besides, Jiang et al. [10] proposed an ensemble learning method based on attribute reduction and self-help sampling. This method uses self-help sampling to disturb the sample space, and uses approximate reduction based on relative decision entropy to disturb the attribute space.

In recent years, the wide application of machine learning and data mining technology in the field of software defect prediction has greatly improved the accuracy of software defect prediction. As an important research direction of machine learning, ensemble learning has also been widely used in the field of software defect prediction. At present, many ensemble learning algorithms have been applied to software defect prediction system [11–13]. For example, Chen et al. [11] proposed a software defect prediction model based on classes overlapping reduction and ensemble unbalanced learning, which combined several generated classifiers with AdaBoost mechanism to establish the software defect prediction model. Abuqaddom et al. [12] proposed a software defect prediction model based on improved hybrid SMOTE ensemble learning method. The model uses cost sensitive learning (CSL) to improve the processing of unbalanced distribution problems. The experimental results show that using cost sensitive learning can improve the performance of software defect prediction. Balogun et al. [13] proposed an ensemble learning method software defect prediction model combining SMOTE with Bagging and Boosting. The model uses decision tree and Bayesian network as the classifier to predict the software defect data set. The experimental results confirm the effectiveness of the model.

Ensemble learning can effectively improve the generalization ability of learning system and keep smaller error. The introduction of ensemble learning method in software defect prediction can still ensure good detection performance in the case of insufficient prior knowledge. However, the existing ensemble learning methods still have some problems in predicting software defects. For example, it is difficult to guarantee the diversity of base learners, and the feature redundancy of software data is always high. Most of the current ensemble learning methods use the method based on training sample disturbance to generate base learners. When dealing with high-dimensional data, these algorithms will face very large computational overhead. Meanwhile, in order to solve the problem of feature redundancy, many scholars have proposed ensemble learning methods based on attribute reduction and applied them to software defect prediction. Most of the existing ensemble learning methods based on attribute reduction use the traditional rough set theory. This kind of methods have to discretize the attribute value before reduction. However, discretization operation may change the data structure, resulting in low accuracy of software defect prediction. Therefore, in order to quickly and accurately detect software defects from massive and high-dimensional software defect data without changing the internal structure of software data, it is necessary to improve the existing ensemble learning algorithm.

Aiming at the problems of existing ensemble learning algorithms, we proposed an ensemble learning algorithm ELNAR based on neighborhood approximate reduction, and use ELNAR to detect software defects. Firstly, we use the attribute reduction technique of neighborhood rough set to reduce the dimension of high-dimensional attribute space, that is, generate multiple neighborhood approximate reductions. Then, we build a base learner on the low-dimensional subspace corresponding to each neighborhood approximate reduction. Finally, by ensembling these base learners, we obtain ELNAR. Building base learner on low-dimensional attribute subspace can effectively reduce the computational cost of software defect prediction system and ensure the efficiency of the system. When generating neighborhood approximate reduction, we firstly select an attribute to the core by random selection, and then select the remaining attributes by heuristic method. In this way, the diversity between different approximate reductions can be guaranteed, so that there is also diversity among the corresponding base learners. In addition, different from the traditional rough set for attribute reduction, the whole process of attribute reduction using neighborhood rough set does not need to discretize the data attribute value, so as to ensure that the internal structure of the data will not be changed.

In order to verify the effect of ELNAR in software defect prediction, we adopt KNN [14] algorithm to train the base learners in this study, and carry out experiments on 20 NASA MDP datasets and Promise datasets [15,16]. The steps of using ELNAR to detect software defects on the data sets are described as follows: 1) Using neighborhood approximate reduction algorithm to generate multiple neighborhood approximate reduction on the training sets. 2) Train a base learner on the attribute subspace corresponding to each neighborhood approximate reduction. 3) Obtain ensemble learner by ensembleing multiple base learners together with majority voting. 4) Predict software defects on the data to be detected by using ensemble learner and return the defect prediction results.

Experimental results show that ELNAR has better performance than six existing ensemble learning algorithms.

2 Preliminaries

In the neighborhood rough set, U is the given domain, A is the set of condition attributes describing U, and D is the decision attributes; And V_a is the set of attribute a, $f : U \times \{C \cup D\} \rightarrow V$ is the mapping function; $\Delta \rightarrow [0, \infty]$ is the distance function; δ is the neighborhood radius parameter and $0 \leq \delta \leq 1$, so called $NDS = \langle U, C, D, V, f, \Delta, \delta \rangle$ neighborhood decision system, abbreviated as $NDS = \langle U, C, D, \delta \rangle$ [17].

Definition 1 (*Neighborhood Relation*). Given a neighborhood decision system $NDS = \langle U, C, D, \delta \rangle$, for any $x, y \in U$ and attribute subset $B \subseteq C$, the neighborhood relation N_B^δ determined by attribute subset B on U is defined as follows [17–19]:

$$N_B^\delta = \{(x, y) \in U \times U \mid \Delta_B(x, y) \leq \delta\} \tag{1}$$

Where $\Delta_B(x, y)$ denotes the distance between object x and y on attribute subset B. For numerical attribute, $\Delta_B(x, y)$ is usually measured by Minkowski distance [18]. For categorical attribute, $\Delta_B(x, y)$ is usually measured by simple matching distance.

Definition 2 (*Neighborhood Class*). Given a neighborhood decision system $NDS = \langle U, C, D, \delta \rangle$, for any attribute subset $B \subseteq C$, assuming that the neighborhood relationship determined by B on U is N_B^δ, for any $x \in U$, the neighborhood class δ_B of x under N_B^δ is defined as follows [17–19]:

$$\delta_B(x) = \{y \in U : (x, y) \in N_B^\delta\} \tag{2}$$

Definition 3 (*Lower and Upper Approximations*). Given a neighborhood decision system $NDS = \langle U, C, D, \delta \rangle$, for any attribute subset $B \subseteq C \cup D$ and sample subset $X \subseteq U$, the *B-lower* and *B-upper approximations* of set X are respectively defned as follows [17–19]:

$$\underline{X}_B = \{x_i \mid \delta_B(x_i) \subseteq X, x_i \in U\} \tag{3}$$

$$\overline{X}_B = \{x_i \mid \delta_B(x_i) \cap X \neq \emptyset, x_i \in U\} \tag{4}$$

Definition 4 (*Relative Positive Region*). Given a neighborhood decision system $NDS = \langle U, C, D, \delta \rangle$, suppose U is divided by D into N equivalence calsses: $X_1, X_2, ..., X_N$, for any attribute subset $B \subseteq C$, the positive region $POS_B(D)$ of decision attribute set D relative to condition attribute subset B is defined as follows [17–19]:

$$POS_B(D) = \bigcup_{j=1}^{N} \underline{X_j}_B \tag{5}$$

Where $\underline{X_{j}}_B$ denotes the *B-lower approximation* of the *jth* equivalent class of U divided by D on attribute subset B.

Definition 5 (*Indiscernibility Attribute*). Given a neighborhood decision system $NDS = \langle U, C, D, \delta \rangle$, for any condition attribute $c \in C$, if $|POS_C(D)| = |POS_{C-\{c\}}(D)|$, called c an indiscernibility attribute of C relative to D. Otherwise, attribute c is a discernibility attribute of C relative to D.

Definition 6 (*Core*). Given a neighborhood decision system $NDS = \langle U, C, D, \delta \rangle$, for any $b \in C$, if $|POS_{C-\{b\}}(D)| \neq |POS_C(D)|$, called b a core attribute of C relative to D [19].

Definition 7 (*Attribute Significance*). Given a neighborhood decision system $NDS = \langle U, C, D, \delta \rangle$, for any $B \subseteq C$ and $c \in C - B$, The significance of attribute C relative to B and D is defined as follows [17–19]:

$$SGF(c, B, D) = \frac{|POS_{B \cup \{c\}}(D)| - |POS_B(D)|}{|U|} \tag{6}$$

3   Neighborhood Approximate Reduction and Ensemble Learning Based on Neighborhood Approximate Reduction

In the definition of neighborhood rough set reduction, given a neighborhood decision system $NDS = \langle U, C, D, \delta \rangle$, if R is a reduction of the initial attribute set C, then R must have exactly the same classification ability as C, that is, $|POS_R(D)| = |POS_C(D)|$, The above requirements on reduction are too strict, resulting in a very small number of reduction on many data sets. In order to ensure enough reduction in each data set, it is necessary to relax the above requirements, that is, if R is a reduction of C, R and C have the same or similar classification ability. Through the above modification, the concept of neighborhood approximate reduction is obtained, which is defined as follows.

Definition 8 (*Neighborhood Approximate Reduction*). Given a neighborhood decision system $NDS = \langle U, C, D, \delta \rangle$, for any $NAR \subset C$, if $|POS_C(D)| \geq |POS_{NAR}(D)| \geq \sigma \times |POS_C(D)|$, then NAR is called a neighborhood approximate reduction of C relative to D, where $\sigma \in (0, 1]$ is a given threshold, called σ the degree of approximation.

According to Definition 8, although the classification ability of neighborhood approximate reduction NAR may be lower than that of initial attribute set C, their classification ability is approximately equal. The approximation degree of the classification ability of NAR and C can be controlled by the threshold σ. if σ is larger, the classification ability of NAR is closer to C. When $\sigma = 1$, the classification ability of NAR is equal to C. At this time, NAR becomes a traditional reduction.

In order to obtain enough reduction to build the base learners, we appropriately slacken the strict requirements of traditional methods for reduction. Although this slackness may lead to the lower classification ability of NAR than C, which may affect the accuracy of the base learners built by NAR, this sacrifice is worth it, because the gap between the classification ability of NAR and C can be controlled by the threshold σ. By setting the δ value reasonably, it can not only ensure the better performance of the base learners built by NAR, but also obtain enough approximate reduction. In addition, ensemble learning not only focuses on the performance of each base learners, but also the diversity of base learners is a key factor for the success of ensemble learning.

The ELNAR algorithm mainly includes the following steps.

Algorithm 1. Calculate Relative Positive Region.

Input: Neighborhood decision system $NDS = \langle U, C, D, \delta \rangle$;
Output: $POS_C(D)$ and set $\{a \in C \mid POS_{C-\{a\}}(D)\}$.
1: **for** each object $x \in U$ **do**
2:     alculate the neighborhood classes $\delta_C(x)$ and $\delta_D(x)$ of x under the neighborhood relation N_C^δ and N_D^δ by using $ball - tree$ method [20];
3: **end for**
4: **for** each attribute $a \in A$ **do**
5:     **for** each object $x \in U$ **do**
6:         calculate the neighborhood classes $\delta_{C-\{a\}}(x)$ under the neighborhood relation $N_{C-\{a\}}^\delta$ by using $ball - tree$ method [20];
7:     **end for**
8:     Calculate relative positive region $POS_{C-\{a\}}(D)$ of D relative to condition attribute subset $C - \{a\}$.
9: **end for**
10: **return** $POS_C(D)$ and set $\{a \in C \mid POS_{C-\{a\}}(D)\}$.

In Algorithm 1, we adopt $ball - tree$ method [20] to calculate the neighborhood classes $\delta_C(x)$ and $\delta_D(x)$ of x under the neighborhood relation N_C^δ and N_D^δ, respectively. The time complexity of steps 1–3 is $O(|A| \times \log |U| \times |U|)$. In steps 4–9, we also use the $ball - tree$ method to calculate the neighborhood class $\delta_{C-\{a\}}(x)$ of x under the neighborhood relationship $N_{C-\{a\}}^\delta$. The time complexity of steps 4–9 is $O\left(|A|^2 \times \log |U| \times |U|\right)$. In conclusion, the time complexity of Algorithm 1 is $O\left(|A|^2 \times \log |U| \times |U|\right)$, and space complexity is $O(|A| + |U|)$.

Based on Algorithm 2, we further propose an ensemble learning algorithm ELNAR based on neighborhood approximate reduction. The detailed description of ELNAR is shown in Algorithm 3.

4   Experimental Results and Analysis

4.1   Experimental Data

To evaluate the performance of ELNAR on software defect prediction, we designed and carried out a series of experiments using 20 datasets collected

---

Algorithm 2. Calculate Neighborhood Approximate Reduction.

Input: $NDS = \langle U, C, D, \delta \rangle$, where $U = \{x_1, ..., x_n\}$, $C = \{a_1, ..., a_m\}$, the number S of approximate reduction, the approximate degree σ;

Output: S approximate reduction sets Set_NAR.

1: Initialization: let $Set\_NAR = \emptyset$, and core $Core_D(C) = \emptyset$;

2: Obtain $POS_C(D)$ and set $\{a \in C \mid POS_{C-\{a\}}(D)\}$ via Algorithm 1;

3: **for** any $a \in C$ **do**

4:    **if** $POS_{C-\{a\}}(D) \neq POS_C(D)$ **then**

5:       $Core_D(C) = Core_D(C) \cup \{a\}$.

6:    **end if**

7: **end for**

8: Let $RASet = C - Core_D(C)$ denote the subset of remaining attributes after removing the core from C.

9: **while** $-Set\_NAR| < S$ **do**

10:    Let $NAR = Core_D(C)$ denote the current neighborhood approximate reduction;

11:    Randomly select an attribute r from $RASet$, and $NAR = \cup \{r\}$ and $RASet = RASet - \{r\}$.

12:    **while** $|POS_{NAR}(D)| < \sigma \times |POS_C(D)|$ **do**

13:       **for** any $c \in RASet$ **do**

14:          calculate the significance $SGF(c, NAR, D)$ of c relative to NAR and D;

15:       **end for**

16:       Select the attribute m with the greatest significance from $RASet$;

17:       $NAR = NAR \cup \{m\}$ and $RASet = RASet - \{m\}$.

18:    **end while**

19:    **if** $NAR \notin Set\_NAR$ **then**

20:       $Set\_NAR = Set\_NAR \cup \{NAR\}$.

21:    **end if**

22: **end while**

23: **return** S different neighborhood approximate reduction sets Set_NAR.

Algorithm 3. ELNAR.

Input: Train set T, the number S of NAR, the approximate degree σ;

Output: Ensemble learner EL.

1: Initialization: set $E = \emptyset$;

2: Obtain the set Set_NAR on train set T via Algorithm 2 according to S and σ.

3: **for** each $NAR \in Set\_NAR$ **do**

4:    Reduce the dimension of the attribute space of T by using NAR, and obtain the reduction training set T_{NAR};

5:    Train the reduction training set T_{NAR} by using given classification algorithm, and obtain a base learner bl_{NAR};

6:    $E = E \cup \{bl_{NAR}\}$.

7: **end for**

8: Ensemble all base learners in set E by majority voting method, and obtain the ensemble learner EL;

9: **return** ensemble learner EL.

---

from real-world software projects. To verify the predictive effect of ELNAR on software defects, we designed and performed a series of experiments using 20 datasets collected from real-world software projects. Where 10 datasets are from NASA MDP repository [15], and the remaining 10 datasets are retrieved from Promise repository [16].

The NASA MDP repository has been widely used in the field of software defect prediction and currently contains 13 publicly available datasets. In experiment, we selected 10 NASA MDP datasets. The description of the 10 NASA MDP datasets is shown in Table 1.

Table 1. Detailed information of the 10 NASA MDP data sets.

Data set	No. of attributes	No. of data	No. of defective data	Ratio of defective data/%
CM1	38	505	48	9.50
KC1	22	2107	325	15.42
KC3	40	458	43	9.39
MC1	39	9466	68	0.72
MW1	38	403	31	7.69
PC1	38	1107	76	6.87
PC2	37	5589	23	0.41
PC3	38	1563	160	10.24
PC4	38	1458	178	12.21
PC5	39	17186	516	3.00

In addition to the 10 NASA MDP datasets, we also choose 10 publicly available datasets from the Promise repository. The description of the 10 Promise datasets is shown in Table 2.

Table 2. Detailed information of the 10 Promise data sets.

Data set	Release	No. of attributes	No. of data	No. of defective data	Ratio of defective data/%
Ant	1.5	38	293	32	10.92
Camel	1.0	22	339	13	3.83
Ivy	1.4	40	241	16	6.64
Jedit	4.3	39	492	11	2.24
Log4j	1.0	38	135	34	25.19
Lucene	2.0	38	195	91	46.67
Poi	2.0	37	314	37	11.78
Synapse	1.0	38	157	16	10.19
Velocity	1.6	38	229	78	34.06
Xerces	1.3	39	453	69	15.23

4.2   Experimental Design

In order to verify the effectiveness of ELNAR algorithm in improving the performance of software defect prediction, we compare ELNAR with the following six ensemble learning methods: (1)Bagging [5]; (2)RSM [6]; (3)Bag(RSM) [21]; (4)SPE [7]; (5)DuBE [9]; (6)DES-MI [8]. Where method (1) is the representative of resampling algorithm, method (2) is the representative of feature subspace algorithm, and method (3) is a multimodal ensemble learning algorithm, which is obtained by the combination of Bagging and RSM methods (3), (4) and (5) are ensemble learning algorithms based on disturbed sample space, method (6) is a dynamic ensemble learning algorithm for multi-class imbalanced datasets.

The parameter settings of different algorithms are as follows: firstly, for the parameter settings of ELNAR algorithm, the neighborhood radius δ, approximation degree σ and the number S of neighborhood approximate reductions is gradually adjusted through many experiments, and select the parameter value that can obtain the optimal experimental results. Eventually, Finally, we set the values of δ and σ to 0.08 and 0.9 respectively, and set the number S of neighborhood approximate reductions to 10. Secondly, for the three comparison algorithms of Bagging, RSM and Bag (RSM), we set the dimension of the random subspace of RSM and Bag (RSM) to 1/2 of the total dimension, and set the ensemble scale of the three algorithms to 10. For the other three comparison algorithms, each parameter of them is set according to the parameter values provided in the relevant literature.

The experimental steps are divided into the following three stages:

1) Data preprocessing. For Bagging, RSM, Bag(RSM) and ELNAR algorithms, we use SMOTE algorithm to deal with the imbalance data sets, and the sample ratio is set to 0.8. Since SPE, DuBE and DEM-MI algorithms are ensemble learning methods based on imbalance processing, there is no need to preprocess the data sets of these three algorithms.
2) Generate base learners. For the , we use KNN classification algorithm to build base learners. After obtaining the prediction results of all base learners, the results of ensemble learning are generated by majority voting rules. This paper focuses on the performance of ensemble learner in software defect prediction, rather than adjusting the parameters of KNN. Therefore, each parameter of KNN is set as the default value.
3) Build ensemble learner and software defect prediction. Ensemble the previously generated base learners together to get an ensemble learner. Finally, the ensemble learner is used to detect software defects on the test sets.

4.3   Experimental Metrics

In order to evaluate the results of software defect prediction by different ensemble learning methods, we use three performance metrics: AUC (Area Under Curve), $F1\text{-}score$ and MCC (Matthews correlation coefficient). Where the AUC value is the area value surrounded by ROC curve and coordinate axis, which can clearly show the classification effect of the classifier. The closer the AUC value is to 1, the better the classification performance is. When its value is less than or equal to 0.5, the worse the classification ability is. $F1\text{-}score$ is the harmonic average of *Precision* and *Recall*. The value range of $F1\text{-}score$ is from 0 to 1. 1 represents the best output of the model and 0 represents the worst output of the model. MCC is a more balanced indicator used to measure the performance of binary classification. The value range of MCC is from -1 to 1. The closer the value is to 1, the better the prediction effect of the tested object is.

4.4   Experimental Results

Tables 3, 4 and 5 shows the software defect prediction results of different ensemble learning methods when using KNN to build the base learner. Where Table 3 shows AUC values of different methods, Table 4 shows $F1$-$score$ values of different methods, and Table 5 shows MCC values of different methods.

Table 3. The AUC values produced by various Ensemble learning methods (KNN).

Data sets	Ensemble learning methods						
	Bagging	RSM	Bag(RSM)	SPE	DuBE	DES-MI	ELNAR
CM1	0.8636	0.8577	0.8659	0.8687	0.8722	0.8494	**0.8897**
KC1	0.8109	0.8133	0.8146	0.8237	0.8257	0.8143	**0.8306**
KC3	0.8906	0.8856	0.8820	0.8962	0.8908	0.8647	**0.9009**
MC1	0.9888	0.9889	0.9912	0.9883	**0.9925**	0.9826	0.9916
MW1	0.8860	0.8917	0.8826	0.9065	0.8964	0.8934	**0.9197**
PC1	0.8946	0.9053	0.9137	0.9115	0.9166	0.8928	**0.9245**
PC2	0.9829	0.9804	0.9862	0.9812	0.9835	0.9753	**0.9864**
PC3	0.8719	0.8755	0.8740	0.8753	0.8798	0.8606	**0.8860**
PC4	0.8771	0.8716	0.8858	0.8771	0.8841	0.8729	**0.8892**
PC5	0.9702	0.9685	0.9732	0.9717	**0.9737**	0.9610	0.9729
ant-1.5	0.8869	0.8826	0.8953	0.9012	0.9052	0.8768	**0.9329**
camel-1.0	0.9203	0.9398	0.9350	**0.9468**	0.9329	0.9332	0.9403
ivy-1.4	0.9109	0.9176	0.9115	0.9130	0.9153	0.9134	**0.9308**
jedit-4.3	0.9357	0.9365	0.9401	0.9488	0.9456	0.9415	**0.9597**
log4j-1.0	0.8360	0.8030	0.8300	0.8378	0.8369	0.8266	**0.9098**
lucene-2.0	0.6321	0.6584	0.6435	0.6655	0.6793	0.6817	**0.7089**
poi-2.0	0.8741	0.8714	0.8618	0.8722	0.8705	0.8782	**0.8902**
synapse-1.0	0.8617	0.8566	0.8403	0.8676	0.8636	0.8764	**0.9236**
velocity-1.6	0.7237	0.7317	0.7280	0.7672	0.7567	0.7530	**0.8274**
xerces-1.3	0.8328	0.8528	0.8460	0.8648	0.8617	0.8555	**0.8939**

In Table 3, the highest values on each dataset are bold. It can be seen from table 3 that when using AUC to evaluate the prediction ability of different ensemble learning methods to software defects, ELNAR is always better than the other six ensemble learning methods. Except for MC1, PC5 and camel-1.0,

ELNAR obtained the highest AUC value on the other 17 data sets. Specifically, the performance of ELNAR is better than RSM and Bagging methods on all 20 data sets. RSM method adopts the random subspace method, which has randomness in the effect of attribute space disturbance, while ELNAR method based on neighborhood approximate reduction can better disturb attribute space. In addition, ELNAR is also better than Bag (RSM), SPE, DuBE and DES-MI. For example, ELNAR is better than DuBE method on 18 data sets. The above results show that ELNAR has good software defect prediction performance from the perspective of AUC.

Table 4. The $F1$-$score$ values produced by various Ensemble learning methods (KNN).

Data sets	Ensemble learning methods						
	Bagging	RSM	Bag(RSM)	SPE	DuBE	DES-MI	ELNAR
CM1	0.8541	0.8460	0.8541	0.8497	0.8583	0.8417	**0.8826**
KC1	0.8064	0.8149	0.8148	0.8191	0.8217	0.8120	**0.8265**
KC3	0.8843	0.8753	0.8754	0.8775	0.8798	0.8573	**0.8930**
MC1	0.9877	0.9882	0.9906	0.9875	**0.9918**	0.9812	0.9908
MW1	0.8848	0.8855	0.8806	0.8954	0.8856	0.8805	**0.9102**
PC1	0.8866	0.8934	0.9084	0.9010	0.9084	0.8794	**0.9150**
PC2	0.9819	0.9787	0.9848	0.9801	0.9829	0.9739	**0.9849**
PC3	0.8600	0.8732	0.8699	0.8697	0.8693	0.8534	**0.8757**
PC4	0.8685	0.8597	0.8784	0.8682	0.8742	0.8668	**0.8786**
PC5	0.9684	0.9668	**0.9721**	0.9701	0.9717	0.9602	0.9716
ant-1.5	0.8862	0.8793	0.8856	0.9006	0.8933	0.8758	**0.9214**
camel-1.0	0.9141	0.9309	0.9317	**0.9432**	0.9258	0.9318	0.9374
ivy-1.4	0.9014	0.9071	0.9091	0.9015	0.9098	0.9093	**0.9254**
jedit-4.3	0.9308	0.9260	0.9384	0.9461	0.9382	0.9378	**0.9534**
log4j-1.0	0.8344	0.7812	0.8328	0.7782	0.8361	0.8428	**0.9090**
lucene-2.0	0.6029	0.6586	0.6449	0.6641	0.6793	0.6488	**0.7033**
poi-2.0	0.8731	0.8666	0.8577	0.8643	0.8667	0.8763	**0.8867**
synapse-1.0	0.8552	0.8411	0.8411	0.8676	0.8421	0.8532	**0.9211**
velocity-1.6	0.7237	0.7315	0.7237	0.7750	0.7560	0.7538	**0.8276**
xerces-1.3	0.8343	0.8452	0.8461	0.8648	0.8554	0.8549	**0.8930**

It can be seen from Table 4 that when $F1$-$score$ is used to evaluate the prediction ability of different ensemble learning methods to software defects, ELNAR is always better than the other six ensemble learning methods. ELNAR obtained the highest $F1$-$score$ on 17 data sets. Specifically, the performance of ELNAR is better than Bagging, RSM and DES-MI methods on all 20 data sets. In addition, ELNAR is better than SPE method on 19 data sets. The above results show that ELNAR has better software defect prediction performance from the perspective of $F1$-*score*.

Table 5. The MCC values produced by various Ensemble learning methods (KNN).

Data sets	Ensemble learning methods						
	Bagging	RSM	Bag(RSM)	SPE	DuBE	DES-MI	ELNAR
CM1	0.7338	0.7110	0.7274	0.7268	0.7477	0.6935	**0.7806**
KC1	0.6183	0.6301	0.6296	0.6422	0.6473	0.6275	**0.6592**
KC3	0.7777	0.7645	0.7596	0.7711	0.7750	0.7234	**0.7951**
MC1	0.9757	0.9765	0.9813	0.9751	**0.9837**	0.9628	0.9817
MW1	0.7826	0.7810	0.7656	0.8069	0.7946	0.7824	**0.8320**
PC1	0.7828	0.7998	0.8258	0.8162	0.8310	0.7775	**0.8392**
PC2	0.9641	0.9577	0.9699	0.9604	0.9659	0.9482	**0.9703**
PC3	0.7315	0.7492	0.7429	0.7438	0.7558	0.7180	**0.7657**
PC4	0.7544	0.7388	0.7668	0.7483	0.7727	0.7413	**0.7764**
PC5	0.9375	0.9341	**0.9444**	0.9406	0.9439	0.9205	0.9436
ant-1.5	0.7782	0.7630	0.7802	0.8015	0.8020	0.7518	**0.8543**
camel-1.0	0.8332	0.8680	0.8673	**0.8925**	0.8590	0.8658	0.8778
ivy-1.4	0.8151	0.8252	0.8193	0.8207	0.8296	0.8218	**0.8539**
jedit-4.3	0.8705	0.8602	0.8786	0.8963	0.8835	0.8775	**0.9109**
log4j-1.0	0.6693	0.5963	0.6708	0.6367	0.6728	0.6930	**0.8185**
lucene-2.0	0.2504	0.3187	0.3112	0.3285	0.3587	0.4073	**0.4105**
poi-2.0	0.7468	0.7445	0.7174	0.7336	0.7411	0.7529	**0.7797**
synapse-1.0	0.7251	0.7013	0.6834	0.7352	0.7273	0.7313	**0.8472**
velocity-1.6	0.4473	0.4640	0.4806	0.5631	0.5128	0.5084	**0.6643**
xerces-1.3	0.6701	0.6983	0.6924	0.7295	0.7187	0.7100	**0.7862**

As can be seen from Table 5, when using MCC to evaluate the prediction ability of different ensemble learning methods to software defects, the performance of ELNAR is always better than the other six ensemble learning methods. ELNAR obtained the highest MCC values on 17 data sets. Specifically, ELNAR is better than Bagging, RSM, and DES-MI methods on all 20 data sets. In addition, the performance of ELNAR is better than both DuBE and SPE methods on 17 data sets.

In order to test whether the performance difference between ELNAR and existing methods is statistically significant, we performed paired t-test [22] on the results listed in Tables 3, 4 and 5, that is, paired t-test was performed on the clustering results generated by the existing initialization methods and ELNAR, where the significance level is 0.05. The results of the paired t-test are shown in Table 6.

Table 6. Paired *t*-test results.

ELNAR and each existing method	*p*-value under AUC	*p*-value under $F1\text{-}score$	*p*-value under MCC
ELNAR vs. Bagging	0.0001	0.0001	0.0001
ELNAR vs. RSM	0.0002	0.0003	0.0002
ELNAR vs. Bag(RSM)	0.0002	0.0005	0.0002
ELNAR vs. SPE	0.0003	0.0018	0.0004
ELNAR vs. DuBE	0.0004	0.0005	0.0008
ELNAR vs. DES-MI	0.0000	0.0000	0.0000

As can be seen from Table 6, under the three metrics AUC, $F1\text{-}score$ and $MCC-$, the *p*-value between ELNAR and each compared method is always less than 0.05. Therefore, the above results demonstrate that the difference between the proposed method and the existing methods is statistically significant.

5   Conclusions

This paper studies ensemble learning from the perspective of attribute space disturbance, defines the concept of neighborhood approximate reduction based on neighborhood rough set, proposes an ensemble learning algorithm ELNAR based on neighborhood approximate reduction, and applies it to software defect prediction, which solves the problems of high feature redundancy and low prediction accuracy in software defect prediction. Instead of traditional rough set attribute reduction integration method, ELNAR does not need to discretize the numerical attributes, and can reduce the mixed data to ensure the stability of the internal structure of the data. In addition, we use attribute reduction technology to divide the attribute space, and use greedy strategy when selecting the remaining attributes. Therefore, ELNAR can not only ensure the diversity of base learners, but also ensure that each base learner has better performance. The next step is to apply different attribute reduction integrated learning methods to software defect prediction to further improve the performance of software defect prediction.

Acknowledgements. This work is supported by the National Natural Science Foundation of China (Grant Nos. 61973180, 62172249, U1806201), and the Shandong Provincial Natural Science Foundation, China (Grant Nos. ZR2022MF326, ZR2021QF074, ZR2018MF007).

References

1. Rajadurai, H., Gandhi, U.D.: A stacked ensemble learning model for intrusion detection in wireless network. In: Neural Computing and Applications **34**, 15387–15395 (2020)
2. Luo, S.Y., Gu, Y.J., Yao, X.X., Wei, F.: Research on text sentiment analysis based on neural network and ensemble learning. Revue d'Intelligence Artificielle **35**(1), 63–70 (2021)

3. Jabbar, M.A.: Breast cancer data classification using ensemble machine learning. Eng. Appl. Sci. Res. **48**(1), 65–72 (2021)
4. Ali, U., Aftab, S., Iqbal, A., Nawaz, Z., Bashir, M.S., Saeed, M.A.: Software defect prediction using variant based ensemble learning and feature selection techniques. Int. J. Modern Educ. Comput. Sci. **12**(5), 29–40 (2020)
5. Bühlmann, P., Yu, B.: Analyzing bagging. Ann. Stat. **30**(4), 927–961 (2002)
6. Ho, T.K.: The random subspace method for constructing decision forests. IEEE Trans. Pattern Anal. Mach. Intell. **20**(8), 832–844 (1998)
7. Liu, Z.N., et al.: Self-paced ensemble for highly imbalanced massive data classification. In: 9th International Proceedings on Data Engineering, pp. 841–852. IEEE, NY (2020)
8. García, S., Zhang, Z.L., Altalhi, A., Alshomrani, S., Herrera, F.: Dynamic ensemble selection for multi-class imbalanced datasets. Inf. Sci. **445–456**, 22–37 (2018)
9. Liu, Z.N., et al.: Towards inter-class and intra-class imbalance in class-imbalanced learning. arXiv preprint arXiv:2111.12791 (2021)
10. Jiang, F., Yu, X., Zhao, H.B., Gong, D.W., Du, J.W.: Ensemble learning based on random super-reduct and resampling. Artif. Intell. Rev. **54**(4), 3115–3140 (2021)
11. Chen, L., Fang, B., Shang, Z.W., Tang, Y.Y.: Tackling class overlap and imbalance problems in software defect prediction. Software Qual. J. **26**(1), 97–125 (2018)
12. Abuqaddom, I., Hudaib, A.: Cost-sensitive learner on hybrid smote-ensemble approach to predict software defects. In: Silhavy, R., Silhavy, P., Prokopova, Z. (eds.) CoMeSySo 2018. AISC, vol. 859, pp. 12–21. Springer, Cham (2019). https://doi.org/10.1007/978-3-030-00211-4_2
13. Balogun, A.O., et al.: SMOTE-based homogeneous ensemble methods for software defect prediction. In: Gervasi, O., et al. (eds.) ICCSA 2020. LNCS, vol. 12254, pp. 615–631. Springer, Cham (2020). https://doi.org/10.1007/978-3-030-58817-5_45
14. Cover, T., Hart, P.: Nearest neighbor pattern classification. IEEE Trans. Inf. Theory **13**(1), 21–27 (1967)
15. MDP Data Repository. http://nasa-softwaredefectdatasets.wikispaces.com/. Accessed 11 Mar 2022
16. PROMISE Data Repository. https://code.google.com/p/promisedata/. Accessed 11 Mar 2022
17. Hu, Q.H., Yu, D.R., Xie, Z.X.: Neighborhood classifiers. Expert Syst. Appl. **34**(2), 866–876 (2008)
18. Hu, Q.H., Yu, D.R., Liu, J.F., Wu, C.X.: Neighborhood rough set based heterogeneous feature subset selection. Inf. Sci. **178**(18), 3577–3594 (2008)
19. Hu, Q.H., Liu, J.F., Yu, D.R.: Mixed feature selection based on granulation and approximation. Knowl.-Based Syst. **21**(4), 294–304 (2008)
20. Dolatshah, M., Hadian, A., Minaei-Bidgoli, B.: Ball*-tree: Efficient spatial indexing for constrained nearest-neighbor search in metric spaces. arXiv preprint arXiv:1511.00628 (2015)
21. Marqués, A.I., García, V., Sánchez, J.S.: Two-level classifier ensembles for credit risk assessment. Expert Syst. Appl. **39**(12), 10916–10922 (2012)
22. Demšar, J.: Statistical comparisons of classifiers over multiple data sets. J. Mach. Learn. Res. **7**, 1–30 (2006)

Granular Computing and Applications

USV Path Planning Based on Adaptive Fuzzy Reward

Zhenhua Duan, Guoyin Wang$^{(\boxtimes)}$, Qun Liu, and Yan Shi

Chongqing Key Laboratory of Computational Intelligence, Chongqing University of Posts and Telecommunications, Chongqing 400065, People's Republic of China
{s200201061,s210231158}@stu.cqupt.edu.cn, {wanggy,liuqun}@cqupt.edu.cn

Abstract. Unmanned surface vehicles (USVs) with autonomous capabilities is the future trend. The capability of path planning is particularly critical to ensure the safety of navigation at sea. The algorithms with known environmental information are no longer suitable for the complex and changeable marine environment. Deep reinforcement learning (DRL) can be better applied to uncertain environments as it obtains optimal policies through the interaction of agents. However, the sparse reward problem of reinforcement learning is more prominent in the path planning task. Agents can not get positive reward in a great number of interactions. To study the path planning problem of USV in uncertain environments, this paper proposes a deep Q-learning (DQN) model based on adaptive fuzzy reward. To address the sparse reward problem in path planning using reinforcement learning, we use fuzzy logic that conforms to human cognition to dynamically adjust the reward for different states so as to improve the performance of DQN algorithm. Through simulation experiments, the validity of our method under different environments is verified. The results show that our model can carry out path planning safely and effectively.

Keywords: Fuzzy logic · USV · Deep reinforcement learning · Path planning

1 Introduction

The exploration and utilization of marine resources is an important developing direction in the future. Therefore, the development of sea surface unmanned technology is an inevitable requirement. However, in terms of autonomous driving, most works focus on the researches of unmanned vehicles. In contrast, there is fewer researches on the unmanned surface vehicle (USV) [13]. For the complex and dangerous marine environment, USV technology can reduce casualties when performing marine missions such as environmental monitoring, search and rescue work, and maritime patrol [17]. Path planning is a core research topic to perform tasks in a complex environment. It is very important to find a safe and collision-free path from the current point to the specified target point. But

© The Author(s), under exclusive license to Springer Nature Switzerland AG 2022
J. Yao et al. (Eds.): IJCRS 2022, LNAI 13633, pp. 117–131, 2022.
https://doi.org/10.1007/978-3-031-21244-4_9

the marine environment is complex and unpredictable, which requires to learn independently and have the anti-interference ability for path planning algorithm.

Traditional path planning algorithms rely heavily on the surrounding environment and can not be apt for complex environments. Meanwhile, with the increasing complexity of application areas, algorithms that only consider the distance and reachability of paths can no longer meet the application requirements. In addition, the unpredictability of ocean wind and waves produces new challenges to the path planning capability of USV. In other words, we need to study more efficient and safer path planning algorithms.

In recent years, with the development of artificial intelligence, reinforcement learning technology has been rapidly applied to robotics, which provides new ideas and directions for path planning of mobile robots in complex environments. Mnih et al. [9] combined convolutional neural network with Q-learning algorithm and proposed a deep Q-network model (DQN) to solve the problem of high-dimensional perception of raw pixels in control decisions. Due to the powerful performance of DQN model, it has been well-applied in many control tasks, including robot path planning. At present, many followed papers try to combine the DQN model with different algorithms to improve the performance of path planning. However, these combined algorithms are only able to improve performance and fail to pay attentions on the sparse reward challenge faced by agents in realistic path planning task scenarios, only few situations in the state space return positive reward signals. To address the above problems, in this paper, we combine deep reinforcement learning (DRL) with fuzzy logic to propose a new path planning method for marine environments.

We summarize the contributions of our work as follows: First, with combined reinforcement learning, we realized path planning without global environmental information and prior knowledge. Secondly, the agent trained by our method can autonomously plan route and avoid obstacles in the simulated ocean environment. Third, by introducing fuzzy logic, we propose adaptive fuzzy reward to solve the problem of sparse reward in reinforcement learning, which is more consistent with human cognitive habits and makes the results more acceptable.

The rest of this paper is organized as follows. Section 2 presents some related works in the area of path planning, deep reinforcement learning and spare reward. In Sect. 3, we introduce the proposed method in detail. Section 4 demonstrates the performance of our model in the simulated environment. At the end, Sect. 5 concludes the work and also discusses future research.

2   Related Work

Compared with the path planning of unmanned vehicles and robots on land, there are few studies on path planning in marine environment. Moreover, by reviewing the relevant literature on optimal path planning for ships, most of them are still traditional algorithms. Singh et al. [14] proposed an A* algorithm that uses the safe distance as a constraint on the generated path to solve the path planning problem of USV in marine environment. Song et al. [15] proposed an improved ant colony algorithm based grid environment model for global path planning method for USV. The method is used to solve the global path planning problem of USV system in complex marine environment where there are a lot of obstacles. Zhang et al. [20] proposed a method based on genetic algorithm and simulated annealing algorithm to plan the optimal path of USV to solve the problems of lack of searching ability and large amount of calculation of traditional genetic algorithm. Although the aforementioned traditional algorithms have advantages in solving the path planning problem of USV, the traditional algorithms rely too much on environmental models and global environmental information, which have great limitations in application scenarios. The complex and volatile ocean environment, which requires the USV to have the ability of autonomous learning.

In recent years, some scholars have applied reinforcement learning to path planning and obstacle avoidance tasks to improve the autonomous learning ability of the algorithm. Lin et al. [6] considered the special requirements of USV navigation, a USV path planning model based on improved Q-learning algorithm is proposed. It reduces the computational complexity of classical Q-learning algorithm and speeds up the speed of path planning. Lei et al. [4] utilized Q-learning algorithm to improve the ability of obstacle avoidance and local planning in dynamic environment. Although the Q-learning algorithm has shown successful performance in the field of path planning, the application scenario is limited to the low dimensional fully observable state space. Zhang et al. [19] proposed an end-to-end path planning model based on deep reinforcement learning, utilized DQN model to solve the problem of high-dimensional observation space. Li et al. [5] investigated the path planning problem of USV in uncertain environments, and integrated deep reinforcement learning with artificial potential field to propose a path planning strategy that complies with the International Regulations for Preventing Collisions at Sea (COLREGs). In complex environments, due to the large environment state space, the sparse reward problem is an inevitable problem for DRL model. In 2017, Pathak et al. [12] proposed curiosity mechanism, which is regarded as an internal reward signal to encourage agents to explore the unknown state space and give a certain degree of reward when discovering new states. Recently, Jin et al. [3] proposed to divide reinforcement learning into exploration phase and planning phase by computing an approximately optimal policy to deal with states that the agent rarely appears in exploration. Similar researches on solving the sparse reward problem include imitation learning [1], auxiliary tasks [11], and reward reshaping [2].

There have been applications of deep reinforcement learning for path planning in marine environments. However, these studies compress the state space without considering the sparse reward problem. In this paper, we designed simulation environments with randomly distributed obstacles with large state spaces. We combine fuzzy logic and the classical DRL model and use adaptive fuzzy reward to give corresponding reward according to different states to solve the sparse reward problem. In conclusion, the proposed method not only effectively solves the sparse reward problem but also can quickly adapt to unknown environments.

3   Fuzzy-DQN Model and Adaptive Fuzzy Reward

In this section, we describe the proposed method in detail. And we first briefly introduce the concept of deep reinforcement learning and fuzzy logic. Then, the overall framework of the DQN model and the design of the fuzzy reward function will be illustrated.

3.1   Problem Formulation

Unlike supervised learning, reinforcement learning is a method that the agent interacts with the environment and learns by trial and error. The goal is to learn state-action mapping from the environment and maximize rewards to obtain optimal policies. The notations involved in reinforcement learning are shown in Table 1.

Table 1. The meaning of notations in reinforcement learning.

Symbol	Meaning
s	The environment state at current time
a	Action performed by the agent
r	The reward after the agent performs an action
π	The policy taken by the agent
γ	Discount factor

Markov decision process is the mathematical description of reinforcement learning. The next state in the Markov process only depends on the current state and action, not the historical state [10]. We use the value function to evaluate the current state. The value is a scalar that represents the expected return of performing an action in the current state for the future. Calculate the value function by using the Bellman equation, the function is shown as follows:

$$V^{\pi}(s) = E_{\pi} \left[R_{t+1} + \gamma V^{\pi} \left(S_{t+1} \right) \mid S_t = s \right] \tag{1}$$

Note that different actions can be performed in each state, we use the Q function to represent the future return, which is defined as follows:

$$Q^{\pi}(s, a) = E_{\pi} \left[R_{t+1} + \gamma Q^{\pi} \left(S_{t+1}, A_{t+1} \right) \mid S_t = s, A_t = a \right] \tag{2}$$

In the path planning problem, there are infinite states in the state space, and it is impossible for traditional reinforcement learning to use a table to record all the values of $Q(s, a)$. Therefore, combined with deep learning, DQN model is proposed and the deep neural network is used to approximate fit $Q(s, a)$. The Q value is calculated according to the following formula:

$$Q\left(s_t, a_t\right) \leftarrow Q\left(s_t, a_t\right) + \alpha \left[R_t + \gamma Q\left(s_{t+1}, a_{t+1}\right) - Q\left(s_t, a_t\right)\right] \tag{3}$$

where α is the learning rate, $Q\left(s_t, a_t\right)$ is an approximation.

The sparse reward problem is widespread in reinforcement learning tasks, especially in path planning missions. The agent can obtain the reward only when it reaches the specified position within the specified time step, and in other intermediate states, it is punished, that is, a negative reward. The sparse reward is shown in Eq. 4. The sparse reward problem causes the agent to fall into a local loop when exploring the environment, which affects the learning of optimal policies.

$$R(x) = \begin{cases} -10, & obstacle\ collision,\ boundary\ collision \\ +10, & arrive\ goal\ position \\ -0.01, & normal\ navigation \end{cases} \tag{4}$$

3.2 Fuzzy Logic

To overcome sparse reward problem, according to the features of the path planning task, we add additional intensive reward information to improve the learning ability of the agent.

The concept of the fuzzy set [18] was first proposed by Lotfi A. Zadeh in 1965. Fuzzy sets have demonstrated superior performance in the solution of network embedding problems [7]. The fuzzy set uses the membership function to express the degree to which the element belongs to the set. The value of membership falls on the interval [0, 1]. In the ocean, due to the influence of wind and waves, the distance and yaw angle between the current position and the target are not fixed values. These values are uncertain. Fuzzy logic is used to imitate human reasoning and cognition, which describes fuzzy concepts through membership function. Therefore, in this work, we design dense reward based on distance and yaw angle by using fuzzy logic.

Navigation at sea is not restricted by road traffic lines, but only requires attention to obstacles and other vessels. Therefore, during navigation, we only focus on the location of the target and the obstacles near USV. Inspired by this characteristic, we use Euclidean distance and yaw angle as the benchmark for navigation. This paper designs a fuzzy controller with dual input and single output. We take the distance between USV and obstacle and the difference of heading angle between USV and target position as input, the output is the fuzzy coefficient to realize the dynamic adjustment of the reward value, then the reasonable dense reward is obtained.

In order to describe the input of fuzzy rules, a coordinate system is established, as shown in Fig. 1. In this figure, XOY indicates the water plane, d_{obs} represents the distance between the USV and the obstacle, α is the heading

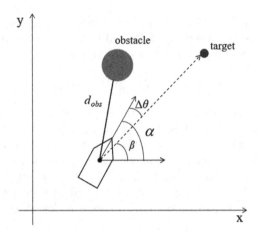

Fig. 1. Positional relationship between USV, obstacle and target.

angle of USV, β indicates the direction angle of the target point, $\Delta\theta$ indicates the difference of heading angle between USV and target position. We take $\Delta\theta$, d_{obs} and $\Delta\theta$, d_{goal} as the input of fuzzy controller respectively, and the output ρ is the coefficient of the reward value. The reward value of each state of the agent is adjusted by the coefficient ρ to make the reward value more reasonable.

In the fuzzy control system, the membership function is used to complete the fuzzification of precise quantity. We adopt the triangular membership function, the universe of discourse for the distance variable is set to $[0, 20]$, and the fuzzy linguistic variable is {VS, S, M, B, VB}. Another input is the angle variable, the universe is set to $[-\pi/2, \pi/2]$, and the fuzzy linguistic variable is {RB, RS, Z, LS, LB}. The output variable is the coefficient of the universe in $[0,1]$, and the linguistic fuzzy sets{VS, S, M, B, VB}. The meaning of the alphabet is {V: Very, S: Small, M: Middle, B: Big, R: Right, Z: Zero, L: Left}. The membership functions of the input variables are shown in Fig. 2. The membership function of coefficient variables is shown in Fig. 3.

According to the input of the fuzzy controller, we can obtain 25 fuzzy rules, respectively. The fuzzy rules are established according to human navigation experience, and the general idea is as follows:

Fuzzy rules about obstacles: the larger the difference $\Delta\theta$ between the agent's heading angle and the direction angle, the smaller the distance d_{obs} between the agent and the obstacle, and the larger the coefficient ρ, that is, the greater the penalty for the agent. On the contrary, the smaller the coefficient ρ. In addition, fuzzy rules about goal are used when the target position is in the detection range of the agent. The closer to the target point, the smaller the angle difference and the larger the coefficient.

All fuzzy rules are shown in the Table 2 and Table 3.

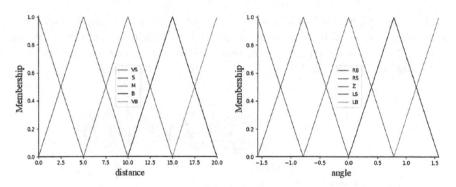

Fig. 2. Membership function of distance and angle.

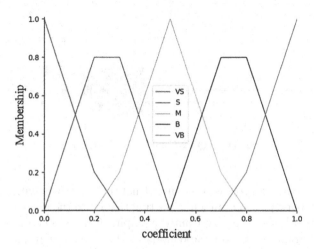

Fig. 3. Membership function of coefficient.

Table 2. Fuzzy rules about obstacles.   **Table 3.** Fuzzy rules about goal.

Distance	Angle				
	RB	RS	Z	LS	LB
VS	VB	VB	B	VB	VB
S	VB	B	B	B	VB
M	B	B	M	B	B
B	B	M	S	M	B
VB	M	S	VS	S	M

Distance	Angle				
	RB	RS	Z	LS	LB
VS	B	B	VB	B	B
S	B	B	VB	B	B
M	M	M	B	M	M
B	M	S	S	S	M
VB	VS	VS	S	VS	VS

3.3   Overall Framework

In this part, we will introduce the overall framework of the model and the fuzzy reward function.

The Deep Q-learning model uses the perception ability of deep learning and the decision-making ability of reinforcement learning, whose structure is shown in Fig. 4. The DQN model consists of two deep neural networks which are used to evaluate the Q-value and a memory buffer to store experience.

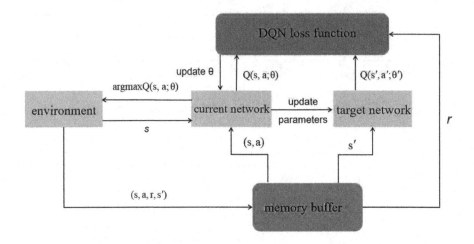

Fig. 4. Deep Q-learning network structure.

During the training process, the current network interacts with the environment to generate empirical data while putting them into the memory buffer, and the target network learns policy from the empirical data. The current network calculates the predicted Q-value and updates the network parameters through gradient descent, the target network calculates the target Q-value, which is used to calculate the loss function. The parameters of the target network are updated through the current network. Ihe state-action values are iteratively updated through the Bellman equation:

$$Q_{t+1}(s, a) = E_{s'} \left[r + \gamma \max_{a'} Q_t (s', a') \right] \tag{5}$$

The DQN approximates the state-action value function $Q(s, a)$ to $Q(s, a; \theta)$, where θ is the neural network parameter. The loss function is the mean square error between the real value and the predicted value. It is utilized to iteratively update the state-action value function as follows:

$$L(\theta) = E_{\pi} \left[\left(r + \gamma \max_{a'} Q (s', a', \theta') - Q(s, a; \theta) \right)^2 \right] \tag{6}$$

To settle the sparse reward problem, a new reward function is designed according to fuzzy rule. The reward function is divided into three parts:

1) normal navigation reward function R_n; 2) collision avoidance reward function R_c; 3) terminal reward R_{end}. The reward function can be written as follows:

$$R(x) = \begin{cases} R_n, & normal\ navigation \\ R_c, & collision\ avoidance \\ R_{end}, & terminal\ state \end{cases} \tag{7}$$

The normal action reward function is calculated based on the distance between the agent and the target point. The agent is in the safe area, that is, there are no obstacles in the detection range. ρ_{goal} represents the fuzzy coefficient about the target. The distance between the initial point of the agent and the target is denoted by d_{max}, the distance between the current position of the agent and the target is represented by d_{goal}. The normal navigation reward function can be written as follows:

$$R_n = \rho_{goal}\frac{d_{\max} - d_{goal}}{d_{\max}} \tag{8}$$

When the agent is within collision range, the reward is calculated by the collision avoidance reward function. ρ_{obs} represents the fuzzy coefficient about the obstacle. The distance between the current position of the agent and the obstacle is represented by d_{obs}. The collision range is a $135°$ semicircle with the radius r_{safe} set in the experiment. In this case, reward function can be written as follows:

$$R_c = \rho_{obs}\frac{d_{obs} - r_{safe}}{r_{safe}} \tag{9}$$

The R_{end} is a constant value, when the agent reaches the target position it is positive, while it is negative while colliding with the boundary or an obstacle.

4 Experiments and Analysis

In this part, according to the method in Sect. 3, we verify the effectiveness of the fuzzy reward function based on fuzzy logic through extensive experiments. The tensorflow framework and Python are used to build the algorithm model, and we build a two-dimensional simulation environment based on python and tkinter. The simulation environment is introduced in Sect. 4.1. The experimental results are presented in Sect. 4.2.

4.1 Environment Design

The simulation environment is designed as a 700×700 2D plane. The initialization environment is shown in Fig. 5. The range of motion of USV is limited to 2D plane space. If the USV moves beyond the range of motion, it is considered a collision and the environment will be initialized. Obstacles are added to the environment to imitate ships and islands on the sea.

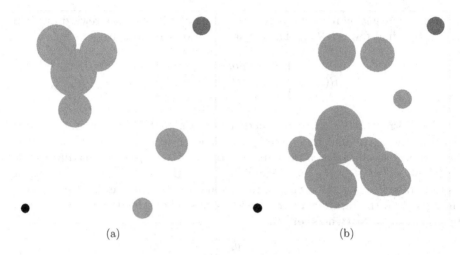

(a)   (b)

Fig. 5. Different simulation maps.

In Fig. 5, overlapping or isolated blue circles are used to regard as obstacles, The black circles and the red circles represent the source and target positions of the USV, respectively. When the USV collides with an obstacle or moves beyond the boundary, it is considered a failed episode, while reaching the target position is a successful episode.

We assume that the initial heading angle of the agent is 45°, and the action space is designed with 5 different directions of movement as:

$$N_{action} = \{-45°, -22.5°, 0°, 22.5°, 45°\} \tag{10}$$

The USV uses radar to perceive surrounding environment information. To simulate the function of radar, we assume that the agent can obtain 180° environmental information within a certain distance, including obstacles and boundary. Finally, the state space consists of environmental information, current position and heading angle, and target distance.

4.2   Training and Result Analysis

The model training is performed on 400 randomly generated maps with different obstacles.

The DQN is composed of two neural networks: the current net and target net with three hidden layers. The structure of the deep neural network is displayed in Fig. 6. The input data is the state of the agent at time t, which is composed of environmental information within a certain distance scanned by the radar, the distance between the USV and the obstacle and the distance and angle between the destination and the location of USV. The data is flattened into one dimension as the input of the fully connected neural network, and the output of the network is the action taken by the agent at the moment t. The parameters of the model in the training process are given in Table 4. The higher reward discount rate γ, the

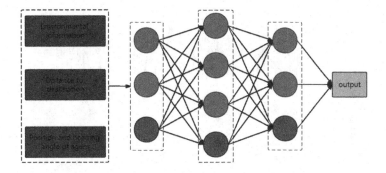

Fig. 6. Neural network structure of the current net and the target net.

more the agent pays attention on future rewards. The learning rate lr determines whether the network can converge. At the beginning of training, the learning rate is set to 1e-3, then, it is adaptively updated by Adam optimizer [21]. To make the agent have a certain exploration ability, the parameter ϵ of the ϵ-greedy algorithm is set to 0.1. The maximum capacity of the experience replay buffer is set to 10000, and the batch size of experience replay learning is set to 32. When the current network is trained C steps, the target network updates the parameters by replicating the parameters of the current network. The update interval is set to 100. Additionally, the agent learns action policies by interacting with the environment. Once the agent is trapped into a loop, the interaction cannot be terminated and the model does not learn an effective strategy. Therefore, except for the termination conditions such as collision, reaching the target position, and crossing the boundary, the maximum step of action in an episode is 100.

Table 4. Hyper parameters of the DQN model.

Hyper parameter	Symbol	Value
Reward discount rate	γ	0.99
Learning rate	lr	1e−3
ϵ-greedy	ϵ	0.1
Experience replay buffer size	M	10000
Replay batch size	B	32
Update interval	C	100

Using the proposed fuzzy reward function, the average reward of episodes is shown in the Fig. 7. To verify the effect of fuzzy reward function on the USV path planning problem, sparse reward, consistent dense reward, and fuzzy dense reward are used for path planning comparison experiments. The experimental results will be visualized in the simulation environment. We tested the path planning problem ability of these three reward functions under the same conditions in different environments, and verified the impact of the reward function

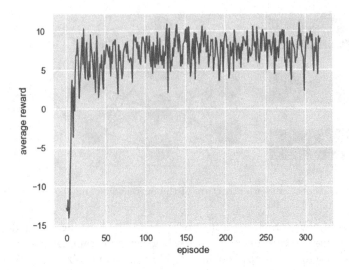

Fig. 7. Average reward for episodes

on the results. We classified the difficulty of tasks based on the complexity of the environment: simple, moderate and complex.

In Fig. 8, the coordinates of the starting point and the target point in the experiment are set in the lower left and upper right corners of the simulation environment, respectively. The purpose of the experimental design is to simulate the path planning task of USV in an unknown environment. The experimental results are displayed in Fig. 8 which includes the route trajectory. We can see that the reward functions from left to right are sparse reward, consistent reward, and fuzzy reward. The complexity of the environment from top to bottom is simple, moderate and complex. We performed path planning 10 rounds in 2000 different scenarios using different reward functions, and exhibited the stable success rate to display the results. The success rates for the three reward functions are shown in the Table 5, and we can observe that the success rate of the dense reward is higher.

Table 5. Success rate of different reward functions.

Reward function	Failure	Success	Success rate
Sparse reward	1412	588	29.40%
Consistent dense reward	312	1688	84.40%
Fuzzy dense reward	239	1761	**88.05%**

The visualization results are presented in the Fig. 8, all three reward functions are available, and the experimental results show that the fuzzy reward method uses the safest route to reach the destination. However, the route chosen by the sparse reward is closer to obstacles and boundaries, it is obviously more dangerous. Although the success rates of the two dense reward function methods are close, it can be seen from the Fig. 8 that in a simple environment, the path selected by the fuzzy reward function is farther from the obstacle, and the route is safer under the same route length. In moderate and complex environments, although the route length of the fuzzy dense reward is increased, route safety is improved. In summary, compared with the spares reward and consistent reward, adaptive fuzzy reward can guide agents to generate more effective and safe path planning trajectories.

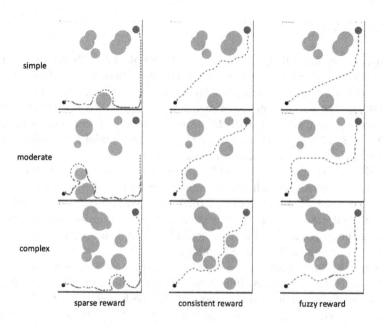

Fig. 8. Routes for different reward functions in different environments.

To illustrate the effectiveness of our proposed adaptive fuzzy reward function, extensive experiments are performed on different deep reinforcement learning algorithms, the experimental results are shown in Table 6. Comparing by using different reward functions on different algorithms, the experimental results show that the success rate of the algorithm combined with fuzzy reward remains relatively stable and demonstrate the effectiveness and stability of the proposed method.

Table 6. Success rate of fuzzy reward on different algorithms.

Algorithm	Success rate
DQN [9]	84.40% ± 1.77%
DDQN [16]	87.10% ± 1.25%
A3C [8]	87.50% ± 1.65%
DQN [9]+Fuzzy reward	88.05% ± 0.54%
DDQN [16]+Fuzzy reward	89.05% ± 0.40%
A3C [8]+Fuzzy reward	89.87% ± 0.42%

5   Conclusion

In this paper, we presented a USV path planning algorithm based on deep reinforcement learning, which uses the fuzzy rule to dynamically adjust the coefficient of reward values to solve the sparse reward problem in the path planning task. In the simulation experiment, we demonstrate the effectiveness of our method by comparing it with the deep reinforcement learning method with sparse reward. Through trial-and-error learning, our model can avoid obstacles and plan a safe path in an unknown environment.

To solve the sparse reward problem, most works on DRL rely on carefully tuned shaping reward through domain knowledge that guides the agent to accomplish the task goal. However, many of these tasks can be fairly easily defined by a terminal goal state, and it is difficult to get an appropriate reward function for solving the problem. In this work, we adaptively tune the reward according to fuzzy rules in line with human cognition.

Although we just apply the fuzzy rule to address the sparse reward problem for USV path planning tasks, it is a useful way to be applied to improve the performance of the agent in other domains. In the future, we intend to further optimize fuzzy rule design and extend it to multi-USV path planning tasks. In addition, at present, our experiment only runs in the simulation environment, we would like to test our model in the real environment.

Acknowledgments. This work is supported by the State Key Program of National Nature Science Foundation of China (61936001), the key cooperation project of Chongqing municipal education commission (HZ2021008), the Natural Science Foundation of Chongqing (cstc2019jcyj-cxttX0002, cstc2021ycjh-bgzxm0013).

References

1. Ho, J., Ermon, S.: Generative adversarial imitation learning. In: Advances in Neural Information Processing Systems, vol. 29 (2016)
2. Hu, Y., Wang, W., Jia, H., Wang, Y., et al.: Learning to utilize shaping rewards: a new approach of reward shaping. Adv. Neural. Inf. Process. Syst. **33**, 15931–15941 (2020)

3. Jin, C., Krishnamurthy, A., Simchowitz, M., Yu, T.: Reward-free exploration for reinforcement learning. In: International Conference on Machine Learning, pp. 4870–4879. PMLR (2020)

4. Lei, X., Zhang, Z., Dong, P.: Dynamic path planning of unknown environment based on deep reinforcement learning. J. Robot. **2018** (2018)

5. Li, L., Wu, D., Huang, Y., Yuan, Z.M.: A path planning strategy unified with a colregs collision avoidance function based on deep reinforcement learning and artificial potential field. Appl. Ocean Res. **113**, 102759 (2021)

6. Lin, X., Guo, R.: Path planning of unmanned surface vehicle based on improved q-learning algorithm. In: 2019 3rd International Conference on Electronic Information Technology and Computer Engineering (EITCE), pp. 302–306. IEEE (2019)

7. Liu, Q., Shu, H., Yuan, M., Wang, G.: Fuzzy hierarchical network embedding fusing structural and neighbor information. Inf. Sci. **603**, 130–148 (2022)

8. Mnih, V., Badia, A.P., Mirza, M., Graves, A., et al.: Asynchronous methods for deep reinforcement learning. In: International Conference on Machine Learning, pp. 1928–1937. PMLR (2016)

9. Mnih, V., Kavukcuoglu, K., Silver, D., Graves, A., et al.: Playing Atari with deep reinforcement learning. arXiv preprint arXiv:1312.5602 (2013)

10. Padakandla, S.: A survey of reinforcement learning algorithms for dynamically varying environments. ACM Comput. Surv. (CSUR) **54**(6), 1–25 (2021)

11. Papoudakis, G., Chatzidimitriou, K.C., Mitkas, P.A.: Deep reinforcement learning for doom using unsupervised auxiliary tasks. arXiv preprint arXiv:1807.01960 (2018)

12. Pathak, D., Agrawal, P., Efros, A.A., Darrell, T.: Curiosity-driven exploration by self-supervised prediction. In: International Conference on Machine Learning, pp. 2778–2787. PMLR (2017)

13. Peng, Y., Yang, Y., Cui, J., Li, X., et al.: Development of the USV 'jinghai-i'and sea trials in the southern yellow sea. Ocean Eng. **131**, 186–196 (2017)

14. Singh, Y., Sharma, S., Sutton, R., Hatton, D., Khan, A.: A constrained a* approach towards optimal path planning for an unmanned surface vehicle in a maritime environment containing dynamic obstacles and ocean currents. Ocean Eng. **169**, 187–201 (2018)

15. Song, C.H.: Global path planning method for USV system based on improved ant colony algorithm. In: Applied Mechanics and Materials, vol. 568, pp. 785–788. Trans Tech Publ (2014)

16. Van Hasselt, H., Guez, A., Silver, D.: Deep reinforcement learning with double q-learning. In: Proceedings of the AAAI Conference on Artificial Intelligence, pp. 2094–2100 (2016)

17. Yan, R.J., Pang, S., Sun, H.B., Pang, Y.J.: Development and missions of unmanned surface vehicle. J. Mar. Sci. Appl. **9**(4), 451–457 (2010)

18. Zadeh, L.A.: Fuzzy sets. In: Fuzzy sets, fuzzy logic, and fuzzy systems: selected papers by Lotfi A Zadeh, pp. 394–432. World Scientific (1996)

19. Zhang, W., Wang, W., Zhai, H., Li, Q.: A deep reinforcement learning method for mobile robot path planning in unknown environments. In: 2021 China Automation Congress (CAC), pp. 5898–5902. IEEE (2021)

20. Zhang, W., Xu, Y., Xie, J.: Path planning of USV based on improved hybrid genetic algorithm. In: 2019 European Navigation Conference (ENC), pp. 1–7. IEEE (2019)

21. Zhang, Z.: Improved Adam optimizer for deep neural networks. In: 2018 IEEE/ACM 26th International Symposium on Quality of Service (IWQoS), pp. 1–2. IEEE (2018)

A Naive Bayes Classifier Based on Neighborhood Granulation

Xingyu Fu[1], Yingyue Chen[2]([✉]), Zhiyuan Yao[2]([✉]), Yumin Chen[1],
and Nianfeng Zeng[3]

[1] School of Computer and Information Engineering, Xiamen University
of Technology, Xiamen 361024, China
[2] School of Economics and Management, Xiamen University of Technology,
Xiamen 361024, China
`{chenyingyue,yaozhiyuan}@xmut.edu.cn`
[3] E-success Information Technology Co., Ltd., Xiamen 361024, China

Abstract. The naive Bayes is a classifier based on probability and statistics theory, which is widely used in the field of text classification. But the assumption of independence between features affects its classification accuracy. To solve this problem, this paper studies the theory of granular computing and proposes a naive Bayes classifier based on neighborhood granulation. The neighborhood discriminant function is introduced to perform single-feature neighborhood granulation for all samples to form neighborhood granules and multiple characteristic granules in a sample form a neighborhood granular vector. The operation rules of granular vector, a prior probability, and the conditional probability of granular vector are defined, and then a naive Bayes classifier based on neighborhood granulation is proposed. Experiments on some UCI data sets, using different neighborhood parameters to compare with the classic naive Bayes classifier, the results show that the method can effectively improve the classification accuracy.

Keywords: Naive Bayes · Classification · Granular computing · Neighborhood granulation · Granular vector

1 Introduction

The naive Bayes (NB) is an important classification algorithm in machine learning [1]. The main idea is divided into two steps: firstly, the conditional probability of the samples to be judged is calculated belongs to each type based on the known prior probability, and then it is judged as the category with the maximum probability. The naive Bayes algorithm has been widely used in many fields, such as text classification [2], multi-label learning [3], imbalanced data processing [4], medical diagnosis [5], and traffic risk management [6]. The naive Bayes is a simple and effective classification algorithm, and many scholars have proposed a variety of improved naive Bayes algorithms from different perspectives [7].

J. Yao et al. (Eds.): IJCRS 2022, LNAI 13633, pp. 132–142, 2022.
https://doi.org/10.1007/978-3-031-21244-4_10

Zadeh, an expert in automatic control theory, proposed the fuzzy set theory [8] and put forward the concept of information granules. Pawlak, a scientist, proposed the rough set theory [9] and expressed that knowledge is granular. Since Professor Lin put forward the concept of granular computing, the research of granular computing has developed rapidly, such as hierarchical classification [10], granular regression [11], aggregation algorithm [12], shadowed sets [13], two-way learning [14], and so on. The classical naive Bayes corresponds to different algorithms when dealing with different data types. Multinomial naive Bayes algorithm is used when dealing with text data, and Gaussian Naive Bayes algorithm is used that deal with continuous data.

The naive Bayes classification algorithm has a good classification effect when dealing with text data, but when dealing with data-type data, the assumption of independence between features will lead to low classification accuracy [15]. To solve this problem, we define the operation and measurement of granules in the single feature classification neighborhood system. We further define the prior probability of the granular vector and the conditional probability of the granular vector. We propose the concept of naive Bayes granular vector, to transform the classification problem into the probability problem of naive Bayes granular vector, and construct the naive Bayes granular classifier model. At the same time, the naive Bayes granular classifier is designed and verified experimentally on several UCI data sets. Theoretical analysis and experimental results show that the naive Bayes classifier based on neighborhood granulation can improve classification accuracy and achieve better classification results with appropriate neighborhood granulation parameters.

2   Granular Representation and Neighborhood Granulation

Elements in a set are not repeatable and have no order. However, granules are ordered and repeatable finite collections and granular vectors are composed of ordered granules. The neighborhood granulation method is to granulate the sample into neighborhood granules, whose values are discrete values of 0 or 1. The neighborhood granular vector is composed of neighborhood granules. The neighborhood granulation on a single feature forms neighborhood granules, and the granules on multiple features are combined into neighborhood granular vectors.

Definition 1. *Given a big data learning system* $U = (X \cup P, C, d)$*, for two samples* $x, y \in X$*, a single feature* $a \in C$*, and a given neighborhood parameter* δ*, the neighborhood discriminant function of the sample* x*,* y *is defined as:*

$$\varphi_a(x,y) = \begin{cases} 0, s_a(x,y) > \delta \\ 1, s_a(x,y) \leq \delta \end{cases} \qquad (1)$$

where $s_a(x,y) = 1 - |v(x,a) - v(y,a)|$ *is the neighborhood metric of* x *and* y *in single feature* a*. When* $\varphi_a(x,y) = 1$*,* x *and* y *are neighbors; When* $\varphi_a(x,y) = 0$*,* x *and* y *are not neighbors.*

Definition 2. *Given a big data learning system* $U = (X \cup P, C, d)$, *for any sample* $x \in X$ *and reference sample set* $P = \{p_1, p_2, ..., p_k\}$, *and any single feature* $c \in C$ *then* x *performs neighborhood granulation on the single feature sample* c, *and the conditional granule formed is defined as:*

$$g = g_c(x) = \{r_j\}_{j=1}^{k} = \{r_1, r_2, ..., r_k\} \tag{2}$$

where $r_j = \varphi_c(x, p_j)$ *is the neighborhood discriminant function of the sample* x, p_j *on the single feature* c.

Definition 3. *Let* $U = (X \cup P, C, d)$ *be a big data learning system, for any sample* $x \in X$, *any feature subset* $A \subseteq C$, *suppose* $A = \{a_1, a_2, ..., a_m\}$, *then the granular vector of* x *on the feature subset* A *is defined as:*

$$G = G_A(x) = (g_{a_1}(x), g_{a_2}(x), ..., g_{a_m}(x))^T \tag{3}$$

where $g_{a_m}(x)$ *is the granule of sample* x *on feature* a_m. *For convenience, the feature set* $A = \{a_1, a_2, ..., a_m\}$ *is marked with an integer, and the particle vector can be expressed as:*

$$G(x) = (g_1(x), g_2(x), ..., g_m(x))^T \tag{4}$$

The granular vector is made up of granules, which in turn are made up of granular nuclei. Therefore, the granular vector can be in the form of a granular core matrix, which is expressed as:

$$G(x) = \begin{bmatrix} g_1(x)_1, g_1(x)_2, ..., g_1(x)_k \\ g_2(x)_1, g_2(x)_2, ..., g_2(x)_k \\ ... \\ g_m(x)_1, g_m(x)_2, ..., g_m(x)_k \end{bmatrix} = \begin{bmatrix} r_{11}, r_{12}, ..., r_{1k} \\ r_{21}, r_{22}, ..., r_{2k} \\ ... \\ r_{m1}, r_{m2}, ..., r_{mk} \end{bmatrix} \tag{5}$$

The granular vector can also be expressed in another form as:

$$G(x) = (g(x)_1, g(x)_2, ..., g(x)_k) \tag{6}$$

where

$$g(x)_j = (g_1(x)_j, g_2(x)_j, ..., g_m(x)_j)^T \tag{7}$$

3   Granular Operations and Measures

The addition, subtraction, multiplication, and division of real numbers are closed on real numbers, and the defined granular operation operators should also be closed on granules.

Definition 4. *Let* $s = \{s_j\}_{j=1}^{n}$ *and* $t = \{t_j\}_{j=1}^{n}$ *be two granules, then the addition operation of two granules is defined as:*

$$s + t = \{s_j + t_j\}_{j=1}^{n} = \{s_1 + t_1, s_2 + t_2, ..., s_n + t_n\} \tag{8}$$

Definition 5. *Let* $s = \{s_j\}_{j=1}^n$ *and* $t = \{t_j\}_{j=1}^n$ *be two granules, then the multiplication operation of two granules is defined as:*

$$s * t = \{s_j * t_j\}_{j=1}^n = \{s_1 * t_1, s_2 * t_2, ..., s_n * t_n\} \tag{9}$$

Definition 6. *Let* $s = \{s_j\}_{j=1}^n$ *and* $t = \{t_j\}_{j=1}^n$ *be two granules, then the division operation of two granules is defined as:*

$$s / t = \{s_j / t_j\}_{j=1}^n = \{s_1/t_1, s_2/t_2, ..., s_n/t_n\} \tag{10}$$

Definition 7. *Let* $o = \{o_j\}_{j=1}^n$ *denote the output granule of the granular classifier and* $y = \{y_j\}_{j=1}^n$ *denote the decision granule, then the 0–1 loss function metric of the two granules is defined as:*

$$L(y, o) = \left\{ \begin{cases} 1, y_j \neq o_j \\ 0, y_j = o_j \end{cases} \right\}_{j=1}^n \tag{11}$$

The loss value is 0 if the elements of the granules are the same, and 1 if they are different.

Definition 8. *Let* $G = (g_1, g_2, ..., g_l)$ *be the input granular vector of the granular classifier, where* $g_i = \{r_j\}_{j=1}^n$ *represents the input granule, then* p_i *is the probability granule represented as:*

$$p_i = \frac{g_i}{\sum_{i=1}^l g_i} \tag{12}$$

4 A Naive Bayes Classifier Model Based on Neighborhood Granulation

The naive Bayes classifier based on neighborhood granulation inputs the granular feature vector, after granular computing, compares the probabilities of multiple granules and outputs the maximum probability granule. Therefore, a naive Bayes classifier based on neighborhood granulation can be used for multi-classification problems.

Definition 9. *Let the granule type be* Y *and* $y = \{c_k\}_{k=1}^n$ *denote the decision granule, then the prior probability of the granular vector is expressed as:*

$$P(Y = c_k) = \frac{\sum_{i=1}^{N_g} I_g(y_i = c_k)}{N_g} \tag{13}$$

where I_g *are the number of* c_k *in the granules and* N_g *are the total number of granules.*

Definition 10. *Let $G = (g_1, g_2, ..., g_j)$ be the input granular vector of the granular classifier, where $g_i = \{r_l\}_{j=1}^n$ represents the input granule and $y = \{c_k\}_{k=1}^n$ represents the decision granule, then the conditional probability of the granular vector is expressed as:*

$$P(G^{(j)} = r_{jl} \mid Y = c_k) = \frac{\sum_{i=1}^N I(g_i^{(j)} = r_{jl}, y_i = c_k)}{\sum_{i=1}^N I(y_i = c_k)} \tag{14}$$

In actual data training, there may be zero probability problems. The zero probability problem is that when calculating the conditional probability of a granular vector, the granules of a certain granular vector have never appeared in the training set, which will cause the calculation result of the conditional probability of the entire granular vector to be 0. To solve this problem, the conditional probability formula of the granular vector is added by 1 through the idea of Laplace smoothing. The value is converted to a value between [1, N + 1]. The smoothing formula for granules can be expressed as:

$$P_i = P_i \times N + 1 \tag{15}$$

Definition 11. *Let $G = (g_1, g_2, ..., g_j)$ be the input granular vector of the granular classifier, where $g_i = \{r_l\}_{j=1}^n$ represents the input granule and $y = \{c_k\}_{k=1}^n$ represents the decision granule, then the posterior probability of the granular vector is expressed as:*

$$P(Y = c_k) \prod_{j=1}^n P(G^{(j)} = g^j \mid Y = c_k) \tag{16}$$

Definition 12. *Let $G = (g_1, g_2, ..., g_j)$ be the input granular vector of the granular classifier, where $g_i = \{r_l\}_{j=1}^n$ represents the input granule and $y = \{c_k\}_{k=1}^n$ represents the decision granule, then the maximum a posterior probability of the granular vector is expressed as:*

$$y = \arg\max_{c_k} P(Y = c_k) \prod_{j=1}^n P(G^{(j)} = g^j \mid Y = c_k) \tag{17}$$

According to the previous theory and principle, the naive Bayes classifier model based on neighborhood granulation is constructed. The granule is a structured representation. The components of granules are independently calculated, and the granules can be separated and combined, which is the essence of the naive Bayes classifier based on neighborhood granulation. The specific naive

Bayes classifier algorithm based on neighborhood granulation is described as shown in Algorithm 1.

Algorithm 1: The naive Bayes classifier algorithm based on neighborhood granulation (NGNB)

Input: A training data set $T = (U, F, L)$, the instance t, and the neighborhood parameter δ

Output: Classification of the instance t

1 Normalizing the data;
2 Circularly executing the step 3 to 5 for each training data $x \in U$;
3 The neighborhood granulation δ is performed on each single characteristic $a_i \in F$ to be $g_{a_i}(x)_\delta$;
4 Form a neighborhood granular vector $G_F(x)_\delta = \{g_{a_1}(x)_\delta, g_{a_2}(x)_\delta, ..., g_{a_m}(x)_\delta\}$ of x;
5 Obtain granular label L_x and construct granular vector rule $R(x) = \langle G_F(x), L_x \rangle$;
6 Compute the prior probability of the granular vector and the conditional probability of the granular vector according to Definition 9 and Definition 10, respectively;
7 The conditional probability of the granular vector is processed according to Eq. 15;
8 Granulation is performed in the training data set for a given instance t, forming an instance granular vector $G_F(t)_\delta$;
9 The granular vector $G_F(t)_\delta$ is calculated according to Definition 11;
10 Determine the class of the instance granular vector t according to Definition 12.

---

5 Experiments and Analysis

In this paper, four data sets in the UCI data set are used as data sources for experimental tests, as shown in Table 1.

Table 1. The description of data sets.

Datasets	Samples	Features	Categories
Iris	150	4	3
Wine	178	13	3
Lymphography	148	18	4
Clean1	476	166	2

Due to the different widths of sample values in the data sets in Table 1, the data sets need to be preprocessed with maximum and minimum normalization. The formula is:

$$X_{norm} = \frac{X - X_{min}}{X_{max} - X_{min}} \tag{18}$$

The data is neighborhood granulated on each single-sample feature to form a neighborhood granular vector. To test the classification accuracy of the naive Bayes classifier based on neighborhood granulation, each data set was randomly divided into a 0.7 training set and a 0.3 testing set.

The neighborhood granulation parameter is a measure of the distance between granules, and the construction of the neighborhood granular vector requires neighborhood parameters, so different neighborhood parameters will affect the accuracy of classification. The experiment adopts the control variable method. The variable is the neighborhood parameter and compares the neighborhood granulation naive Bayes algorithm (NGNB) with the classical naive Bayes algorithm (NB). The experiment set the values of neighborhood parameters from 0.05 to 0.95 and set the interval to 0.05. The classification results of the four UCI data sets are shown in Fig. 1, 2, 3 and Fig. 4.

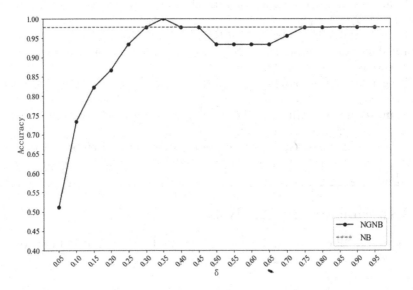

Fig. 1. Classification accuracy of different neighborhood parameters on Iris data set.

From the analysis of Fig. 1 and Fig. 2, it can be seen that the accuracy curve of the naive Bayes classifier based on neighborhood granulation is similar to the curve of the logarithmic function. In the Iris data set, the NGNB and the NB have 8 times the same classification accuracy, but when the neighborhood parameter is 0.35, the classification accuracy of NGNB reaches the maximum value of 1, exceeding the classification accuracy of NB. In the Wine data set, starting from the neighborhood parameter of 0.35, the classification accuracy of NGNB is not lower than that of NB, and 8 of them are higher than the classification accuracy of NB, and the classification accuracy reaches the maximum value of 1 when the neighborhood parameter is 0.85.

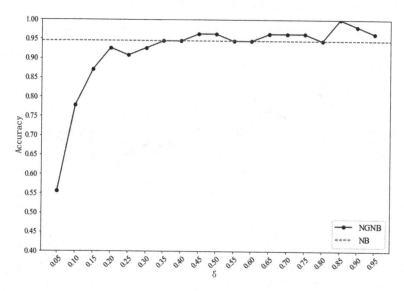

Fig. 2. Classification accuracy of different neighborhood parameters on Wine data set.

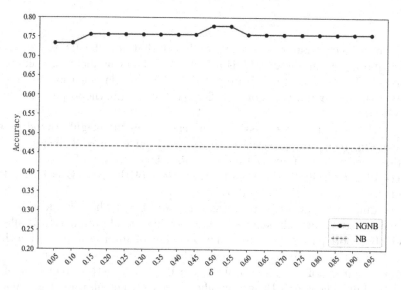

Fig. 3. Classification accuracy of different neighborhood parameters on Lymphography data set.

From the analysis in Fig. 3, for the Lymphography data set, the NGNB algorithm outperforms the NB algorithm in all neighborhood parameters. The classification accuracy of the NGNB algorithm reaches the maximum value of 0.7778 when the neighborhood parameters are 0.50 and 0.55. And it is not sensitive to the neighborhood parameters, and the accuracy is relatively stable.

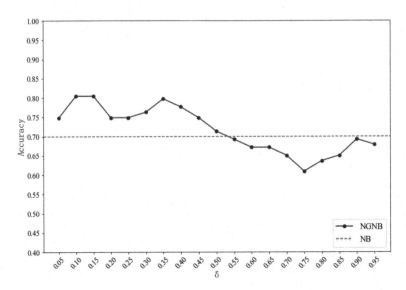

Fig. 4. Classification accuracy of different neighborhood parameters on Lymphography data set.

It can be seen from the analysis of Fig. 4 that for the Clean1 data set, the neighborhood parameter 0.55 is a boundary. The classification accuracy of NGNB is higher than that of NB from 0.05 to 0.5. The classification accuracy of NGNB is the maximum value of 0.8042 when the neighborhood parameters are 0.1 and 0.15.

The value with the best classification accuracy in the neighborhood parameters experiment is compared with the standard naive Bayes. The specific classification results of the four UCI data sets are shown in Table 2.

According to Table 2, the classification effect of the NGNB is better than that of the NB under the optimal parameters.

According to the analysis from Fig. 1, 2, 3 and Fig. 4, the classification accuracy of the naive Bayes classifier based on neighborhood granulation is affected by neighborhood parameters, and the selection of appropriate neighborhood parameters is the key to the accuracy of classification. In most cases, the classification accuracy of the NGNB algorithm is slightly better than that of the NB algorithm. The naive Bayes classifier based on neighborhood granulation is different from the classical naive Bayes classifier. NGNB uses neighborhood granulation technology to improve the structure so that the data can better meet the requirements of the algorithm and improve the accuracy of the algorithm on different data sets.

Table 2. The specific experiment results.

Datasets	NGNB	NB
Iris	1.0000	0.9778
Wine	1.0000	0.9444
Lymphography	0.7778	0.4667
Clean1	0.8042	0.6993

6   Conclusions

Starting from the study of granular computing, this paper proposes a naive Bayes classifier based on neighborhood granulation. First, the concept of neighborhood rough set is introduced for neighborhood granulation. Granular vectors are constructed in the classification system, the prior probability and conditional probability of the granular vectors are defined, and the naive Bayes classifier based on neighborhood granulation is designed. Through experiments on some UCI data sets and comparison experiments with the classical naive Bayes algorithm, the effectiveness of the algorithm is verified, and better classification performance can be achieved with appropriate granulation parameters. In future work, we will further explore the use of more granulation methods in naive Bayes classifiers, using rotational granulation, for the construction of naive Bayes classifiers based on rotational granulation.

Acknowledgement. This work is supported by the National Natural Science Foundation of China (No. 61976183), the Joint Project of Production, Teaching and Research of Xiamen (No. 2022CXY04028) and the School-Level Teaching Reform Project of Xiamen University of Technology (No. JG2021028).

References

1. Blanquero, R., Carrizosa, E., Ramirez-Cobo, P., Sillero-Denamiel, M.: Variable selection for Naive Bayes classification. Comput. Oper. Res. **135**, 105456 (2021)
2. Ruan, S., Chen, B., Song, K., Li, H.: Weighted Naive Bayes text classification algorithm based on improved distance correlation coefficient. Neural. Comput. Appl. **34**(4), 2729–2738 (2022)
3. Kim, H., Park, J., Kim, D., Lee, J.: Multilabel Naive Bayes classification considering label dependence. Pattern Recognit. Lett. **136**, 279–285 (2020)
4. Blanquero, R., Carrizosa, E., Ramirez-Cobo, P., Sillero-Denamiel, M.: Constrained Naive Bayes with application to unbalanced data classification. Cent. Eur. J. Oper. Res. **30**, 1403–1425 (2021)
5. Xiong, Y., Ye, M., Wu, C.: Cancer classification with a cost-sensitive Naive Bayes stacking ensemble. Comput. Math. Methods Med. **2021**, 5556992 (2021)
6. Chen, H., Hu, S., Hua, R., Zhao, X.: Improved naive Bayes classification algorithm for traffic risk management. EURASIP J. Adv. Sig. Process. **2021**(1), 1–12 (2021). https://doi.org/10.1186/s13634-021-00742-6

7. Jiang, L., Zhang, L., Yu, L., Wang, D.: Class-specific attribute weighted Naive Bayes. Pattern Recognit. **88**, 321–330 (2019)
8. Zadeh, L.A.: Fuzzy sets. Inf. Control **8**, 338–353 (1965)
9. Pawlak, Z.: Rough sets. Int. J. Comput. Inf. Sci. **11**, 341–356 (1982)
10. Guo, S., Zhao, H.: Hierarchical classification with multi-path selection based on granular computing. Artif. Intell. Rev. **54**(3), 2067–2089 (2021)
11. Chen, Y., Miao, D.: Granular regression with a gradient descent method. Inf. Sci. **537**, 246–260 (2020)
12. Liu, N., Xu, Z., Wu, H., Ren, P.: Conversion-based aggregation algorithms for linear ordinal rankings combined with granular computing. Knowl. Based Syst. **219**, 106880 (2021)
13. Zhou, J., Lai, Z., Miao, D., Gao, C., Yue, X.: Multigranulation rough-fuzzy clustering based on shadowed sets. Inf. Sci. **507**, 553–573 (2020)
14. Xu, W., Li, W.: Granular computing approach to two-way learning based on formal concept analysis in fuzzy datasets. IEEE Trans. Cybern. **46**(2), 336–379 (2016)
15. Jing, L., Li, C., Wang, S., Zhang, L.: Deep feature weighting for Naive Bayes and its application to text classification. Eng. Appl. Artif. Intell. **52**, 26–39 (2016)

Matrix Representations and Interdependency on an *L*-fuzzy Covering-Based Rough Set

Wei Li and Bin Yang[✉]

College of Science, Northwest A & F University,
Yangling, Xianyang 712100, People's Republic of China
binyang0906@nwsuaf.edu.cn

Abstract. In this paper, the matrix representations and interdependency of a pair of *L*-fuzzy covering-based approximation operators are investigated. The aim of matrix representations of lower and upper approximation operators is to make calculation more valid by means of operations on matrices. Furthermore, in accordance with the concept of β-base, we give a necessary and sufficient condition under what two *L*-fuzzy β-coverings can generate the same lower and upper approximation operations.

Keywords: Fuzzy covering rough set · *L*-fuzzy β-covering · *L*-fuzzy β-neighborhood · Residuated lattice · Matrix representation · Interdependency

1 Introduction

The classical rough set theory was proposed by Pawlak [18], in which the relationship of objects were built on equivalence relation. However, the equivalence relation imposes restrictions on the actual application process, so many scholars have carried out generalized researches on rough sets from other aspects [1,2,5,19,20,22,26]. One of the generalized researches is the covering-based rough set model by relaxing the partitions (which constituted by the equivalence class based on equivalence relation) to coverings. In the frame of covering-based rough set, Pomykala [19,20] proposed two pairs of dual rough approximation operators. In 1998, Yao [30] studied a type of covering-based rough set model in terms of the neighborhood operators. In addition, Zhu et al. [32] discussed three kinds of covering-based rough set to deal with the vagueness and granularity in information systems. In accordance with neighborhood system, Yao and Yao [31] introduced a framework to further explore the covering-based approximations. By the concept of complementary neighborhood, Ma [12] in 2012 considered some types of neighborhood-related covering-based rough set. Furthermore, Ma [13] defined the twin approximation operators by the notions of neighborhood and complementary neighborhood. Later, Han [7] generalized the (finite) covering approximation space to locally finite covering approximation space.

However, the covering-based rough set faces stringent limitation when dealing with real-valued data sets, which can only process the qualitative (discrete) data [8]. Fuzzy covering-based rough set was constructed by fuzzy rough set [4,16,17,21] based on fuzzy covering, which can be seen as a bridge linking covering rough set theory and

J. Yao et al. (Eds.): IJCRS 2022, LNAI 13633, pp. 143–157, 2022.
https://doi.org/10.1007/978-3-031-21244-4_11

fuzzy rough set theory. From the perspective of model structure, the fuzzy covering-based rough set model can be seen as a model by using fuzzy neighborhood operator based on fuzzy coverings instead of the general fuzzy binary relation in the fuzzy rough set model. Further, Ma [14] in 2016 generalized the notion of fuzzy covering to fuzzy β-covering and proposed the concept of fuzzy β-neighborhood. So the fuzzy β-covering-based rough set can be viewed as an extension of the fuzzy covering approximation space.

In the meantime, some lattice structures were discussed to replace the interval $[0, 1]$ as the truth table for membership degrees, among which residuated lattices [25] play a significant role. In general, the residuated lattice-valued fuzzy approximation operators are discussed from two directions: L-fuzzy relation-based rough approximation operators and L-fuzzy β-covering-based approximation operators. In 2007, Deng et al. [3] explored a pair of L-fuzzy covering-based rough approximation operators. Later, Li et al. [10] introduced another two pairs of L-fuzzy covering-based approximation operators when $L = [0, 1]$. In addition, Jin and Li [9] gave two L-fuzzy covering-based rough approximation operators with the condition of completely distributive complete lattice.

Starting from the axiomatic approaches of fuzzy operators, Li et al. [11] recently studied three pairs of L-fuzzy covering-based approximation operators, which are presented in [3, 10], and further analysed the differences between the axiom sets of L-fuzzy covering-based approximation operators and their crisp counterparts. In addition, Yang and Hu [28] studied the matrix representations and interdependency of the three pairs of L-fuzzy covering-based approximation operators. Based on the work of Ma [14], Yang [27] introduced the notion of L-fuzzy β-covering and proposed a pair of L-fuzzy β-covering-based approximation operators. Further, the basic properties and axiomatic representation are studied. However, the matrix representations and interdependency of this pair of L-fuzzy β-covering-based approximation operators are not considered.

In this paper, we give the matrix representations of a pair of L-fuzzy β-covering-based approximation operators, which discussed in [27]. In addition, to solve the issue that under what two L-fuzzy β-coverings can generate the same L-fuzzy β-covering-based approximation operators, a necessary and sufficient condition is proposed by redefining the β-independent element.

The organization of this paper is introduced as follows: In Sect. 2, we review some basic concepts of residuated lattice, L-fuzzy β-covering and L-fuzzy β-neighborhood. In Sect. 3, the matrix representations of L-fuzzy β-covering-based approximation operators are proposed. Section 4 explores the necessary and sufficient condition under what two L-fuzzy β-coverings can generate the same L-fuzzy β-covering-based approximation operators. Section 5 concludes this paper.

2 Preliminaries

This section recapitulates some well-known concepts that shall be used in the sequel.

A complete residuated lattice [25] is a pair $L = (L, \otimes)$ subject to the following conditions:

(1) L is a complete lattice with a top element 1 and a bottom element 0;
(2) $(L, \otimes, 1)$ is a commutative monoid;

(3) $a \otimes \bigvee_{j \in J} b_j = \bigvee_{j \in J} (a \otimes b_j)$ for all $a \in L$ and $\{b_j : j \in J\} \subseteq L$.

The binary operation $\otimes$ induces another binary operation $\rightarrow$ on L via the adjoint property:

$$a \otimes b \leq c \Longleftrightarrow b \leq a \rightarrow c.$$

In this paper, if not otherwise specified, $L = (L, \wedge, \vee, \otimes, \rightarrow, 0, 1)$ is always a complete residuated lattice. In addition, a function $A : U \longrightarrow L$ is an L-fuzzy set in U. L^U is denoted as the set of all L-fuzzy sets in U and called the L-fuzzy power set on U. The operators $\vee, \wedge, \otimes, \rightarrow$ on L can be translated onto L^U in a pointed wise, i.e.,

$$A \leq B \Longleftrightarrow A(x) \leq B(x),$$

$$(\bigwedge_{t \in T} A_t)(x) = \bigwedge_{t \in T} A_t(x), \ (\bigvee_{t \in T} A_t)(x) = \bigvee_{t \in T} A_t(x),$$

$$(A \otimes B)(x) = A(x) \otimes B(x), \ (A \rightarrow B)(x) = A(x) \rightarrow B(x).$$

where $A, B, A_t \ (t \in T) \in L^U$ and $x \in U$. For a crisp subset $A \subseteq U$, let 1_A be the characteristic function, i.e., $1_A(x) = 1$ if $x \in A$ and $1_A(x) = 0$ if $x \notin A$. Clearly, the characteristic function 1_A of a subset $A \subseteq U$ can be regarded as an L-fuzzy set in U. Thus, when $L = \{0, 1\}$, the set L^U degenerates into the power set $\mathscr{P}(U)$ of U if we make no difference between a subset of U and its characteristic function.

The following lemma shows some basic properties of residuated lattices.

Lemma 1. *[6, 15, 23] Suppose that $L = (L, \wedge, \vee, \otimes, \rightarrow, 0, 1)$ is a residuated lattice. For any $x, y, z \in L$, $\{x_i\}_{i \in I}$ and $\{y_i\}_{i \in I} \subseteq L$, the following statements hold.*

(1) $x \otimes (y \rightarrow z) \leq (x \rightarrow y) \rightarrow z$.
(2) $\bigvee_{i \in I} (x_i \rightarrow y) \leq \bigwedge_{i \in I} x_i \rightarrow y$.
(3) If $x_1 \leq x_2$*, then* $x_2 \rightarrow y \leq x_1 \rightarrow y$ *and* $y \rightarrow x_1 \leq y \rightarrow x_2$.
(4) $y \otimes (\bigwedge_{i \in I} x_i) \leq \bigwedge_{i \in I} (y \otimes x_i)$.
(5) $y \otimes (\bigvee_{i \in I} x_i) = \bigvee_{i \in I} (y \otimes x_i)$.
(6) $x \rightarrow (y \rightarrow z) = (x \otimes y) \rightarrow z = y \rightarrow (x \rightarrow z)$.
(7) $\bigvee_{i \in I} x_i \rightarrow y = \bigwedge_{i \in I} (x_i \rightarrow y)$.
(8) $y \rightarrow \bigwedge_{i \in I} x_i = \bigwedge_{i \in I} (y \rightarrow x_i)$.
(9) $\bigwedge_{y \in L} ((x \rightarrow y) \rightarrow y) = x$.
(10) $y \rightarrow \bigvee_{i \in I} x_i \geq \bigvee_{i \in I} (y \rightarrow x_i)$.
(11) $0 \otimes x = 0$.
(12) $(x \otimes y) \leq (x \wedge y)$.
(13) $(x \rightarrow (x \otimes y)) \geq y$.
(14) $(x \otimes (x \rightarrow y)) \leq y$.
(15) $(1 \rightarrow x) = x, (x \rightarrow x) = 1, (0 \rightarrow 1) = 1, (0 \rightarrow 0) = 1$.

(16) $(x \to y) = 1 \Longleftrightarrow x \le y.$
(17) $\neg\neg x \ge x.$
(18) $(x \to y) \le (\neg y \to \neg x).$
(19) $((x \to y) \otimes (y \to z)) \le (x \to z).$
(20) $(x \to \neg y) = \neg(x \otimes y).$
(21) $\neg(\bigvee_{i \in I} x_i) = \bigwedge_{i \in I}(\neg x_i).$
(22) $\neg(\bigwedge_{i \in I} x_i) \ge \bigvee_{i \in I}(\neg x_i).$
(23) $(\bigwedge_{i \in I} x_i) \to (\bigvee_{i \in I} y_i) \ge \bigvee_{i \in I}(x_i \to y_i).$

Next, some related concepts about *L-fuzzy β-covering* and *L-fuzzy β-neighborhood* are given below.

Definition 1. *[27] Let $L = (L, \bigwedge, \bigvee, \otimes, \to, 0, 1)$ be a residuated lattice and U be a non-empty finite set. For any $\beta > 0$ and $\beta \in L$, we call $\mathscr{C} = \{C_1, C_2, \ldots, C_m\}$ with $C_i \in L^U$ $(i = 1, 2, \ldots, m)$, L-fuzzy β-covering of U, if $(\bigcup_{i=1}^{m} C_i)(x) \ge \beta$ for any $x \in U$. Further, $(U, \mathscr{C})$ is called an L-fuzzy β-covering approximation space.*

Note that *L-fuzzy covering* [3] is one of the special cases of *L-fuzzy β-covering* when $\beta = 1$.

Definition 2. *[29] Let $(U, \mathscr{C})$ be an L-fuzzy β-covering approximation space with $\mathscr{C} = \{C_1, C_2, \ldots, C_m\}$ being an L-fuzzy β-covering of U for $\beta > 0$ $(\beta \in L)$. For each $x \in U$,*

$$N_{\mathscr{C}}^{\beta}(x) = \{C_i \in \mathscr{C} : C_i(x) \ge \beta\}$$

is called the L-fuzzy β-neighborhood system of x.

Definition 3. *Let $(U, \mathscr{C})$ be an L-fuzzy β-covering approximation space with $\mathscr{C} = \{C_1, C_2, \ldots, C_m\}$ being an L-fuzzy β-covering of U for $\beta > 0$ $(\beta \in L)$. For each $x \in U$,*

$$N_x^{\beta} = \bigcap\{C_i \in \mathscr{C} : C_i(x) \ge \beta\} = \bigcap N_{\mathscr{C}}^{\beta}(x)$$

is called the L-fuzzy β-neighborhood of x.

Note that $(N_x^{\beta})'(y)$ is called *L-fuzzy β-neighborhood* of x induced by *L-fuzzy β-covering* $\mathscr{C}'$, i.e.,

$$(N_x^{\beta})'(y) = \left(\bigcap\{C_i \in \mathscr{C}' : C_i(x) \ge \beta\}\right)(y) = \left(\bigcap N_{\mathscr{C}'}^{\beta}(x)\right)(y).$$

Based on the concept of *L-fuzzy β-neighborhood*, Yang introduced a pair of *L-fuzzy covering-based approximation operators* as follows in [27].

Definition 4. *[27] Let $(U, \mathscr{C})$ be an L-fuzzy β-covering approximation space. For any $X \in L^U$, the L-fuzzy covering-based lower approximation operator $\underline{\mathscr{C}}(X)$ and upper approximation operator $\overline{\mathscr{C}}(X) : L^U \longrightarrow L^U$ are defined as follows:*

$$\underline{\mathscr{C}}(X)(x) = \bigwedge_{y \in U} [N_x^{\beta}(y) \to X(y)], \tag{1}$$

$$\overline{\mathscr{C}}(X)(x) = \bigvee_{y \in U} [N_x^\beta(y) \otimes X(y)], \tag{2}$$

where $x \in U$. If $\underline{\mathscr{C}}(X) \neq \overline{\mathscr{C}}(X)$, then X is denoted as an L-fuzzy covering-based rough set.

3   The Matrix Representations of L-fuzzy β-covering-based Approximation Operators

In this section, we present the matrix representations of L-fuzzy β-covering-based approximation operators defined in Definition 4. Firstly, we propose some relevant notions and notations.

Definition 5. Let $\mathscr{C} = \{C_1, C_2, \ldots, C_m\}$ be an L-fuzzy β-covering of $U = \{x_1, x_2, \ldots, x_n\}$ and $X \in L^U$. We denote the matrix representation of $\mathscr{C}$ by $M_{\mathscr{C}} = (C_j(x_i))_{n \times m}$, and X by $M_X = (X(x_i))_{n \times 1}$.

Further, let $M = (a_{ij})_{s \times t}$ be a matrix, and its transpose is denoted as $M^T = (a_{ji})_{t \times s}$ in the following discussions.

Example 1. Let $U = \{x_1, x_2, x_3, x_4, x_5, x_6\}$ and $L = ([0, 1], \otimes, \rightarrow, \bigvee, \bigwedge, 0, 1)$ be a Gödel-residuated lattice, where

$$x \otimes y = x \bigwedge y \text{ and } x \rightarrow y = \begin{cases} 1, & x \leq y, \\ y, & x > y. \end{cases} (\forall x, y \in L)$$

A family $\mathscr{C} = \{C_1, C_2, C_3, C_4\}$ of L-fuzzy subsets of U is listed below.

$$C_1 = \frac{0.7}{x_1} + \frac{0.5}{x_2} + \frac{0.4}{x_3} + \frac{0.6}{x_4} + \frac{0.3}{x_5} + \frac{0.2}{x_6}, C_2 = \frac{0.5}{x_1} + \frac{0.4}{x_2} + \frac{0.6}{x_3} + \frac{0.5}{x_4} + \frac{0.7}{x_5} + \frac{0.3}{x_6},$$

$$C_3 = \frac{0.3}{x_1} + \frac{0.7}{x_2} + \frac{0.1}{x_3} + \frac{0.5}{x_4} + \frac{0.6}{x_5} + \frac{0.3}{x_6}, C_4 = \frac{0.6}{x_1} + \frac{0.6}{x_2} + \frac{0.2}{x_3} + \frac{0.3}{x_4} + \frac{0.5}{x_5} + \frac{0.7}{x_6}.$$

According to Definition 1, $\mathscr{C}$ is an L-fuzzy β-covering of U for any $\beta \in (0, 0.6]$, then the matrix representations of $M_{\mathscr{C}}$ and M_{C_3} can be expressed as follows.

$$M_{\mathscr{C}} = \begin{pmatrix} 0.7 \ 0.5 \ 0.3 \ 0.6 \\ 0.5 \ 0.4 \ 0.7 \ 0.6 \\ 0.4 \ 0.6 \ 0.1 \ 0.2 \\ 0.6 \ 0.5 \ 0.5 \ 0.3 \\ 0.3 \ 0.7 \ 0.6 \ 0.5 \\ 0.2 \ 0.3 \ 0.3 \ 0.7 \end{pmatrix} \text{ and } M_{C_3} = \begin{pmatrix} 0.3 \\ 0.7 \\ 0.1 \\ 0.5 \\ 0.6 \\ 0.3 \end{pmatrix}.$$

Definition 6. Let $A = (a_{ik})_{n \times m}$ and $B = (b_{kj})_{m \times l}$ be two lattice valued matrices. Then, $C = A \circ B = (c_{ij})_{n \times l}$ and $D = A \bullet B = (d_{ij})_{n \times l}$ can be stipulated as follows, respectively,

$$c_{ij} = \bigvee_{k=1}^{m} (a_{ik} \otimes b_{kj}), i = 1, 2, \ldots, n, j = 1, 2, \ldots, l,$$

$$d_{ij} = \bigwedge_{k=1}^{m} (a_{ik} \rightarrow b_{kj}), i = 1, 2, \ldots, n, j = 1, 2, \ldots, l.$$

Definition 7. *Let* $U = \{x_1, x_2, \ldots, x_n\}$ *be a finite universe and* $\mathscr{C} = \{C_1, C_2, \ldots, C_m\}$ *be an L-fuzzy β-covering of U.* $M_{\mathscr{C}} = (C_j(x_i))_{n \times m}$ *is denoted as a matrix representation of* $\mathscr{C}$, *and the Boolean matrix* $M_\beta = (t_{ij})_{n \times m}$ *is denoted as a β-matrix representation of* $\mathscr{C}$, *where*

$$t_{ij} = \begin{cases} 1, C_j(x_i) \geq \beta, \\ 0, C_j(x_i) < \beta. \end{cases}$$

Example 2. Let $(U, \mathscr{C})$ be the L-fuzzy β-covering approximation space in Example 1. For two orders $\{C_1, C_2, C_3, C_4\}$ and $\{C_1, C_4, C_3, C_2\}$ of $\mathscr{C}$, $M_{\mathscr{C}}, N_{\mathscr{C}}$ are both matrix representations of $\mathscr{C}$, while $M_{0.6}, N_{0.6}$ are both 0.6-matrix representations of $\mathscr{C}$, where

$$M_{\mathscr{C}} = \begin{pmatrix} 0.7 \ 0.5 \ 0.3 \ 0.6 \\ 0.5 \ 0.4 \ 0.7 \ 0.6 \\ 0.4 \ 0.6 \ 0.1 \ 0.2 \\ 0.6 \ 0.5 \ 0.5 \ 0.3 \\ 0.3 \ 0.7 \ 0.6 \ 0.5 \\ 0.2 \ 0.3 \ 0.3 \ 0.7 \end{pmatrix}, \quad N_{\mathscr{C}} = \begin{pmatrix} 0.7 \ 0.6 \ 0.3 \ 0.5 \\ 0.5 \ 0.6 \ 0.7 \ 0.4 \\ 0.4 \ 0.2 \ 0.1 \ 0.6 \\ 0.6 \ 0.3 \ 0.5 \ 0.5 \\ 0.3 \ 0.5 \ 0.6 \ 0.7 \\ 0.2 \ 0.7 \ 0.3 \ 0.3 \end{pmatrix},$$

and

$$M_{0.6} = \begin{pmatrix} 1 \ 0 \ 0 \ 1 \\ 0 \ 0 \ 1 \ 1 \\ 0 \ 1 \ 0 \ 0 \\ 1 \ 0 \ 0 \ 0 \\ 0 \ 1 \ 1 \ 0 \\ 0 \ 0 \ 0 \ 1 \end{pmatrix}, \quad N_{0.6} = \begin{pmatrix} 1 \ 1 \ 0 \ 0 \\ 0 \ 1 \ 1 \ 0 \\ 0 \ 0 \ 0 \ 1 \\ 1 \ 0 \ 0 \ 0 \\ 0 \ 0 \ 1 \ 1 \\ 0 \ 1 \ 0 \ 0 \end{pmatrix}.$$

As we can see from the above example that for a fixed order of all elements of U, different orders of $\mathscr{C}$ lead to different matrix representations, and the matrices can be transformed into each other by list exchanges. Next, the relationship between matrix representations under different order of $\mathscr{C}$ are given.

Proposition 1. *Let* $U = \{x_1, x_2, \ldots, x_n\}$ *be a finite universe of which the order of elements is given, and* $\mathscr{C} = \{C_1, C_2, \ldots, C_m\}$ *be an L-fuzzy β-covering of U. Based on two different orders of* $\mathscr{C}$, $M_{\mathscr{C}}, N_{\mathscr{C}}$ *are two matrix representations of* $\mathscr{C}$ *and* M_β, N_β *are two β-matrix representations of* $\mathscr{C}$, *respectively. Furthermore,*

$$M_\beta \bullet M_{\mathscr{C}}^T = N_\beta \bullet N_{\mathscr{C}}^T \quad and \quad M_\beta \bullet M_\beta^T = N_\beta \bullet N_\beta^T.$$

Proof. Since $M_{\mathscr{C}}, N_{\mathscr{C}}$ are two matrix representations of $\mathscr{C}$, $M_{\mathscr{C}}$ and $N_{\mathscr{C}}$ can be converted to each other by list exchanges, so do M_β and N_β. Without loss of generality, we suppose that they can be represented through block matrix columns as

$$M_{\mathscr{C}} = \{\alpha_1, \ldots, \alpha_p, \ldots, \alpha_q, \ldots, \alpha_m\}, N_{\mathscr{C}} = \{\alpha_1, \ldots, \alpha_q, \ldots, \alpha_p, \ldots, \alpha_m\},$$
$$M_\beta = \{t_1, \ldots, t_p, \ldots, t_q, \ldots, t_m\}, N_\beta = \{t_1, \ldots, t_q, \ldots, t_p, \ldots, t_m\}.$$

Further, denote that $M_\beta \bullet M_{\mathscr{C}}^T = (a_{ij})_{n \times n}$ and $N_\beta \bullet N_{\mathscr{C}}^T = (b_{ij})_{n \times n}$. Then

$$a_{ij} = (t_{j1}, \dots, t_{jp}, \dots, t_{jq}, \dots, t_{jm}) \bullet (C_1(x_i), \dots, C_p(x_i), \dots, C_q(x_i), \dots, C_m(x_i))^T$$

$$= \bigwedge_{k=1}^m [t_{jk} \to C_k(x_i)]$$

$$= (t_{j1}, \dots, t_{jq}, \dots, t_{jp}, \dots, t_{jm}) \bullet (C_1(x_i), \dots, C_q(x_i), \dots, C_p(x_i), \dots, C_m(x_i))^T$$

$$= b_{ij},$$

where $i, j \in \{1, 2, \dots, m\}$. Thus, it holds that $M_\beta \bullet M_{\mathscr{C}}^T = N_\beta \bullet N_{\mathscr{C}}^T$. In addition, $M_\beta \bullet M_\beta^T = N_\beta \bullet N_\beta^T$ can be proven in a similar way.

Proposition 2. *Let* $U = \{x_1, x_2, \dots, x_n\}$ *be a finite universe of which the order of elements is given, and* $\mathscr{C} = \{C_1, C_2, \dots, C_m\}$ *be an L-fuzzy β-covering of U. If* $M_{\mathscr{C}}$ *is a matrix representation of* $\mathscr{C}$, *and* M_β *is a β-matrix representation of* $\mathscr{C}$, *then*

$$\left(N_{x_i}^\beta(x_j)\right)_{n \times n} = M_\beta \bullet M_{\mathscr{C}}^T.$$

Proof. Since $\mathscr{C}$ is an L-fuzzy β-covering of U, and $M_\beta = (t_{ik})_{n \times m}$ is a β-matrix representation of $\mathscr{C}$, for each i $(1 \le i \le n)$, there exists k $(1 \le k \le m)$ such that $t_{ik} = 1$. Further, assume that $M_\beta \bullet M_{\mathscr{C}}^T = (c_{ij})_{n \times n}$, then

$$c_{ij} = \bigwedge_{k=1}^m [t_{ik} \to C_k(x_j)]$$

$$= \bigwedge_{t_{ik}=1} C_k(x_j)$$

$$= \bigwedge_{C_k(x_i) \ge \beta} C_k(x_j)$$

$$= \left(\bigcap_{C_k(x_i) \ge \beta} C_k \right)(x_j) = N_{x_i}^\beta(x_j).$$

Hence, $\left(N_{x_i}^\beta(x_j)\right)_{n \times n} = M_\beta \bullet M_{\mathscr{C}}^T$.

Example 3. We can calculate $M_\beta \bullet M_{\mathscr{C}}^T$ and $N_\beta \bullet N_{\mathscr{C}}^T$ in Example 1 as follows.

$$M_\beta \bullet M_{\mathscr{C}}^T = N_\beta \bullet N_{\mathscr{C}}^T = \begin{pmatrix} 0.6 & 0.5 & 0.2 & 0.3 & 0.3 & 0.2 \\ 0.3 & 0.6 & 0.1 & 0.3 & 0.5 & 0.3 \\ 0.5 & 0.4 & 0.6 & 0.5 & 0.7 & 0.3 \\ 0.7 & 0.5 & 0.4 & 0.6 & 0.3 & 0.2 \\ 0.3 & 0.4 & 0.1 & 0.5 & 0.6 & 0.3 \\ 0.6 & 0.6 & 0.2 & 0.3 & 0.5 & 0.7 \end{pmatrix} = \left(N_{x_i}^\beta(x_j)\right)_{n \times n}.$$

Next, we represent the lower approximation operator $\underline{\mathscr{C}}(X)$ and upper approximation operator $\overline{\mathscr{C}}(X)$ of L-fuzzy set X by operations on matrices.

Proposition 3. *Let* $U = \{x_1, x_2, \ldots, x_n\}$ *be a finite universe of which the order of elements is given, and* $\mathscr{C} = \{C_1, C_2, \ldots, C_m\}$ *be an L-fuzzy β-covering of* U. *Suppose that* $M_{\mathscr{C}}$ *is a matrix representation of* $\mathscr{C}$, *and* M_β *is a β-matrix representation of* $\mathscr{C}$, *then for any* $X \in L^U$, *we can obtain that*

$$\underline{\mathscr{C}}(X) = (M_\beta \bullet M_{\mathscr{C}}^T) \bullet M_X \ and \ \overline{\mathscr{C}}(X) = (M_\beta \bullet M_{\mathscr{C}}^T) \circ M_X.$$

Proof. It follows Definition 4 that for any i $(1 \leq i \leq n)$,

$$(M_\beta \bullet M_{\mathscr{C}}^T) \bullet M_X(x_i) = \bigwedge_{k=1}^{m} [N_{x_i}^\beta(x_k) \to X(x_k)] = \underline{\mathscr{C}}(X)(x_i)$$

and

$$(M_\beta \bullet M_{\mathscr{C}}^T) \circ M_X(x_i) = \bigvee_{k=1}^{m} [N_{x_i}^\beta(x_k) \otimes X(x_k)] = \overline{\mathscr{C}}(X)(x_i).$$

Hence, $\underline{\mathscr{C}}(X) = (M_\beta \bullet M_{\mathscr{C}}^T) \bullet M_X$ and $\overline{\mathscr{C}}(X) = (M_\beta \bullet M_{\mathscr{C}}^T) \circ M_X$ can be obtained.

Example 4. The *L*-fuzzy β-covering-based lower and upper approximation operators $\underline{\mathscr{C}}(X)$ and $\overline{\mathscr{C}}(X)$ in Example 1 can be calculated as follows. Then for

$$X = \frac{0.6}{x_1} + \frac{0.4}{x_2} + \frac{0.3}{x_3} + \frac{0.5}{x_4} + \frac{0.5}{x_5} + \frac{0.4}{x_6},$$

it holds that

$$(M_\beta \bullet M_{\mathscr{C}}^T) \bullet M_X = \begin{pmatrix} 0.6\ 0.5\ 0.2\ 0.3\ 0.3\ 0.2 \\ 0.3\ 0.6\ 0.1\ 0.3\ 0.5\ 0.3 \\ 0.5\ 0.4\ 0.6\ 0.5\ 0.7\ 0.3 \\ 0.7\ 0.5\ 0.4\ 0.6\ 0.3\ 0.2 \\ 0.3\ 0.4\ 0.1\ 0.5\ 0.6\ 0.3 \\ 0.6\ 0.6\ 0.2\ 0.3\ 0.5\ 0.7 \end{pmatrix} \bullet \begin{pmatrix} 0.6 \\ 0.4 \\ 0.3 \\ 0.5 \\ 0.5 \\ 0.4 \end{pmatrix} = \begin{pmatrix} 0.4 \\ 0.4 \\ 0.3 \\ 0.3 \\ 0.5 \\ 0.4 \end{pmatrix} = M_{\underline{\mathscr{C}}(X)},$$

$$(M_\beta \bullet M_{\mathscr{C}}^T) \circ M_X = \begin{pmatrix} 0.6\ 0.5\ 0.2\ 0.3\ 0.3\ 0.2 \\ 0.3\ 0.6\ 0.1\ 0.3\ 0.5\ 0.3 \\ 0.5\ 0.4\ 0.6\ 0.5\ 0.7\ 0.3 \\ 0.7\ 0.5\ 0.4\ 0.6\ 0.3\ 0.2 \\ 0.3\ 0.4\ 0.1\ 0.5\ 0.6\ 0.3 \\ 0.6\ 0.6\ 0.2\ 0.3\ 0.5\ 0.7 \end{pmatrix} \circ \begin{pmatrix} 0.6 \\ 0.4 \\ 0.3 \\ 0.5 \\ 0.5 \\ 0.4 \end{pmatrix} = \begin{pmatrix} 0.6 \\ 0.5 \\ 0.5 \\ 0.6 \\ 0.5 \\ 0.6 \end{pmatrix} = M_{\overline{\mathscr{C}}(X)}.$$

Hence, we obtain that

$$\underline{\mathscr{C}}(X) = \frac{0.4}{x_1} + \frac{0.4}{x_2} + \frac{0.3}{x_3} + \frac{0.3}{x_4} + \frac{0.5}{x_5} + \frac{0.4}{x_6},$$

$$\overline{\mathscr{C}}(X) = \frac{0.6}{x_1} + \frac{0.5}{x_2} + \frac{0.5}{x_3} + \frac{0.6}{x_4} + \frac{0.5}{x_5} + \frac{0.6}{x_6}.$$

4 Interdependency of *L*-fuzzy β-covering-based Approximation Operators

Let $\mathscr{C}_1$, $\mathscr{C}_2$ be two *L*-fuzzy β-coverings on U and $\beta > 0$ ($\beta \in L$). When $N_{\mathscr{C}_1}^{\beta}(x) = N_{\mathscr{C}_2}^{\beta}(x)$ for any $x \in U$, $\mathscr{C}_1$ is not necessarily equal to $\mathscr{C}_2$. To illustrate this conclusion, we show the following example.

Example 5. Let $U = \{x_1, x_2, x_3, x_4, x_5, x_6\}$, $\mathscr{C} = \{C_1, C_2, C_3, C_4\}$ and $L = ([0,1], \otimes, \rightarrow, \bigvee, \bigwedge, 0, 1)$ be a Gödel-residuated lattice in Example 1. $\mathscr{C}' = \{C_1, C_2, C_3, C_4, C_5\}$ is a family of fuzzy sets of U, where

$$C_5 = \frac{0.4}{x_1} + \frac{0.5}{x_2} + \frac{0.2}{x_3} + \frac{0.1}{x_4} + \frac{0.3}{x_5} + \frac{0.2}{x_6}.$$

It is easy to see that $\mathscr{C}$ and $\mathscr{C}'$ are two *L*-fuzzy β-coverings on U for any $\beta \in (0, 0.6]$. In general, we take $\beta = 0.6$, then $N_{\mathscr{C}}^{0.6}(x_i)$ and $N_{\mathscr{C}'}^{0.6}(x_i)$ ($i = 1, 2, 3, 4, 5, 6$) can be listed in Table 1. We can easy to see that

$$\left(N_{x_i}^{0.6}\right)(y) = \left(\bigcap N_{\mathscr{C}}^{0.6}(x_i)\right)(y) = \left(\bigcap N_{\mathscr{C}'}^{0.6}(x_i)\right)(y) = \left(N_{x_i}^{0.6}\right)'(y),$$

where $y \in U$ and $i \in \{1, 2, 3, 4, 5, 6\}$. However, $\mathscr{C} \neq \mathscr{C}'$.

Table 1. $N_{\mathscr{C}}^{0.6}(x_i)$ and $N_{\mathscr{C}'}^{0.6}(x_i)$.

U	x_1	x_2	x_3	x_4	x_5	x_6
$N_{\mathscr{C}}^{0.6}(x_i)$	$\{C_1, C_4\}$	$\{C_3, C_4\}$	$\{C_2\}$	$\{C_1\}$	$\{C_2, C_3\}$	$\{C_4\}$
$N_{\mathscr{C}'}^{0.6}(x_i)$	$\{C_1, C_4\}$	$\{C_3, C_4\}$	$\{C_2\}$	$\{C_1\}$	$\{C_2, C_3\}$	$\{C_4\}$

Next, we consider the condition under what two *L*-fuzzy β-coverings can generate the same *L*-fuzzy β-neighborhood system of any elements on U. At first, we introduce some basic concepts.

Definition 8. *Let* $(U, \mathscr{C})$ *be an L-fuzzy β-covering approximation space and* $C \in \mathscr{C}$. *If one of the following statements holds:*

(1) $C(x) < \beta$ *for all* $x \in U$.
(2) For $x \in U$, $C(x) \geq \beta$ *implies that there exists* $C' \in \mathscr{C} - \{C\}$ *such that* $C' \subseteq C$ *and* $C'(x) \geq \beta$.

then C is called a β-independent element of $\mathscr{C}$; if not, C is called a β-dependent element of $\mathscr{C}$.

Example 6. Let $\mathscr{C}, \mathscr{C}'$ be two L-fuzzy β-coverings in Example 5. On the one hand, for any $x \in U$, $C_5(x) < 0.6$, then C_5 is a 0.6-independent element of $\mathscr{C}'$ and C_i ($i = 1, 2, 3, 4$) are 0.6-dependent elements of $\mathscr{C}'$.

On the other hand, let $\mathscr{C}'' = \mathscr{C} \bigcup C_6$, where

$$C_6 = \frac{0.7}{x_1} + \frac{0.5}{x_2} + \frac{0.4}{x_3} + \frac{0.8}{x_4} + \frac{0.4}{x_5} + \frac{0.3}{x_6}.$$

It is easy to seen that $\mathscr{C}''$ is an L-fuzzy β-covering on U for any $\beta \in (0, 0.6]$. In general, we take $\beta = 0.6$, then $N_{\mathscr{C}''}^{0.6}(x_i)$ ($i = 1, 2, 3, 4, 5, 6$) can be listed in Table 2.

Table 2. $N_{\mathscr{C}}^{0.6}(x_i)$ and $N_{\mathscr{C}''}^{0.6}(x_i)$.

U	x_1	x_2	x_3	x_4	x_5	x_6
$N_{\mathscr{C}}^{0.6}(x_i)$	$\{C_1, C_4\}$	$\{C_3, C_4\}$	$\{C_2\}$	$\{C_1\}$	$\{C_2, C_3\}$	$\{C_4\}$
$N_{\mathscr{C}''}^{0.6}(x_i)$	$\{C_1, C_4, C_6\}$	$\{C_3, C_4\}$	$\{C_2\}$	$\{C_1, C_6\}$	$\{C_2, C_3\}$	$\{C_4\}$

Then, for $x_i \in U$, $C_6(x_i) \geq \beta$ ($i = 1, 4$), there exists $C_1 \subseteq C_6$ and $C_1(x_i) \geq \beta$ ($i = 1, 4$), further, it can be easy to obtain that

$$\left(N_{x_i}^{0.6}\right)(y) = \left(\bigcap N_{\mathscr{C}}^{0.6}(x_i)\right)(y)$$

$$= \bigwedge_{C_i(x_i) \geq \beta, C_i \in \mathscr{C}} C_i(y)$$

$$= \left(\bigwedge_{C_i(x_i) \geq \beta, C_i \in \mathscr{C}} C_i(y)\right) \bigwedge \left(\bigwedge_{C_6(x_i) \geq \beta, C_6 \in \mathscr{C}''} C_6(y)\right)$$

$$= \bigwedge_{C_i(x_i) \geq \beta, C_i \in \mathscr{C}''} C_i(y)$$

$$= \left(\bigcap N_{\mathscr{C}''}^{0.6}(x_i)\right)(y)$$

$$= \left(N_{x_i}^{0.6}\right)''(y),$$

where $y \in U$ and $i \in \{1, 2, 3, 4, 5, 6\}$. Hence, C_6 is a 0.6-independent element of $\mathscr{C}''$, and C_i ($i = 1, 2, 3, 4$) are a 0.6-dependent elements of $\mathscr{C}''$. The L-fuzzy β-covering $\mathscr{C}$ and $\mathscr{C}''$ can generate the same L-fuzzy β-neighborhood of any $x_i \in U$.

Proposition 4. *Let $(U, \mathscr{C})$ be an L-fuzzy β-covering approximation space. If C is a β-independent element of $\mathscr{C}$, then $\mathscr{C} - \{C\}$ is still an L-fuzzy β-covering of U.*

Proof. Suppose that $\mathscr{C} = \{C, C_1, C_2, \ldots, C_m\}$, where $C, C_i \in L^U$ ($i = 1, 2, \ldots, m$), and C is a β-independent element.

Case(1). For any $x \in U$, $C(x) < \beta$, we can conclude that

$$\left(\bigcup_{k=1}^{m} C_i\right)(x) = \bigvee_{k=1}^{m} C_i(x) = \left(\bigvee_{k=1}^{m} C_i(x)\right) \bigvee C(x) \geq \beta.$$

Case(2). For $x \in U$, if $C(x) > \beta$, then there exists $C_r \in \{C_1, C_2, \ldots, C_m\}$ such that $C_r \subseteq C$ and $C_r(x) \geq \beta$, i.e., $\left(\bigcup_{k=1}^{m} C_i \right)(x) \geq C_r(x) \geq \beta$.

Thus, $\mathscr{C} - \{C\}$ is an L-fuzzy β-covering of U.

Proposition 5. *Let $(U, \mathscr{C})$ be an L-fuzzy β-covering approximation space, C be a β-independent element of $\mathscr{C}$ and $C_1 \in \mathscr{C} - \{C\}$. Then C_1 is a β-independent element of $\mathscr{C}$ if and only if it is a β-independent element of $\mathscr{C} - \{C\}$.*

Proof. $\Longleftarrow$): It is obvious.

$\Longrightarrow$): Suppose $\mathscr{C} = \{C, C_1, C_2, \ldots, C_m\}$, where $C, C_i \in L^U$ $(i = 1, 2, \ldots, m)$. If C is a β-independent element, then we have the following four cases:

For any $x \in U$, $C(x) < \beta$:

Case(1). $C_1(x) < \beta$ for any $x \in U$. Since $C_1 \in \mathscr{C} - \{C\}$, C_1 is a β-independent element of $\mathscr{C} - \{C\}$.

Case(2). For $y \in U$, since $C_1(y) \geq \beta$, there exists $C' \in \mathscr{C}$ such that $C' \subseteq C_1$ and $C'(y) \geq \beta$. If $C' = C$, it contradicts $C(x) < \beta$ for any $x \in U$. If $C' \neq C$, then C_1 is a β-independent element of $\mathscr{C} - \{C\}$.

For $x \in U$, $C(x) > \beta$, then there exists $C_r \in \{C_1, C_2, \ldots, C_m\}$ such that $C_r \subseteq C$ and $C_r(x) \geq \beta$.

Case(3). $C_1(x) < \beta$ for any $x \in U$. It is easy to see that C_1 is a β-independent element of $\mathscr{C} - \{C\}$.

Case(4). For $y \in U$, since $C_1(y) \geq \beta$, there exists $C' \in \mathscr{C} - \{C_1\}$ such that $C' \subseteq C_1$ and $C'(y) \geq \beta$. If $C' = C$, then there exists $C_r \in \{C_1, C_2, \ldots, C_m\}$ such that $C_r \subseteq C' \subseteq C_1$ and $C_r(y) \geq \beta$, then C_1 is a β-independent element of $\mathscr{C} - \{C\}$. If $C' \neq C$, then it is easy to obtain that C_1 is a β-independent element of $\mathscr{C} - \{C\}$.

Proposition 4 ensures that after a β-independent element is removed from an L-fuzzy β-covering, it remains an L-fuzzy β-covering. And Proposition 5 emphasizes that deleting a β-independent element from an L-fuzzy β-covering will not affect the existence of other β-independents, that is, it will not create any new β-independent elements, and will not make the original β-independent elements into β-dependent elements. Therefore, the β-base of an L-fuzzy β-covering is calculated by directly deleting all the β-independent elements simultaneously or one by one.

Definition 9. *Let $(U, \mathscr{C})$ be an L-fuzzy β-covering approximation space and $\mathscr{B}$ be a subset of $\mathscr{C}$. If $\mathscr{C} - \mathscr{B}$ is the set of all β-independent elements of $\mathscr{C}$, then $\mathscr{B}$ is called the β-base of $\mathscr{C}$, which is denoted as $\mathscr{B}^{\beta}(\mathscr{C})$.*

Definition 10. *Let $(U, \mathscr{C})$ be an L-fuzzy β-covering approximation space. If each element of U is a β-dependent element, i.e., $\mathscr{B}^{\beta}(\mathscr{C}) = \mathscr{C}$, then $\mathscr{C}$ is β-dependent; if not, $\mathscr{C}$ is β-independent.*

Example 7. Let $(U, \mathscr{C}')$ and $(U, \mathscr{C})$ be two L-fuzzy β-covering approximation spaces in Example 5. Then $\mathscr{C}'$ is β-independent because $\mathscr{B}^{0.6}(\mathscr{C}') = \{C_1, C_2, C_3, C_4\} \neq \mathscr{C}'$ and $\mathscr{C}$ is β-dependent due to $\mathscr{B}^{0.6}(\mathscr{C}) = \{C_1, C_2, C_3, C_4\} = \mathscr{C}$.

The following proposition shows that deleting β-independent elements has no effect on the β-neighborhood of any elements on U.

Proposition 6. *Let $(U, \mathscr{C})$ be an L-fuzzy β-covering approximation space. For any $x \in U$, the following statement holds.*

$$\bigcap N_{\mathscr{C}}^{\beta}(x) = \bigcap N_{\mathscr{B}^{\beta}(\mathscr{C})}^{\beta}(x).$$

Proof. Suppose that $\mathscr{C} = \{C, C_1, C_2, \ldots, C_m\}$, where $C, C_i \in L^U$ $(i = 1, 2, \ldots, m)$ and C is a β-independent element of $\mathscr{C}$. For each $x \in U$, we denote that the L-fuzzy β-neighborhood of x induced by L-fuzzy β-covering of $\mathscr{C}$ as $\bigcap N_{\mathscr{C}}^{\beta}(x)$ and the L-fuzzy β-neighborhood of x induced by L-fuzzy β-covering of $\mathscr{C} - \{C\}$ as $\bigcap N_{\mathscr{C}-\{C\}}^{\beta}(x)$. Then, there exists two cases:

Case(1). For any $x \in U$, $C(x) < \beta$, then

$$\bigcap N_{\mathscr{C}}^{\beta}(x) = \bigcap\{C_i \in \mathscr{C} : C_i(x) \geq \beta\} = \bigcap\{C_i \in \mathscr{C} - \{C\} : C_i(x) \geq \beta\} = \bigcap N_{\mathscr{C}-\{C\}}^{\beta}(x).$$

Case(2). For $y \in U$, $C(y) \geq \beta$, there exists $C' \in \{C_1, C_2, \ldots, C_m\}$ such that $C' \subseteq C$ and $C'(y) \geq \beta$. It is obvious that $C \notin \bigcap N_{\mathscr{C}}^{\beta}(x)$. Thus, $\bigcap N_{\mathscr{C}}^{\beta}(x) = \bigcap N_{\mathscr{C}-\{C\}}^{\beta}(x)$ holds.

Furthermore, if $\mathscr{C} - \{C\}$ is β-dependent, then we have that $\mathscr{B}^{\beta}(\mathscr{C}) = \mathscr{C} - \{C\} = \mathscr{B}^{\beta}(\mathscr{C} - \{C\})$ and $\bigcap N_{\mathscr{C}}^{\beta}(x) = \bigcap N_{\mathscr{C}-\{C\}}^{\beta}(x) = \bigcap N_{\mathscr{B}^{\beta}(\mathscr{C}-\{C\})}^{\beta}(x) = \bigcap N_{\mathscr{B}^{\beta}(\mathscr{C})}^{\beta}(x)$ for any $x \in U$. If $\mathscr{C} - \{C\}$ is β-independent, then there exists $\{C_{i_1}, C_{i_2}, \ldots, C_{i_s}\} \in \mathscr{C} - \{C\}$ $(i_1, i_2, \ldots, i_s \in \{1, 2, \ldots, m\})$ such that $C_{i_1}, C_{i_2}, \ldots, C_{i_s}$ are all of β-independent elements of $\mathscr{C} - \{C\}$, then $\mathscr{C} - \{C, C_{i_1}, C_{i_2}, \ldots, C_{i_s}\}$ is β-dependent. Thus, $\mathscr{B}^{\beta}(\mathscr{C} - \{C\}) = \mathscr{C} - \{C, C_{i_1}, C_{i_2}, \ldots, C_{i_s}\}$. In summary, it holds that

$$\bigcap N_{\mathscr{C}}^{\beta}(x) = \bigcap N_{\mathscr{C}-\{C\}}^{\beta}(x) = \bigcap N_{\mathscr{C}-\{C,C_{i_1}\}}^{\beta}(x) = \cdots$$

$$= \bigcap N_{\mathscr{C}-\{C,C_{i_1},C_{i_2},\ldots,C_{i_s}\}}^{\beta}(x) = \bigcap N_{\mathscr{B}^{\beta}(\mathscr{C}-\{C\})}^{\beta}(x) = \bigcap N_{\mathscr{B}^{\beta}(\mathscr{C})}^{\beta}(x).$$

Hence, $\bigcap N_{\mathscr{C}}^{\beta}(x) = \bigcap N_{\mathscr{B}^{\beta}(\mathscr{C})}^{\beta}(x)$ holds for any $x \in U$.

Proposition 7. *Let $\mathscr{C}_1$, $\mathscr{C}_2$ be two L-fuzzy β-coverings of U. For any $x \in U$, $\bigcap N_{\mathscr{C}_1}^{\beta}(x) = \bigcap N_{\mathscr{C}_2}^{\beta}(x)$ if and only if $\mathscr{B}^{\beta}(\mathscr{C}_1) = \mathscr{B}^{\beta}(\mathscr{C}_2)$.*

Proof. $\Longleftarrow$): It follows from Proposition 6.

$\Longrightarrow$): It can be obtained by Definition 3.

The above proposition states the necessary and sufficient condition under what two L-fuzzy β-coverings can generate the same β-neighborhood of any elements on U is that their β-bases are equal.

Corollary 1. *Let $\mathscr{C}_1$, $\mathscr{C}_2$ be two β-dependent L-fuzzy β-coverings of U. For any $x \in U$, $\bigcap N_{\mathscr{C}_1}^{\beta}(x) = \bigcap N_{\mathscr{C}_2}^{\beta}(x)$ if and only if $\mathscr{C}_1 = \mathscr{C}_2$.*

Theorem 1. *Let $(U, \mathscr{C})$ be an L-fuzzy β-covering approximation space. For any $X \in L^U$, the following statements hold:*

$$\bigwedge_{y \in U}\left[\left(\bigcap N_{\mathscr{C}}^{\beta}(x)\right)(y) \rightarrow X(y)\right] = \bigwedge_{y \in U}\left[\left(\bigcap N_{\mathscr{B}^{\beta}(\mathscr{C})}^{\beta}(x)\right)(y) \rightarrow X(y)\right],$$

$$\bigvee_{y \in U}\left[\left(\bigcap N_{\mathscr{C}}^{\beta}(x)\right)(y) \otimes X(y)\right] = \bigvee_{y \in U}\left[\left(\bigcap N_{\mathscr{B}^{\beta}(\mathscr{C})}^{\beta}(x)\right)(y) \otimes X(y)\right].$$

Proof. It follows from Proposition 6.

According to Definition 4, Proposition 7 and Theorem 1, we can conclude the following two propositions.

Proposition 8. *Let $\mathscr{C}_1$, $\mathscr{C}_2$ be two L-fuzzy β-coverings of U. For any $X \in L^U$, $\mathscr{C}_1$ and $\mathscr{C}_2$ generate the same L-fuzzy β-covering-based lower approximation of X if and only if $\mathscr{B}^\beta(\mathscr{C}_1) = \mathscr{B}^\beta(\mathscr{C}_2)$.*

Proposition 9. *Let $\mathscr{C}_1$, $\mathscr{C}_2$ be two L-fuzzy β-coverings of U. For any $X \in L^U$, $\mathscr{C}_1$ and $\mathscr{C}_2$ generate the same L-fuzzy β-covering-based upper approximation of X if and only if $\mathscr{B}^\beta(\mathscr{C}_1) = \mathscr{B}^\beta(\mathscr{C}_2)$.*

Propositions 8 and 9 show a necessary and sufficient condition under what two L-fuzzy covering-based approximations of an L-subset is that their β-base are equal.

Corollary 2. *Let $\mathscr{C}_1$, $\mathscr{C}_2$ be two L-fuzzy β-coverings of U. For any $X \in L^U$, $\mathscr{C}_1$ and $\mathscr{C}_2$ generate the same L-fuzzy β-covering-based lower approximation of X if and only if $\mathscr{C}_1$ and $\mathscr{C}_2$ generate the same L-fuzzy β-covering-based upper approximation of X.*

Example 8. Let $\mathscr{C}''$ be an L-fuzzy β-coverings in Example 6 for any $\beta \in (0, 0.6]$. In general, we take $\beta = 0.6$ and

$$X = \frac{0.6}{x_1} + \frac{0.4}{x_2} + \frac{0.3}{x_3} + \frac{0.5}{x_4} + \frac{0.5}{x_5} + \frac{0.4}{x_6}.$$

It follows Definition 9 that $\mathscr{B}^\beta(\mathscr{C}'') = \{C_1, C_2, C_3, C_4\} = \mathscr{C}$. Then, combining the value of $M_{\underline{\mathscr{C}}(X)}$ and $M_{\overline{\mathscr{C}}(X)}$ in Example 4, it holds that

$$M_{\underline{\mathscr{C}''}(X)} = (M_\beta \bullet M_{\mathscr{C}''}^T) \bullet M_X$$

$$= \begin{pmatrix} 0.6\ 0.5\ 0.2\ 0.3\ 0.3\ 0.2 \\ 0.3\ 0.6\ 0.1\ 0.3\ 0.5\ 0.3 \\ 0.5\ 0.4\ 0.6\ 0.5\ 0.7\ 0.3 \\ 0.7\ 0.5\ 0.4\ 0.6\ 0.3\ 0.2 \\ 0.3\ 0.4\ 0.1\ 0.5\ 0.6\ 0.3 \\ 0.6\ 0.6\ 0.2\ 0.3\ 0.5\ 0.7 \end{pmatrix} \bullet \begin{pmatrix} 0.6 \\ 0.4 \\ 0.3 \\ 0.5 \\ 0.5 \\ 0.4 \end{pmatrix} = \begin{pmatrix} 0.4 \\ 0.4 \\ 0.3 \\ 0.3 \\ 0.5 \\ 0.4 \end{pmatrix}$$

$$= M_{\mathscr{B}^\beta(\mathscr{C}'')(X)} = M_{\underline{\mathscr{C}}(X)},$$

$$M_{\overline{\mathscr{C}''}(X)} = (M_\beta \bullet M_{\mathscr{C}''}^T) \circ M_X$$

$$= \begin{pmatrix} 0.6\ 0.5\ 0.2\ 0.3\ 0.3\ 0.2 \\ 0.3\ 0.6\ 0.1\ 0.3\ 0.5\ 0.3 \\ 0.5\ 0.4\ 0.6\ 0.5\ 0.7\ 0.3 \\ 0.7\ 0.5\ 0.4\ 0.6\ 0.3\ 0.2 \\ 0.3\ 0.4\ 0.1\ 0.5\ 0.6\ 0.3 \\ 0.6\ 0.6\ 0.2\ 0.3\ 0.5\ 0.7 \end{pmatrix} \circ \begin{pmatrix} 0.6 \\ 0.4 \\ 0.3 \\ 0.5 \\ 0.5 \\ 0.4 \end{pmatrix} = \begin{pmatrix} 0.6 \\ 0.5 \\ 0.5 \\ 0.6 \\ 0.5 \\ 0.6 \end{pmatrix}$$

$$= M_{\overline{\mathscr{B}^\beta(\mathscr{C}'')}(X)} = M_{\overline{\mathscr{C}}(X)}.$$

5   Conclusion

The main conclusions of this paper and continuous work to do are listed as follows.

(1) The matrix representations of the L-fuzzy β-covering-based lower and upper approximations are proposed, which can make the calculations more valid through operations on matrices.
(2) The interdependency of the L-fuzzy β-covering-based rough approximations is discussed. Using the concept of β-independent element, we give a necessary and sufficient condition under what the two L-fuzzy β-coverings can generate the same approximation operations.
(3) The L-fuzzy covering-based rough sets should be further researched from the perspective of topological properties [15, 24, 24].

Acknowledgments. This research was supported by the National Natural Science Foundation of China (Grant no. 12101500) and the Chinese Universities Scientific Fund (Grant nos. 2452018054 and 2452022370).

References

1. Bonikowski, Z., Bryniarski, E., Wybraniec-Skardowska, U.: Extensions and intentions in the rough set theory. Inf. Sci. **107**, 149–167 (1998)
2. Degang, C., Changzhong, W., Qinghua, H.: A new approach to attribute reduction of consistent and inconsistent covering decision systems with covering rough sets. Inf. Sci. **177**, 3500–3518 (2007)
3. Deng, T., Chen, Y., Xu, W., Dai, Q.: A novel approach to fuzzy rough sets based on a fuzzy covering. Inf. Sci. **177**, 2308–2326 (2007)
4. Dubois, D., Prade, H.: Rough fuzzy sets and fuzzy rough sets. Int. J. Gen. Syst. **17**, 191–209 (1990)
5. Greco, S., Matarazzo, B., Slowinski, R.: Rough approximation by dominance relations. Int. J. Intell. Syst. **17**, 153–171 (2002)
6. Hájek, P.: Metamathematics of fuzzy logic. Springer (2013). https://doi.org/10.1007/978-94-011-5300-3
7. Han, S.E.: Covering rough set structures for a locally finite covering approximation space. Inf. Sci. **480**, 420–437 (2019)
8. Jensen, R., Shen, Q.: Fuzzy-rough attribute reduction with application to web categorization. Fuzzy Sets Syst. **141**, 469–485 (2004)
9. Jin, Q., Li, L.: On the second type of L-fuzzy covering rough sets. Int. Inf. Inst. (Tokyo). Information **16**, 1101–1106 (2013)
10. Li, T.J., Leung, Y., Zhang, W.X.: Generalized fuzzy rough approximation operators based on fuzzy coverings. Int. J. Approx. Reason. **48**(3), 836–856 (2008)
11. Li, L.Q., Jin, Q., Hu, K., Zhao, F.F.: The axiomatic characterizations on L-fuzzy covering-based approximation operators. Int. J. Gen Syst **46**, 332–353 (2017)
12. Ma, L.: On some types of neighborhood-related covering rough sets. Int. J. Approximate Reasoning **53**, 901–911 (2012)
13. Ma, L.: Some twin approximation operators on covering approximation spaces. Int. J. Approximate Reasoning **56**, 59–70 (2015)
14. Ma, L.: Two fuzzy covering rough set models and their generalizations over fuzzy lattices. Fuzzy Sets Syst. **294**, 1–17 (2016)

15. Ma, Z.M., Hu, B.Q.: Topological and lattice structures of L-fuzzy rough sets determined by lower and upper sets. Inf. Sci. **218**, 194–204 (2013)
16. Mi, J.S., Zhang, W.X.: An axiomatic characterization of a fuzzy generalization of rough sets. Inf. Sci. **160**, 235–249 (2004)
17. Morsi, N.N., Yakout, M.M.: Axiomatics for fuzzy rough sets. Fuzzy Sets Syst. **100**, 327–342 (1998)
18. Pawlak, Z.: Rough sets. Int. J. Comput. Inform. Sci. **11**, 341–356 (1982)
19. Pomykala, J.A.: Approximation operations in approximation space. Bullet. Polish Acad. Sci. Math. **35**, 653–662 (1987)
20. Pomykała, J.: On definability in the nondeterministic information system. Bullet. Polish Acad. Sci. Math. **36**, 193–210 (1988)
21. Radzikowska, A.M., Kerre, E.E.: A comparative study of fuzzy rough sets. Fuzzy Sets Syst. **126**, 137–155 (2002)
22. Slowinski, R., Vanderpooten, D.: A generalized definition of rough approximations based on similarity. IEEE Trans. Knowl. Data Eng. **12**, 331–336 (2000)
23. Turunen, E.: Mathematics behind fuzzy logic. Physica-Verlag Heidelberg (1999)
24. Wang, C.Y.: Topological characterizations of generalized fuzzy rough sets. Fuzzy Sets Syst. **312**, 109–125 (2017)
25. Ward, M., Dilworth, R.P.: Residuated lattices. Trans. Am. Math. Soc. **45**, 335–354 (1939)
26. Wu, W.Z., Zhang, W.X.: Neighborhood operator systems and approximations. Inf. Sci. **144**, 201–217 (2002)
27. Yang, B.: A fuzzy covering-based rough set model on residuated lattice. Fuzzy Syst. Math. **34**, 26–42 (2020)
28. Yang, B., Hu, B.Q.: Matrix representations and interdependency on L-fuzzy covering-based approximation operators. Int. J. Approximate Reasoning **96**, 57–77 (2018)
29. Yang, B., Hu, B.Q.: Fuzzy neighborhood operators and derived fuzzy coverings. Fuzzy Sets Syst. **370**, 1–33 (2019)
30. Yao, Y.: Relational interpretations of neighborhood operators and rough set approximation operators. Inf. Sci. **111**, 239–259 (1998)
31. Yao, Y., Yao, B.: Covering based rough set approximations. Inf. Sci. **200**, 91–107 (2012)
32. Zhu, W., Wang, F.Y.: On three types of covering-based rough sets. IEEE Trans. Knowl. Data Eng. **19**, 1131–1144 (2007)

Classification and Deep Learning

Uncertainty-Aware Deep Open-Set Object Detection

Qi Hang[1,2] , Zihao Li[1] , Yudi Dong[1] , and Xiaodong Yue[1,2(✉)]

[1] School of Computer Engineering and Science, Shanghai University,
Shanghai 200444, China
{hangqi,zihao,dyd,yswantfly}@shu.edu.cn
[2] Artificial Intelligence Institute of Shanghai University, Shanghai, China

Abstract. Open-set object detection better simulates the real world compared with close-set object detection. Besides the classes of interest, it also pays attention to unknown objects in the environment. We extend the previous concept of open-set object detection, aiming to detect both known and unknown objects. Because unknown objects have different textural features from known classes and the background, we assume that detecting unknown instances will generate high uncertainty. Therefore, in this paper, we propose an uncertainty-aware open-set object detection framework based on faster R-CNN. We introduce evidential deep learning to the field of object detection to estimate the uncertainty of the predictions and perform more accurate classification in open-set conditions. The obtained uncertainty will be utilized to pseudo-label unknown instances in the training data. We also introduce a contrastive clustering module to separate the feature representations of each class during the training phase. We set an uncertainty-based unknown identifier at the inference phase to enhance the generalization of the detector. We conduct experiments on three different data splits, and our method outperforms the recent SOTA method. We also demonstrate each component in our method is effective and indispensable in our ablation studies.

Keywords: Uncertainty · Evidential deep learning · Open-set object detection

1 Introduction

As one of the most important research fields in computer vision, object detection aims to recognize and localize objects of interest while ignoring the others. Thanks to the development of deep learning, object detection has made significant advances in the past few years and achieved remarkable performance [8,10,25,39]. Most existing detectors are under a strong close-set assumption that detectors only encounter objects that are from the classes available during the training phase. This assumption simply classifies objects into known classes and scene background, while actually, detectors will inevitably meet objects from unknown classes.

J. Yao et al. (Eds.): IJCRS 2022, LNAI 13633, pp. 161–175, 2022.
https://doi.org/10.1007/978-3-031-21244-4_12

Open-set object detection [4,19,20,32] believes the detectors are essentially deployed in open-set conditions. Objects not seen in the training phase can not be simply regarded as background. We extend this concept, requiring detectors capable of detecting known and unknown objects. Figure 1 illustrates the detection results of our open-set detector. It can be seen that both known and unknown objects in example images are detected.

Fig. 1. Examples of open-set object detection. Detectors will counter unknown objects at inference because of lacking sufficient information. Our open-set object detection requires detectors to detect known and unknown objects correctly.

Open-set object detection problem is crucial, especially for safely applying object detection to the ever-changing real world, such as autonomous driving and autonomous robot. Open-set detector has the ability to judge whether the condition is too complex. Therefore, we can obtain a more trustworthy detector and avoid serious consequences.

For performance evaluation under the close-set assumption, test sets contain lots of labeled objects from the classes of interest, and detectors are encouraged to improve the performance of detecting these objects. However, when applied to open-set situations, there are no annotations of unknown objects in existing test sets. We follow the experimental setup in [12,21] to evaluate the open-set detection performance.

Since Bendale et al. [30] formally put out the open-set recognition problem, there has been a great amount of research on this problem, and it has made significant progress [1,7,23,29]. However, these advances cannot be directly adapted to open-set object detection because of the effect of the background. For close-set object detection, it is natural to regard unlabeled things in images as the background. When it comes to open-set object detection, unlabeled unknown objects in training images will harm unknown detection if they are not properly handled. Joseph et al. [12] also noticed this problem and proposed a self-supervised method to solve it. Their work effectively reduces the impact of unlabeled objects in the training set.

The texture of the unknown objects is neither similar to known classes nor the background. Thus, we propose a hypothesis that for well-trained object detectors, the detection of unknown objects will generate higher uncertainty. Miller et al. [19] proposed a similar assumption, they introduced dropout sampling [5] to estimate the uncertainty of predictions derived from SSD [16] attempting to avoid identifying unknown objects as known.

Evidential deep learning [31] can efficiently estimate the uncertainty of the predictions of neural networks. To the best of our knowledge, we are the first to introduce it into the field of object detection. Correctly modeling the uncertainty of predictions is extremely important because we believe uncertainty is the key to open-set object detection. More details about this technique will be discussed in Sect. 3.1.

There are some challenges in open-set object detection: (i) for model training, there are no annotations of unknown objects in any existing training sets; (ii) unlabeled unknown objects in the training data may confuse detecting unknown objects; (iii) unknown class includes many different classes thus it is challenging to form similar feature representations for unknown objects. In this paper, we propose an uncertainty-aware open-set object detection method, which can effectively address the above issues. Our contributions are as follows:

- We introduce evidential deep learning to object detection to quantify the uncertainty of predictions and obtain calibrated classification probability.
- An uncertainty-aware pseudo-labeling method is proposed to label instances with high uncertainty as unknown objects during the training phase. These annotations are utilized to learn an open-set object detector and to mitigate the confusion caused by the background.
- We introduce the contrastive clustering module to optimize feature representation for every class. With this module, the detector can produce similar feature representations for unknown objects and make sure these representations are different from known objects.
- To improve the generalizability, we construct an uncertainty-based identifier that only works at the inference phase.

Section 2 explains related work. Every component of our method is elaborated in detail in Sect. 3. Section 4 presents the experiment setup and comparison of results, and Sect. 5 summarizes our work.

2   Related Work

2.1   Open-Set Object Detection

Closed-set object detection assumes all objects are from classes available during training, i.e. Known classes or the Background. In the real world, however, detectors only get limited information about Known classes during the training phase while meeting various novel classes at inference. Open-set object detection better simulates the real world, which requires detecting both known and unknown objects in the environment.

There have been some works on open-set object detection. Dhamija et al. [4] formalized the problem of open-set object detection and evaluated the performance of some popular object detectors with their proposed open-set object detection protocols. These detectors sometimes classify unknown objects as known with high confidence. Besides, the use of the background cannot effectively alleviate this problem. Miller et al. [19] adapted uncertainty estimation to object detection using Monte Carlo Dropout sampling [5] in an SSD detector to estimate the uncertainty of the detections and regard those with high uncertainty as unknown objects. However, as the first exploration of open-set object detection, this work only rejects unknown objects rather than recognizes them. Meanwhile, it relies on sampling operations, which reduce the efficiency of the object detector. Joseph et al. [12] proposed a method based on faster R-CNN [27] and named it ORE. ORE achieves state-of-the-art on the open-set object detection problem although designed for the open-world task. The open-world task asks the model to perform incremental learning when knowledge of unknown classes is provided. ORE utilizes RPN to generate pseudo unknown labels for proposals that do not overlap the ground-truth but have high 'objectness' scores. These proposals are then regarded as the ground-truth of unknown objects to train the model. To improve detection performance, ORE also models the unknown class and each known class in the latent space to make the feature representations of different classes separable. In addition, it uses an energy-based identifier to determine whether an object belongs to known classes or the unknown class by calculating its energy value. ORE suffers from relying on a validation set containing manually labeled annotations of unknown objects to train the energy distribution. Simultaneously, ORE relies on training data to train unknown detection, while in open-set conditions, detectors face many more unknown classes that do not exist in the training data, its auto-labeling will thus result in poor generalization.

2.2   Deep Learning with Uncertainty

The reliability of deep learning has been drawing more attention with its wide application in the real world. Evaluating the uncertainty of neural networks is critical for obtaining safe predictions and calibrated confidence. Research on uncertainty estimation can be classified into four types [6]: single deterministic methods [18,24,31], Bayesian methods [5,28,34], ensemble methods [14,33,36], and test-time augmentation methods [22]. Evidential neural network (ENN) [31] is one of the single deterministic methods. It is based on the Dempster-Shafer Theory of Evidence (DST) [3] and the theory of Subjective Logic (SL) [13]. The uncertainty estimate can be derived by one single forward pass with a deterministic network. There have been researches based on evidential neural network [9,17,37,38]. We adapt uncertainty estimate to object detection to guide the detector to identify unknown objects in the training data.

2.3   Contrastive Clustering

In computer vision, contrastive learning aims to learn an encoder network that can generate similar feature representations for images from the same class and force feature representations from diverse classes far from each other [11]. ORE [12] contains contrastive clustering, a cluster-based contrastive learning module.

By contrastive clustering, the detector can generate separate feature representations for objects from different classes, while the feature representations of objects from the same class are close to each other.

To perform contrastive clustering, we maintain a class prototype set: $P = \{p_0 \dots p_C\}$ and a feature store $F_{store} = \{q_0 \dots q_c\}$ storing the class-specific features. p_i is the prototype vector for each class $i \in C$, while q_i is a fixed-length queue that stores the features of objects belonging to class i. At the training phase, we compute the class-wise mean of F_{store} as P_{new} every I_P iteration and update P with a momentum parameter $\eta : P = \eta P + (1 - \eta)P_{new}$. The contrastive loss is defined as:

$$\mathcal{L}_{CC}\left(\boldsymbol{f}_c\right) = \sum_{i=0} \ell\left(\boldsymbol{f}_c, p_i\right), \text{ where,}$$

$$\ell\left(\boldsymbol{f}_c, p_i\right) = \begin{cases} \mathcal{D}\left(\boldsymbol{f}_c, p_i\right) & i = c \\ \max\left\{0, \Delta - \mathcal{D}\left(\boldsymbol{f}_c, p_i\right)\right\} & \text{otherwise} \end{cases} \quad (1)$$

where f_c is the feature of an object from class c, $\mathcal{D}$ is the distance function and δ is the minimum distance between the feature vector and the prototype vectors of the other classes. As the prototype vectors continuously evolve during training, the detector will model each class in latent space and finally generate separatable feature representations for objects from different classes.

3   Method

We aim to propose an object detection approach that can detect novel instances while having a comparable performance of known detection to other detectors designed under the close-set assumption. We choose Faster R-CNN as the base detector because it outperforms other common detectors [15, 26] in open-set situations [4]. Figure 2 shows the overview architecture of our method. We introduce ENN to the classification head, aiming to estimate the label uncertainty of the predictions. According to our hypothesis, regions with high uncertainty will be labeled as unknown objects. The obtained uncertainty is utilized to pseudo-label regions generated by RPN. The unknown class contains many different classes, such that the feature representations of unknown objects can be various. We introduce a contrastive clustering module to optimize feature representations generated by the Region of Interest (RoI) head. It models unknown and each known class in the latent feature space by constructing a set of prototype vectors. Details on contrastive clustering can be found in Sect. 2.2. An identifier based on uncertainty is introduced as a supplement to improve the performance of unknown detection.

3.1   Object Detection with Evidential DNN

Existing deep learning models usually use the softmax operator to obtain class probabilities for classification tasks. Our method regards the label uncertainty

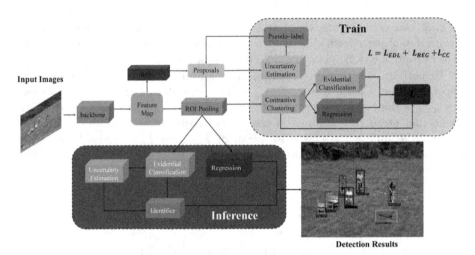

Fig. 2. The overall architecture. Our method is based on the standard faster R-CNN framework. We introduce the evidential classification head to conduct the uncertainty estimate. In the training phase, pseudo-labeling module is introduced to detect unknown instances in training data and utilize them to train the detector. We also introduce the contrastive clustering module to optimize feature representations for each class. The above two modules are only available during training. At the inference phase, we introduce an uncertainty-based identifier to determine whether an object is unknown or not.

of predictions as the key to open-set object detection. However, label uncertainty cannot be derived from such models. The softmax-based detectors can be interpreted as Multinomial distributions parameterized by the neural network's output. For a the K-class classification problem, the likelihood function for a sample x and its label y is

$$P(y|x,\theta) = Mult(y|\sigma(f_1(x,\theta)),\ldots,\sigma(f_k(x,\theta))) \qquad (2)$$

where $Mult()$ is a multinomial mass function, $f()$ is a neural network parametrized by θ, $f_i(x,\theta)$ is the output of channel i, and $\sigma()$ is the softmax function. Minimizing the negative log-likelihood $-logP(y|x,\theta)$ is equivalent to the cross-entropy loss function. Therefore, softmax essentially conducts a point estimation to the predictive distribution, which cannot obtain the variance of the predictive distribution. Meanwhile, the softmax-based classifiers could produce over-confident predictions for unknown objects [2,31,35]. Evidential deep learning (EDL) [31] is proposed to overcome these drawbacks. For K-class classification problem, EDL provides a belief mass bi for each class i, $1 \leq i \leq K$, and an overall uncertainty mass u. These mass values are non-negative and sum up to one: $u + \sum_{i=1}^{K} b_i = 1$. EDL regards $e_i \geq 0$ as the evidence derived from the network, supporting classifying sample x into class i, and the total evidence S is computed as: $S = \sum_{i=1}^{K}(e_i + 1)$. Then the belief mass b_i can be computed as: $b_i = e_k/S$, and the overall uncertainty is $u = K/S$. SL assumes that the probability density of the probability mass function **p** of the class probability follows a Dirichlet distribution, the conjugate prior distribution of the multinomial distribution. For a K-class classification task, and the Dirichlet distribution is characterized by K parameters $\boldsymbol{\alpha} = [\alpha_1, \ldots, \alpha_k]$ and given by

$$Dir(\mathbf{p}|\boldsymbol{\alpha}) = \frac{1}{B(\boldsymbol{\alpha})} \sum_{i=1}^{K} p_i^{\alpha_i - 1} \tag{3}$$

where $B()$ is a K-dimensional multinomial beta function, $\alpha_i = e_i + 1$. EDL treat α as the output of neural network: $\alpha = f(g(x)) + 1$, where $g()$ is the network and $f()$ is the evidence activation function which can ensure $\alpha_i \geq 1, 1 \leq i \leq K$. The probability that sample x belongs to class i is: $p_i = \alpha_i / S$. Then we can train the network by minimizing negative log-marginal likelihood, let y be a one-hot vector encoding the ground-truth of the input sample x, the eventual loss function is given by

$$
\begin{aligned}
L_{EDL} &= -\log\left(\int \prod_{j=1}^{K} p_j^{y_i} \frac{1}{B(\alpha_i)} \prod_{j=1}^{K} p_j^{\alpha_j - 1} dP\right) \\
&= \prod_{j=1}^{K} y_i(\log(S) - \log(\alpha_j))
\end{aligned}
\tag{4}
$$

We introduce EDL to our classification head, and the uncertainty can be obtained with one single forward pass. Uncertainty estimate is utilized to label unknown objects in the training phase and identity unknown objects in the inference phase. EDL can alleviate over-confident predictions and regarding unknown objects as known.

3.2   Model Training of Evidential DNN for Uncertain Object Detection

Existing datasets only contain the annotations of objects from classes of interest, while there are numerous unlabeled unknown objects in the training images. To avoid being confused by these unknown objects without labels, closed-set object detection introduces the background and forces these objects to be classified as background. Although this method has a degree of limitations, as the texture characteristics of unknown objects are different from the background, it has been proven effective in closed-set object detection.

In open-set object detection, with the demand of detecting unknown objects, ignoring the unlabeled unknown objects in the training data can cause unimaginable harm to the performance of the unknown detection because, in this way, the detector may recognize objects with similar features as the background. It is natural to solve this problem by adding the annotations of these unknown instances into the training set such that the detector can learn to detect unknown objects in open-set conditions. However, the cost of labeling these objects is unacceptable, especially in large-scale datasets.

To address this issue, we propose an uncertainty-based self-supervised approach. The Region Proposal Network (RPN) in the faster R-CNN framework can propose a set of regions that might contain objects. With the introduced evidential DNN, the detector can obtain the label uncertainty of these regions. After sorting these regions by uncertainty, the top K regions are selected, which are highly uncertain predictions. According to the hypothesis in Sect. 1, we believe that these regions are more likely to contain unknown objects. As a result, instances with high uncertainty in the training images are pseudo-labeled as unknown objects. This way, we can provide training data for open-set object detection. Meanwhile, the harm caused by regarding the unknown instances without a label in the training set as the background is effectively alleviated. Figure 3 illustrates the process of pseudo-labeling unknown objects.

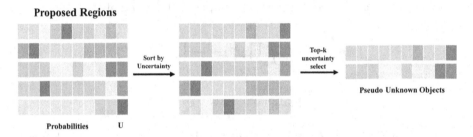

Fig. 3. The process of pseudo-labeling unknown objects. Each row represents the prediction result for a region. The grey grids represent the probability of an object belonging to a class, and the blue grids represent the prediction uncertainty of the object. Darker colors indicate higher values. We sort regions by uncertainty and label the top k regions. (Color figure online)

Though we obtain the feature representations of unknown instances, we have to deal with the problem that the feature representations of the unknown class can be various. Any class that does not exist in the annotations of the training set can be regarded as an unknown class, so the obtained features of the unknown class may be various from each other. To generate similar feature representations for the unknown class, we introduce the contrastive clustering module. The above pseudo-labeled instances will be employed in the contrastive clustering task to construct the prototype vector for the unknown class. More details on contrastive clustering can be found in Sect. 2.2. After comparative clustering, we can provide a prototype vector for each class, and the features of objects from different classes are separable.

In the end, with the original regression loss function in faster R-CNN unchanged, our training loss is:

$$L = L_{EDL} + L_{REG} + L_{CC} \tag{5}$$

3.3 Identify Unknown Objects with Uncertainty

Our pseudo-labeling effectively avoids being confused by unlabeled unknown objects in the training data. But we have to notice that the detector will encounter lots of objects which do not exist in training images at inference. The unknown objects in the training set can be regarded as a small subset of the whole unknown objects. The detector will lack generalizability when deployed in the real world. Based on our hypothesis, it is natural to identify unknown objects by their uncertainty during inference. We can set a certain uncertainty threshold, and the class of an object is unknown if its uncertainty is higher than the threshold.

Then we need to deal with the threshold setting problem. Manually setting a fixed threshold lacks mathematical interpretability. For the tricky hyperparameter setting problem, inspired by [2], we use a validation set that only contains annotations of known instances and compute their logarithmic total evidence ($logS$). We set a group of class-wise thresholds according to these $logS$ to avoid manually setting a threshold.

Specifically, let the threshold vector of unknown $\delta \in R^K$. For Class i, $1 \leq i \leq K$, record the $logS$ when an instance is classified to class i, such that we have $\Omega_i = \{logS : y = i\}$. The calculation of S is explained in Sect. 3.1. For each class, we fit a Gaussian distribution by its recorded $logS$ and obtain the class-specific mean μ and standard deviation σ. According to the "three-sigma" rule, for class i, we set $\delta_i = \mu_i - 2 * \alpha_i$, which is because the $logS$ of only 2.3% of objects is less than δ_i, when classified as class i. These objects can be considered as having insufficient evidence, so they are more likely to be unknown objects. We formalize the above operation as:

$$y = \begin{cases} i, & logS \geq \delta_i, i = \arg\max_{1 \leq c \leq K} \alpha_c \\ unknown, & logS < \delta_i, i = \arg\max_{1 \leq c \leq K} \alpha_c \end{cases} \qquad (6)$$

4 Experiments

4.1 Experimental Settings

Data Treatment. We choose Pascal VOC and MS-COCO to conduct our experiments. We set three tasks to evaluate the performance of the detectors in simple and complex open-set situations. In task 1, we regard the first 15 classes in Pascal VOC as known and the left five classes as unknown. We use the original test set, val set and training set of Pascal VOC 2007. In task 2, we regard the 20 classes in Pascal VOC as known classes and the other 60 classes in MS-COCO as unknown. To achieve this, we group the Pascal VOC training set and the training images in MS-COCO that contain objects of VOC classes as the training set. In task 3, we add 20 more classes in MS-COCO as known classes and the remaining 40 classes as unknown. As for evaluation, we group the test set of Pascal VOC and the val set of MS-COCO to test the performance of detecting known and unknown objects, respectively. We use the val set of Pascal

VOC to fit the category-wise Gaussian distribution to determine the uncertainty threshold. Note that the unknown annotations are removed in the training and validation phase.

Evaluation Protocols. We use the universal mean average precision (mAP) to evaluate the detecting ability for known classes. We only have limited unknown annotations in the test set, so we choose recall as the detection metric for unknown objects. Recall shows how many annotated unknown objects can be detected.

Implementation Details. Our method extends the standard faster R-CNN with a ResNet-50 backbone. We set the batch size as 16 and the initial learning rate as 0.01. We choose the exp function as the activation function. We set the clustering momentum as 0.99, the length of queues in F_{store} as 20, and the clustering distance Δ as 10.

4.2   Comparative Studies

Although specially designed for open-world object detection, ORE outperforms the prior work [19] on open-set object detection. As ORE relies on a validation set containing labeled unknown instances to learn the energy-based unknown identifier (EBUI), for a fair comparison, we choose ORE-EBUI as the baseline to compare the performance with our method. Note that we change the original classification head of ORE to adapt to the open-set setting. We train a faster R-CNN to examine whether the unknown detection could cause a severe drop in detection performance for known classes. Table 1 shows the results comparison of our method with baseline and standard faster R-CNN on three tasks. Known mAP indicates the ability to detect known objects. Unknown Recall quantifies the ability to retrieve unknown objects, i.e., how many labeled unknown objects have been detected. The results of task 1 show that even with a slight performance drop on known detection, our method can achieve much better capability on unknown detection. The results of task 2 and task 3 show that our method performs better when applied in complex situations, thanks to our uncertainty-aware design.

Table 1. The detection results on our data splits. Known mAP indicates the ability to detect known and Unknown Recall measures how well the detector could detect labeled unknown objects in the test set. As shown in Table 1, our uncertainty-aware method outperforms the baseline on known and unknown detecting. It results in a comparable performance of detecting known classes to a closed-set detector.

Task ID →	Task 1		Task 2		Task 3	
Method ↓	Known mAP (↑)	Unknown recall (↑)	Known mAP (↑)	Unknown recall (↑)	Known mAP (↑)	Unknown recall (↑)
faster R-CNN	70.69	–	57.79	–	51.40	–
ORE-EBUI	69.54	19.66	56.44	6.08	51.25	3.11
Ours	67.21	27.21	56.67	9.90	52.16	3.90

Figure 4 shows the comparison of our method, faster R-CNN and ORE on example images. It can be seen that our method achieves satisfactory performance in detecting known objects. Our method can mitigate the impact of the presence of unknown objects on known object detection (shown in Fig. 4b, 4g). The false negative (shown in Fig. 4i–4j) and the false positive detections (shown in Fig. 4j–4k) indicate the limitation of our method.

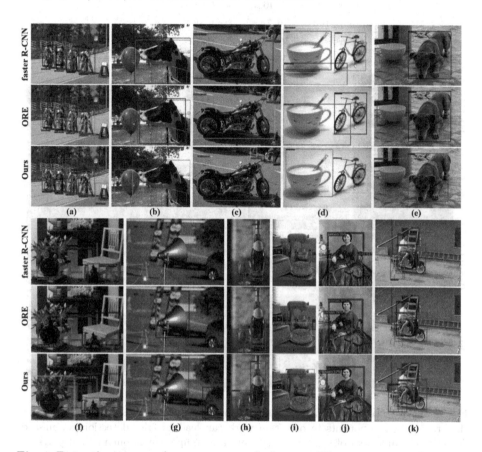

Fig. 4. Detection comparisons on example images. We compare some detection results of faster R-CNN, ORE and our method in the above three tasks. In each of above images, our method performs better in detecting unknown objects than ORE. As for known instances, the detection performance of our method is comparable to faster R-CNN and ORE.

We also perform ablation experiments on task 2 to study the effect of each component of our method, including Contrastive Clustering (CC) and Identify Unknown with Uncertainty (IUU). Table 2 shows the ablation results. The absence of CC or IUU will cause a performance drop. Each component of our architecture contributes to open-set object detection and plays an irreplaceable role (Fig. 5).

Table 2. Results of ablation study of our method. It can be seen that each component in our architecture has an irreplaceable effect on open-set detection. CC refers to "Contrastive Clustering", and IUU refers to "Identify Unknown with Uncertainty".

Method	Known mAP	Unknown recall
ORE-EBUI	56.44	6.08
Ours-CC	53.46	6.58
Ours-IUU	**56.74**	6.38
Ours	56.67	**9.90**

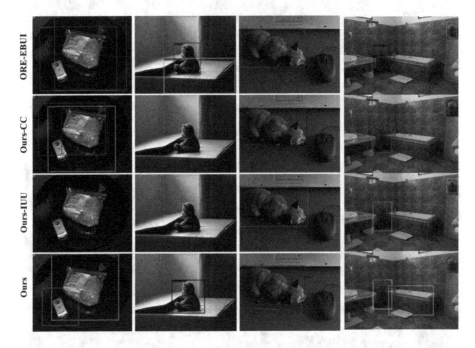

Fig. 5. Results of ablation experiments on example images. The results show that removing components in our framework can lead to false detections or missed detections of unknown objects, proving that each component of our method plays an important role.

5   Conclusion

In this paper, we propose a novel uncertainty-aware open-set object detection framework. We hypothesize that object detectors will generate higher uncertainty for unknown objects and design the model architecture based on this assumption. We aim to detect known and unknown objects in the open-set conditions, which better simulates the real world than the close-set setting that regards unknown objects as background. The key of our framework is to obtain training data for detecting unknown objects through uncertainty estimation.

Experiments are conducted on three data splits based on Pascal VOC and MS-COCO. The results show that our uncertainty-aware method outperforms the baseline on the open-set object detection task. This paper focuses on label uncertainty. In future work we will explore the effect of spatial uncertainty in open-set object detection.

Acknowledgments. This work was supported by National Natural Science Foundation of China (Serial Nos. 61976134, 61991410, 61991415), Natural Science Foundation of Shanghai (Serial No. 21ZR1423900) and Open Project Foundation of Intelligent Information Processing Key Laboratory of Shanxi Province, China (No. CICIP2021001).

References

1. Bendale, A., Boult, T.E.: Towards open set deep networks. In: Proceedings of the IEEE Conference on Computer Vision and Pattern Recognition, pp. 1563–1572 (2016)
2. Chen, L., Lou, Y., He, J., Bai, T., Deng, M.: Evidential neighborhood contrastive learning for universal domain adaptation (2022)
3. Dempster, A.P.: A generalization of Bayesian inference. J. Roy. Stat. Soc.: Ser. B (Methodol.) **30**(2), 205–232 (1968)
4. Dhamija, A., Gunther, M., Ventura, J., Boult, T.: The overlooked elephant of object detection: open set. In: Proceedings of the IEEE/CVF Winter Conference on Applications of Computer Vision, pp. 1021–1030 (2020)
5. Gal, Y., Ghahramani, Z.: Dropout as a Bayesian approximation: representing model uncertainty in deep learning. In: International Conference on Machine Learning, pp. 1050–1059. PMLR (2016)
6. Gawlikowski, J., et al.: A survey of uncertainty in deep neural networks. arXiv preprint arXiv:2107.03342 (2021)
7. Geng, C., Huang, S.J., Chen, S.: Recent advances in open set recognition: a survey. IEEE Trans. Pattern Anal. Mach. Intell. **43**(10), 3614–3631 (2020)
8. Girshick, R.: Fast R-CNN. In: Proceedings of the IEEE International Conference on Computer Vision, pp. 1440–1448 (2015)
9. Han, Z., Zhang, C., Fu, H., Zhou, J.T.: Trusted multi-view classification. arXiv preprint arXiv:2102.02051 (2021)
10. He, K., Gkioxari, G., Dollár, P., Girshick, R.: Mask R-CNN. In: Proceedings of the IEEE International Conference on Computer Vision, pp. 2961–2969 (2017)
11. Jaiswal, A., Babu, A.R., Zadeh, M.Z., Banerjee, D., Makedon, F.: A survey on contrastive self-supervised learning. Technologies **9**(1), 2 (2020)
12. Joseph, K., Khan, S., Khan, F.S., Balasubramanian, V.N.: Towards open world object detection. In: Proceedings of the IEEE/CVF Conference on Computer Vision and Pattern Recognition, pp. 5830–5840 (2021)
13. Jøsang, A.: Subjective Logic: A Formalism for Reasoning Under Uncertainty. Springer, Cham (2016). https://doi.org/10.1007/978-3-319-42337-1
14. Lakshminarayanan, B., Pritzel, A., Blundell, C.: Simple and scalable predictive uncertainty estimation using deep ensembles. In: Advances in Neural Information Processing Systems, vol. 30 (2017)
15. Lin, T.Y., Goyal, P., Girshick, R., He, K., Dollár, P.: Focal loss for dense object detection. In: Proceedings of the IEEE International Conference on Computer Vision, pp. 2980–2988 (2017)

16. Liu, W., et al.: SSD: single shot multibox detector. In: Leibe, B., Matas, J., Sebe, N., Welling, M. (eds.) ECCV 2016. LNCS, vol. 9905, pp. 21–37. Springer, Cham (2016). https://doi.org/10.1007/978-3-319-46448-0_2

17. Liu, W., Yue, X., Chen, Y., Denoeux, T.: Trusted multi-view deep learning with opinion aggregation. In: The 36th AAAI Conference on Artificial Intelligence (AAAI-2022), vol. 36, pp. 7585–7593 (2022)

18. Malinin, A., Gales, M.: Predictive uncertainty estimation via prior networks. In: Advances in Neural Information Processing Systems, vol. 31 (2018)

19. Miller, D., Nicholson, L., Dayoub, F., Sünderhauf, N.: Dropout sampling for robust object detection in open-set conditions. In: 2018 IEEE International Conference on Robotics and Automation (ICRA), pp. 3243–3249. IEEE (2018)

20. Miller, D., Sünderhauf, N., Milford, M., Dayoub, F.: Uncertainty for identifying open-set errors in visual object detection. IEEE Robot. Autom. Lett. **7**(1), 215–222 (2022). https://doi.org/10.1109/LRA.2021.3123374

21. Miller, D., Sünderhauf, N., Milford, M., Dayoub, F.: Uncertainty for identifying open-set errors in visual object detection. IEEE Robot. Autom. Lett. **7**(1), 215–222 (2021)

22. Molchanov, D., Lyzhov, A., Molchanova, Y., Ashukha, A., Vetrov, D.: Greedy policy search: a simple baseline for learnable test-time augmentation. arXiv preprint arXiv:2002.09103 (2020)

23. Neal, L., Olson, M., Fern, X., Wong, W.-K., Li, F.: Open set learning with counterfactual images. In: Ferrari, V., Hebert, M., Sminchisescu, C., Weiss, Y. (eds.) ECCV 2018. LNCS, vol. 11210, pp. 620–635. Springer, Cham (2018). https://doi.org/10.1007/978-3-030-01231-1_38

24. Oberdiek, P., Rottmann, M., Gottschalk, H.: Classification uncertainty of deep neural networks based on gradient information. In: Pancioni, L., Schwenker, F., Trentin, E. (eds.) ANNPR 2018. LNCS (LNAI), vol. 11081, pp. 113–125. Springer, Cham (2018). https://doi.org/10.1007/978-3-319-99978-4_9

25. Redmon, J., Divvala, S., Girshick, R., Farhadi, A.: You only look once: unified, real-time object detection. In: Proceedings of the IEEE Conference on Computer Vision and Pattern Recognition, pp. 779–788 (2016)

26. Redmon, J., Farhadi, A.: YOLO9000: better, faster, stronger. In: Proceedings of the IEEE Conference on Computer Vision and Pattern Recognition, pp. 7263–7271 (2017)

27. Ren, S., He, K., Girshick, R., Sun, J.: Faster R-CNN: towards real-time object detection with region proposal networks. In: Advances in Neural Information Processing Systems, vol. 28 (2015)

28. Salimans, T., Kingma, D.P.: Weight normalization: a simple reparameterization to accelerate training of deep neural networks. In: Advances in Neural Information Processing Systems, vol. 29 (2016)

29. Scheirer, W.J., Jain, L.P., Boult, T.E.: Probability models for open set recognition. IEEE Trans. Pattern Anal. Mach. Intell. **36**(11), 2317–2324 (2014)

30. Scheirer, W.J., de Rezende Rocha, A., Sapkota, A., Boult, T.E.: Toward open set recognition. IEEE Trans. Pattern Anal. Mach. Intell. **35**(7), 1757–1772 (2012)

31. Sensoy, M., Kaplan, L., Kandemir, M.: Evidential deep learning to quantify classification uncertainty. In: Advances in Neural Information Processing Systems, vol. 31 (2018)

32. Stevens, W.: Efficient uncertainty estimation for open-set object detection. In: Epistemic Uncertainty Estimation for Object Detection in Open-Set Conditions, p. 91 (2021)

33. Valdenegro-Toro, M.: Deep sub-ensembles for fast uncertainty estimation in image classification. arXiv preprint arXiv:1910.08168 (2019)
34. Welling, M., Teh, Y.W.: Bayesian learning via stochastic gradient Langevin dynamics. In: Proceedings of the 28th International Conference on Machine Learning (ICML-2011), pp. 681–688 (2011)
35. Wen, Y., Liu, W., Weller, A., Raj, B., Singh, R.: SphereFace2: binary classification is all you need for deep face recognition. arXiv preprint arXiv:2108.01513 (2021)
36. Wen, Y., Tran, D., Ba, J.: BatchEnsemble: an alternative approach to efficient ensemble and lifelong learning. arXiv preprint arXiv:2002.06715 (2020)
37. Yue, X., Chen, Y., Yuan, B., Lv, Y.: Three-way image classification with evidential deep convolutional neural networks. Cogn. Comput. 1–13 (2021). https://doi.org/10.1007/s12559-021-09869-y
38. Zhou, X., Yue, X., Xu, Z., Denoeux, T., Chen, Y.: Deep neural networks with prior evidence for bladder cancer staging. In: 2021 IEEE International Conference on Bioinformatics and Biomedicine (BIBM), pp. 1221–1226. IEEE (2021)
39. Zhu, X., Su, W., Lu, L., Li, B., Wang, X., Dai, J.: Deformable DETR: deformable transformers for end-to-end object detection. arXiv preprint arXiv:2010.04159 (2020)

Rule Acquisition in Generalized One-Sided Decision Systems

Zhiyong Hu[1], Mingwen Shao[2]([✉]), and Meishe Liang[3]

[1] College of Control Science and Engineering, China University of Petroleum
(East China), Qingdao, China
[2] College of Computer Science and Technology, China University of Petroleum
(East China), Qingdao, China
smw278@126.com
[3] Department of Mathematics and Physics, Shijiazhuang Tiedao University,
Shijiazhuang, China

Abstract. As a major direction of data mining, rule acquisition has
been applied and researched widely. However, there are few studies on
association rules determined by ordered attributes. In this paper, we
propose methods of mining decision rules based on situations where only
condition attributes are ordered (generalized one-sided formal decision
context) and both the condition and decision attributes are ordered (generalized one-sided ordered formal decision context), to discuss the influence of ordered attributes on decision rules. Moreover, the discussion
is carried out to explain relations between decision rules based on generalized one-sided formal decision contexts and decision rules based on
generalized one-sided ordered formal decision contexts. Furthermore, the
rules based on ordered decision attributes pay more attention to global
information while the rules constructed by disordered decision attributes
pay more attention to local information through the experimental results
and theoretical derivation.

Keywords: Concept lattice · Generalized one-sided formal context ·
Rule acquisition

1 Introduction

Data mining [1] aims to extract interesting knowledge from a big data set. As
a means of data processing, granular computing was proposed by Lin [2]. The
main idea of granular computing is to simplify a complex problem into several
simple ones. The similar elements are processed as a class [3,4] in the research.

This work was supported by the grants from the National Natural Science Foundation
of China (Nos. 61673396, 61976245), National Key Research and development Program
of China (2021YFA1000102) and the Fundamental Research Funds for the Central
Universities (18CX02140A).

J. Yao et al. (Eds.): IJCRS 2022, LNAI 13633, pp. 176–190, 2022.
https://doi.org/10.1007/978-3-031-21244-4_13

This class is regarded as an information granule or particle. One of the advanced researches field in granular computing is the rough set [5].

As a method of knowledge representation and data mining, formal concept analysis [6] was first proposed by Wille. Formal concepts [7–9] are basic research objects for formal concept analysis and reflect the interdependence between attributes and objects. Since its inception, formal concept analysis has become a rapidly developing field in applied information science and theoretical information science [10–12].

Multi-valued ordered attributes commonly exist in practical applications. And generalized one-sided concept lattices [13] analyze object-attribute models with different structures for truth values of attributes. On this basis, Shao et al. [14] gave a method for attribute reduction in the generalized one-sided formal context to reduce information while keeping the original lattice unchanged. Besides, Liang et al. [15] extended the generalized one-sided formal context to an intuitionistic fuzzy information system to make the attribute reduction. Furthermore, Shao et al. [16] combined generalized one-sided concept lattices and multi-granularity concept lattices to explore a transformation algorithm of lattice structures between different granularities.

In this paper, we present algorithms to extract quantitative association rules in multi-valued ordered formal decision contexts directly. Compared with existing algorithms, the algorithms in this paper needn't transform the original multi-valued context into a binary context. Moreover, the influence of multi-valued ordered attributes on mining decision rules is explored in the contexts where only condition attributes are ordered and both the condition and decision attributes are ordered respectively. Furthermore, the relation between decision rules in the decision context and decision rules in the order context is studied.

2   Preliminaries

In this section, we briefly recall some previous results about generalized one-sided formal contexts [13] and traditional association rules [1].

2.1   Generalized One-Sided Formal Context

Generally, a formal context can be expressed as an ordered 3-tuple (U, A, R), where $U = \{x_1, x_2, ..., x_n\}$ is a set constructed by objects and $A = \{a_1, a_2, ..., a_m\}$ is a set constructed by attributes. $R \subseteq U \times A$ is a relation between U and A. Generally, attribute values in (U, A, R) aren't considered as ordered, which is contrary to human cognition. Thus, Butka et al. [13] proposed generalized one-sided formal contexts to deal with the situation.

Definition 1 [13]. *Let a 4-tuple $(U, A, \mathcal{L}, R)$ be a generalized one-sided formal context, if the following conditions are satisfied:*

(1) U is a non-empty set constructed by objects, and A is a non-empty set constructed by attributes.

(2) $\pounds : A \to CL$ is a mapping between an attribute set and a class of complete lattices. Here CL is a class of complete lattices about the attribute value of each attribute. Then $\pounds(a)$ is seen as the structure of truth values for any attribute $a \in A$.

(3) R is a generalized incidence relation, i.e., for any $x \in U$ and $a \in A$, $R(x, a) \in \pounds(a)$ represents the degree, from the structure $\pounds(a)$, that the object $x \in U$ possesses the attribute a.

Normally, the structure of truth values $\pounds(a_i)$ is considered to be the order of attribute values for each a_i.

Example 1. The generalized one-sided formal context $(U, A, \pounds, R)$ is shown as Table 1, where the set of objects is $U = \{x_1, x_2, x_3, x_4\}$, the set of attributes is $A = \{a_1, a_2, a_3, a_4\}$. $\pounds(a_1), \pounds(a_2), \pounds(a_3), \pounds(a_4)$ represent the corresponding structure of truth values for a_1, a_2, a_3, a_4 which are shown in Fig. 1. Table 1 also illustrates the generalized incidence relation R.

Table 1. The generalized one-sided formal context.

	a_1	a_2	a_3	a_4
x_1	2	1	1	1
x_2	3	2	2	3
x_3	1	2	1	2
x_4	3	3	2	3

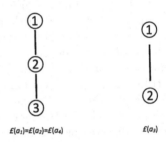

$\pounds(a_1)=\pounds(a_2)=\pounds(a_4)$ $\pounds(a_3)$

Fig. 1. Structure of truth values.

Let $(U, A, \pounds, R)$ be a generalized one-sided formal context. The power set of U is $P(U)$. $\prod_{a \in A} \pounds(a)$ is the direct product for the structure of truth values. The mappings $\uparrow : P(U) \to \prod_{a \in A} \pounds(a)$ and $\downarrow : \prod_{a \in A} \pounds(a) \to P(U)$ are given below:

$$\uparrow (X)(a) = \bigwedge_{x \in X} R(x, a), \downarrow (g) = \{x \in U \mid \forall a \in A, g(a) \leq R(x, a)\},$$

where $X \subseteq U, a \in A, g \in \prod_{a \in A} \pounds(a)$.

Let $\varphi : P \to Q$ and $\psi : Q \to P$ be the mappings between the partially ordered set $(P, \leq)$ and the partially ordered set $(Q, \leq)$. If the pair (ψ', φ') satisfies the conditions: (1) $p_1 \leq p_2 \Rightarrow \varphi' p_2 \leq \varphi' p_1$, (2) $q_1 \leq q_2 \Rightarrow \psi' q_2 \leq \psi' q_1$, (3) $p \leq \psi' \varphi' p, q \leq \psi' \varphi' q$, (ψ', φ') forms a Galois connection between P and Q. And then the pair of mappings $(\uparrow, \downarrow)$ constitutes a Galois connection in the generalized one-sided formal context $(U, A, \pounds, R)$. $c_{\uparrow \downarrow} : \prod_{a_i \in A} \pounds(a_i) \to \prod_{a_i \in A} \pounds(a_i)$ is a compound operation in the generalized one-sided formal context $(U, A, \pounds, R)$, which can be calculated by $c_{\uparrow \downarrow}(g) = \uparrow (\downarrow (g))$ for $g \in \prod_{a_i \in A} \pounds(a_i)$. And the equation $\| \downarrow (g) \| = \| \downarrow (c_{\uparrow \downarrow}(g)) \|$ can be derived from the properties of Galois connections.

Let $(U, A, \mathcal{L}, R)$ be a generalized one-sided formal context. For $X \subseteq U$, $g \in \prod_{a \in A} \mathcal{L}(a)$, the pair (X, g) is described as a generalized one-sided formal concept, if the following conditions are satisfied:

$$\uparrow (X) = g, \downarrow (g) = X.$$

X and g are the extent and intent of the generalized one-sided formal concept, respectively. $B(U, A, \mathcal{L}, R)$ is a set containing all generalized one-sided formal concepts of $(U, A, \mathcal{L}, R)$.

(X_1, g_1) and (X_2, g_2) are generalized one-sided formal concepts in $B(U, A, \mathcal{L}, R)$. If $X_1 \subseteq X_2$ or $g_2 \leq g_1$, then $(X_1, g_1) \leq (X_2, g_2)$. $\leq$ is a partial order on $B(U, A, \mathcal{L}, R)$. All concepts in $B(U, A, \mathcal{L}, R)$ structured by partial order $\leq$ constitutes a complete lattice called a generalized one-sided concept lattice, denoting as $\mathbf{B}(U, A, \mathcal{L}, R)$.

Example 2. A generalized one-sided concept lattice based on the generalized one-sided formal context (see Table 1) is shown in Fig. 2.

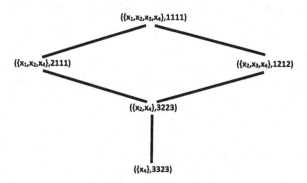

Fig. 2. The generalized one-sided concept lattice.

2.2   Traditional Association Rules

The background of rule acquisition is made by A^*, T and U^*. $A^* = \{a_1^*, a_2^*, ..., a_m^*\}$ is a set including m items. $\forall X \subseteq A^*$ is called an itemset. t_j is a transaction consisting of items, $t_j \subseteq A^*$. $T = \{t_1, t_2, ..., t_n\}$ is a transaction data set. Each t_j has a particular identifier x_j, and all identifiers make a set U^*. Taking the classic shopping problem as an example, $A^* = \{milk, sugar, bread, egg\}$, $T = \{\{milk, sugar, bread\}, \{milk, egg\}, \{milk, bread, egg\}, \{sugar, bread\}\}$, $U^* = \{x_1, x_2, x_3, x_4\}$. Table 2 shows a background of rule acquisition.

Table 2. The background of rule acquisition.

U^*	T
x_1	Milk, sugar, bread
x_2	Milk, egg
x_3	Milk, bread, egg
x_4	Sugar, bread

Table 3. The shopping formal context.

	Milk	Sugar	Bread	Egg
x_1	1	1	1	0
x_2	1	0	0	1
x_3	1	1	0	1
x_4	0	1	1	0

Definition 2 [1]. *Let* $A^* = \{a_1^*, a_2^*, ..., a_m^*\}$ *be a set of items.* $T = \{t_1, t_2, ..., t_n\}$ *is a transaction data set. For* $B \subseteq A^*$, *the support of* B *is calculated by* $\sigma(B) = \| \{t_j | B \subseteq t_j, t_j \in T\} \|$. *If* $\sigma(B)$ *is not less than the preset threshold* p *which is called the support threshold of the itemset, then* B *is a frequent itemset in the background.*

Here the operation $\| \cdot \|$ is known as the cardinality operation.

Definition 3 [1]. *Let the transaction data set* T, *the set of items* A^* *and the set of identifiers* U^* *be preset sets in the background of rule acquisition. An association rule in the background is expressed as an implicit formula*

$$M \Rightarrow N,$$

where $M \bigcap N \neq \phi$. $M \subseteq A^*$ *and* $N \subseteq A^*$ *are called the antecedent and consequent of the rule, respectively. The support of the rule is calculated by*

$$s = \frac{\sigma(M \bigcup N)}{\| T \|}.$$

The confidence of the rule is calculated by

$$c = \frac{\sigma(M \bigcup N)}{\sigma(M)}.$$

Let h *be the support threshold of the rule and* w *be the confidence threshold of the rule. If* $s \geq h, c \geq w$, *the rule is an* (h, w) *rule.*

If the confidence of a rule is 1, then the rule is called a certain rule, otherwise called a possible rule. Generally, a background of rule acquisition can be transformed into a formal context. The set of items A^* is represented by the attribute set A. The object set U becomes a set of identifiers U^*. And the relation R between U and A turns to a transaction data set T. If the item a_q^* is possessed by the transaction t_j, then $(t_j, a_q^*) = 1$, otherwise $(t_j, a_q^*) = 0$. For example, the shopping background (Table 2) becomes the formal context (Table 3).

3   Generalized One-Sided Decision Rule

Let $(U^+, A^+, \pounds^+, R^+, D^+, I^+)$ be a generalized one-sided formal decision context (GOFDC), where $(U^+, A^+, \pounds^+, R^+)$ makes a generalized one-sided formal context, D^+ is a non-empty finite set of decision attributes, $A^+ \cap D^+ = \phi$, and I^+ is a decision incidence relation between U^+ and D^+.

Example 3. Table 4 shows a $GOFDC$ $(U^+, A^+, \pounds^+, R^+, D^+, I^+)$, where $U^+ = \{x_1, x_2, x_3, x_4, x_5, x_6\}$ is a set of objects, $A^+ = \{a_1, a_2\}$ is a set of condition attributes. The set of decision attributes is $D^+ = \{d\}$. $\parallel \pounds^+(a_1) \parallel = 3, \parallel \pounds^+(a_2) \parallel = 4$. The decision incidence relation in GOFDC is represented in Table 5.

Table 4. $GOFDC$

	a_1	a_2	d
x_1	1	1	1
x_2	1	1	1
x_3	3	4	2
x_4	2	3	2
x_5	2	2	2
x_6	2	2	2

Table 5. The decision incidence relation.

	x_1	x_2	x_3	x_4	x_5	x_6
d	1	1	2	2	2	2

Definition 4. *Let* $(U^+, A^+, \pounds^+, R^+, D^+, I^+)$ *be a GOFDC.* $\frac{U^+}{I^+} = \{P_1, P_2, ..., P_t\}$ *is a partition of* U^+, *where* $P_i = [x]_{D^+} = \{y | I^+(x) = I^+(y), x, y \in U^+\}$ *is a decision equivalence class,* $P_i \cap P_j = \phi$, $i, j \in \{1, 2, ..., t\}$. *If the cardinality of* P_i *is greater than the support threshold* p *of the itemset, then* $I^+(P_i)$ *as the value of* P_i *is a decision frequent itemset.*

Definition 5. *Let* $(U^+, A^+, \pounds^+, R^+, D^+, I^+)$ *be a GOFDC. The association rule in the context is known as an implication*

$$g \Rightarrow I^+(P_i),$$

where $g \in \prod_{a \in A^+} \pounds^+(a)$ *is the antecedent of the rule and* $I^+(P_i)$ *is the consequent of the rule,* $P_i \in \frac{U^+}{I^+}$. *The decision association rules under the GOFDC are named generalized one-sided decision rules.*

$$s = \frac{\parallel \downarrow (g) \cap P_i \parallel}{\parallel U \parallel},$$

is the support of the rule.

$$c = \frac{\parallel \downarrow (g) \cap P_i \parallel}{\parallel \downarrow (g) \parallel},$$

is the confidence of the rule. Let h be the support threshold of the rule and w be the confidence threshold of the rule. If $s \geq h, c \geq w$, the decision rule is an (h, w) generalized one-sided decision rule.

The generalized one-sided decision rule is an implication between a combination of values for condition attributes and corresponding values of a decision class.

Proposition 1. *Let $(U^+, A^+, \pounds^+, R^+, D^+, I^+)$ be a GOFDC. For $g \in \prod_{a \in A^+} \pounds^+(a)$ and $P_i \in \frac{U^+}{I^+}$, $g \Rightarrow I^+(P_i)$ and $c_{\uparrow\downarrow}(g) \Rightarrow I^+(P_i)$ have the same support and confidence.*

Proof. It follows directly by applying the equation $\| \downarrow (g) \| = \| \downarrow (c_{\uparrow\downarrow}(g)) \|$.

From Proposition 1, the support and confidence of the rule are fixed whether an itemset is functioned by closure operation or not. Therefore, mining generalized one-sided decision rules can be converted to discuss relationships between generalized one-sided formal concept lattices and decision partitions.

Proposition 2. *Let $(U^+, A^+, \pounds^+, R^+, D^+, I^+)$ be a GOFDC, $g_1, g_2 \in \prod_{a \in A^+} \pounds^+(a)$, $g_1 \leq g_2$. P_i is a decision class in this context, $P_i \subseteq \downarrow (g_2)$. If $g_1 \Rightarrow I^+(P_i)$ is an (h, w) generalized one-sided decision rule, then $g_2 \Rightarrow I^+(P_i)$ is an (h, w) generalized one-sided decision rule.*

Proof. Due to $g_1 \leq g_2$, the formula $\downarrow (g_2) \subseteq \downarrow (g_1)$ can be obtained. And then $P_i \subseteq \downarrow (g_1)$. Therefore, the result can be derived from Definition 5.

The antecedents of generalized one-sided decision rules are monotonous.

Proposition 3. *Let $(U^+, A^+, \pounds^+, R^+, D^+, I^+)$ be a GOFDC, $g_1, g_2 \in \prod_{a \in A^+} \pounds^+(a)$, $g_1 \leq g_2$. P_i is a decision class in this context, $P_i \subseteq \downarrow (g_2)$. If $g_1 \Rightarrow I^+(P_i)$ is an (h, w) generalized one-sided decision rule, then $c_{\uparrow\downarrow}(g_2) \Rightarrow I^+(P_i)$ is an (h, w) generalized one-sided decision rule.*

Proof. Follows directly by applying Proposition 2 and $\downarrow (g_2) = \downarrow (c_{\uparrow\downarrow}(g_2))$.

Example 4. In the generalized one-sided formal decision context (Table 4), the rules constructed by the generalized one-sided condition concept lattice and the decision partition are shown in Table 6.

Table 6. GOFDC.

Rule	s	c	Rule	s	c
$11 \Rightarrow I(P_1)$	1/3	1/3	$11 \Rightarrow I(P_2)$	2/3	2/3
$22 \Rightarrow I(P_1)$	0	0	$22 \Rightarrow I(P_2)$	2/3	1
$23 \Rightarrow I(P_1)$	0	0	$23 \Rightarrow I(P_2)$	1/3	1
$34 \Rightarrow I(P_1)$	0	0	$34 \Rightarrow I(P_2)$	1/6	1

Theorem 1. *Let $(U^+, A^+, \pounds^+, R^+, D^+, I^+)$ be a GOFDC. $\{P_1, P_2, ..., P_t\}$ is a partition in the context. $g \in \prod_{a \in A^+} \pounds^+(a)$ is an itemset represented by condition attributes. c_i is the confidence of $g \Rightarrow I^+(P_i), i = 1, 2..., t$. There is an equation $\sum_{i=1}^{t} c_i = 1$.*

Proof. Since $\{P_1, P_2, ..., P_t\}$ is a partition, $U^+ = \bigcup_{i=1}^{t} P_i$, $P_i \bigcap P_j = \phi$. Then $\parallel P_i \bigcup P_j \parallel = \parallel P_i \parallel + \parallel P_j \parallel$. From the definition about the confidence of the rule, equation $\sum_{i=1}^{t} c_i = \sum_{i=1}^{t} \frac{\parallel \downarrow(g) \bigcap P_i \parallel}{\parallel \downarrow(g) \parallel}$ can be found. Therefore, $\sum_{i=1}^{t} c_i = \frac{\parallel \downarrow(g) \bigcap U^+ \parallel}{\parallel \downarrow(g) \parallel} = \frac{\parallel \downarrow(g) \parallel}{\parallel \downarrow(g) \parallel} = 1$.

Nextly, the generalized one-sided decision rule mining algorithm is studied. Before studying the decision association rule mining algorithm, finding the method of mining multi-valued frequent itemsets is necessary. The novel frequent itemset mining algorithm directly extracts multi-valued itemsets in a multi-valued context while traditional algorithms normally convert the original multi-valued context into a Boolean context.

Let $(U^+, A^+, \pounds^+, R^+, D^+, I^+)$ be a GOFDC. The frequent itemsets existing in the condition part of the formal decision context can be generated by the set $JI = \{l_{11}, ..., l_{1j_{(a1)}}, ..., l_{i1}, ..., l_{ij_{(ai)}}\}$, where $l = \prod_{a_i \in A} min\pounds(a_i)$, $o_{ij_{(ai)}} \in \pounds(a_i)$, $l_{ij_{(ai)}} = l \bigvee o_{ij_{(ai)}}$. i represents the index of the attribute, and $j_{(ai)}$ illustrates the index of the value for attribute a_i. Specifically, the $j_{(ai)}$ th value of the attribute a_i is $o_{ij_{(ai)}}$. For the common understanding in data mining, an itemset may be frequent only if the itemset used to generate it is frequent. On this basis, the mining algorithm of multi-valued frequent itemsets is shown as Algorithm 1.

Algorithm 1. Mining Frequent Itemsets

Input: generalized one-sided formal context: K
support threshold of itemset: minSup
Output: frequent itemset: FrIs
1: produce the generating base JI
2: **for** all $l \in JI$ **do**
3:     **if** $\sigma(l) \geq$ minSup **then**
4:         $B' = [l, B']$
5: $FrIs = B'$
6: **for** $i = 1 : size(K, 2)$ **do**
7:     **for** all $a \in FrIs$ **do**
8:         **for** all $c \in B'$ **do**
9:             $d = a \bigvee c$
10:             **if** $\sigma(d) \geq$ minSup **then**
11:                 $FrIs = [b, FrIs]$
12:                 $FrIs = unique(FrIs,' rows')$
13: **return** $FrIs$

Finally, the generalized one-sided decision rule mining algorithm is given as Algorithm 2.

Algorithm 2. Mining Generalized One-sided Decision Rule

Input: generalized one-sided formal decision context: $K = (U^+, A^+, \pounds^+, R^+, D^+, I^+)$
support threshold: minSup
confidence threshold: minCon
Output: decision rule: (h,w) rule

1: $Fr = \phi, DF = \phi, Rule = \phi$;
2: $Fr = \{all\ condition\ frequent\ itemsets\ in\ K\}$
3: $DF = \{all\ decision\ classes\ in\ K\}$
4: $l = size(Fr, 1); o = size(DF, 1)$;
5: **for** t=1 to l **do**
6:    $Ac = Fr(t, ;)$;
7:    **for** e=1 to o **do**
8:       $Co = DF(e, ;)$;
9:       **if** $\frac{\|\downarrow(Ac) \bigcup Co\|}{\|U\|} \geq$ minSup **then**
10:          **if** $\frac{\|\downarrow(Ac) \bigcup Co\|}{\|\downarrow(Ac)\|} \geq$ minCon **then**
11:             $Rule = Rule \bigcup \{Ac \Rightarrow I^+(Co)\}$;
12: **return** $Rule$

4   Generalized One-Sided Ordered Decision Rule

For practical applications, decision attributes are also considered to be ordered in many cases, such as judging the grade of wine by known conditions. The decision attribute "grade" is ordered.

Let the context $(U', A', \pounds', R', D', \pounds'_D, I')$ be a generalized one-sided ordered formal decision context, where A' is a non-empty finite set of condition attributes, D' is a non-empty finite set of decision attributes and $A' \bigcap D' = \phi$. $(U', A', \pounds', R')$ and $(U', D', \pounds'_D, I')$ are generalized one-sided formal contexts induced by the condition part and decision part of the context respectively. $(\uparrow'_C, \downarrow'_C)$ is a Galois connection in $(U', A', \pounds', R')$.

$$\uparrow'_C (X)(a) = \bigwedge_{x \in X} R'(x, a), \downarrow'_C (g) = \{x \in U' \mid \forall a \in A', g(a) \leq R'(x, a)\},$$

where $X \in P(U'), g \in \prod_{a \in A'} \pounds'(a)$. $(\uparrow'_D, \downarrow'_D)$ is a Galois connection in $(U', D', \pounds'_D, I')$.

$$\uparrow'_D (Y)(d) = \bigwedge_{y \in Y} I'(y, d), \downarrow'_D (e) = \{y \in U' \mid \forall d \in D', e(d) \leq I'(y, d)\},$$

where $Y \in P(U'), e \in \prod_{d \in D'} \pounds'_D(d)$.

Definition 6. *Let $(U', A', \pounds', R', D', \pounds'_D, I')$ be a generalized one-sided ordered formal decision context. For $g \in \prod_{a \in A'} \pounds'(a)$ induced by the condition part of the context, the support of the itemset is expressed by*

$$\sigma'_C(g) = \| \downarrow'_C (g) \| .$$

If $\sigma'_C(g)$ is larger than the support threshold p of the itemset, then g is a frequent condition itemset. For $e \in \prod_{d \in D'} \pounds'_D(d)$ induced by the decision part of the context, the support of the itemset is expressed by

$$\sigma'_D(e) = \| \downarrow'_D (e) \| .$$

If $\sigma'_D(e)$ is larger than the preset threshold p', then e is a frequent decision itemset.

Definition 7. Let $(U', A', \pounds', R', D', \pounds'_D, I')$ be the generalized one-sided ordered formal decision context. The decision association rule is expressed as

$$g \Rightarrow e,$$

where $g \in \prod_{a \in A'} \pounds(a)$ is the antecedent of the rule and $e \in \prod_{d \in D'} \pounds'_D(d)$ is the consequent of the rule. A decision association rule in the ordered formal decision context is made a generalized one-sided ordered decision rule.

$$s = \frac{\| \downarrow'_C (g) \cap \downarrow'_D (e) \|}{\| U' \|},$$

is the support of the rule.

$$c = \frac{\| \downarrow'_C (g) \cap \downarrow'_D (e) \|}{\| \downarrow'_C (g) \|},$$

is the confidence of the rule. For a preset pair (h, w), if $h \leq s$, $w \leq c$ and $h, w \in [0, 1]$, then the rule is known as an (h, w) generalized one-sided ordered rule.

Proposition 4. Let $(U', A', \pounds', R', D', \pounds'_D, I')$ be a generalized one-sided ordered formal decision context, $g \in \prod_{a \in A'} \pounds(a)$, $e \in \prod_{d \in D'} \pounds'_D(d)$. $c_{\uparrow'_C \downarrow'_C}(g) = \uparrow'_C (\downarrow'_C (g))$, $c_{\uparrow'_D \downarrow'_D}(e) = \uparrow'_D (\downarrow'_D (e))$.

(1) The support and confidence of $g \Rightarrow e$ and $c_{\uparrow'_C \downarrow'_C}(g) \Rightarrow e$ are equal respectively.
(2) The support and confidence of $g \Rightarrow e$ and $g \Rightarrow c_{\uparrow'_D \downarrow'_D}(e)$ are equal respectively.
(3) The support and confidence of $g \Rightarrow e$ and $c_{\uparrow'_C \downarrow'_C}(g) \Rightarrow c_{\uparrow'_D \downarrow'_D}(e)$ are equal respectively.

Proof. It can be easily proven according to the formula $\| \downarrow (g) \| = \| \downarrow (c_{\uparrow \downarrow}(g)) \|$, so it is omitted.

The concept lattices $CL_{A'}$ and $CL_{D'}$ are constructed under the condition part and decision part of the generalized one-sided ordered formal decision context $(U', A', \pounds', R', D', \pounds'_D, I')$ respectively. The generalized one-sided ordered decision rules in the context reflect the relationship between $CL_{A'}$ and $CL_{D'}$. The antecedent of the rule corresponds to the node $(\downarrow'_C (g), c_{\uparrow'_C \downarrow'_C}(g)) \in CL_{A'}$, and the consequent of the rule corresponds to the node $(\downarrow'_D (e), c_{\uparrow'_D \downarrow'_D}(e)) \in CL_{D'}$, where $g \in \prod_{a \in A'} \pounds'(a), e \in \prod_{d \in D'} \pounds'_D(d)$.

Proposition 5. *Let $(U', A', \mathcal{L}', R', D', \mathcal{L}'_D, I')$ be a generalized one-sided ordered formal decision context. $g_1, g_2 \in \prod_{a \in A'} \mathcal{L}'(a)$ are itemsets represented by condition attributes, $g_1 \leq g_2$. $e \in \prod_{d \in D'} \mathcal{L}'_D(d)$ is an itemset represented by decision attributes, $\downarrow'_D (e) \subseteq \downarrow'_C (g_2)$. If $g_1 \Rightarrow e$ is an (h, w) generalized one-sided ordered decision rule, then $g_2 \Rightarrow e$ is an (h, w) generalized one-sided ordered decision rule.*

Proof. Since $g_1 \leq g_2$, there is $\downarrow'_C (g_2) \subseteq \downarrow'_C (g_1)$. And for $\downarrow'_D (e) \subseteq \downarrow'_C (g_2)$, $\downarrow'_D (e) \subseteq \downarrow'_C (g_1)$ can be obtained. From the (h, w) generalized one-sided ordered decision rule $g_1 \Rightarrow e$,

$$s_1 = \frac{\| \downarrow'_C (g_1) \bigcap \downarrow'_D (e) \|}{\| U' \|} \geq h, c_1 = \frac{\| \downarrow'_C (g_1) \bigcap \downarrow'_D (e) \|}{\| \downarrow'_C (g_1) \|} \geq w.$$

Due to $\downarrow'_C (g_2) \subseteq \downarrow'_C (g_1)$, there is $\| \downarrow'_C (g_2) \| \leq \| \downarrow'_C (g_1) \|$. So $s_2 = s_1 \geq h$, $c_2 \geq c_1 \geq w$. Therefore, $g_2 \Rightarrow e$ is an (h, w) generalized one-sided ordered decision rule.

The antecedents of generalized one-sided ordered decision rules are also monotonous.

Proposition 6. *Let $(U', A', \mathcal{L}', R', D', \mathcal{L}'_D, I')$ be a generalized one-sided ordered formal decision context. $e_1, e_2 \in \prod_{d \in D'} \mathcal{L}'_D(d)$ are itemsets represented by decision attributes, $e_1 \leq e_2$. $g \in \prod_{a \in A'} \mathcal{L}(a)$ is an itemset represented by condition attributes. If $g \Rightarrow e_2$ is an (h, w) generalized one-sided ordered decision rule, then $g \Rightarrow e_1$ is an (h, w) generalized one-sided ordered decision rule.*

Proof. Since $e_1 \leq e_2$, $\downarrow'_D (e_2) \subseteq \downarrow'_D (e_1)$. The proposition can be easily proved by the formula and the original definition.

Proposition 7. *Let $(U', A', \mathcal{L}', R', D', \mathcal{L}'_D, I')$ be a generalized one-sided ordered formal decision context. $g_1, g_2 \in \prod_{a \in A'} \mathcal{L}(a)$ are itemsets about condition attributes, $g_1 \leq g_2$. $e_1, e_2 \in \prod_{d \in D'} \mathcal{L}'_D(d)$ are itemsets about decision attributes, $e_1 \leq e_2$. Under the premise of $\downarrow'_D (e_1) \subseteq \downarrow'_C (g_2)$, if $g_1 \Rightarrow e_2$ is an (h, w) generalized one-sided ordered decision rule, then $g_2 \Rightarrow e_1$ is also an (h, w) generalized one-sided ordered decision rule.*

Proof. The proof is similar to that of Proposition 3, so it is omitted.

In order to reduce the generating rules, the effect of extending these properties to generalized one-sided concept lattices is discussed.

Proposition 8. *Let $(U', A', \mathcal{L}', R', D', \mathcal{L}'_D, I')$ be a generalized one-sided ordered formal decision context. $g_1, g_2 \in \prod_{a \in A'} \mathcal{L}'(a)$ are condition attribute itemsets in the context, $g_1 \leq g_2$. $e \in \prod_{d \in D'} \mathcal{L}'_D(d)$ is a decision attribute itemset, $\downarrow'_D (e) \subseteq \downarrow'_C (g_2)$. If $g_1 \Rightarrow e$ is an (h, w) generalized one-sided ordered decision rule, then $c_{\uparrow'_C \downarrow'_C}(g_2) \Rightarrow e$ is an (h, w) generalized one-sided ordered decision rule.*

Proof. Easy to prove by $\| \downarrow'_C (g) \| = \| \downarrow'_C (c_{\uparrow'_C \downarrow'_C}(g)) \|$.

Proposition 9. *Let* $(U', A', \pounds', R', D', \pounds'_D, I')$ *be a generalized one-sided ordered formal decision context.* $e_1, e_2 \in \prod_{d \in D'} \pounds'_D(d)$ *are decision attribute itemsets in the context,* $e_1 \le e_2$. $g \in \prod_{a \in A'} \pounds(a)$ *is a condition attribute itemset. If* $g \Rightarrow e_2$ *is an* (h, w) *generalized one-sided ordered decision rule, then* $c_{\uparrow'_C \downarrow'_C}(g) \Rightarrow c_{\uparrow'_D \downarrow'_D}(e_1)$ *is an* (h, w) *generalized one-sided ordered decision rule.*

Proof. Easy to prove by $\| \downarrow'_C (g) \| = \| \downarrow'_C (c_{\uparrow'_C \downarrow'_C}(g)) \|$, $\| \downarrow'_D (e) \| = \| \downarrow'_D (c_{\uparrow'_D \downarrow'_D}(e)) \|$.

Example 5. Table 4 shows a generalized one-sided ordered formal decision context. The generalized one-sided ordered decision rules constructed by the generalized one-sided concept lattices based on the condition part and decision part of the formal decision context are shown in Table 7.

Table 7. Generalized one-sided ordered decision rules.

Rule	s	c	Rule	s	c
$11 \Rightarrow 1$	1	1	$11 \Rightarrow 2$	2/3	2/3
$22 \Rightarrow 1$	2/3	1	$22 \Rightarrow 2$	2/3	1
$23 \Rightarrow 1$	1/3	1	$23 \Rightarrow 2$	1/3	1
$34 \Rightarrow 1$	1/6	1	$34 \Rightarrow 2$	1/6	1

Nextly, the generalized one-sided ordered decision rule mining algorithm (Algorithm 3) is designed. Decision attribute frequent itemsets and condition attribute frequent itemsets in the related context can also be mined by applying Algorithm 1.

5 Decision Rule V.S. Ordered Decision Rule

Theorem 2. *Let* $(U, A, \pounds, R, D, I)$ *be a generalized one-sided formal decision context.* $g \Rightarrow I(P_i)$ *is a generalized one-sided decision rule in the context, where* $g \in \prod_{a \in A} \pounds(a)$, $I(P_i)$ *is the value of* P_i. s_1 *and* c_1 *are the support and confidence of the rule, respectively.* $g \Rightarrow \uparrow (P_i)$ *is a generalized one-sided ordered decision rule in the context.* s_2 *and* c_2 *are the support and confidence of the ordered rule, respectively. There are the formulas* $s_1 \le s_2$ *and* $c_1 \le c_2$.

Proof. The theorem can be proved by the formula $P_i \subseteq \downarrow (\uparrow (P_i))$.

The validity of Theorem 2 can be verified by Example 4 and Example 5. The generalized one-sided ordered decision rule has larger support and confidence than the generalized one-sided decision rule. The reason is that generalized one-sided ordered decision rules pay more attention to the global information, and generalized one-sided decision rules pay more attention to the local information. In terms of algorithm complexity, the generalized one-sided decision rule has low complexity.

Algorithm 3. Mining Generalized One-sided Ordered Decision Rule

---

   Input: generalized one-sided ordered formal decision context: K = $(U', A', \mathcal{L}', R', D', \mathcal{L}'_D, I')$
support threshold: minSup
confidence threshold: minCon
  Output: decision rule: (h,w) rule
1: $Fr = \phi, DF = \phi, Rule = \phi$;
2: $Fr = \{all\ condition\ frequent\ itemsets\ in\ K\}$
3: $DF = \{all\ decision\ frequent\ itemsets\ in\ K\}$
4: $l = size(Fr, 1); o = size(DF, 1)$;
5: **for** t=1 to l **do**
6:    $Ac = Fr(t, ;)$;
7:    **for** e=1 to o **do**
8:       $Co = DF(e, ;)$;
9:       **if** $\frac{\|\downarrow'_C(Ac) \cup \downarrow'_D(Co)\|}{\|U\|} \geq$ minSup **then**
10:          **if** $\frac{\|\downarrow'_C(Ac) \cup \downarrow'_D(Co)\|}{\|\downarrow'_C(Ac)\|} \geq$ minCon **then**
11:            $Rule = Rule \cup \{Ac \Rightarrow Co\}$;
12: **return** $Rule$

---

6 Numerical Experiment

In this section, we conduct some numerical experiments to explain the effect of the proposed methods. The experiments are run in Matlab 2018a configured on a PC with Windows 2010 64-bit operating system, CPU is Intel(R) Core(TM) i7-9750H CPU @2.60 GHz, 16.00 GB memory. Six data sets are chosen from UCI Machine Learning Repository (http://archive.ics.uci.edu/ml/datasets.php) for target domains. The relevant information about the data sets is described as Table 8.

Table 8. The datasets for experiments.

Data	U	C	D
Balance scale	625	4	3
Harberman	306	3	2
Exasen	399	3	4
Soybean	47	35	4
Zoo	101	16	7
Car	1728	6	4

In the beginning, the method for mining generalized one-sided decision rules shown as Algorithm 2 is operated on the chosen data sets. Table 9 shows the related results in detail, where h and w are the support and confidence thresholds used in the experiment respectively. The number of extraction rules and the time consumption are also exhibited in the table.

Table 9. Generalized one-sided decision rules.

Data	h	w	Number	Time(s)
Balance scale	25/625	0.4	600	217.22
Harberman	40/306	0.4	756	9.84
Exasen	150/399	0.4	5	0.99
Soybean	17/47	0.4	384	2.81
Zoo	30/101	0.4	184	3.91
Car	45/1728	0.4	1383	209.27

In addition, the method for mining generalized one-sided ordered decision rules shown as Algorithm 3 is also operated on the chosen data sets. The corresponding results are depicted in Table 10. We use the same parameters as Algorithm 2.

Table 10. Generalized one-sided ordered decision rules.

Data	h	w	Number	Time(s)
Balance scale	25/625	0.4	1150	69.22
Harberman	40/306	0.4	1482	95.78
Exasen	150/399	0.4	113	1.36
Soybean	17/47	0.4	2796	2590.15
Zoo	30/101	0.4	250	11.84
Car	45/1728	0.4	1799	209.46

Comparing Table 9 and Table 10, Algorithm 3 can extract more rules than Algorithm 2 with the same conditions. Moreover, Algorithm 2 requires less running time in most cases.

7   Conclusion

This paper studies rule acquisition methods in generalized one-sided formal contexts with decision attributes to investigate the effect of ordered attributes on rule acquisition. Different from classical formal contexts, each attribute in the generalized one-sided formal context has an ordered structure. To this end, this paper defines two formal decision contexts: generalized one-sided formal decision context and generalized one-sided ordered formal decision context. Moreover, we explore corresponding decision rule acquisition algorithms in these contexts to mine related decision rules. Related theoretical derivations are also carried out to explain the relations between decision rules. Compared with generalized one-sided formal decision contexts, rules in generalized one-sided ordered formal decision contexts have larger support and confidence. That is, for the same level of support and confidence thresholds, more decision rules are extracted from the generalized one-sided ordered formal decision context.

References

1. Han, J., Kamber, M., Pei, J., et al.: Data Mining: Concepts and Techniques, 3rd edn. China Machine Press, Beijing (2012)
2. Lin, T.: Granular computing on binary relations II: rough set representations and belief functions. In: Rough Sets in Knowledge Discovery, vol. 1, pp. 121–140 (1998)
3. Hu, Q., Pan, W., An, S., Ma, P., Wei, J.: An efficient gene selection technique for cancer recognition based on neighborhood mutual information. Int. J. Mach. Learn. Cybern. **1**, 63–74 (2010). https://doi.org/10.1007/s13042-010-0008-6
4. Pawlak, Z.: Rough Sets: Theoretical Aspects of Reasoning About Data. Kluwer Academic Publishers, Boston (1991)
5. Qian, Y., Liang, J., Yao, Y.: MGRS: a multi-granulation rough set. Inf. Sci. **180**(6), 949–970 (2010)
6. Ganter, B., Wille, R.: Formal Concept Analysis: Mathematical Foundations. Springer, Berlin (1999). https://doi.org/10.1007/978-3-642-59830-2
7. Shao, M., Wu, W., Wang, X., Wang, C.: Knowledge reduction methods of covering approximate spaces based on concept lattice. Knowl.-Based Syst. **191**, 105269 (2020)
8. Hu, Q., Qin, K., Yang, L.: The updating methods of object-induced three-way concept in dynamic formal contexts. Appl. Intell. (2022). https://doi.org/10.1007/s10489-022-03646-6
9. Chen, D., Li, J., Li, R.: Formal concept analysis of multi-scale formal context. J. Ambient. Intell. Humaniz. Comput. **11**(11), 5315–5327 (2020). https://doi.org/10.1007/s12652-020-01867-6
10. Yang, W.: Analysis and application of formal concept in information science. J. Phys. Conf. Ser. **1624**(3), 032032 (2020)
11. Gligorijević, M.F., Bogdanovic, M., Veljkovic, N., Stoimenov, L.: Open data categorization based on formal concept analysis. IEEE Trans. Emerg. Top. Comput. **9**(2), 571–581 (2021)
12. Yao, J., Medina, J., Zhang, Y., Ślezak, D.: Formal concept analysis, rough sets, and three-way decisions. Int. J. Approx. Reason. **140**, 1–6 (2022)
13. Butka, P., Pócs, J.: Generalization of one-sided concept lattices. Comput. Inform. **32**(2), 355–370 (2013)
14. Shao, M., Li, K.: Attribute reduction in generalized one-sided formal contexts. Inf. Sci. **378**, 317–327 (2017)
15. Liang, M., Mi, J., Feng, T., Jin, C.: Attribute reduction in intuitionistic fuzzy formal concepts. J. Intell. Fuzzy Syst. **43**(3), 3561–3573 (2022)
16. Shao, M., Lv, M., Li, K., Wang, C.: The construction of attribute (object)-oriented multi-granularity concept lattices. Int. J. Mach. Learn. Cybern. **11**, 1017–1032 (2020). https://doi.org/10.1007/s13042-019-00955-0

Density Peak Clustering Based Split-and-Merge

Jixiang Lu and Caiming Zhong$^{(\boxtimes)}$

College of Science and Technology, Ningbo University,
Ningbo 315300, Zhejiang, China
{lujixiang,zhongcaiming}@nbu.edu.cn

Abstract. Density peak clustering (DPC) defines cluster centers to be
the objects with the highest density in their neighborhoods and far away
from the objects that are with higher densities, where density of each
object is determined by a pre-specified parameter. This definition may
find the number of clusters and detect clusters with arbitrary shapes.
However, as a drawback, it may also lead to multiple centers for one cluster and hence degrade the clustering quality. Some work in the literature
try to remedy it, but this drawback can be regarded as the advantage
in split-and-merge strategy-based clustering. In this paper, we make use
of a modified DPC to partition a dataset into a large number of small
clusters by defining a refined K nearest neighbors, and design a merge
criterion to combine the small clusters. The experimental results on 6
synthetic and 6 real datasets demonstrate the proposed algorithm outperforms some traditional clustering algorithms.

Keywords: Density peak clustering · Split-and-merge · K nearest
neighbors

1 Introduction

Clustering aims to partition a dataset into a number of subsets in terms of a
similarity measure of data points, and points in the same subset are similar while
those in different subsets are dissimilar. Although there exist a large number of
clustering algorithms in the literature [2,5,8,11,14,15,18], most of them such
as K-means, Single-linkage can only deal with datasets of specific cluster structures. For example, K-means favors sphere like clusters and Single-linkage can
only detect clusters consisting of points of high connectivity. We regard them as
algorithms of less universality. Since users usually have not any priori knowledge
about their datasets, it is difficult for them to select a suitable clustering algorithm. This implicates that for a clustering algorithm the higher universality the
better usability.

Density-based algorithms are capable of disclosing clusters with arbitrary
shapes and sizes, and are with relatively high universality. One of the well known

---

Supported by NSFC 62172242.

J. Yao et al. (Eds.): IJCRS 2022, LNAI 13633, pp. 191–202, 2022.
https://doi.org/10.1007/978-3-031-21244-4_14

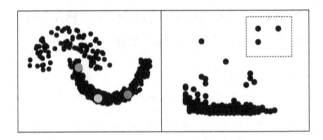

Fig. 1. Inappropriate cluster centers of a two clusters dataset determined by DPC. The three large circles in the left are the cluster centers corresponding to the three points in the dashed rectangle from the right decision graph (The cutoff distance d_c here is set by 2.5% of nearest distances).

paradigms is DBSCAN [4]. However, it suffers from two drawbacks: the density threshold is fixed and two parameters to estimate the density are required to be specified by users (Fig. 1).

Apart from density-based algorithms, multiobjective clustering [1,9], clustering ensemble [6,7] and split-and-merge strategy based clustering [10,12,16] are also of high universality.

Multiobjective clustering optimizes multiple objective functions simultaneously so that datasets with complex cluster structure can be processed. As in most cases heuristic search is used to figure out the optimal solution, the process is complicated. Clustering ensemble is a robust method which can produce clustering with high quality. It contains two steps, generating base partitions and combining them into a single partition. Although the majority of the studies on clustering ensemble focus on the combination process, the base partitions is certainly the primary factor for the clustering result. If base partitions are of high homogeneity, which means a pair of points in a base cluster belong to the same true cluster, the ensemble result is of high quality, or vice versa. Split-and-merge strategy based clustering partitions the dataset into a large number of small clusters, and then merges them into the requested number of final clusters. Comparing with clustering ensemble, the homogeneity of the small clusters is more crucial, as one pair of heterogeneous points may dramatically degrade the clustering result.

In this paper, we propose a split-and-merge based clustering algorithm, in which density peak based clustering [13] is used in split stage to produce small clusters of high homogeneity. In the merge stage, a simple yet effective merge criterion is designed and accordingly the small clusters are combined into the final clustering.

2 The Proposed Method

2.1 Refinement of KNN

In pattern recognition community, KNN is widely used for classification, density estimation etc. The number of nearest neighbors K is crucial to the final

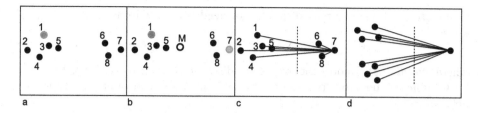

Fig. 2. Dissimilarity and partition in [17]. In a, point 1 has 7 neighbors. In b, M is center of 8 points, and point 7 is the furthest point from M. In c, the right, the 8 points are partitioned into two groups. In d, an isolated point is achieved.

application results. However, it is always a tough problem for users to select the suitable K. We employ the dissimilarity defined by furthest reference point [17] to refine the K nearest neighbors.

For a data point, its neighbors must be nearer to itself than those non-neighbors. Normally, when K is specified by users, some non-neighbors may be included. Our intuition here is to exclude those non-neighbors.

In Fig. 2, we suppose K is 7. Point 1 has 7 neighbors $\{2, 3, 4, 5, 6, 7, 8\}$, and point M is the center of point 1 and its 7 neighbors. The furthest point 7 from the center M is selected as the reference point. The reason behind this is that the furthest point has better ability to discriminate neighbors than other points. The distances from each point to point 7 are computed as $d_{7,1}, d_{7,2}, \cdots, d_{7,8}$, and sorted as $\langle d_{7,7}, d_{7,8}, d_{7,6}, d_{7,5}, \cdots, d_{7,2} \rangle$. In the sorted list, as the neighbor pair $\langle d_{7,6}, d_{7,5} \rangle$ has the biggest difference, $d_{7,5}$–$d_{7,6}$, the list is partitioned into two sublists from the middle of this pair: $\langle d_{7,7}, d_{7,8}, d_{7,6} \rangle$ and $\langle d_{7,5}, d_{7,3}, \cdots, d_{7,2} \rangle$. Accordingly, the 8 points are partitioned into two groups: $\{1, 2, 3, 4, 5\}$ and $\{6, 7, 8\}$.

Although K is 7, we can say that point $\{2, 3, 4, 5\}$ are more eligible to be neighbors of point 1 than $\{6, 7, 8\}$, and thus the number of neighbors of point 1 prefers to 4. We denote the number of neighbors of x by $K_p(x)$ from the view of the furthest reference point.

Suppose $NN_K(x, X)$ is x's K nearest neighbors from X, and r is the furthest point from the center of $NN_K(x, X) \cup \{x\}$. Let $L(x)$ be the sorted list of distances from r to $NN_K(x, X) \cup \{x\}$:

$$L(x) = \langle d_{r,i_1}, d_{r,i_2}, \cdots, d_{r,i_{K+1}} \rangle \tag{1}$$

where $d_{r,i_j} < d_{r,i_{j+1}}, 1 \leq j \leq K$. Suppose i_k is the cut point:

$$d_{r,i_{k+1}} - d_{r,i_k} \geq d_{r,i_{j+1}} - d_{r,i_j} \tag{2}$$

Then, $K_p(x) = k - 1$.

However, the above furthest reference point based partition may lead to an isolated point, for example in Fig. 2(d), the rightmost point has not any neighbor. As in this paper we use KNN to estimate the point density, we consider the average of distances from x to its neighborhood as the remedy.

Let $K_a(x)$ be the potential number of neighbors of x from the view of the average:

$$K_a(x) = |\{y | y \in NN_K(x, X) \wedge d_{x,y} < avg\_dist(x)\}| \qquad (3)$$

where $avg\_dist(x)$ denotes the average of distances from x to $NN_K(x, X)$.

Combining the above two situations, the refined number of neighbors of x is defined as:

$$K_c(x) = \max(K_p(x), K_a(x)) \qquad (4)$$

For a dataset X, the refined number of neighbors K_X is defined as:

$$K_X = ceil(\frac{1}{|X|} \sum_{x \in X} K_c(x)) \qquad (5)$$

where $ceil(f)$ rounds the decimal f to the nearest integer greater than or equal to f.

2.2   Partition the Dataset into Small and Homogeneous Clusters

In DPC, the density of a data point is defined as:

$$\rho_i = \sum_j \varphi(d_{x_i, x_j} - d_c) \qquad (6)$$

where $\varphi(a)$ is 1 if $a < 0$ and 0 otherwise, d_{x_i, x_j} denotes the distance between data point x_i and x_j, and d_c is a cutoff distance, which is determined by:

$$d_c = dist(round(|X| * percent)) \qquad (7)$$

where $dist(i)$ is the ith element in the sequence of all pairs of distances of data points arranged in ascending order, $round(a)$ returns the the nearest integer of number a, and $percent$ is a percentage.

In this paper, we simply define the density of a data point as the reciprocal of the average neighbor distance:

$$\rho_i' = \frac{1}{\frac{1}{K_X} \sum_{x_j \in NN_{K_X}(x_i, X)} d_{x_i, x_j}} \qquad (8)$$

With the above density definition, the procedure to produce the initial clustering is as follows.

Repeatedly taking the data point with the most density, say x_i, from X, if there exists point x_j which has been assigned a label and $x_j \in NN_{K_X}(x_i, X) \wedge \rho_i' < \rho_j'$, then it has the same label, otherwise it is assigned a new label.

In Fig. 3, the five data points are assigned the same label. The label assignment process is described as follows. If the number of neighbors is 2, point a, b and c are neighbors and with the highest density, then they are assigned the same label. While c is one of neighbors of d, d is assigned the same label with c. Similarly, e is assigned the same label with d.

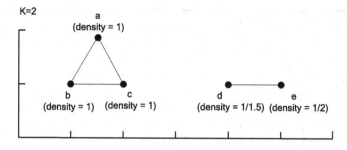

Fig. 3. Label assignment.

The split procedure is described in Algorithm 1. From line 2 to line 9, the number of neighbors of each point is computed. In line 4, mean(S) is to compute the mean point of dataset S. In line 5, array(S) denotes S is stored in an array and sort($\|r - \text{array}(S)\|$) sorts in ascending order the distances between r and the elements of the array and return the indices of the array. In line 6, array(S, j) means the jth element of the array. In line 13, sortdescent($array$) sorts in descending order the elements in $array$ and return the indices of the array.

2.3   Combine the Initial Clustering

After the dataset is partitioned into a number of small homogeneous clusters, split-and-merge clustering scheme then combines those clusters into user-required number of clusters. To iteratively combine the small clusters generated in the previous step, the pairwise similarities are needed. When designing the similarity measure, two kinds of information are crucial to definition of a cluster: density and distance.

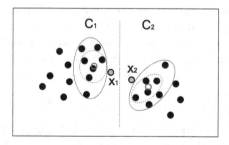

Fig. 4. Cluster similarity. The two solid ellipses enclose the K_X nearest neighbors of x_1 and x_2, respectively. The broken ellipses denote the refined nearest neighborhood. The two hollow circles are the centers of the refined nearest neighbors.

For example, DBSCAN [4] detects a cluster with a concept of density reachable, which means that the data points within the same cluster are density-

---

Algorithm 1: Split procedure

Input: X - Data set, K - Initial number of nearest neighbors
Output: $label$ - The labels of a partition of X

1 $label(1..|X|) \leftarrow null$
2 **for** each point $x \in X$ **do**
3 $\quad S \leftarrow NN_K(x, X) \cup \{x\}$
4 $\quad r \leftarrow \text{argmax}_{y \in S}\, d_{\text{mean}(S), y}$
5 $\quad idx \leftarrow \text{sort}(\|r - array(S)\|)$
6 $\quad k \leftarrow \text{argmax}_i (d_{r, \text{array}(S, idx(i+1))} - d_{r, \text{array}(S, idx(i))})$
7 $\quad K_p(x) \leftarrow k - 1$
8 $\quad K_a(x) \leftarrow |\{y | y \in NN_K(x, X) \wedge d_{x,y} < \text{avg\_dist(x)}\}|$
9 $\quad K_c(x) \leftarrow \max(K_p(x), K_a(x))$
10 $K_X \leftarrow \text{ceil}(\frac{1}{|X|} \sum_{x \in X} K_c(x))$
11 **for** $x_i \in X$ **do**
12 $\quad \rho'(i) \leftarrow \dfrac{1}{\frac{1}{K_X} \sum_{x_j \in NN_{K_X}(x_i, X)} d_{ij}}$
13 $idx \leftarrow \text{sortdescent}(\rho')$
14 $lab \leftarrow 1$
15 **for** $i \leftarrow idx(1 : |X|)$ **do**
16 $\quad$ **if** $\exists j(\rho'(j) \geq \rho'(i) \wedge x_j \in NN_K(x_i, X) \wedge (label(j) \neq null))$ **then**
17 $\quad\quad label(i) \leftarrow label(j)$
18 $\quad$ **else**
19 $\quad\quad label(i) \leftarrow lab$
20 $\quad\quad lab \leftarrow lab + 1$

---

reachable. This concept simultaneously considers density and distance information. In [3], the robustness property comes from incorporation of density information into distance information. In the decision graph of [13], the two dimensions are density and distance information of a data point, respectively.

With the above understanding in the mind, we define a cluster dissimilarity as follows and the main components are illustrated in Fig. 4.

For a given pair of clusters to be merged, say C_1 and C_2, suppose data point x_1 and x_2 are the nearest pair from the two clusters respectively, namely, $x_1 \in C_1$, $x_2 \in C_2$ and the following holds:

$$\forall(y_1 \in C_1, y_2 \in C_2), d_{y_1,y_2} \geq d_{x_1,x_2} \tag{9}$$

Compute the average distance of x_1. Since at this stage the dataset has been partitioned into a number of clusters, we consider the neighbors that have the same cluster label with x_1. The reciprocal of the density is defined as:

$$den(x_1) = \frac{1}{|NN_{K_X}(x_1, C_1)|} \sum_{y \in NN_{K_X}(x_1, C_1)} d_{x_1, y} \tag{10}$$

where $NN_{K_X}(x_1, C_1)$ denotes the K_X nearest neighbors of x_1 from C_1.

Algorithm 2: Merge procedure

Input: *label* - Initial clustering, k - The required number of clusters, X
Output: *label* - Final clustering

1   $n \leftarrow$ getnum($label$)
2   $Dis(1..n, 1..n) \leftarrow INF$
3   $C_1, C_2, \cdots, C_n \leftarrow$ getcluster($X, label$)
4   **for** each cluster pair $\langle C_i, C_j \rangle$, if $i > j$ **do**
5    $\lfloor$ $Dis(i, j) \leftarrow dis_{final}(C_i, C_j)$

6   **while** $|\{C_i | C_i \neq \varnothing, 1 \leq i \leq n\}| > k$ **do**
7     $i, j \leftarrow \text{argmin}_{i,j \in \{1..n\}} Dis(i, j)$
8     $C_j \leftarrow C_i \cup C_j$
9     $label(\text{index}(C_i)) \leftarrow \text{lab}(C_j)$
10    $C_i \leftarrow \varnothing$
11    $Dis(i, 1..i - 1) \leftarrow INF$
12    $Dis(i + 1..n, i) \leftarrow INF$
13   **if** $Dis(j, 1..j - 1)$ is not INF **then**
14     $\lfloor$ $Dis(j, 1..j - 1) \leftarrow dis_{final}(C_j, C_{1..j-1})$
15   **if** $Dis(j + 1..n, j)$ is not INF **then**
16     $\lfloor$ $Dis(j + 1..n, j) \leftarrow dis_{final}(C_{j+1..n}, C_j)$

The above distance implies the density of x_1 in C_1. Similarly, the average distance of x_2, $den(x_2)$ is defined.

Refine the neighborhood of x_1. From the Eq. 10, although the neighbors of x_1 are from C_1, some neighbors may be far away from x_1 as the number of neighbors is fixed to K_X. The neighborhood is refined as:

$$NN(x_1) = \begin{cases} T = \{y | d_{x_1, y} < den(x_1)\}, & \text{if } |T| \geq \frac{K_X}{2} \\ NN_{\frac{K_X}{2}}(x_1, C_1), & \text{otherwise} \end{cases} \tag{11}$$

where $y \in NN_{K_X}(x_1, C_1)$ and $NN_{\frac{K_X}{2}}(x_1, C_1)$ denotes the $\frac{K_X}{2}$ nearest neighbors of x_1 from C_1.

Similarly, the refined neighborhood of x_2 from C_2, $NN(x_2)$, can be defined. The average distance of $NN(x_1)$ is defined:

$$denAvg(x_1) = \frac{1}{|NN(x_1)|K_X} \sum_{y \in |NN(x_1)|} \sum_{z \in NN_{K_X}(y, C_1)} d_{y, z} \tag{12}$$

This average distance expresses the average density of data points in $NN(x_1)$. Similarly, the average distance of $NN(x_2)$, $denAvg(x_2)$ is defined.

Let m_1 and m_2 be the centers of $NN(x_1)$ and $NN(x_2)$, respectively.

The cluster dissimilarity is defined as:

$$dis(C_1, C_2) = \text{abs}(den(x_1) - d_{m_1,m_2})$$
$$+ \text{abs}(den(x_2) - d_{m_1,m_2})$$
$$+ \text{abs}(den(x_1) - den(x_2))$$
$$+ \text{abs}(denAvg(x_1) - denAvg(x_2)) \qquad (13)$$

where $\text{abs}(x)$ returns the absolute of x.

The first two items in Eq. 13 indicates that the larger the differences between d_{m_1,m_2} and $den(x_1)$, $den(x_2)$, the bigger the dissimilarity. The third item is the difference between the densities of x_1 and x_2, while the forth item is the difference between the average densities of the refined neighborhood of x_1 and x_2. These two differences are positively correlated to the dissimilarity.

The defined dissimilarity measure has one drawback: Its effectiveness is mainly decided by the x_1 and x_2. At the same time, the dissimilarity measure does not consider the size of clusters. Intuitively, under the same condition a small cluster has the priority to be merged. Therefore, we improve the dissimilarity measure as follows.

We define a variant of the measure, in which we remove x_1 and x_2, and then re-compute the measure denoted as $dis'(C_1, C_2)$.

$$dis'(C_1, C_2) = \begin{cases} dis(C_1', C_2'), \text{if } \min(|C_1|, |C_2|) > \alpha\frac{N}{k} \\ dis(C_1', C_2') * \frac{\min(|C_1|, |C_2|)}{N/k}, \text{otherwise} \end{cases} \qquad (14)$$

where $C_1' = C_1\backslash\{x_1\}$, $C_2' = C_2\backslash\{x_2\}$, k is the user-specified number of clusters and $0 < \alpha \leq 1$.

The final dissimilarity measure, dis_{final}, is defined as:

$$dis_{final}(C_1, C_2) = dis(C_1, C_2) + dis'(C_1, C_2) \qquad (15)$$

The merge procedure is described in Algorithm 2. In line 1, getnum($label$) returns the number of clusters from $label$. $Dis(i, j)$ denotes the dissimilarity of cluster C_i and C_j in line 2. Function getcluster($X, label$) in line 3 returns the n clusters. In line 9, function index(C_i) denotes the indices of points in C_i and lab(C_j) denotes the cluster label of C_j.

3 Experimental Results and Discussion

3.1 Datasets and Parameter Setting

To demonstrate the performance of the proposed method, we test it with 16 datasets, of which 8 datasets are synthetic and the others are real. The 8 synthetic datasets, DS1-DS8, are available at http://cs.uef.fi/sipu/datasets. DS1-DS5 are of complex cluster structures, while DS6-DS8 are composed of clusters which are in Gaussian distribution. The 8 real datasets, Iris, Ionosphere, Wine, Diabetes, Segmentation, Glass, WDBC, WPBC are available at http://www.ics. uci.edu/~mlearn/MLRepository.html (Fig. 5).

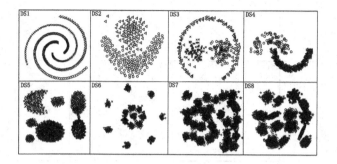

Fig. 5. Eight synthetic datasets.

The proposed method has two parameters which are to be set by users: the number of initial nearest neighbors K and the coefficient α in Eq. 14. In all of the experiments, we set K to 10 and α to 0.33, and we will discuss the two parameters in the next sub-section.

3.2   Split-and-Merge Methods to Be Compared

We compare the proposed method with two split-and-merge based methods: CSM [12] and Chameleon [10].

In CSM, K-means is employed to partition the dataset into a number of small clusters, which are then merged into required number of clusters by a merge criterion. This criterion is defined according of the minimum conditional probability of a point belongs to a pair of clusters. When the probabilities are estimated, each cluster is supposed to be under Gaussian distribution. The number of initial clusters is set to $\sqrt{N}$. As the initial clustering of CSM is produced by K-means and hence not unique, we use the best one of 10 runs for comparison.

In Chameleon, a KNN graph is constructed and partitioned into the initial clustering such that the edge cut is minimized. In the merge stage, a relative interconnectivity and a relative closeness are defined to form a merge criterion. The two parameters, the number of the nearest neighbors and the importance coefficient α, are set to 10 and 0.3, respectively.

3.3   Experimental Results

The clustering on the 8 synthetic datasets are illustrated in Fig. 6. From the figure, one can see the proposed method can find the true clusters of the 8 datasets. Although CSM also find the true clusters, they are the best results selected from 10 runs on each dataset. While for Chameleon, it can not detect the true cluster structure of DS1 and DS3.

The proposed method

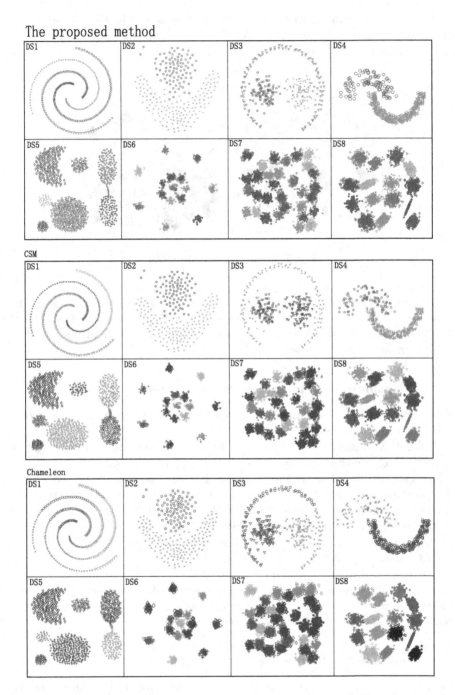

Fig. 6. The clustering results of the proposed method, CSM and Chameleon on the eight synthetic datasets.

For the 8 real datasets, CA and ARI are employed to measure the clustering qualities. The results are in Table 1. It is clear that the proposed method has the best performance on 5 datasets, while CSM has the best performance on 4 datasets and Chameleon on 3 datasets. Taking into account the clustering results of all of the 16 datasets, the proposed method outperforms CSM and Chameleon (Table 2).

Table 1. The comparison of clustering qualities measured by CA and ARI on 8 real datasets.

Dataset	Proposed method		CSM		Chameleon	
	CA	ARI	CA	ARI	CA	ARI
Iris	**0.9667**	**0.9038**	0.9000	0.7455	0.7267	0.5506
Wine	**0.8933**	**0.6989**	0.5843	0.1249	0.6854	0.3663
Segmentation	0.5667	0.3238	**0.5762**	**0.3843**	0.4714	0.1634
WDBC	**0.9315**	**0.7423**	0.7170	0.1508	0.9104	0.6684
WPBC	**0.7629**	**0.0507**	**0.7629**	**0.0507**	**0.7629**	**0.0507**
Ionosphere	0.6410	0.0242	0.7464	0.2103	**0.8519**	**0.2513**
Diabetes	**0.6510**	**0.0102**	**0.6510**	**0.0102**	**0.6510**	**0.0102**
Glass	0.5140	0.1094	**0.6121**	**0.2193**	0.5654	0.2639

Table 2. The average clustering qualities measured by CA and corresponding standard deviations on the 16 datasets.

Dataset	Averaged-CA		STD	
	Ks refined	Ks not refined	Ks refined	Ks not refined
DS1	**0.8627**	0.7566	**0.1231**	0.1251
DS2	1.0000	1.0000	0.0000	0.0000
DS3	**0.9769**	0.9739	**0.0039**	0.0042
DS4	**0.9870**	0.9556	**0.0177**	0.0359
DS5	0.9967	0.9967	0.0000	0.0000
DS6	0.9948	0.9948	0.0090	0.0090
DS7	**0.9950**	0.9949	**0.0004**	0.0005
DS8	**0.9497**	0.9344	**0.0154**	0.0234
Iris	**0.9230**	0.9085	0.0418	**0.0374**
Wine	0.8933	0.8933	0.0000	0.0000
Segmentation	0.6095	**0.6416**	**0.0475**	0.0510
WDBC	**0.8747**	0.8636	**0.0471**	0.0873
WPBC	0.7629	0.7629	0.0000	0.0000
Ionosphere	0.6410	0.6410	0.0000	0.0000
Diabetes	0.6510	0.6510	0.0000	0.0000
Glass	0.5391	**0.5493**	**0.0071**	0.0443

4 Conclusion

In this paper, a DPC-based split-and-merge clustering method is proposed. It takes advantage of the characteristics of DPC, which can split a cluster into some small local groups, and defines a merge criterion to combine the small groups. When a KNN is used in split and merge steps, K nearest neighbor are refined, and this may lead to good clustering result and good stability with respect to different initial Ks.

References

1. Armano, G., Farmani, M.: Multiobjective clustering analysis using particle swarm optimization. Expert Syst. Appl. **55**, 184–193 (2016)
2. Bakoben, M., Bellotti, A., Adams, N.: Improving clustering performance by incorporating uncertainty. Pattern Recogn. Lett. **77**, 28–34 (2016)
3. Chang, H., Yeung, D.: Robust path-based spectral clustering. Pattern Recogn. **41**, 191–203 (2008)
4. Ester, M., Peter, K., Hans, S., Xu, X.: A density-based algorithm for discovering clusters in large spatial databases with noise. In: Proceedings of the 2nd KDD. AAAI Press (1996)
5. Hou, J., Zhang, A.: Enhanced dominant sets clustering by cluster expansion. IEEE Access **6**, 8916–8924 (2018)
6. Huang, D., Lai, J., Wang, C.: Ensemble clustering using factor graph. Pattern Recogn. **50**, 131–142 (2016)
7. Huang, D., Lai, J.H., Wang, C.D.: Robust ensemble clustering using probability trajectories. IEEE Trans. Knowl. Data Eng. **28**, 1312–1326 (2016)
8. Jain, A.: Data clustering: 50 years beyond K-means. Pattern Recogn. Lett. **31**, 651–666 (2010)
9. Julia, H., Joshua, K.: An evolutionary approach to multiobjective clustering. IEEE Trans. Evol. Comput. **11**, 56–76 (2007)
10. Karypis, G., Han, E., Kumar, V.: CHANELEON: a hierarchical clustering algorithm using dynamic modeling. IEEE Comput. **32**, 68–75 (1999)
11. Leski, J., Kotas, M.: Hierarchical clustering with planar segments as prototypes. Pattern Recogn. Lett. **54**, 1–10 (2015)
12. Lin, C., Chen, M.: Combining partitional and hierarchical algorithms for robust and efficient data clustering with cohesion self-merging. IEEE Trans. Knowl. Data Eng. **17**, 145–159 (2005)
13. Rodriguez, A., Laio, A.: Clustering by fast search and find of density peaks. Science **344**, 1492–1496 (2014)
14. Wang, J., Zhu, C., Zhou, Y., Zhu, X., Wang, Y., Zhang, W.: From partition-based clustering to density-based clustering: fast find clusters with diverse shapes and densities in spatial databases. IEEE Access **6**, 1718–1729 (2018)
15. Zhao, Q., Shi, Y., Liu, Q., Fränti, P.: A grid-growing clustering algorithm for geospatial data. Pattern Recogn. Lett. **53**, 77–84 (2015)
16. Zhong, C., Miao, D., Fränti, P.: Minimum spanning tree based split-and-merge: a hierarchical clustering method. Inf. Sci. **181**, 3397–3410 (2011)
17. Zhong, C., Miao, D., Wang, R., Zhou, X.: DIVFRP: an automatic divisive hierarchical clustering method based on the furthest reference points. Pattern Recogn. Lett. **29**, 2067–2077 (2008)
18. Zhuo, Z., Zhang, X., Niu, W., Yang, G., Zhang, J.: Improving data field hierarchical clustering using Barnes-Hut algorithm. Pattern Recogn. Lett. **80**, 113–120 (2016)

Multi-label Feature Extraction With Distance-Based Graph Attention Network

Yue Peng[1] , Kun Qian[1,2] , Guojie Song[2] , and Fan Min[1,2,3(✉)]

[1] School of Computer Science, Southwest Petroleum University,
Chengdu 610500, China
[2] School of Sciences, Southwest Petroleum University, Chengdu 610500, China
[3] Institute for Artificial Intelligence, Southwest Petroleum University,
Chengdu 610500, China
minfan@swpu.edu.cn

Abstract. Feature extraction deals with information redundancy in data with a large number of features. Existing feature extraction approaches to multi-label data usually consider label correlations, while rarely consider sample correlations. In this paper, we propose a multi-label feature extraction with the distance-based graph attention network (DBGAT) algorithm. First, to easily extract the neighbors of the sample later, we construct an adjacency matrix according to the distance between samples and the number of neighbors specified by the user. Second, to obtain the importance of neighbor features to instances, we get the weight coefficients of each instance and its neighbors through the attention network. Third, a new representation for each instance is obtained by weighted summation of neighboring instances. The difference in the weight coefficient reflects the degree of influence of different neighbors on the new feature. We tested the new algorithm and eight other popular algorithms on twelve datasets. Experiments show that this method improves the accuracy of multi-label classification.

Keywords: Attention mechanism · Feature extraction · Multi-label learning · Neural networks · Weight coefficient

1 Introduction

High-dimensional data makes the sample distribution sparse, which increases the computational complexity and reduces the performance of the classification model. Therefore, many dimension reduction [1,5] methods have been proposed. To reduce the computational complexity, feature extraction [6,11] algorithms project original data into a low-dimensional feature space, reducing the number of features. Principal component analysis (PCA) [18] maps high-dimensional features to low-dimensional features. It is a new orthogonal feature reconstructed

This work is in part supported by the Central Government Funds of Guiding Local Scientific and Technological Development (No. 2021ZYD0003).

on the basis of the original feature. Feature extraction can achieve information gain to improve the classification accuracy. Singular value decomposition (SVD) [2] directly decomposes the original matrix into three matrices. It takes values in order on the diagonal matrix, multiplying three matrices to get one matrix. This new matrix will contain most of the information of the original matrix, thus eliminating redundant information. Therefore, feature extraction is an essential step in the data processing.

Feature extraction can be applied to multi-label learning [23] problems. Popular feature extraction algorithms include unsupervised principal component analysis (PCA) and supervised dependence maximization. Unsupervised algorithms treat all data equally in terms of dimension reduction. For supervised feature extraction algorithms, they can be roughly divided into two categories: problem transformation (PT) and algorithm adaptation (AA). PT methods transform the multi-label learning problem into a series of single-label learning problems. Such algorithms are applied to multi-label data. AA methods can be applied to multi-label feature extraction (MLFE) [15,19,20] algorithms by modifying the existing single-label feature extraction algorithm.

There are many common AA methods applied to MLFE. Multi-Label Informed Latent Semantic Indexing (MLSI) [21] maps input features to a new feature space. This captures the output dimensional dependencies while preserving the information of the original input. This process is an optimization problem of linear projection. Multi-label Linear Discriminant Analysis (MLDA) [17] uses label correlations to build class-wise between-class scatter matrix. It avoids ambiguity about how much data points with multiple labels contribute to the scatter matrix. It can also incorporate label correlations at the same time. It can also contain tag dependencies at the same time. These algorithms enrich MLFE methods.

In this paper, we propose a multi-label feature extraction algorithm with distance-based graph attention network. Figure 1 shows the framework of the algorithm. First, to easily extract the neighbors of the sample later, we construct an adjacency matrix based on the distance between samples and the number of neighbors specified by the user. The adjacency matrix can help us get the graph structure. Second, to obtain the importance of neighbor features to instances, we get the weight coefficients of each instance's neighbors through the attention network. The network needs information about neighbors, so we need to use the adjacency matrix information obtained in the first step. Using an adjacency matrix, we can find neighbors quickly and easily. Third, a new representation for each instance is obtained by weighted summation of neighboring instances. We compute the weight of each neighbor then sum it weighted. This result will serve as the new representation of the node.

There are two main contributions of this paper. On the one hand, we construct an adjacency matrix with the idea of k-nearest neighbors based on Euclidean distance. Different datasets have different adjacency matrices. On the other hand, we introduce the constructed adjacency matrix as a graph structure into the attention mechanism. In order to obtain the importance of adjacent fea-

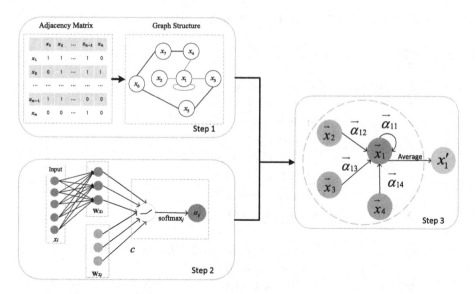

Fig. 1. The framework of DBGAT. Step 1 computes the adjacency matrix from the original data. Step 2 computes the attention coefficient using the generalized GAT. Step 3 obtains a new representation of the final node by the weighted summation of its neighbors.

tures to the instance, we calculate the attention coefficient. Each neighbor has its own attention coefficient. It indicates that different neighbors have different importance to the new representation of the node.

The rest of this paper is organized as follows. Section 2 presents the data model of MLFE, and reviews popular approaches to MLFE. The details of the proposed DBGAT algorithm is presented in Sect. 3. Section 4 reports the experimental results and analysis. Finally, Sect. 5 draws out conclusion.

2 Related Work

This section first introduces the data model of MLFE, then discusses popular MLFE methods. The comparison algorithm used in the analysis will be focused on.

2.1 Data Model

Table 1 lists some important notations used throughout the paper. The data is represented as follows. Let $\mathbf{X} = [\mathbf{x}_1, \mathbf{x}_2, \ldots, \mathbf{x}_N]^\mathsf{T} \in \mathbb{R}^{N \times M}$ denotes the M-dimensional feature space of instances, and $\mathbf{Y} = [\mathbf{y}_1, \mathbf{y}_2, \ldots, \mathbf{y}_N]^\mathsf{T} \in \{0, 1\}^{N \times L}$ denote the label space with L labels, where $[\cdot]^\mathsf{T}$ denotes matrix transpose. Here, we have $\mathbf{y}_i = [y_{i1}, y_{i2}, \ldots, y_{iL}]^\mathsf{T}$, $y_{ij} = 1$ indicates $\mathbf{x}_i$ has the label, while $y_{ij} = 0$ indicates no.

Table 1. Notations.

Notation	Meaning
$\mathbf{X} = [x_{ij}]_{N \times M}$	Data matrix
$\mathbf{x}_i = [x_{i1}, \ldots, x_{iM}]$	The i-th instance
$\mathbf{Y} = [y_{ik}]_{N \times L}$	Label matrix
$\mathbf{y}_i = [y_{i1}, \ldots, y_{iL}]$	Labels of $\mathbf{x}_i$
N	Number of instances
M	Number of features
L	Number of labels
$\mathbf{D} \in \mathbb{R}^{N \times N}$	Distance matrix
$\mathbf{A} \in \{0, 1\}^{N \times N}$	Adjacency matrix
$\mathcal{A}_i$	Neighbors of node i
F	Number of features after dimension reduction
k	Number of neighbors
$\mathbf{W} \in \mathbb{R}^{F \times M}$	A weight matrix
$\mathbf{c} \in \mathbb{R}^{2F}$	A weight vector
c	A single-layer feedforward neural network
α_{ij}	Attention mechanism coefficients

2.2   Multi-label Feature Extraction methods

There are many methods for multi-label feature extraction, some of which are mainly introduced here. The first is some unsupervised methods. It works by minimizing the loss of information. The classic PCA algorithm is unsupervised. Non-negative matrix factorization (NMF) [23] is a matrix factorization method under the constraint that all elements in the matrix are non-negative. It extends two-matrix factorization to three-matrix factorization. Independent component analysis (ICA) [8] is to find a linear representation of non-Gaussian data such that the components are statistically independent, or as independent as possible. Compared with singular value decomposition (SVD) and principal component analysis (PCA), ICA is an analysis method based on higher-order statistical properties. In many applications, the analysis of higher-order statistical properties is more practical.

The second is some supervised methods. It works by maximizing the difference between classes. Multi-Label Dimensionality Reduction via Dependence Maximization (MDDM) [24] attempts to project the original data into a low-dimensional feature space based on the Hilbert-Schmidt independence criterion. It maximizing the dependencies between the original feature description and the relevant class labels. The basic idea of linear discriminant analysis (LDA) [9] model implementation is the same as that of PCA. LDA considers the factors of categories on the basis of dimension reduction. Hoping that the variance within the projected projection is the smallest, and the variance between classes is the largest. MLSI [21] is a multi-label extension of Latent Semantic Indexing (LSI)

[3] that extends multi-label learning by treating each concept combination as a class. General Graph Embedding (GGE) unifies dimension reduction methods in a common framework. It overcomes the limitations of traditional linear discriminant analysis algorithms in terms of data distribution assumptions and available projection directions. Canonical Correlation Analysis (CCA) [14] projects two sets of variables into the low-dimensional space where they are most correlated. CCA in the multi-label case can be formulated as a least squares problem.

3 DBGAT Method

In this section, to obtain the graph structure, we construct an adjacency matrix based on the distances between samples and the number of neighbors specified by the user. Then, we obtain the weight coefficients of each instance relative to its neighbors. Finally, a new representation for each instance is obtained by weighted summation of neighboring instances.

3.1 Method Description

Graph Attention Network (GAT) [16] is one of the mainstream algorithms in Graph Neural Networks (GNN) [12]. GAT is a continuation of Graph Convolutional Network (GCN) [10] with the introduction of the attention mechanism. GCN assigns the same weights to all adjacent nodes during the convolution process, which limits its power. To address this issue, GAT adds masked self-attention layers to learn different weights for neighbor nodes.

First, in order to build a graph structure to facilitate the extraction of neighbors of samples later, we need to process the data to construct an adjacency matrix. We calculate the distance of each example feature by Euclidean distance. Therefore, we construct a distance matrix $\mathbf{D} = [d_{ij}]_{N \times N} \in \mathbb{R}^{N \times N}$. N is the number of instances, M is the number of features, and x_{ij} represents the j-th feature of the i-th instance. Where

$$d_{ij} = \left(\sum_{l=1}^{M} (x_{il} - x_{jl})^2 \right)^{1/2}. \tag{1}$$

We construct an adjacency matrix $\mathbf{A} = [a_{ij}]_{N \times N} \in \{0, 1\}^{N \times N}$ according to the distance between samples and the number of neighbors specified by the user. Let k be the number of neighbors.

$$a_{ij} = \begin{cases} 1, & \text{if } |\{l | d_{il} < d_{ij}\}| < k; \\ 0, & \text{otherwise.} \end{cases} \tag{2}$$

The smallest top k value in each row is selected from the distance matrix $\mathbf{D}$. In other words, there are exactly k 1s in each row of $\mathbf{A}$ if the distance between any instance pair is different.

Second, to obtain the importance of neighbor features to instances, we inject the adjacency matrix into the mechanism by performing masked attention. The data matrix $\mathbf{X}$ is input to the attention layer. $\mathbf{X} = \{\mathbf{x}_1, \ldots, \mathbf{x}_N\}$, $\mathbf{x}_i \in \mathbb{R}^M$, where N is the number of nodes, and M is the number of features in each node. In order to obtain more expressive features, we need to do a linear transformation. To share a linear transformation, we parameterize and apply a weight matrix $\mathbf{W} \in \mathbb{R}^{F \times M}$ to each node. Here F is the number of features after dimension reduction. Then we use the shared attention mechanism $c : \mathbb{R}^F \times \mathbb{R}^F \to \mathbb{R}$ to perform self-attention on the node to calculate the attention coefficient,

$$e_{ij} = c\left(\mathbf{W}\mathbf{x}_i, \mathbf{W}\mathbf{x}_j\right) \tag{3}$$

that indicate the importance of the features of node j to node i. We compute e_{ij} for nodes $j \in \mathbf{A}_i$, where $\mathbf{A}_i$ is some neighborhood of node i in the adjacency matrix. To make coefficients convenient to compare across nodes, we normalize all j options using the softmax function,

$$\alpha_{ij} = \text{softmax}_j\left(e_{ij}\right) = \frac{\exp\left(e_{ij}\right)}{\sum_{k \in \mathbf{A}_i} \exp\left(e_{ik}\right)}. \tag{4}$$

The attention mechanism c is a feedforward neural network, parameterized by a weight vector $\mathbf{c} \in \mathbb{R}^{2F}$, and applying the LeakyReLU nonlinearity. When fully expanded, it can be expressed as:

$$\alpha_{ij} = \frac{\exp\left(\text{LeakyReLU}\left(\mathbf{c}^{\mathrm{T}}\left[\mathbf{W}\mathbf{x}_i \| \mathbf{W}\mathbf{x}_j\right]\right)\right)}{\sum_{k \in \mathbf{A}_i} \exp\left(\text{LeakyReLU}\left(\mathbf{c}^{\mathrm{T}}\left[\mathbf{W}\mathbf{x}_i \| \mathbf{W}\mathbf{x}_k\right]\right)\right)}, \tag{5}$$

where $\cdot^{\mathrm{T}}$ represents transposition, $\|$ is the concatenation operation.

Third, the output feature of each node is calculated by the normalized attention coefficient to calculate the linear combination of the corresponding features. This gets a new representation for each instance.

$$\mathbf{x}_i' = \sigma\left(\sum_{j \in \mathbf{A}_i} \alpha_{ij} \mathbf{W}\mathbf{x}_j\right), \tag{6}$$

where $\sigma(\cdot)$ is the nonlinear activation function.

3.2   Algorithm Description

Algorithm 1 illustrates our algorithm. Line 1 initializes the network. Line 2 through 3 are the process of building the adjacency matrix. Line 2 uses Euclidean distance to calculate the distance between each instance according to Eq. (1). Line 3 obtains the adjacency matrix $\mathbf{A}$ according to Eq. (2). Line 4 performs self-attention on the node and calculates the attention coefficient e_{ij} according to Eq. (3). Line 5 normalizes the adjacent node coefficients. Line 6 to obtain a fully expanded, the attention mechanism a is parameterizes by a weights vector $\overrightarrow{\mathbf{c}}$. Line 7 obtains a new representation of a node by weighted summation according to the attention coefficients between the node and its neighbors.

In summary, DBGAT constructs the adjacency matrix in the initial stage, and learns new features through the attention mechanism in the main learning process.

Algorithm 1. Multi-label feature extraction with the distance-based graph attention network.

Input: data matrix $\mathbf{X} : N \times M$, label matrix $\mathbf{Y} : N \times L$, number of neighbors: k;
Output: $\mathbf{X}'$: New representation of the instance;
  1: Initialize the network;
     // Stage 1. Construct adjacency matrix.
  2: Calculate the distance between i instances and other instances in data matrix $\mathbf{X}$ according to Eq. (1);
  3: To get adjacency matrix $\mathbf{A}$, it is calculated by Eq. (2);
     // Stage 2. Main learning process.
  4: Computes attention coefficients e_{ij} according to Eq. (3);
  5: Get attention coefficients α_{ij} according to Eq. (4);
  6: Fully expanded attention coefficient according to Eq. (5);
  7: Get a new representation of instance i, x_i' according to Eq. (6) ;
  8: Get new representation of the instance $\mathbf{X}'$;

---

4   Experiments

In this section we conduct experiments to analyze the effectiveness of the DBGAT algorithm. The experiment includes comparison and analysis with other 6 methods from the aspects of evaluation metrics and parameter setting.

Experiments are performed on 12 benchmark datasets to evaluate the effectiveness of the DBGAT algorithm. The number of samples ranges from 194 to 16105, and the features of samples range from 14 to 34, 096. We compared other 8 feature processing methods. DBGAT is compared with 8 algorithm on the multi-label dimension reduction (MLDR) task. The source code of DBGAT is available online[1].

4.1   Datasets

Table 2 presents the characteristics of each experimental dataset. It includes the number of samples N, the number of features M, the number of class labels L, and the label cardinality. They can be downloaded for free from Mulan[2].

4.2   Evaluation Metrics

In general classification tasks, the classification effect is judged according to the evaluation metrics. Single-label classification usually uses accuracy, precision, recall, F_1 and other metrics to evaluate the effect of the model. However, multi-label classification is different from single-label classification. The evaluation metrics of multi-label classification is relatively more complicated. Table 3 shows the evaluation metrics used in our paper.

---

[1] https://gitee.com/pengyyuu/dbgat.
[2] http://mulan.sourceforge.net/datasets-mlc.html .

Table 2. Datasets.

Dataset	N	M	L	Cardinality	Density
Arts	7,484	23,146	26	1.654	0.064
Bibtex	7,395	1,836	159	2.402	0.015
Business	11,214	21,924	30	1.599	0.053
CAL500	502	68	174	26.044	0.150
Computers	12,444	34,096	33	1.507	0.046
Corel5k	5,000	499	374	3.522	0.009
Emotions	593	72	6	1.869	0.311
Enron	1,702	1,001	53	3.378	0.064
Flags	194	14	12	3.392	0.485
Recreation	12,828	30,324	22	1.429	0.065
Medical	978	1,449	45	1.245	0.028
Yeast	2,417	103	14	4.237	0.303

Table 3. Definitions of six multi-label Evaluation metrics. '↓' means the lower the better, '↑' means the higher the better.

Evaluation metrics	Fomulation
hamming loss ↓	$\frac{1}{NL} \sum_{i=1}^{N} \sum_{j=1}^{L} \mathbb{I}\left[h_{ij} \neq y_{ij}\right]$
one-error ↓	$\frac{1}{N} \sum_{i=1}^{N} \mathbb{I}\left[\arg\max f\left(\mathbf{x}_i\right) \notin Y_i^+\right]$
coverage ↓	$\frac{1}{NL} \sum_{i=1}^{N} \mathbb{I}\left[\max_{j \in Y_i^+} \operatorname{rank}_f\left(\mathbf{x}_i, j\right) - 1\right]$
ranking loss ↓	$\frac{1}{N} \sum_{i=1}^{N} \frac{\left|S_{\text{rank}}^i\right|}{\left|Y_i^+\|Y_{i.}^-\right|}$
average precision ↑	$\frac{1}{N} \sum_{i=1}^{N} \frac{1}{\left|Y_i^+\right|} \sum_{j \in Y_i^+} \frac{\left|S_{\text{precision}}^{ij}\right|}{\operatorname{rank}_f(\mathbf{x}_i, j)}$
macro-AUC ↑	$\frac{1}{L} \sum_{j=1}^{L} \frac{\left|S_{\text{macro}}^j\right|}{\left|Y_{\cdot j}^+\|Y_{\cdot j}^-\right|}$

4.3 Comparison with Dimension Reduction algorithms

To demonstrate the effectiveness of our proposed DBGAT algorithm, we train it using public datasets.

After dimension reduction of the dataset, we employ linear layers for multi-label classification. We use 6 evaluation metrics to evaluate the classification results. Table 4, 5, 6, 7, 8 and 9 list the comparison of 8 dimension reduction algorithms. To be fair, all dimension reduction methods reduce to the same dimension. The brief characterization and parameter settings of these algorithms are as follows.

PCA [18] is a popular and typical feature extraction method.

Table 4. Hamming loss (↓) comparison with MLDR algorithms.

Approach	DBGAT	PCA	MDDM	MLSI	MCLS	MDFS	Original	KPCA	GAT
Arts	0.063	0.064	0.062●	0.309	0.063	0.063	0.546	0.064	0.121
Bibtex	0.015	0.013●	0.045	0.031	0.015	0.015	0.334	0.013●	0.015
Bussiness	0.028●	0.030	0.030	0.428	0.030	0.030	0.517	0.030	0.044
CAL500	0.136	0.138	0.629	0.139	0.132●	0.132●	0.191	0.138	0.219
Computers	0.037●	0.042	0.038	0.043	0.043	0.042	0.485	0.042	/
Corel5k	0.009●	0.100	0.009●	0.011	0.009●	0.009●	0.437	0.010	0.009●
Emotions	0.202●	0.232	0.476	0.236	0.349	0.270	0.236	0.232	0.323
Enron	0.050●	0.071	0.069	0.198	0.064	0.063	0.653	0.072	0.093
Flags	0.165●	0.232	0.414	0.235	0.178	0.274	0.224	0.219	0.276
Medical	0.023	0.020●	0.030	0.026	0.027	0.029	0.467	0.020●	0.028
Recreation	0.062●	0.063	0.068	0.064	0.064	0.065	0.519	0.063	/
Yeast	0.187●	0.213	0.319	0.214	0.229	0.228	0.214	0.213	0.229
Mean rank	**1.583**	3.333	5.75	5.917	3.667	3.833	7.417	3.083	5.167

Table 5. One error (↓) comparison with MLDR algorithms.

Approach	DBGAT	PCA	MDDM	MLSI	MCLS	MDFS	Original	KPCA	GAT
Arts	0.755	0.713	0.601●	0.904	0.739	0.744	0.768	0.713	0.910
Bibtex	0.601	0.602	0.667	0.748	0.833	0.844	0.613	0.598●	0.861
Bussiness	0.142	0.143	0.140●	0.582	0.140●	0.142	0.434	0.143	0.142
CAL500	0.090	0.130	0.460	0.130	0.080●	0.110	0.410	0.130	0.660
Computers	0.472	0.463●	0.488	0.473	0.471	0.472	0.521	0.463●	/
Corel5k	0.742	0.741	0.640●	0.825	0.754	0.755	0.875	0.749	0.743
Emotions	0.288●	0.356	0.508	0.364	0.542	0.407	0.356	0.364	0.703
Enron	0.476	0.488	0.409●	0.915	0.597	0.591	0.815	0.506	0.679
Flags	0.105●	0.158	0.474	0.158	0.184	0.132	0.158	0.158	0.158
Medical	0.492	0.385●	0.764	0.472	0.646	0.656	0.615	0.385●	0.672
Recreation	0.788	0.720	0.726	0.804	0.801	0.789	0.713●	0.718	/
Yeast	0.220●	0.238	0.586	0.236	0.255	0.255	0.269	0.238	0.251
Mean rank	2.833	**2.75**	5.167	5.833	5.25	5.25	5.667	2.917	5.583

Table 6. Coverage (↓) comparison with MLDR algorithms.

Approach	DBGAT	PCA	MDDM	MLSI	MCLS	MDFS	Original	KPCA	GAT
Arts	0.239	0.238	0.199●	0.575	0.237	0.234	0.426	0.238	0.517
Bibtex	0.462	0.198	0.288	0.374	0.450	0.436	0.320	0.197●	0.510
Bussiness	0.089	0.085●	0.097	0.528	0.092	0.092	0.550	0.085●	0.241
CAL500	0.735●	0.764	0.962	0.764	0.739	0.754	0.873	0.762	0.905
Computers	0.140	0.138●	0.158	0.139	0.141	0.142	0.434	0.138●	/
Corel5k	0.427	0.498	0.359 ●	0.605	0.434	0.437	0.751	0.493	0.492
Emotions	0.322●	0.338	0.565	0.338	0.469	0.383	0.360	0.338	0.500
Enron	0.221●	0.346	0.385	0.524	0.338	0.337	0.719	0.349	0.422
Flags	0.491	0.511	0.650	0.534	0.489●	0.553	0.531	0.518	0.500
Medical	0.120	0.070	0.262	0.141	0.151	0.161	0.297	0.069●	0.292
Recreation	0.231	0.239●	0.235	0.258	0.253	0.252	0.414	0.239	/
Yeast	0.454●	0.465	0.615	0.467	0.483	0.483	0.477	0.466	0.484
Mean rank	**2.75**	3	5.75	5.75	4.333	4.833	7.083	3	5.917

Table 7. Ranking loss (↓) comparison with MLDR algorithms.

Approach	DBGAT	PCA	MDDM	MLSI	MCLS	MDFS	Original	KPCA	GAT
Arts	0.182	0.179	0.141●	0.580	0.179	0.176	0.588	0.179	0.874
Bibtex	0.330	0.127●	0.193	0.247	0.306	0.301	0.217	0.127●	0.352
Bussiness	0.050	0.047●	0.058	0.492	0.050	0.052	0.551	0.047●	0.223
CAL500	0.180●	0.197	0.576	0.197	0.185	0.184	0.276	0.197	0.281
Computers	0.100●	0.100●	0.113	0.101	0.102	0.103	0.526	0.100●	/
Corel5k	0.197	0.237	0.163●	0.315	0.202	0.198	0.443	0.238	0.222
Emotions	0.191●	0.199	0.449	0.202	0.367	0.252	0.220	0.200	0.463
Enron	0.134●	0.146	0.166	0.31	0.147	0.144	0.609	0.149	0.201
Flags	0.143	0.167	0.408	0.169	0.129●	0.197	0.179	0.170	0.161
Medical	0.092	0.055	0.226	0.120	0.123	0.134	0.453	0.053●	0.262
Recreation	0.214	0.201	0.192●	0.220	0.217	0.216	0.558	0.200	/
Yeast	0.171●	0.180	0.328	0.181	0.202	0.204	0.189	0.180	0.216
Mean rank	**2.667**	2.75	5.583	5.583	4.583	4.75	7.25	3	6

Table 8. Average precision (↑) comparison with MLDR algorithms.

Approach	DBGAT	PCA	MDDM	MLSI	MCLS	MDFS	Original	KPCA	GAT
Arts	0.426	0.446	0.535●	0.257	0.436	0.440	0.336	0.447	0.202
Bibtex	0.191	0.371	0.310	0.238	0.162	0.159	0.396●	0.374	0.140
Bussiness	0.854	0.857●	0.851	0.201	0.856	0.854	0.436	0.857	0.946
CAL500	0.504●	0.484	0.169	0.484	0.493	0.498	0.368	0.484	0.350
Computers	0.606	0.614	0.572	0.606	0.609	0.608	0.400	0.615●	/
Corel5k	0.209	0.203	0.290●	0.161	0.208	0.208	0.12	0.202	0.203
Emotions	0.767●	0.756	0.591	0.753	0.633	0.717	0.749	0.755	0.547
Enron	0.495	0.505	0.553●	0.223	0.471	0.478	0.269	0.498	0.373
Flags	0.829●	0.790	0.648	0.790	0.824	0.817	0.78	0.794	0.812
Medical	0.605	0.706●	0.362	0.596	0.454	0.458	0.445	0.705	0.403
Recreation	0.395	0.442	0.454●	0.379	0.383	0.390	0.392	0.443	/
Yeast	0.791●	0.742	0.584	0.742	0.715	0.709	0.735	0.744	0.694
Mean rank	3.25	3.333	5.5	6.083	5.083	4.583	6.5	**2.917**	6.8

Table 9. AUC (↑) comparison with MLDR algorithms.

Approach	DBGAT	PCA	MDDM	MLSI	MCLS	MDFS	Original	KPCA	GAT
Arts	0.913	0.541	0.590	0.538	0.593	0.609	0.690	0.542	0.923●
Bibtex	0.967	0.838	0.796	0.721	0.601	0.830	0.676	0.839	0.997●
Bussiness	0.632	0.586	0.495	0.606	0.748	0.817●	0.694	0.584	0.933●
CAL500	0.505	0.492	0.499	0.491	0.459	0.497	0.488	0.492	0.799●
Computers	0.939	0.556	0.593	0.939	0.870	0.752	0.753	0.555	/
Corel5k	0.727	0.506	0.515	0.405	0.728●	0.718	0.482	0.507	0.785●
Emotions	0.805	0.796	0.751	0.794	0.566	0.749	0.785	0.796	0.863●
Enron	0.491	0.593	0.563	0.571	0.597	0.621●	0.607	0.595	0.607●
Flags	0.737	0.721	0.478	0.720	0.783●	0.558	0.716	0.734	1
Medical	0.742	0.614	0.361	0.545	0.708	0.495	0.625	0.613	0.744●
Recreation	0.984	0.584	0.639	0.100	0.803	0.751	0.776	0.586	/
Yeast	0.668	0.645	0.632	0.645	0.584	0.577	0.650	0.645	0.982●
Mean rank	2.833	5.583	6.75	6.083	4.75	5.083	5.333	5.333	**2.333**

MDDM[3] [24] attempts to project the original data into a low-dimensional feature space maximizing the dependence between the original feature description and the associated class labels.

MLSI [21], the parameter β is set to 0.5 as recommended in paper.

MCLS [7] is a feature selection method named manifold-based constraint Laplacian score. The settings are as follows: $knear = 5$.

MDFS [22] is an embedded multi-label feature selection method with manifold regularization. It seeks discriminative features across multiple class labels. The settings are as follows: $mu = 0.5, dim\_para = 10$.

DBGAT is our new algorithm. We use the following settings: the number of neighbors we set $k = 0.05N$. In the network we use the sigmoid activation function.

Original indicates that no feature extraction is applied.

KPCA [13] (Kernel Principal Component Analysis) can achieve nonlinear dimension reduction of data for processing linearly inseparable datasets.

GAT is a degenerate version of DBGAT. The GAT algorithm is designed in the way of steps 2 and 3 of Fig. 1 in this paper.

For each datasets, the best measurement is shown with black dots. The last row lists the average ranking of each method on the 12 datasets.

From the results we observe that:

1) DBGAT can be compared with popular MLDR methods. Note that we have not fine-tuned the number of neighbors yet.
2) DBGAT runs on different types of datasets. At the same time it performs well on both large and small datasets, proving its adaptability.

Figure 2 shows the results of the Bonferroni-Dunn test for further analysis of pairwise comparisons [4]. The performance of two methods is significantly different if the corresponding average ranks differ by at least the critical difference (CD = 3.05) for $\alpha = 0.05$. If there is no significant difference between the two algorithms, the solid line is connected, and vice versa. This test is illustrated in Fig. 2, where groups of methods that are not significantly different are connected. That shows DBGAT performs well.

---

[3] http://www.lamda.nju.edu.cn/code_MDDM.ashx.

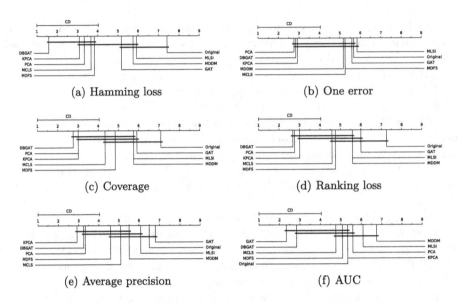

Fig. 2. Performance comparison of DBGAT algorithm against the others with the Bonferroni-Dunn test.

4.4   Ablation Learning

1) In this paper, we set different k values as the number of neighbors. N is the number of instances. Table 10 shows our classification performance after fine-tuning the number of neighbors.

Table 10. Set different k values, the results on emotions.

k	Hamming loss	One error	Coverage	Ranking loss	Average precision	AUC
$0.05N$	0.208	0.305	0.302	0.164	0.8	0.835
$0.1N$	0.209	0.322	0.314	0.175	0.785	0.826
$0.15N$	0.218	0.331	0.299	0.167	0.794	0.819
$0.2N$	0.23	0.356	0.331	0.194	0.757	0.806

2) There are many methods for calculating the distance between instances, such as Cosmic similarity, Euclidean distance, Pearson correlation, etc. In this experiment, we calculated several distance methods. The final results were similar. Therefore, in our paper, we use a relatively common Euclidean distance calculation method.

5 Conclusions

In our paper, the multi-label learning of graph feature representation is studied. Considering the correlation between instances, a neighbor representation update method is adopted to learn new node representations. Experiments show that this method has new research significance for multi-label classification.

References

1. Cunningham, P.: Dimension reduction. In: Cord, M., Cunningham, P. (eds.) Machine Learning Techniques for Multimedia, pp. 91–112. Springer, Berlin (2008). https://doi.org/10.1007/978-3-540-75171-7_4
2. De Lathauwer, L., De Moor, B., Vandewalle, J.: A multilinear singular value decomposition. SIAM J. Matrix Anal. Appl. **21**(4), 1253–1278 (2000)
3. Deerwester, S., Dumais, S.T., Furnas, G.W., Landauer, T.K., Harshman, R.: Indexing by latent semantic analysis. J. Am. Soc. Inf. Sci. **41**(6), 391–407 (1990)
4. Demšar, J.: Statistical comparisons of classifiers over multiple data sets. J. Mach. Learn. Res. **7**, 1–30 (2006)
5. Fodor, I.K.: A survey of dimension reduction techniques. Technical report, Lawrence Livermore National Lab., CA (US) (2002)
6. Guyon, I., Gunn, S., Nikravesh, M., Zadeh, L.A.: Feature Extraction: Foundations And Applications, STUDFUZZ, vol. 207. Springer (2008). https://doi.org/10.1007/978-3-540-35488-8
7. Huang, R., Jiang, W., Sun, G.: Manifold-based constraint laplacian score for multi-label feature selection. Patt. Recogn. Lett. **112**, 346–352 (2018)
8. Hyvärinen, A., Oja, E.: Independent component analysis: algorithms and applications. Neural Networks **13**(4–5), 411–430 (2000)
9. Izenman, A.J.: Linear discriminant analysis. In: Modern Multivariate Statistical Techniques, pp. 237–280. Springer, New York (2013). https://doi.org/10.1007/978-0-387-78189-1_8
10. Kipf, T.N., Welling, M.: Semi-supervised classification with graph convolutional networks. In: ICLR (Poster) (2016)
11. Nixon, M., Aguado, A.: Feature Extraction and Image Processing for Computer Vision. Academic Press (2019)
12. Scarselli, F., Gori, M., Tsoi, A.C., Hagenbuchner, M., Monfardini, G.: The graph neural network model. IEEE Trans. Neural Networks **20**(1), 61–80 (2008)
13. Schölkopf, B., Smola, A., Müller, K.R.: Kernel principal component analysis. In: Gerstner, W., Germond, A., Hasler, M., Nicoud, JD. (eds.) International Conference on Artificial Neural Networks. LNCS, vol. 1327, pp. 583–588. Springer,Berlin (1997). https://doi.org/10.1007/BFb0020217
14. Sun, L., Ji, S., Ye, J.: Canonical correlation analysis for multilabel classification: a least-squares formulation, extensions, and analysis. IEEE Trans. Patt. Anal. Mach. Intell. **33**(1), 194–200 (2010)
15. Tsoumakas, G., Katakis, I., Vlahavas, I.: Mining multi-label data. In: Maimon, O., Rokach, L. (eds.) Data Mining and Knowledge Discovery Handbook, pp. 667–685. Springer, Boston (2009). https://doi.org/10.1007/978-0-387-09823-4_34
16. Veličković, P., Cucurull, G., Casanova, A., Romero, A., Liò, P., Bengio, Y.: Graph attention networks. In: International Conference on Learning Representations (2018)

17. Wang, H., Ding, C., Huang, H.: Multi-label linear discriminant analysis. In: Dani-ilidis, K., Maragos, P., Paragios, N. (eds.) European Conference on Computer Vision. LNIP, vol. 6316, pp. 126–139. Springer, Berlin (2010). https://doi.org/10.1007/978-3-642-15567-3_10

18. Wold, S., Esbensen, K., Geladi, P.: Principal component analysis. Chemometrics Intell. Lab. Syst. **2**(1–3), 37–52 (1987)

19. Xu, J.: A weighted linear discriminant analysis framework for multi-label feature extraction. Neurocomputing **275**, 107–120 (2018)

20. Xu, J., Liu, J., Yin, J., Sun, C.: A multi-label feature extraction algorithm via maximizing feature variance and feature-label dependence simultaneously. Knowl. Based Syst. **98**, 172–184 (2016)

21. Yu, K., Yu, S., Tresp, V.: Multi-label informed latent semantic indexing. In: Proceedings of the 28th Annual International ACM SIGIR Conference on Research and Development in Information Retrieval, pp. 258–265 (2005)

22. Zhang, J., Luo, Z., Li, C., Zhou, C., Li, S.: Manifold regularized discriminative feature selection for multi-label learning. Pattern Recogn. **95**, 136–150 (2019)

23. Zhang, M.L., Zhou, Z.H.: A review on multi-label learning algorithms. IEEE Trans. Knowl. Data Eng. **26**(8), 1819–1837 (2013)

24. Zhang, Y., Zhou, Z.H.: Multilabel dimensionality reduction via dependence maximization. TKDD **4**(3), 1–21 (2010)

Multi-scale Subgraph Contrastive Learning for Link Prediction

Shilin Sun, Zehua Zhang[(✉)], Runze Wang, and Hua Tian

College of Information and Computer, Taiyuan University of Technology,
Jinzhong 030600, China
zehua_zhang@163.com

Abstract. Existing works about link prediction rely mainly on pooling operations which cause loss of edge information or similarity assumptions, so that they are limited in specific networks, and mainly supervised learning methods. We propose a Multi-scale Subgraph Contrastive Learning (MSCL) method. To adapt to networks of different sizes and make direct use of edge information, MSCL converts a sampled subgraph centered on the target link into a line graph as a node-scale to represent links, and mines deep representations by combining two scales information, subgraph-scale and line graph node-scale. After learning the information of the two subgraphs separately by encoders, we use contrastive learning to balance the information of two scales to alleviate the over-reliance of the model on labels and enhance the model's robustness. MSCL outperforms a set of state-of-the-art graph representation learning solutions on link prediction task in a variety of graphs including biology networks and social networks.

Keywords: Multi-scale · Contrastive learning · Subgraph · Link prediction

1 Introduction

Link prediction methods combined with specific applications can be considered to predict the presence of interactions between pairs of nodes or the type of interactions. And link prediction methods have many specific applications in the real world, such as drug interaction prediction, community discovery, etc.

Traditional methods for link prediction score links by calculating nodes similarity. Based on the assumption that similar nodes are more inclined to connect, these heuristic methods use known node information in the network to score nodes. Some methods consider local information about the network, such as Common Neighbors (CN). Other methods take a global perspective, for example, Katz [5] and Pagerank [1] directly aggregate over all the paths between node pairs to score links. Although such methods improve performance, they lack universal applicability to various networks, for example, similar proteins do not tend to interact [7].

© The Author(s), under exclusive license to Springer Nature Switzerland AG 2022
J. Yao et al. (Eds.): IJCRS 2022, LNAI 13633, pp. 217–223, 2022.
https://doi.org/10.1007/978-3-031-21244-4_16

To design a universal model, based on the fact that subgraphs already contain enough information and are suitable for networks of different sizes, Weisfeiler-Lehman Neural Machine(WLNM) [9] extracts local subgraphs around the links, and learns the subgraphs corresponding to link existence through the fully connected layer to achieve better performance. To enhance the graph feature learning capability and incorporate latent features, SEAL [10] uses a graph neural network to replace the fully connected neural network in WLNM. However, the information loss of pooling operations with subgraphs is not negligible. LGLP [2] converts subgraphs into line graphs to obtain a unique node to represent every link directly for more edge information. Although it works well, only a single scale of information is considered in LGLP and it is based on a supervised learning approach. So the performance improvement is limited by the noise from the growing edges of the line graph and over-reliance on labels.

In this paper, we propose a **M**ulti-scale **S**ubgraph **C**ontrastive **L**earning (MSCL) method for the link prediction task. Firstly, the subgraphs are extracted centered on the links, and then the subgraphs are converted into line graph subgraphs so that each link has its corresponding node representation. Thus, a new view is obtained. Secondly, the original subgraph and the line graph subgraph are learned by different encoders to obtain link representation, and finally, multi-scale information is balanced using contrastive learning. Our work is summarized as follows:

- We propose a multi-scale subgraph contrastive learning framework, which converts subgraphs into line graphs to improve the efficiency of aggregating information and reduce the information loss of subgraph pooling.
- We adopt a contrastive learning component to achieve multi-scale learning and balance information of two views. So the robustness of the model is enhanced and the dependence of models on labels is decreased.
- The multi-scale information is learned using different encoders, and the model achieves state-of-the-art results on two public available datasets.

2   Proposed Method

2.1   Problem Formulation

Through multi-scale contrastive learning, the model integrates line graph and subgraph information. The line graph node transformed from the subgraph of the target link is the positive sample g^+, and the node of the line graph corresponding to the other link is negative sample g^-, and the anchor g is the subgraph of target link. The model then trains the mapping function using contrastive learning to enhance model performance. Our contrastive learning component is shown in Eq. (1). $score(\cdot)$ is a function to calculate the similarity of two representations for a target link. And $f(\cdot)$ denotes graph encoder.

$$score(f(g), f(g^+)) >> score(f(g), f(g^-)) \tag{1}$$

2.2   Overview

Figure 1 illustrates the overall framework of MSCL. The line graph transformation component samples the original graph for subgraphs and transforms them to line graphs. Next, the encoder encodes the subgraph and line graph respectively. Finally, the contrastive learning component balances multi-scale information.

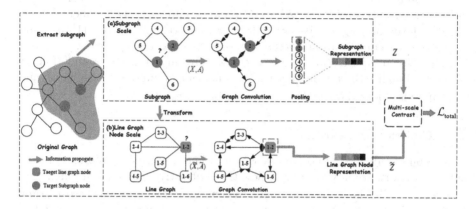

Fig. 1. Overview of the MSCL framework

2.3   Line Graph Transformation

MSCL mainly uses the different scale information of subgraph $G(V, E, X, A)$ for link prediction, and its core point is to balance the scale information of line graph nodes by contrastive learning.

Line Graph Transformation: To transform a subgraph into a line graph, we transform the subgraph's edges to the line graph's set of nodes, as illustrated in Eq. (2). If any two nodes of a line graph correspond to two edges of a subgraph that share a common node, then these two nodes form an edge [2]. $\widetilde{V}$ is the line graph node identity matrix, and $\widetilde{A}$ is the set of edges of the line graph, which is identical to the representation of the adjacency matrix. Equation (2) is used to generate the line graph representation $\widetilde{G}(\widetilde{V}, \widetilde{E}, \widetilde{X}, \widetilde{A})$.

$$\widetilde{V} = \{e\}, \forall e \in E$$

$$\widetilde{X} = \{\text{concate}(x_i, x_j) | \forall e_{(v_i, v_j)} \in \widetilde{V}\}$$

$$\widetilde{E} = \{l_{(e_i, e_j)} | e_i \cap e_j \neq \emptyset\}, \forall e_i, e_j \in \widetilde{V} \tag{2}$$

$$\text{Num}(\widetilde{E}) = \frac{1}{2} \sum_{i=1}^{m=\text{Num}(V)} \text{Deg}(v_i)^2 - \text{Num}(E), \forall v \in V \tag{3}$$

As shown in Eq. (3), the number of edges in the line graph increases exponentially compared to the number of edges in the original graph. $Deg(\cdot)$ is a function to calculate the degree of nodes.

2.4 Graph Encoder

The graph encoder $f(\cdot)$ is separated into two parts: a subgraph encoder $f_S(\cdot)$, a line graph encoder $f_L(\cdot)$. And the graph encoder handles the feature and adjacency matrices of the two views. The representation of subgraphs and line graphs is conducted independently as $\boldsymbol{Z}, \widetilde{\boldsymbol{Z}}$.

The graph encoder extracts substructure features by stacking multiple graph convolution layers as $\boldsymbol{Z}^{1:h} := [\boldsymbol{Z}^1 \dots \boldsymbol{Z}^h]$, and h represents the number of layer. Finally, the obtained graph is pooled through the SortPooling layer [11] to obtain a graph-scale representation, thereby obtaining a cross entropy loss $\mathcal{L}$.

The design of the line graph encoder convolution layer mainly follows GCN [6]. Finally, the pooling layer selects nodes to represent target links, and the supervised loss $\widetilde{\mathcal{L}}$ of the line graph can be obtained.

2.5 Contrastive Learning

To combine multi-scale information which is line graph node information and subgraph information, we mainly follow the contrastive model in GraphCL [8].

Take $z^{(n)}, \widetilde{z}^{(n)}$ in $\boldsymbol{Z}, \widetilde{\boldsymbol{Z}}$ respectively, to denote the two views of the nth graph in the small batch. Negative samples are generated from the other $n - 1$ line graphs. $sim(\cdot)$ is cosine similarity function, and contrastive loss $\mathcal{L}_{CON}$ is defined as:

$$\mathcal{L}_{\text{CON}} = \frac{1}{|T|} \sum_{n=1}^{|T|} - \log \frac{\exp(\text{sim}(z^{(n)}, \widetilde{z}^{(n)})/\tau)}{\sum_{n'=1, n' \neq n}^{N} \exp(\text{sim}(z^{(n)}, \widetilde{z}^{(n')})/\tau)} \tag{4}$$

$|T|$ denotes the number of links in the training set and τ is a hyperparameter.

Finally, the total loss function $\mathcal{L}_{total}$ is obtained by combining the self-supervised task loss with the supervised loss of both views.

$$\mathcal{L}_{\text{total}} = \alpha \mathcal{L} + \beta \widetilde{\mathcal{L}} + \lambda \mathcal{L}_{\text{CON}} \tag{5}$$

α, β, λ are the hyperparameters used to balance the different losses.

3 Experiment

During the training process of the model, we use the Adam optimizer with a learning rate of 0.05. To evaluate the effectiveness of link prediction, we use Area Under the Curve(AUC) and Average Precision(AP) as evaluation metrics.

3.1 Datasets and Baseline Models

We conduct experiments on two datasets in different areas, HPD, ADV [2] to verify MSCL's effectiveness. In this work, we compare three network similarity methods, including Katz [5], PageRank (PR) [1] and SimRank (SR) [4]. Also, the network embedding methods Node2vec(N2V) [3] and graph representation learning methods SEAL [10] and LGLP [2] (Table 1).

Table 1. Summary of datasets used in our experiments.

Datasets	Nodes	Links	Degree	Area
HPD	8756	32331	7.38	Biology
ADV	5155	39285	15.24	Social network

3.2 Comparison With Baselines

Figure 2 shows that the performance of MSCL method has been improved compared with the other three types of methods on 80% training percentage of links. The graph representation learning method can learn deeper feature information and topological information than other methods, so the performance is significantly improved. MSCL combines multi-scale information and therefore has the best performance among all methods.

3.3 Model Robustness Analysis

To explore the robustness of the model to network sparsity, experiments are conducted on edge datasets of different sizes from 30% to 80%. Figure 3 shows the robustness of MSCL to network sparsity. MSCL outperforms at all assignments with various levels of network sparsity. The performance of Katz and PageRank, which are based on network similarity, is poor compared to other methods due to their assumptions, and the gap becomes larger as the density increases. MSCL alleviates the model's over-reliance on labels and the line graph noise problem.

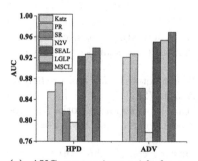
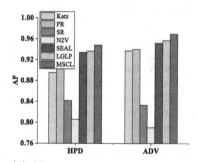

(a) AUC comparison with baselines

(b) AP comparison with baselines

Fig. 2. Performances of MSCL and baselines

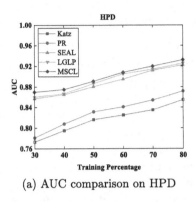
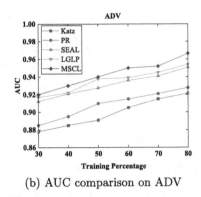

(a) AUC comparison on HPD   (b) AUC comparison on ADV

Fig. 3. Robustness analysis on HPD and ADV with different training percentage

4   Conclusions

In this paper, we propose a method for balancing line graph node scale and subgraph scale information by contrastive learning. Final experiments show that MSCL performs well. Future work will continue to explore the information yield of graphs at different scales for diverse tasks.

Acknowledgements. This work was supported by the National Natural Science Foundation of China (61702356), Industry-University Cooperation Education Program of the Ministry of Education, and Shanxi Scholarship Council of China.

References

1. Brin, S., Page, L.: The anatomy of a large-scale hypertextual web search engine. Comput. Networks ISDN Syst. **30**(1), 107–117 (1998) (proceedings of the Seventh International World Wide Web Conference)
2. Cai, L., Li, J., Wang, J., Ji, S.: Line graph neural networks for link prediction. IEEE Trans. Patt. Anal. Mach. Intell. **44**, 5103–5113 (2021)
3. Grover, A., Leskovec, J.: node2vec: scalable feature learning for networks. In: Proceedings of the 22nd ACM SIGKDD International Conference on Knowledge Discovery and Data Mining, pp. 855–864 (2016)
4. Jeh, G., Widom, J.: Simrank: a measure of structural-context similarity. In: Proceedings of the Eighth ACM SIGKDD International Conference on Knowledge Discovery and Data Mining, pp. 538–543 (2002)
5. Katz, L.: A new status index derived from sociometric analysis. Psychometrika **18**(1), 39–43 (1953)
6. Kipf, T.N., Welling, M.: Semi-supervised classification with graph convolutional networks. arXiv preprint arXiv:1609.02907 (2016)
7. Kovács, I.A., et al.: Network-based prediction of protein interactions. Nature Commun. **10**(1), 1–8 (2019)
8. You, Y., Chen, T., Sui, Y., Chen, T., Wang, Z., Shen, Y.: Graph contrastive learning with augmentations. Adv. Neural Inf. Process. Syst. **33**, 5812–5823 (2020)

9. Zhang, M., Chen, Y.: Weisfeiler-lehman neural machine for link prediction. In: Proceedings of the 23rd ACM SIGKDD International Conference on Knowledge Discovery and Data Mining, pp. 575–583 (2017)
10. Zhang, M., Chen, Y.: Link prediction based on graph neural networks. Adv. Neural Inf. Process. Syst. **31**, 5171–5181 (2018)
11. Zhang, M., Cui, Z., Neumann, M., Chen, Y.: An end-to-end deep learning architecture for graph classification. In: Proceedings of the AAAI Conference on Artificial Intelligence. vol. 32 (2018)

Video Object Detection with MeanShift Tracking

Shuai Zhang[1,3], Wei Liu[2], Haijie Fu[1], and Xiaodong Yue[1,3(✉)]

[1] School of Computer Engineering and Science, Shanghai University,
Shanghai 200444, China
{wszs1998,fhttsfhj,yswantfly}@shu.edu.cn
[2] College of Electronics and Information Engineering, Tongji University,
Shanghai 200444, China
[3] Artificial Intelligence Institute of Shanghai University, Shanghai, China

Abstract. Video object detection, a basic task in the computer vision, is rapidly evolving and widely used in various real-world applications. Recently, with the success of deep learning, deep video object detection has become an important research direction. Although existing deep video object detection methods have achieved excellent results compared with those of traditional methods, they ignore the motion laws of objects and are hard to improve the detection performance of the fast moving objects suffering from deteriorated problems such as the motion blur, video defocus, object occlusion and rare poses. To address this limitation, we add the object trajectory information into the process of the video object detection and devise a novel deep video object detection method which utilizes the MeanShift algorithm to guide the deep neural networks to enhance the video object detection performance. The experiments on ImageNet VID dataset validate that the proposed method can improve the recognition performance of fast moving objects with taking into account the motion laws of objects.

Keywords: Video object detection · Deep neural networks · MeanShift

1 Introduction

Object detection, which tries to locate the object of interest according to the input image and give the category information, is one of the key methodologies in computer vision research [38]. With the explosion of video data close to our daily life such as video surveillance, face recognition, autonomous driving, and robot vision, the research on video target detection has greater practical research significance and application value. Although there have been a significant progress in object detection in still images, directly applying these detection methods to videos faces great challenges, such as the unaffordable computational cost and low recognition performance [35], which promotes the development of the video object detection.

© The Author(s), under exclusive license to Springer Nature Switzerland AG 2022
J. Yao et al. (Eds.): IJCRS 2022, LNAI 13633, pp. 224–237, 2022.
https://doi.org/10.1007/978-3-031-21244-4_17

video defocus

motion blur

rare poses

object occlusion

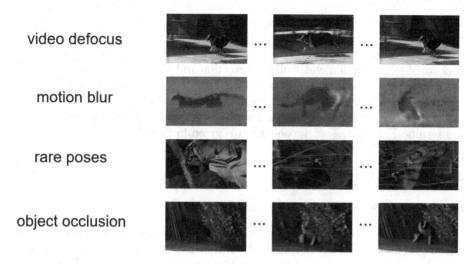

Fig. 1. The fast moving objects suffering from different deteriorated problems in video.

Earlier methods in video object detection relied on handcrafted features, e.g., [16,23], which produce lower accuracies in video object detection. Recently, joining the success of deep learning, deep video object detection, which aims to locate and recognize the object in video with deep neural networks (DNNs) [13, 25,33,34], has become an important research direction. The common attempts in deep video object detection involved performing deep object detection methods on each image frame to promote the recognition accuracy in video.

However, although existing deep video object detection methods have achieved excellent results compared with those of traditional methods, they ignore the motion laws of objects and cannot take into consideration both spatial and temporal correlations between image frames, which are hard to improve the detection performance of the video fast moving objects suffering from deteriorated problems such as the motion blur, video defocus, object occlusion and rare poses (shown in Fig. 1). *To address this limitation, inspired by the work of [17,32], we introduce the trajectory information into the detection process and devise a novel deep video object detection method from the perspective of object tracking, which utilizes the MeanShift algorithm to guide the deep neural networks for video object detection.* The MeanShift tracking algorithm, a non-parametric mode-seeking method for density functions, is a popular algorithm for object tracking since it is fast, robust and easy to implement. The MeanShift algorithm tracks by minimizing a distance between two probability density functions represented by a reference and candidate histograms. Since the histogram distance (or, equivalently, similarity) does not depend on spatial structure of the search window, the method is suitable for deformable objects caused by fast moving [30]. As to the characteristics of MeanShift for tracking of fast moving objects suffering from deteriorated problems in video, we apply it to deep video object detection for adding the object trajectory information into the process of the video object detection and thereby enhancing the video detection performance.

The rest of this paper is organized as follows. Section 2 introduces the related works of object detection, video object detection and MeanShift methods, respectively. Section 3 describes the proposed method in detail, which consists of the workflow of deep video object detection and constructed Object Motion Law Prediction (OMLP) module. Section 4 describes our specific experimental settings and results. Finally, the work conclusion is given in Sect. 5.

2   Related Work

2.1   Object Detection in Still Images

In general, object detection methodologies can be grouped into two major categories: (1) one-stage object detection algorithms and (2) Two-stage object detection algorithms. One-stage object detection algorithms which are trained by optimizing classification-loss and localization-loss simultaneously are often more computationally efficient than two-stage detection methods which first generate a limited number of object box proposals and then classify those proposals to achieve the detection task. However, two-stage object detection methods can produce higher accuracies compared to one-stage object detection algorithms. Specifically, in one-stage detectors, OveFeat [27] is one of the first CNN-based one-stage detect detection method. Thereafter, different designs of one-stage detectors are proposed, including SSD [20], YOLO [24], DSSD [9] and DSOD [28]. In two-stage detectors, Faster-RCNN [13,25] utilizes two fully connected layers as the RoI heads. CascadeRCNN [1] consists of a sequence of detectors trained with higher IoU thresholds to get a high quality object detector. TSD [29] decouples the classification and localization branches for each RoI.

2.2   Video Object Detection

Although the object detection methods for images have achieved significant progress, it is hard to directly apply these detectors to videos. Initially, video object detection approaches have relied on handcrafted features, e.g., [16,18,21,23]. Joining the success of deep learning, A number of deep-learning based video object detection approaches were proposed after the ImageNet Large Scale Visual Recognition Challenge (ILSVRC2015). DFF [37] runs the expensive convolutional sub-network only on sparse key video frames and propagates their deep feature maps to other frames via a flow field to boost the recognition accuracy. FGFA [36] proposes a flow-guided feature aggregation framework to leverage the pre-frame features by aggregation of nearby features along the motion paths for improving the video object detection accuracy. RDN [8] proposes a new architecture that novelly aggregates and propagates object relation to augment object features for detection. MEGA [2] Introduces a memory enhanced global-local aggregation network to take full consideration of both global and local

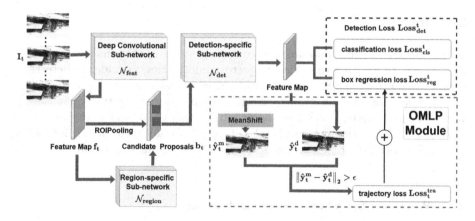

Fig. 2. The framework of video object detection with MeanShift.

information. HVR-Net [12] designs a Hierarchical Video Relation Network by integrating intra-video and inter-video proposal relations in a hierarchical fashion. The work of [11] proposes a novel Temporal RoI Align operator to extract features from other frames feature maps for current frame proposals by utilizing feature similarity. SLTnet FPN-X101 [5] presents a new network architecture to take advantage of spatio-temporal information available in videos to boost object detection precision. Although these methods achieve excellent results on video object detection, they ignore the object trajectory information in detection and are hard to recognize the fast moving objects. In contrast, our method adds the motion laws of objects into the learning process, which can promote the video object detection performance of the fast moving objectives suffering from deteriorated problems.

2.3   MeanShift Tracking

The MeanShift algorithm proposed by Fukunaga and Hostetler [10] is a non-parametric mode-seeking method for density functions. It was firstly introduced to computer vision by Comaniciu et al. [4] and became a popular object tracking algorithm since it is fast, robust, easy to implement and performs well in a range of conditions. The MeanShift algorithm tracks by minimizing a distance between two probability density functions represented by a reference and candidate histograms. Since the histogram distance (or, equivalently, similarity) does not depend on spatial structure of the search window, the method is suitable for deformable objects [3, 22, 30, 31]. In this paper, we use the MeanShift algorithm to extract the object trajectory information and guide the deep neural network to improve the recognition performance of fast moving objectives in videos.

3   Method

In this section, we will elaborate how we devise our model that enables the whole architecture to fully use the object trajectory information to guide the detection neural networks for improving the performance of fast moving objects suffering from the deteriorated problems. In specific, compared with the traditional single frame image detection model, we construct an Object Motion Law Prediction (OMLP) module that extracts the object trajectory information and integrate it into the workflow of video object detection. The overall architecture is shown in Fig. 2.

3.1   Workflow for Video Object Detection

Here we will first introduce the workflow of video object detection in our work. Given a video dataset that contains a set of T frames $\{I_t\}_{t=1}^{T}$, our goal is to give detection results for each frame I_t of video for video object detection task. Typically, the video object detection process contains three main steps: (1) A deep convolutional subnetwork $\mathcal{N}_{\text{feat}}$ is applied to each frame I_t to produce feature maps $f_t = \mathcal{N}_{\text{feat}}(I_t)$; (2) A shallow region-specific sub-network, $\mathcal{N}_{\text{region}}$ (i.e., Region Proposal Networks (RPN) [25]) is applied on the feature maps f_t to generate the candidate proposals (RoIs), $b_t = \mathcal{N}_{\text{region}}(f_t)$, and (3) then extracts the RoI feature x_t of each candidate proposal b_t and outputs the detection loss $Loss_{\text{det}}^{t}$ that consists of the cross-entropy loss $Loss_{\text{cls}}^{t}$ for per-region classification score and bounding box regression loss $Loss_{\text{reg}}^{t}$ in terms of the detection-specific sub-network, $\mathcal{N}_{\text{det}}$ (i.e., Region-based Fully Convolutional Network (R-FCN) [6]).

3.2   Integrate OMLP Module into Video Object Detection

In the Sect. 3.1, we have introduced a common workflow for video object detection, although this workflow can recognize the object in video with deep neural networks, it is hard to improve the detection performance of fast moving objects suffering from deteriorated problems in video. Therefore, we integrate the object trajectory information to guide the training direction of deep object detection neural network.

In specific, our main goal is to optimize the overall loss $Loss^t$ that consists of the object detection loss $Loss_{\text{det}}^{t}$ and trajectory loss $Loss_{\text{tra}}^{t}$ as follows

$$Loss^t = Loss_{\text{det}}^{t} + \lambda Loss_{\text{tra}}^{t}. \tag{1}$$

where $Loss_{\text{det}}^{t} = Loss_{\text{cls}}^{t} + Loss_{\text{reg}}^{t}$, similar to the work of [25], and λ is a balance factor.

To achieve this goal, we construct our OMLP module to extract the object trajectory information to improve the performance of video object detection. Our OMLP module acts on the RoI feature x_t and each candidate proposal b_t produced by detection-specific sub-network $\mathcal{N}_{\text{det}}$. We use the MeanShift algorithm to calculate the trajectory center point of an object detected in the tth

frame, and compare with the center point of the candidate proposal predicted by the $\mathcal{N}_{\text{det}}$. If the distance between the two points exceeds the threshold ϵ, which means we need to add the object trajectory information into the detection networks to guide the training direction. In detail, the proposed OMLP module based on the color histogram can be divided into three steps [19].

First, on the base of color histogram and kernel density estimation, we use the initial center candidate y_1 of the candidate proposal b_1 and its corresponding RoI feature $\boldsymbol{x}_1 = \left\{x_1^i\right\}_{i=1}^N$, N represents the size of the feature, to calculate the target kernel function histogram $\hat{q}_u$ and the candidate kernel function histogram $\hat{p}_u$.

$$\hat{q}_u = C \sum_{i=1}^N k\left(\left\|x_1^i\right\|^2\right) \delta\left[b\left(x_1^i\right) - u\right] u = 1, 2, \ldots, m, \tag{2}$$

$$\hat{p}_u = C \sum_{i=1}^N k\left(\left\|\frac{x_1^i - y_1}{h}\right\|^2\right) \delta\left[b\left(x_1^i\right) - u\right] u = 1, 2, \ldots, m, \tag{3}$$

where $k\left(\cdot\right)$ is a kernel function (i.e., Epanechnikov), $b\left(\cdot\right)$ is the bin number $(1, \ldots, m)$ associated with $\boldsymbol{x}_1^i$, h is the size of kernel, C is the normalization factor.

Then, we should maximize the similarity between these two histograms $\hat{q}_u$ and $\hat{p}_u$. Similar to the Bhattacharyya coefficient, the similarity metric is defined as:

$$\hat{\rho}(y) = \sum_{u=1}^m \sqrt{\hat{p}_u(y)\hat{q}_u}, \tag{4}$$

and using Taylor expansion, Eq. 4 can be reduced to

$$\hat{\rho}(y) = \sum_{u=1}^m \sqrt{\hat{p}_u(y)\hat{q}_u} \approx \frac{1}{2} \sum_{u=1}^m \sqrt{\hat{p}_u(y)\hat{q}_u} \\ + \frac{1}{2} \sum_{i=1}^N w_i k\left(\left|\frac{x^i - y}{h}\right|^2\right), \tag{5}$$

where

$$w_i = \sum_{u=1}^m \delta\left[b\left(y_i\right) - u\right] \sqrt{\frac{\hat{q}_u}{\hat{p}_u\left(\hat{y}_0\right)}}, \tag{6}$$

and y indicates the arbitrary location and $\hat{y}_0$ indicates the estimated target location in the previous frame.

Finally, the estimated target location $\hat{y}_t$ in the tth frame can be calculated by an iterative MeanShift procedure below

$$\hat{y}_t = \frac{\sum_{i=1}^N x_1^i w_i g\left(\left\|\frac{\hat{y}_{t-1} - x_1^i}{h}\right\|^2\right)}{\sum_{i=1}^N w_i g\left(\left\|\frac{\hat{y}_{t-1} - x_1^i}{h}\right\|^2\right)}. \tag{7}$$

After getting the estimated target location $\hat{y}_t^m$ inferred by MeanShift and the predicted coordinate $\hat{y}_t^d$ inferred by detection-specific sub-network $\mathcal{N}_{\text{det}}$, we compute the trajectory loss of the current frame t as follows

$$Loss_{\text{tra}}^t = \left\| \hat{y}_t^m - \hat{y}_t^d \right\|_2. \tag{8}$$

According to the Eq. 1, we can add the object trajectory information into the process of the video object detection and thereby devise a novel deep video object detection method which utilizes the MeanShift algorithm to guide the deep neural networks to enhance the video object detection performance.

4 Experiments

In this section, we extensively evaluate the proposed method on real-world video dataset and compare it with existing deep video object detection methods. Experimental results show that our algorithm achieves the state-of-the-art results for the video object detection.

4.1 Experimental Settings

Dataset

In experiments, we test our video object detection algorithm on the ImageNet VID dataset [26], which is the most commonly used dataset in video object detection. This dataset is divided into training set and validation set, including 3862 video clips and 555 video clips, respectively. Video streaming in each frame rate is 25 or 30 frames per second (fps). In addition, the ImageNet VID dataset contains 30 object categories, which are a subset of categories in the ImageNet DET dataset [7], the detailed categories information of ImageNet VID dataset are shown in Table 1.

Table 1. Categories of ImageNet VID dataset

airplane	antelope	bear	bicycle	bird	bus
car	cattle	dog	domestic_cat	elephant	fox
giant_panda	hamster	horse	lion	lizard	monkey
motorcycle	rabbit	red_panda	sheep	snake	squirrel
tiger	train	turtle	watercraft	whale	zebra

Compared Methods

We compare our method with existing state-of-the-art video object detection methods, which are listed as follows:

- **Baseline:** Faster-RCNN [13,25] utilizes two fully connected layers as the RoI heads to conduct object detection.
- **DFF:** Deep Feature Flow [37] runs the expensive convolutional sub-network only on sparse key video frames and propagates their deep feature maps to other frames via a flow field to boost the recognition accuracy.
- **FGFA:** Flow-Guided Feature Aggregation [36] proposes a flow-guided feature aggregation framework to leverage the pre-frame features by aggregation of nearby features along the motion paths for improving the video object detection accuracy.
- **RDN-base:** Relation Distillation Networks with relation only in basic stage [8] proposes a new architecture that novelly aggregates and propagates object relation to augment object features for detection.
- **RDN:** Relation Distillation Networks with advanced stage [8] models object relation across frames to boost video object detection performance.
- **MEGA:** Memory Enhanced Global-local Aggregation network [2] takes full consideration of both global and local information to improve the performance of video object detection.

Evaluation Metric

Following protocols widely adopted in [2,36,37], we evaluate our method on the validation set and use mean average precision (mAP) as the evaluation metric which is usually used in conventional object detection and provides a method performance evaluation in terms of regression and classification accuracies. Moreover, similar to the [36], we also divide the ground truth objects into three categories according to their motion speed, including slow, medium and fast motion, respectively. An object's speed is measured by its averaged intersection-over-union (IoU) scores with its corresponding instances in the nearby frames (± 10 frames). The lower the motion IoU is, the faster the object moves. In general, the speed consists of slow (IoU score > 0.9), medium (IoU score $\in [0.7, 0.9]$), fast (IoU score <0.7). Then we report the mAP scores over the slow, medium, and fast groups, respectively, denoted as mAP (slow), mAP (medium), mAP (fast), which provides a more detailed analysis and in-depth understanding.

Network Architecture

Feature Extractor. We adopt the state-of-the-art ResNet-101 and ResNet-50 [15] as the feature extractors. As common practice in [2], we enlarge the resolution of the feature maps by changing the stride of the first convolution block in last stage of convolution from 2 to 1. Further, the dilation of convolutional layers is set as 2 to retain the receptive field size.

Detection Network. We adopt the Faster-RCNN [25] as our detection network. The RPN head is added on the top of conv4 stage. RPN relies on a sliding window on the shared feature map to generate 9 target boxes (anchors) for each location. The area of these 9 anchors are 128×128, 256×256, 512×512, and the aspect ratios are 1:1, 1:2, 2:1. About 300 candidate boxes are generated per frame during training and inference, and we set the threshold for NMS to 0.7. After getting candidate boxes, we use RoI-Align [14] module and a FC layer after conv5 layer to obtain RoI feature for each candidate boxes. Moreover, we also use the MEGA module at training and inference stages, the detailed setting is the same as the work of [2].

Implementation Details

We resize the input image as a shorter side of 600 pixels. We set the number of training iterations for our model to be 120K. The learning rate is set as 10^{-3} and 10^{-4} in the first 80K and in the last 40K iterations, respectively. During inference, we adopt NMS with a threshold of 0.5 IoU in order to suppress duplicate detection boxes. Experiments are conducted on 2 NVIDIA TITAN Xp with GPU of 12 GB memory. Each GPU holds one mini-batch and each mini-batch contains one set of images or frames.

4.2   Experimental Results

In this section, we report the main experimental results of our model on the ImageNet VID dataset to verify the effectiveness of our method, including the comparison results with existing state-of-the-art video object detection methods based on ResNet-101 and ResNet-50, respectively and a detailed ablation study that consists of average detection accuracy for each category and intuitive display for detection results of the fast moving example,

Table 2. Comparison results on ResNet-101

Model	Backbone	mAP (%)	mAP (%) (fast)	mAP (%) (med)	mAP (%) (slow)
Baseline	ResNet-101	76.7	52.3	74.1	84.9
DFF	ResNet-101	75.0	48.3	73.5	84.5
FGFA	ResNet-101	78.0	55.3	76.9	85.6
RDN-base	ResNet-101	81.1	60.2	79.4	87.7
RDN	ResNet-101	**81.7**	59.5	80.0	**89.0**
MEGA	ResNet-101	81.06	61.86	80.59	86.61
Ours	ResNet-101	81.45	**62.0**	**80.6**	87.2

Comparative Results with Video Object Detection Methods

In this section, we overall compare our model with existing state-of-the-art video object detection methods shown in Sect. 4.1 to verify the improved performance of our method. Table 2 and Table 3 show the mAP results of different video object detection methods based on the ResNet-101 and ResNet-50, respectively. In the Table 2 and Table 3, it is obvious that the overall mAP evaluation of our method is just slightly worse than the RDN method, but we have the highest mAP on the objects of fast and medium motion speed compared with other methods. Taking the results on fast moving objects as examples, our method improves the mAP by about 0.2% compared to the second-best model and improves the mAP by about 10% compared to the baseline on Resnet-101, which overall evaluates the effectiveness of our method and verifies that adding the object trajectory information extracted by constructed OMLP module into object detection process can improve the performance of the fast moving object to address the different deteriorated problems in video.

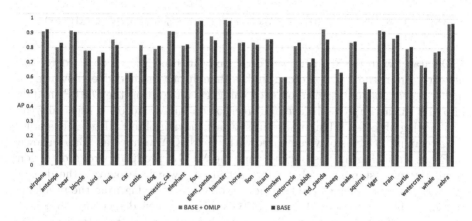

Fig. 3. The average detection precision for each category on the ImageNet VID dataset.

Ablation Study

In this part, we conduct a detailed ablation study to clearly demonstrate the effectiveness of our major technical component, OMLP module. We first compare our method with the applied base detection network MEGA for each category on the ImageNet VID dataset to provide more detailed analysis. Table 1 gives a used categories information of ImageNet VID dataset in our experiment. Figure 3 shows the average detection accuracy for each category on the ImageNet

Table 3. Comparison results on ResNet-50

Model	Backbone	mAP (%)	mAP (%) (fast)	mAP (%) (med)	mAP (%) (slow)
Baseline	ResNet-50	71.8	47.2	69.2	80.6
DFF	ResNet-50	70.4	43.6	68.9	80.8
FGFA	ResNet-50	74.3	50.6	72.3	84.0
RDN-base	ResNet-50	**76.7**	53.8	74.8	**85.4**
Ours	ResNet-50	76.0	**55.2**	**74.8**	83.3

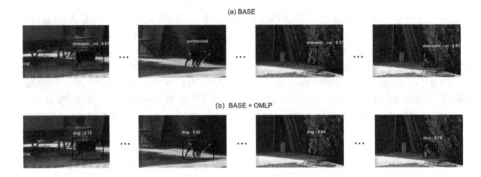

Fig. 4. Comparison with the video object detection neural network (MEGA).

VID dataset with ResNet-101, the "Base" indicates the MEGA object detection network. From the Fig. 3, we can find, although our method does not achieve higher detection performance on each category, integrating the OMLP module into MEGA can improve the average precision on many categories, especially the fast moving objects. Taking the results of fast moving object (dog) as example, we give an intuitive display compared with the MEGA, termed as "Base" shown in Fig. 4. From Fig. 4, it is obvious that the MEGA produces wrong detection result on the fast moving object for each frame, which recognizes the dog as the cat. Moreover, due to the fast moving object is suffering from the problems of motion blur and video defocus, MEGA can not even detect any object on the second frame. In contrast, our method has more accurate detection result on the fast moving object for each frame, even if the object exists the motion blur, video defocus (shown in the second frame) and object occlusion (shown in the third and fourth frames) problems, which further verifies that adding the motion laws of objects is effective for the improvement of video object detection performance.

5   Conclusion

In this work, we propose a novel video object detection framework, which utilizes the object trajectory information extracted by constructed OMLP module to improve the performance of video object detection neural network. Through adding the motion laws of objects in the process of detection, our method effectively mitigates the problems of motion blur, video defocus, object occlusion and rare locations that exist on the fast moving objects in video frames to achieve a better result for detection. Experiments conducted on ImageNet VID dataset validate the effectiveness of our method.

Acknowledgment. This work was supported by National Natural Science Foundation of China (Serial Nos. 61976134, 61991410, 61991415), Natural Science Foundation of Shanghai (Serial No. 21ZR1423900) and Open Project Foundation of Intelligent Information Processing Key Laboratory of Shanxi Province, China (No. CICIP2021001).

References

1. Cai, Z., Vasconcelos, N.: Cascade R-CNN: delving into high quality object detection. In: Proceedings of the IEEE Conference on Computer Vision and Pattern Recognition, pp. 6154–6162 (2018)
2. Chen, Y., Cao, Y., Hu, H., Wang, L.: Memory enhanced global-local aggregation for video object detection. In: Proceedings of the IEEE/CVF Conference on Computer Vision and Pattern Recognition, pp. 10337–10346 (2020)
3. Comaniciu, D., Ramesh, V., Meer, P.: Real-time tracking of non-rigid objects using mean shift. In: Proceedings IEEE Conference on Computer Vision and Pattern Recognition, CVPR 2000 (Cat. No. PR00662), vol. 2, pp. 142–149. IEEE (2000)
4. Comaniciu, D., Ramesh, V., Meer, P.: Kernel-based object tracking. IEEE Trans. Pattern Anal. Mach. Intell. **25**(5), 564–577 (2003)
5. Cores, D., Brea, V.M., Mucientes, M.: Short-term anchor linking and long-term self-guided attention for video object detection. Image Vis. Comput. **110**, 104179 (2021)
6. Dai, J., Li, Y., He, K., Sun, J.: R-FCN: object detection via region-based fully convolutional networks. In: Advances in Neural Information Processing Systems, vol. 29 (2016)
7. Deng, J., Dong, W., Socher, R., Li, L., Li, K., Fei-Fei, L.: ImageNet: a large-scale hierarchical image database. In: Proceedings of the IEEE Conference on Computer Vision and Pattern Recognition, pp. 248–255. IEEE, Piscataway (2009)
8. Deng, J., Pan, Y., Yao, T., Zhou, W., Li, H., Mei, T.: Relation distillation networks for video object detection. In: Proceedings of the IEEE/CVF International Conference on Computer Vision, pp. 7023–7032 (2019)
9. Fu, C.Y., Liu, W., Ranga, A., Tyagi, A., Berg, A.C.: DSSD: deconvolutional single shot detector. arXiv preprint arXiv:1701.06659 (2017)
10. Fukunaga, K., Hostetler, L.: The estimation of the gradient of a density function, with applications in pattern recognition. IEEE Trans. Inf. Theory **21**(1), 32–40 (1975)
11. Gong, T., et al.: Temporal ROI align for video object recognition. In: Proceedings of the AAAI Conference on Artificial Intelligence, vol. 35, pp. 1442–1450 (2021)
12. Han, M., Wang, Y., Chang, X., Qiao, Yu.: Mining inter-video proposal relations for video object detection. In: Vedaldi, A., Bischof, H., Brox, T., Frahm, J.-M. (eds.) ECCV 2020. LNCS, vol. 12366, pp. 431–446. Springer, Cham (2020). https://doi.org/10.1007/978-3-030-58589-1_26
13. He, K., Gkioxari, G., Dollár, P., Girshick, R.: Mask R-CNN. In: Proceedings of the IEEE International Conference on Computer Vision, pp. 2961–2969 (2017)
14. He, K., Gkioxari, G., Dollár, P., Girshick, R.: Mask R-CNN. IEEE Trans. Pattern Anal. Mach. Intell. (2017)
15. He, K., Zhang, X., Ren, S., Sun, J.: Deep residual learning for image recognition. In: Proceedings of the IEEE Conference on Computer Vision and Pattern Recognition, pp. 770–778 (2016)
16. Horn, B.K., Schunck, B.G.: Determining optical flow. Artif. Intell. **17**(1–3), 185–203 (1981)
17. Huang, X., Yue, X., Xu, Z., Chen, Y.: Integrating general and specific priors into deep convolutional neural networks for bladder tumor segmentation. In: 2021 International Joint Conference on Neural Networks (IJCNN), pp. 1–8. IEEE (2021)

18. Jayabalan, E., Krishnan, A.: Object detection and tracking in videos using snake and optical flow approach. In: Das, V.V., Stephen, J., Chaba, Y. (eds.) CNC 2011. CCIS, vol. 142, pp. 299–301. Springer, Heidelberg (2011). https://doi.org/10.1007/978-3-642-19542-6_52

19. Leichter, I., Lindenbaum, M., Rivlin, E.: Mean shift tracking with multiple reference color histograms. Comput. Vis. Image Underst. **114**(3), 400–408 (2010)

20. Liu, W., et al.: SSD: single shot multibox detector. In: Leibe, B., Matas, J., Sebe, N., Welling, M. (eds.) ECCV 2016. LNCS, vol. 9905, pp. 21–37. Springer, Cham (2016). https://doi.org/10.1007/978-3-319-46448-0_2

21. Nguyen, H.T., Worring, M., Dev, A.: Detection of moving objects in video using a robust motion similarity measure. IEEE Trans. Image Process. **9**(1), 137–141 (2000)

22. Ning, J., Zhang, L., Zhang, D., Wu, C.: Scale and orientation adaptive mean shift tracking. IET Comput. Vision **6**(1), 52–61 (2012)

23. Oreifej, O., Li, X., Shah, M.: Simultaneous video stabilization and moving object detection in turbulence. IEEE Trans. Pattern Anal. Mach. Intell. **35**(2), 450–462 (2012)

24. Redmon, J., Divvala, S., Girshick, R., Farhadi, A.: You only look once: unified, real-time object detection. In: Proceedings of the IEEE Conference on Computer Vision and Pattern Recognition, pp. 779–788 (2016)

25. Ren, S., He, K., Girshick, R., Sun, J.: Faster R-CNN: towards real-time object detection with region proposal networks. In: Advances in Neural Information Processing Systems, vol. 28 (2015)

26. Russakovsky, O., et al.: ImageNet large scale visual recognition challenge. Int. J. Comput. Vision **115**(3), 211–252 (2015)

27. Sermanet, P., Eigen, D., Zhang, X., Mathieu, M., Fergus, R., LeCun, Y.: OverFeat: integrated recognition, localization and detection using convolutional networks. arXiv preprint arXiv:1312.6229 (2013)

28. Shen, Z., Liu, Z., Li, J., Jiang, Y.G., Chen, Y., Xue, X.: DSOD: learning deeply supervised object detectors from scratch. In: Proceedings of the IEEE International Conference on Computer Vision, pp. 1919–1927 (2017)

29. Song, G., Liu, Y., Wang, X.: Revisiting the sibling head in object detector. In: Proceedings of the IEEE/CVF Conference on Computer Vision and Pattern Recognition, pp. 11563–11572 (2020)

30. Vojir, T., Noskova, J., Matas, J.: Robust scale-adaptive mean-shift for tracking. Pattern Recogn. Lett. **49**, 250–258 (2014)

31. Yang, C., Duraiswami, R., Davis, L.: Efficient mean-shift tracking via a new similarity measure. In: 2005 IEEE Computer Society Conference on Computer Vision and Pattern Recognition (CVPR 2005), vol. 1, pp. 176–183. IEEE (2005)

32. Zhang, C., Yue, X., Chen, Y., Lv, Y.: Integrating diagnosis rules into deep neural networks for bladder cancer staging. In: Proceedings of the 29th ACM International Conference on Information & Knowledge Management, pp. 2301–2304 (2020)

33. Zhang, R., Miao, Z., Zhang, Q., Hao, S., Wang, S.: Video object detection by aggregating features across adjacent frames. In: Journal of Physics: Conference Series, vol. 1229, pp. 012–039. IOP Publishing (2019)

34. Zhu, H., Wei, H., Li, B., Yuan, X., Kehtarnavaz, N.: A review of video object detection: datasets, metrics and methods. Appl. Sci. **10**(21), 7834 (2020)

35. Zhu, X., Dai, J., Yuan, L., Wei, Y.: Towards high performance video object detection. In: Proceedings of the IEEE Conference on Computer Vision and Pattern Recognition, pp. 7210–7218 (2018)

36. Zhu, X., Wang, Y., Dai, J., Yuan, L., Wei, Y.: Flow-guided feature aggregation for video object detection. In: Proceedings of the IEEE International Conference on Computer Vision, pp. 408–417 (2017)
37. Zhu, X., Xiong, Y., Dai, J., Yuan, L., Wei, Y.: Deep feature flow for video recognition. In: Proceedings of the IEEE Conference on Computer Vision and Pattern Recognition, pp. 2349–2358 (2017)
38. Zou, Z., Shi, Z., Guo, Y., Ye, J.: Object detection in 20 years: a survey. arXiv preprint arXiv:1905.05055 (2019)

MP-KMeans: K-Means with Missing Pattern for Data of Missing Not at Random

Ruifeng Zhou and Hong Yu[✉]

Chongqing Key Laboratory of Computational Intelligence, Chongqing University of Posts and Telecommunications, Chongqing 400065, China
yuhong@cqupt.edu.cn

Abstract. K-Means is one of the most popular clustering algorithm. It aims to minimize the sum of pair-wise distance within a cluster. It has been widely used in data analysis, image recognition and many other fields. However, traditional K-Means cannot handle missing values, which greatly limits its application scenarios. Missing values are ubiquitous in the real world due to sensor failure, high cost, and privacy protection. The appearance of missing values leads to useful information lost in the information system, and makes it difficult to perform data mining. Currently, improvements of K-Means for missing values generally based on data completion and partial distance strategy. Above methods achieve satisfied performance with random missing values, but they will fail when data is missing not at random (MNAR). Considering the effect of missing mechanism, this paper proposes an improved method of traditional K-Means for data of missing not at random, which integrating missing pattern in the distance measurement to assist clustering process. The experiment results on public datasets show that the proposed method outperforms data completion-based K-Means and partial distance-based K-Means.

Keywords: K-Means · Missing not at random · Missing mechanism · Missing pattern

1 Introduction

Cluster analysis or clustering is an important data mining approach. The goal of clustering is to unsupervisedly divide data into different clusters, to make data within the same cluster are as similar as possible and data in different clusters are as dissimilar as possible. As one of the most popular and widely used clustering method, K-Means [8] has been applied in data analysis, image recognition, social network and other fields. One of the disadvantages of traditional K-Means is the incapability in handling missing values. Missing values is ubiquitous in the real world: data can be lost due to network fault or sensor failure, sensitive user information will be omitted for privacy protection, etc. The appearance of missing values poses challenges to studies of clustering.

© The Author(s), under exclusive license to Springer Nature Switzerland AG 2022
J. Yao et al. (Eds.): IJCRS 2022, LNAI 13633, pp. 238–249, 2022.
https://doi.org/10.1007/978-3-031-21244-4_18

Furthermore, according to Rubin's research [11], the type of missing values can be divided into missing at completely random (MCAR), missing at random (MAR) and missing not at random (MNAR). Existing improved versions of K-Means for missing values include: complete the data and then perform traditional K-Means, adopt partial distance strategy [9] to calculate pair-wise distance, and put data completion and clustering into a unified framework [15]. The above methods can achieve fair results when the data is missing completely at random. When the data is missing not at random, these methods will obtain biased distances between samples, which leads to the decline in clustering performance.

Assuming that data is missing not at random, an improved version of K-Means (MP-KMeans) is proposed in this paper, which incorporates missing pattern in the traditional K-means algorithm, thus making full use of the information provided by incomplete dataset to assist the clustering process. The experimental results on public dataset verify the effectiveness of the proposed method.

The rest of this paper is organized as follows: related works are reviewed in Sect. 2; the proposed method is formulated in Sect. 3; comparative experiment and experimental results are provided in Sect. 4; this paper is concluded in Sect. 5.

2   Related Work

2.1   Notations

In this paper, a missing value is denoted by NaN (not a number). Mathematical scalar, vector and matrix are denoted as lowercase letter, bold lowercase letter and bold uppercase letter, respectively. Given an incomplete dataset $\mathbf{X} = \{\mathbf{X}_{obs}, \mathbf{X}_{miss}\} \in \mathbb{R}^{n \times d}$ with n samples and d dimensions, where $\mathbf{X}_{obs}$ is the set of observed data and $\mathbf{X}_{miss}$ is the set of missing data. Given a matrix $\mathbf{A}$, its i-th row and j-th column are written as $\mathbf{A}_{i:}$ and $\mathbf{A}_{:j}$, and the (i, j)-th element of $\mathbf{A}$ is denoted as $\mathbf{A}_{ij}$ (Table 1).

Table 1. Notations.

Notation	Definition
NaN	Not a number (a missing value)
$x, \mathbf{x}$ and $\mathbf{X}$	Scalar, vector and matrix
$\mathbf{X}_{miss}$	The set of observed data
$\mathbf{X}_{obs}$	The set of missing data
$\mathbf{A}_{i:}$ and $\mathbf{A}_{:j}$	The i-th row and j-th column of $\mathbf{A}$
$\mathbf{A}_{ij}$	The (i, j)-th element of $\mathbf{A}$

2.2   K-Means Clustering Method

K-Means is one of the most widely applied clustering method in machine learning, which seeks a partition of data with the minimized sum of pair-wise distance within a cluster. The objective function of K-Means is written as follows:

$$\min_{\mathbf{H}} \sum_{i=1}^{n} \sum_{c=1}^{k} \mathbf{H}_{ic} \left\| \mathbf{x}_i - \boldsymbol{\mu}_c \right\|^2, \ s.t. \sum_{c=1}^{k} \mathbf{H}_{ic} = 1, \mathbf{H} \in \{0,1\}^{n \times k}. \tag{1}$$

where $\mathbf{x}_i$ and $\boldsymbol{\mu}_c$ represent i-th sample and c-th cluster center, respectively. $\mathbf{H}$ is the class indicator matrix, $\mathbf{H}_{ic} = 1$ when i-th sample belongs to c-th cluster. The constraint term ensures that a sample point belongs to only one cluster, hence K-Means is a typical hard clustering method.

Optimization problem in Eq. (1) is a NP-hard problem since minimizing squared distance in a cluster [3]. So far, the most popular solution of K-Means is randomly initializing k cluster centers and then finding the local optimal minimum. The detail approximate solution of K-Means is presented in Algorithm 1. Therefore, K-Means is sensitive to the initialization of cluster centers. In order to avoid occasionality, K-Means algorithm usually needs to be run multiple times and get the mean value.

Algorithm 1. K-Means

Input: Dataset $\mathbf{X}$, number of clusters k and max iteration number $maxIter$.
Initialize: $\mathbf{H} = 0$, k cluster centers: $\boldsymbol{\mu}_1, \boldsymbol{\mu}_2, ..., \boldsymbol{\mu}_k$.

1: **for** $iter = 1$ **to** $maxIter$ **do**
2:      **for** $i = 1$ **to** n **do**
3:          $c = min_c \left\| \mathbf{x}_i - \boldsymbol{\mu}_c \right\|^2$ for $c = 1, .., k$;
4:          Assign $\mathbf{x}_i$ to c-th cluster, i.e., $\mathbf{H}_{ic} = 1$;
5:      **end for**
6:      **if** $\mathbf{H}$ is no longer change **then**
7:          **break**;
8:      **end if**
9:      Update cluster centers;
10: **end for**

Output: Class indicator matrix $\mathbf{H}$.

Moreover, K-Means can not effectively deal with missing values, which limits its application scenarios. At present, various improvements of K-Means for missing values usually complete the dataset first or use local distance strategy. Data completion methods include simple deletion, mean filling, k-nearest neighbor (KNN) filling, etc. The partial distance strategy directly ignores variables with missing values when calculating the pair-wise distance. When the missing rate is low and the data is missing completely at random, these methods can perform well. When the data is missing not at random, methods based on data completion or partial distance will get biased pair-wise distance, which brings significant deviation to the clustering results.

2.3   Missing Mechanism

Definition 1. *In dataset* **X**, *variables that do not contain missing values are complete variables, and variables that contain missing values are incomplete variables.*

Rubin D. B. [11] proposed missing mechanism in 1976 to illustrate the causality of the missingness. According to Rubin's taxonomy, the missing mechanism is classified as missing at completely random (MCAR), missing at random (MAR) and missing not at random (MNAR).

Given an incomplete dataset $\mathbf{X} = \{\mathbf{X}_{obs}, \mathbf{X}_{miss}\} \in \mathbb{R}^{n \times d}$ with n samples and d dimensions. And the missing indicator matrix corresponding to $\mathbf{X}$ is denoted as $\mathbf{M} \in \mathbb{R}^{n \times d}$. $\mathbf{M}_{ij} = 1$ when $\mathbf{X}_{ij} = \text{NaN}$ and $\mathbf{M}_{ij} = 0$ when $\mathbf{X}_{ij} \neq \text{NaN}$. $\mathbf{M}$ can be regarded as the direct embodiment of missing pattern, and it can be considered that the missing indicator matrix and the missing pattern refer as the same in the rest of this article. Figure 1 shows the missing indicator matrix $\mathbf{M}$ of an example dataset $\mathbf{X}$.

Definition 2. *Missing at completely random (MCAR): The probability of missing data has nothing to do with observed data and missing data.*

$$P(\mathbf{M} = 1 \mid \mathbf{X}, \xi) = P(\mathbf{M} = 1 \mid \xi), \tag{2}$$

where ξ represents additional parameter, for example, missing ratio.

Definition 3. *Missing at random (MAR): The probability of missing data is related to observed data.*

$$P(\mathbf{M} = 1 \mid \mathbf{X}, \xi) = P(\mathbf{M} = 1 \mid \mathbf{X}_{obs}, \xi). \tag{3}$$

Definition 4. *Missing not at random (MNAR): The probability of missing data is related to missing data themselves.*

$$P(\mathbf{M} = 1 \mid \mathbf{X}, \xi) = P(\mathbf{M} = 1 \mid \mathbf{X}_{miss}, \xi), \tag{4}$$

$$P(\mathbf{M} = 1 \mid \mathbf{X}, \xi) = P(\mathbf{M} = 1 \mid \mathbf{X}_{obs}, \mathbf{X}_{miss}, \xi). \tag{5}$$

Here is a running example to help readers better understand missing mechanism: adolescent tobacco use study [12]. This case includes two variables, i.e., "age" and "number of cigarettes". The "age" is a complete variable, and the "number of cigarettes" is an incomplete variable. In MCAR, the missingness of "number of cigarettes" only depends on parameter ξ. In the case of MAR, younger participants may less likely to disclose the specific smoked cigarettes number because they know underage smoking is a misbehavior. Namely, the missingness of "number of cigarettes" depends on the complete variable "age". At last, in MNAR, some teenagers may not willing to report their daily used cigarette number because they are heavy smokers. In other words, the missingness of "number of cigarettes" depends on the incomplete variable "number of cigarettes".

According to the above definition of missing mechanisms and missing indicator matrix, it is apparent that:

V1	V2	V3	V4	V5	V6	V7
2	1	NaN	3	1	NaN	NaN
2	2	NaN	2	1	NaN	NaN
2	0	0	2	1	2	1
2	2	NaN	2	1	NaN	NaN
2	1	NaN	2	1	NaN	NaN
1	1	0	1	1	1	0

X

V1	V2	V3	V4	V5	V6	V7
0	0	1	0	0	1	1
0	0	1	0	0	1	1
0	0	0	0	0	0	0
0	0	1	0	0	1	1
0	0	1	0	0	1	1
0	0	0	0	0	0	0

M

Fig. 1. Missing indicator matrix (missing pattern) **M** of original data **X**.

1. In MCAR, missing patterns cannot provide any useful information about the data.
2. In MAR, the similarity of missing patterns indicates the similarity of the values of complete variables.
3. In MNAR, the similarity of missing patterns indicates the similarity of the values of incomplete variables.

In fact, MCAR is difficult to be met, most of them are caused by accident; MAR and MNAR are more common in practical research and engineering, therefore research on MAR and MNAR data is more relevant.

2.4   Clustering Method for Incomplete Data

In the past decades, many clustering methods for incomplete data have been proposed, and quite a lot of works are based on Fuzzy C-Means (FCM), three-way decision and subspace learning.

FCM [2] is a classic soft clustering method. In comparison to hard clustering, soft clustering does not require that a sample must belong to only one cluster. Hathaway et al. [9] proposed four strategies to enable FCM to handle incomplete data: whole data strategy (WDS), partial distance strategy (PDS), optimal completion strategy (OCS) and nearest prototype strategy (NPS). Zhang et al. [18] treat missing values adhering to a certain Gaussian distribution as probabilistic information granules based on the nearest neighbors of incomplete data, and incorporate probabilistic information granules into FCM by maximum likelihood criterion. Li et al. [10] developed a robust fuzzy c-means clustering algorithm for incomplete data, which represents missing feature values by intervals and adopts a min-max optimization model to reduce noises.

Derived from rough set theory and Bayesian decision theory, three-way decision [16] is an ideal approach to cope with uncertain and incomplete data. Three-way decision introduces a third decision state when information is insufficient, i.e., delay decision. Yu et al. [17] proposed a three-way decisions clustering algorithm for incomplete data, which jointly considers attribute significance and missing rate. Afridi et al. [1] introduced game-theory rough set (GTRS) to automatically determine the thresholds that are used in three-way clustering, and proposed a method based on GTRS for data with missing entries.

There are also a couple of subspace-based clustering methods for incomplete data. Gunnemann et al. [7] develop a general fault tolerance definition that are allowed a certain number of missing values in the subspace clustering model to obtain high quality results. Elhamifar et al. [4] mentioned one can fill missing values with random values when the missing rate is relatively low, and then solve sparse subspace clustering (SSC) problem to obtain the subspace representation coefficients. Besides, Fan et al. [5] proposed a framework by integrating high-rank matrix completion and subspace clustering, which achieves satisfied result on random missing data.

However, above methods have the common limitation, that is they overlooked the function of missing mechanisms, which causes poor clustering results when the data is missing not at random.

3   Proposed Method

3.1   The Relationship Between Missing Mechanism and Clustering

As discussed in Sect. 2.3, in order to enhance the performance of clustering on data of missing not at random, missing mechanism should be concerned in the clustering process.

Unfortunately, it is difficult to determine the missing mechanism of dataset without any prior knowledge. In this case, it is unrealistic to fuse missing mechanism into clustering process. But the missing pattern is off-the-shelf, which describes the location of missing data in the dataset. More importantly, missing pattern can reflect the intrinsic missing mechanism.

According to the definition of missing mechanism, we can get the relationship between missing mechanism and clustering:

- When the missing mechanism is MCAR, the missing data has nothing to do with its real value, so the missing pattern can not provide any useful information about the data.
- When the missing mechanism is MAR, the value of incomplete variables is related to the value of complete variables. The similarity of missing pattern indicates the similarity of the value of complete variables. Thus, the information reflected by the missing pattern is consistent with that reflected by the partial distance.
- When the missing mechanism is MNAR, the value of incomplete variables is related to incomplete variables themselves. The similarity of the missing pattern indicates the similarity of the incomplete variables. The partial distance is not able to express this type of similarity, so the similarity of missing patterns is the supplement to the partial distance.

3.2 MP-KMeans

Figure 2 demonstrates the missing indicator matrix and true label of Soybean dataset, the first 35 columns represent the missing indicator matrix, the dark green indicates corresponding entry is observed, and the light green indicates the corresponding entry is missing; the last column is the true label, and each color represent a class. It is obvious that Soybean dataset has MNAR missing values.

In addition, Fig. 2 also shows that data with the same missing patterns are more likely belong to the same groups, and data with different missing patterns are more likely to come from different groups, which is consistent with Wang's view point [14]. As a result, integrating missing patterns as auxiliary information in cluster analysis can improve the clustering accuracy of MNAR missing data.

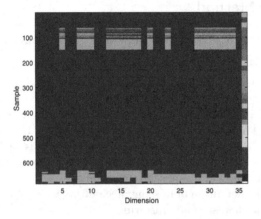

Fig. 2. Visualization of missing indicator matrix and true label of Soybean dataset.

Traditional K-Means directly calculates the Euclidean distance between data samples and cluster centers. Given that data is missing not at random, according to the analysis in Sect. 3.1, the Euclidean distance can be replaced by the joint representation of partial distance and pattern distance when the data is missing not at random. Therefore, the objective function of MP-KMeans is given as follows:

$$\min_{H} \sum_{i=1}^{n} \sum_{c=1}^{k} \mathbf{H}_{ic} Dist(\mathbf{x}_i, \boldsymbol{\mu}_c), \ s.t. \sum_{c=1}^{k} \mathbf{H}_{ic} = 1, \mathbf{H} \in \{0,1\}^{n \times k}. \tag{6}$$

$Dist$ is the proposed distance measurement: given two samples $\mathbf{x}_i$ and $\mathbf{x}_j$, the distance between $\mathbf{x}_i$ and $\mathbf{x}_j$ is redefined in Eq. (11):

$$Dist_{ij} = ParDist_{ij} + PatDist_{ij}. \tag{7}$$

ParDist and *PatDist* represent partial distance and pattern distance, respectively, and their definitions are given in Eq. (8) and Eq. (9).

$$ParDist_{ij} = \frac{d}{min(I_i, I_j)} \sum_{l=1}^{d} (x_{il} - x_{jl})^2 \mathbf{M}_{il} \mathbf{M}_{jl}; \tag{8}$$

where $1 \leq l \leq d$ and $1 \leq i \leq k$ and $I_i = \sum_{l=1}^{d} \mathbf{M}_{il}$. Since missing pattern of a sample is a binary vector, so we utilize Hamming distance to calculate $PatDist_{ij}$ of two samples as follows.

$$PatDist_{ij} = HammingDistance(\mathbf{M}_{i:}, \mathbf{M}_{j:}). \tag{9}$$

In order to determine the importance of partial distance and pattern distance to the total distance, a weight parameter ω is introduced in the objective function. When the missing degree of the sample $\mathbf{x}_i$ is high, pattern distance will play a major role in distance measurement, namely, the ω is large. The ω can be calculated as follows:

$$\omega = max(\frac{\sum_{l=1}^{d} \mathbf{M}_{il}}{d}, \frac{\sum_{l=1}^{d} \mathbf{M}_{jl}}{d}). \tag{10}$$

Accordingly, the formulation of $Dist_{ij}$ can be updated as:

$$Dist_{ij} = (1 - \omega)ParDist_{ij} + \omega PatDist_{ij}; \tag{11}$$

In the case that cluster boundaries are not explicit, the traditional K-means algorithm will assign all samples into the same cluster. Therefore, we refer to literature [13] and introduce a penalty term in the objective function, thereby making any two cluster centers as far as possible. The objective function of MP-KMeans is rewritten as Eq. (12).

$$\min_{H} \sum_{i=1}^{n} \sum_{c=1}^{k} \mathbf{H}_{ic} Dist(\mathbf{x}_i - \boldsymbol{\mu}_c) + \sum_{i=1}^{k} \sum_{j=1, j \neq i}^{k} \frac{1}{Dist(\boldsymbol{\mu}_i, \boldsymbol{\mu}_j)},$$
$$s.t. \sum_{c=1}^{k} \mathbf{H}_{ic} = 1, \mathbf{H} \in \{0, 1\}^{n \times k}. \tag{12}$$

Finally, we can get the following rule of updating cluster center $\boldsymbol{\mu}_c$:

$$\boldsymbol{\mu}_c = \frac{\sum_{i=1}^{n} I_i \mathbf{x}_i - \sum_{j=1, j \neq i}^{k} Dist(\boldsymbol{\mu}_c, \boldsymbol{\mu}_j) \boldsymbol{\mu}_j}{\sum_{i=1}^{n} I_i - \sum_{j=1, j \neq i}^{k} Dist(\boldsymbol{\mu}_c, \boldsymbol{\mu}_j)}. \tag{13}$$

The pseudo code of MP-KMeans is summarized in Algorithm 2. The major difference of Algorithm 1 and Algorithm 2 are missing pattern integration and pair-wise distance measurement. The goal of MP-KMeans is to address the defect that K-Means cannot handle missing values.

Algorithm 2. MP-KMeans

Input: Incomplete dataset $\mathbf{X}$, number of clusters k and max iteration number $maxIter$.
Initialize: $\mathbf{H} = \mathbf{0}$, k cluster centers: $\boldsymbol{\mu}_1, \boldsymbol{\mu}_2, ..., \boldsymbol{\mu}_k$.

1: Construct missing indicator matrix $\mathbf{M}$;
2: **for** $iter = 1$ **to** $maxIter$ **do**
3:     **for** $i = 1$ **to** n **do**
4:         $c = min_c Dist(\mathbf{x}_i - \boldsymbol{\mu}_c)$, $c = 1, .., k$
5:         Assign $\mathbf{x}_i$ to c-th cluster, i.e., $\mathbf{H}_{ic} = 1$;
6:     **end for**
7:     **if** $\mathbf{H}$ is no longer change **then**
8:         **break**;
9:     **end if**
10:     Update cluster centers using Eq. (13).
11: **end for**

Output: Class indicator matrix $\mathbf{H}$.

4 Experiment

4.1 Dataset

Soybean[1] is used for soybean disease diagnosis, which has 19 classes and 35 features, it contains non-random missing values.
Iris[2] is the most popular dataset in UCI repository, which contains 3 classes and each class has 50 instances.
Glass[3] is used for classification of glass types, which includes 7 glass types and 219 glass samples.

The original Iris and Glass dataset are complete, so we refer to the literature [12] to generate MNAR missing values artificially. Detailed information of datasets is summarized in Table 2. The missing mechanism of Iris and Glass are specified in Table 3 and Table 4, respectively.

Table 2. Characteristic of datasets.

Name	# Instances	# Dimensions	# Missing values	# Clusters
Soybean	683	35	2337	19
Iris	150	4	194	3
Glass	214	9	288	7

---

[1] http://archive.ics.uci.edu/ml/datasets/Soybean+%28Large%29.
[2] https://archive.ics.uci.edu/ml/datasets/Iris.
[3] https://archive.ics.uci.edu/ml/datasets/Glass+Identification.

Table 3. Missing mechanism of Iris dataset.

No.	Missing mechanism
1	$P(\text{pedal width} = \text{NaN} \mid \text{pedal width} > 6) = 0.95$
2	$P(\text{sepal width} = \text{NaN} \mid \text{sepal width} > 3.2) = 0.95$
3	$P(\text{pedal length} = \text{NaN} \mid \text{pedal length} > 5) = 1$
4	$P(\text{petal width} = \text{NaN} \mid \text{petal width} < 1) = 1$

Table 4. Missing mechanism of Glass dataset.

No.	Missing mechanism
1	$P(\text{Na} = \text{NaN} \mid \text{Na} > 14) = 0.95$
2	$P(\text{Mg} = \text{NaN} \mid \text{Mg} < 2) = 0.9$
3	$P(\text{Al} = \text{NaN} \mid \text{Al} > 2) = 0.9$
4	$P(\text{Si} = \text{NaN} \mid \text{Si} < 72) = 0.95$
5	$P(\text{K} = \text{NaN} \mid \text{K} < 1) = 0.9$
6	$P(\text{Ca} = \text{NaN} \mid \text{Ca} < 1) = 0.95$

4.2 Compared Methods

In the experiment, we adopt traditional K-Means as the baseline. K-Means cannot directly handle missing values, so there are four strategies utilized in the experiment:

- K-Means$_{zero}$: This method fills missing values with zeros.
- K-Means$_{mean}$: This algorithm imputes missing values with mean values of observed values of corresponding dimension.
- K-Means$_{knn}$ [6]: This method fills missing values with mean values of K-nearest neighbor.
- K-Means$_{pd}$: Partial distance [9] is implemented in traditional K-Means so that incomplete variables are ignored in the similarity calculation.

4.3 Experimental Results

The dataset used in experiment contains true label, so we utilize three external clustering evaluation metrics to verify clustering performance: accuracy (ACC), normalized mutual information (NMI) and adjusted rand index (ARI). For all metrics, higher value indicates better performance. For each method, we repeat 30 times and report the mean value and standard error. The experiment results are presented in Table 5, and the best values are highlighted in bold.

From Table 5, when the data is missing not at random, the proposed MP-KMeans has advantages over data completion based K-Means. Integrating missing pattern to distance measurement could effectively mitigate the impact of data deviation brought by partial distance strategy and enhance clustering performance.

Table 5. Clustering results of different methods.

Dataset	Method	ACC	NMI	ARI
Soybean	K-Means$_{zero}$	0.562 ± 0.030	0.686 ± 0.017	0.425 ± 0.029
	K-Means$_{mean}$	0.568 ± 0.045	0.685 ± 0.027	0.413 ± 0.048
	K-Means$_{knn}$	0.577 ± 0.044	0.709 ± 0.022	0.436 ± 0.040
	K-Means$_{pd}$	0.570 ± 0.039	0.693 ± 0.020	0.428 ± 0.034
	Proposed	$\mathbf{0.632 \pm 0.029}$	$\mathbf{0.719 \pm 0.017}$	$\mathbf{0.452 \pm 0.030}$
Iris	K-Means$_{zero}$	0.783 ± 0.151	0.586 ± 0.144	0.568 ± 0.194
	K-Means$_{mean}$	0.716 ± 0.187	0.606 ± 0.167	0.544 ± 0.231
	K-Means$_{knn}$	0.742 ± 0.113	0.525 ± 0.122	0.457 ± 0.136
	K-Means$_{pd}$	0.800 ± 0.139	0.600 ± 0.129	0.587 ± 0.178
	Proposed	$\mathbf{0.827 \pm 0.146}$	$\mathbf{0.625 \pm 0.138}$	$\mathbf{0.619 \pm 0.194}$
Glass	K-Means$_{zero}$	0.383 ± 0.023	0.250 ± 0.030	0.117 ± 0.026
	K-Means$_{mean}$	0.377 ± 0.022	0.234 ± 0.029	0.098 ± 0.018
	K-Means$_{knn}$	0.394 ± 0.024	0.241 ± 0.025	0.102 ± 0.019
	K-Means$_{pd}$	0.387 ± 0.021	0.258 ± 0.026	0.124 ± 0.023
	Proposed	$\mathbf{0.410 \pm 0.024}$	$\mathbf{0.268 \pm 0.030}$	$\mathbf{0.132 \pm 0.025}$

5   Conclusion

Most of the existing clustering methods for incomplete data do not consider the function of missing mechanism on clustering. By analyzing the relationship between missing mechanism and clustering, as a manifestation of the missing mechanism, missing patterns can assist clustering of incomplete data. With regard to data of missing not at random, this paper proposes an improved K-Means clustering method with missing patterns. Experiments show that the clustering performance of the proposed method on data of missing not at random is superior to compared methods. In the future work, we will further explore the relationship between missing mechanism and clustering, and research on clustering method for multi-view data with MNAR missing values.

Acknowledgements. This work was jointly supported by the National Natural Science Foundation of China (62136002, 61876027), and the Natural Science Foundation of Chongqing (cstc2022ycjh-bgzxm0004).

References

1. Afridi, M.K., Azam, N., Yao, J., Alanazi, E.: A three-way clustering approach for handling missing data using GTRS. Int. J. Approx. Reason. **98**, 11–24 (2018)
2. Bezdek, J.C., Ehrlich, R., Full, W.: FCM: the fuzzy c-means clustering algorithm. Comput. Geosci. **10**(2–3), 191–203 (1984)

3. Drineas, P., Frieze, A., Kannan, R., Vempala, S., Vinay, V.: Clustering large graphs via the singular value decomposition. Mach. Learn. **56**(1), 9–33 (2004)

4. Elhamifar, E., Vidal, R.: Sparse subspace clustering: algorithm, theory, and applications. IEEE Trans. Pattern Anal. Mach. Intell. **35**(11), 2765–2781 (2013)

5. Fan, J., Chow, T.W.: Sparse subspace clustering for data with missing entries and high-rank matrix completion. Neural Netw. **93**, 36–44 (2017)

6. García-Laencina, P.J., Sancho-Gómez, J.L., Figueiras-Vidal, A.R., Verleysen, M.: K nearest neighbours with mutual information for simultaneous classification and missing data imputation. Neurocomputing **72**(7–9), 1483–1493 (2009)

7. Gunnemann, S., Muller, E., Raubach, S., Seidl, T.: Flexible fault tolerant subspace clustering for data with missing values. In: 2011 IEEE 11th International Conference on Data Mining, pp. 231–240. IEEE (2011)

8. Hartigan, J.A., Wong, M.A.: Algorithm as 136: a k-means clustering algorithm. J. R. Stat. Soc. Ser. C (Appl. Stat.) **28**(1), 100–108 (1979)

9. Hathaway, R.J., Bezdek, J.C.: Fuzzy c-means clustering of incomplete data. IEEE Trans. Syst. Man Cybern. Part B (Cybern.) **31**(5), 735–744 (2001)

10. Li, J., Song, S., Zhang, Y., Zhou, Z.: Robust k-median and k-means clustering algorithms for incomplete data. Math. Probl. Eng. **2016** (2016)

11. Rubin, D.B.: Inference and missing data. Biometrika **63**(3), 581–592 (1976)

12. Santos, M.S., Pereira, R.C., Costa, A.F., Soares, J.P., Santos, J., Abreu, P.H.: Generating synthetic missing data: a review by missing mechanism. IEEE Access **7**, 11651–11667 (2019)

13. Vassilvitskii, S., Arthur, D.: k-means++: the advantages of careful seeding. In: Proceedings of the eighteenth annual ACM-SIAM Symposium on Discrete Algorithms, pp. 1027–1035 (2006)

14. Wang, H., Wang, S.: Discovering patterns of missing data in survey databases: an application of rough sets. Expert Syst. Appl. **36**(3), 6256–6260 (2009)

15. Wang, S., et al.: K-means clustering with incomplete data. IEEE Access **7**, 69162–69171 (2019)

16. Yao, Y.: Three-way decision: an interpretation of rules in rough set theory. In: Wen, P., Li, Y., Polkowski, L., Yao, Y., Tsumoto, S., Wang, G. (eds.) RSKT 2009. LNCS (LNAI), vol. 5589, pp. 642–649. Springer, Heidelberg (2009). https://doi.org/10.1007/978-3-642-02962-2_81

17. Yu, H., Su, T., Zeng, X.: A three-way decisions clustering algorithm for incomplete data. In: Miao, D., Pedrycz, W., Ślęzak, D., Peters, G., Hu, Q., Wang, R. (eds.) RSKT 2014. LNCS (LNAI), vol. 8818, pp. 765–776. Springer, Cham (2014). https://doi.org/10.1007/978-3-319-11740-9_70

18. Zhang, L., Lu, W., Liu, X., Pedrycz, W., Zhong, C.: Fuzzy C-means clustering of incomplete data based on probabilistic information granules of missing values. Knowl.-Based Syst. **99**, 51–70 (2016)

Conceptual Knowledge Discovery and Machine Learning Based on Three-Way Decisions and Granular Computing

Using User's Expression Propensity for Sarcasm Detection Based on Sequential Three-Way Decision

Jie Chen[1,2,3], Jinpeng Chen[1,2,3], Shu Zhao[1,2,3]([✉]), and Yanping Zhang[1,2,3]

[1] Key Laboratory of Intelligent Computing and Signal Processing,
Ministry of Education, Hefei 230601, Anhui, People's Republic of China
[2] School of Computer Science and Technology, Anhui University, Hefei 230601,
Anhui, People's Republic of China
[3] Information Materials and Intelligent Sensing Laboratory of Anhui Province,
Hefei 230601, Anhui, People's Republic of China
zhaoshuzs2002@hotmail.com

Abstract. Sarcasm detection is mainly to distinguish whether the target comment is sarcasm that can help identify the actual sentiment. The previous sarcasm detection mainly focused on text features using vocabulary, grammar, and semantics. But users' expression propensity is ignored which is helpful to distinguish some comments with uncertain sarcasm polarity in sarcasm detection. However, how to use the user's expression propensity for sarcasm detection effectively is a challenge. Based on the ideas of granular computing and three-way decisions, we propose a sarcasm detection model based on the sequential three-way decision (S3WD) to integrate text features and users' expression propensity. The S3WD divides the comments into the sarcasm (SAR) region, non-sarcasm (NSAR) region, and boundary region (BND), and then gradually divides the uncertain BND region into a clear SAR region and NSAR region. We firstly construct a sequential structure through analysis sentiment of comments' chunks. Second, text features and users' expression propensity are fed into different sequential layers for fusion that can guide the comment classification more effectively. Finally, contextual information is further applied to consider sentiment context during sarcasm detection. The experimental results on a large Reddit corpus show that our model improves sarcasm classification performance effectively.

Keywords: Sarcasm detection · Sequential three-way decision · User's expression propensity

1 Introduction

Sarcasm detection in social networks has drawn much attention in recent years. Many people often use sarcasm to implicitly express their stronger emotions, especially on controversial topics, which increases the difficulty of sentiment analysis. Consider the following example: Tom: 'How to spot a Linux user?', Bob:

J. Yao et al. (Eds.): IJCRS 2022, LNAI 13633, pp. 253–264, 2022.
https://doi.org/10.1007/978-3-031-21244-4_19

'Actually, if they used Linux they wouldn't have gotten their network connection working yet, so they wouldn't be able to post on the web'.

Existing sarcasm detection methods identify the sarcasm as it mainly focused on text features. One important basis of text feature is lexical(such as specific words and punctuation) found in target comment (Kreuz et al. [1]) to judge the sarcasm polarity. Another is the semantic-level analysis of the text. Rohanian et al. [2] and Tay et al. [3] identify sarcasm according to the contrastive semantics in the target comments. When such text features are present in comments, sarcasm detection can achieve high precision. As a sarcastic example from Reddit, "I say go for it." sarcasm is sometimes expressed implicitly, which means without the presence of such text features, sarcasm detection can not be completed. Thus, Hazarika et al. [4] and Kolchinski et al. [5] utilize another way that encode stylometric and personality features of users' to solve such problems. But they both have the problem of established users personalities that cannot be changed. However, The users' personality is not always constant, and even for the user with a tendency to be sarcastic, it may still make non-sarcastic comments. So, users' expression propensity needs to be judged along with the text of the comment itself. Thus, how to fusion text features and users' expression propensity is still a challenge.

The three-way decision theory was first proposed by professor Yao [7] to process uncertain data. By introducing granular computing, Yao et al. [8] continues to present a sequential three-way decision (S3WD) model for achieving multi-division of the boundary regions. When the current information does not support the decision, the object can be divided into boundary domains, and the object can be divided after obtaining more sufficient information in finer sequential layers. Due to its effectiveness, the S3WD model is suitable for making decisions after the fusion of multiple information.

In this paper, we use users' expression propensity for sarcasm detection based on the S3WD model (UEP-S3WD). First, it constructs a sequential structure through analysis sentiment of comments' chunks. Second, text features and users' expression propensity are fed into different sequential layers for fusion that can guide the comment classification more effectively. Finally, contextual information is further applied to consider sentiment context during sarcasm detection. Our contributions are as follows:

1) We construct the UEP-S3WD model which analysis the sarcasm comments and fuses different attributes by multi-level sequential structure to guide classification.
2) Users' expression propensity obtained by comparing and analyzing users' historical sarcastic and non-sarcastic comments are used to enhance the detection of sarcasm.
3) The experimental results demonstrate that our model achieves a good classification performance on a large Reddit corpus.

The remainder of this work is summarized as follows: Sect. 2 lists related works; Sect. 3 explains the detailed design of the proposed sarcasm detection model; Sect. 4 presents experimentation details of the model and result in analysis; Sect. 5 draws conclusions.

2 Related Work

2.1 Sarcasm Detection

With the development of sentiment analysis technology, sarcasm detection has also drawn much attention from the natural language processing area. By reviewing existing studies on sarcasm detection, previous works can be classified into two main categories: textual feature-based and contextual feature-based sarcasm detection models.

Textual Feature-Based Models. Naturally, many networks model the problem of sarcasm detection task as a standard text classification problem and try to find lexical and semantics to identify sarcasm. Kreuz et al. [1] studied the text features used to detect sarcasm and found that words such as parenthesis and punctuation were useful. Carvalho et al. [9] use linguistic features to identify sarcasm, like positive predicates, interjections, emoticons, quotation marks, etc. Felbo, B et al. [10] also study the use of emoticons.

At a semantic level, Riloff et al. [11] propagate the contrast theme forward, presenting an algorithm strongly based on the intuition that sarcasm arises from a juxtaposition of positive and negative situations. The algorithm expands the list of positive verbs and negative phrases by iterating over the dataset. Joshi et al. [12] use multiple features comprising lexical, pragmatics, implicit and explicit context incongruity. Tay et al. [3] utilized a multi-dimensional intra-attention component to overcome the limitations of sequential models and capture words' incongruities by leveraging intra-sentence relationships. Although the text feature-based method for detecting sarcasm is useful, it does not work well when there is no specific satire marked in the sentence.

Contextual Feature-Based Models. Contextual models utilize both contextual and user information. Texts found in the discussion are plagued by grammatical inaccuracies and contain information that is contextual, thus mining linguistic information becomes relatively inefficient. Wallace et al. [13] claim that when human graders attempted to mark comments as sarcastic or not sarcastic, they needed additional context thus capturing previous and following comments on Reddit increases classification performance. Ghosh et al. [14] detected sarcasm by adding the conversation context to the LSTM model while adding an attention mechanism to show which part of the context triggered the sarcasm reply.

Every user has different attitudes towards different news, and some people are used to using sarcasm to express their opinions. Khattri et al. [15] try to discover users' sentiments towards entities in their histories to find contrasting evidence. Amir's et al. [16] merges all historical tweets of a user into one historical document and models a representation of that document. Kolchinski et al. [5] use a Bayesian method that captures only an author's raw propensity for sarcasm. Hazarika et al. [4] proposed CASCADE, by adopting a hybrid approach of both contents and context-driven modeling for sarcasm detection where they model stylometric and personality details of users along with discourse features of discussion forums to learn informative contextual representations. Du Y et al. [6] propose a dual-channel convolutional neural network that analyzes not only the semantics of the target text but also its sentimental context. The attention mechanism is then applied to take the user's expression habits into account. There is a growing emphasis on textual and contextual information, but how to fuse the two well is still a challenge. Our research is mostly related to this line of work. In particular, we fuse the two kinds of information to process sarcasm detection through the sequential three-way decision.

2.2   Sequential Three-Way Decision

As a useful tool to solve the human problem and process information, Yao [17] thought the basic notion of three-way decisions can be interpreted as a two-step approach. The first step with trisecting is to divide the objects into three pairwise disjoint regions, denoted as positive (POS) region, negative (NEG) region, and boundary region (BND), respectively. The second step is to divide objects among three regions by appropriate strategies. Jia et al. [18] propose a three-way decisions-based feature fusion method for Chinese irony detection in the microblog. It starts only from the text to find multiple features for three-way decision fusion. However, it does not consider both contextual and user information, so it cannot fuse more important features using a multilevel framework for sequential three-way decision. Yao [23] proposed Tri-level thinking, which gives further elaboration on a three-way decision. Tri-level thinking is to divide the whole task into three levels and ask different questions at different levels according to the natural order of the three levels. It can integrate the three relatively simple levels into the complex whole and improve the overall understanding of the whole. Similarly, we improve the understanding of the sarcastic text by integrating multiple features into a complex whole through a multi-level framework of the sequential three-way decision model.

Yao et al. [8,21] thought that granular computing is a network of interacting granules that can be used to establish multiple levels of describing the universe. Thus, by introducing granular computing, three-way decision models with the sequential strategy were presented for achieving multi-division of the boundary regions. Due to the significant advantages of sequential three-way decision, more and more researchers are conducting in-depth studies. Savchenkoa [24] reduces the computation of neural networks using sequential three-way decision to speed up inference in convolutional neural networks. XU Y et al. [25] propose that decisions can be made from multiple views and multiple levels simultaneously based on two different search methods to enhance the effectiveness of the sequential three-way decision. Zhang et al. [22] proposed a cost-sensitive combination technique using sequential three-way decisions, which is designed for strong base classifiers in sentiment classification tasks. But, it only uses the results of different classifiers and misses textual and contextual information.

Previous research used text feature-based or context-based models for detecting complex sarcasm. To tackle the irrationality of the simple combination of all features, our work effectively improves the performance of sarcasm detection by using sequential three-way decisions to fuse text features and users' expression propensity.

3   Proposed Method

This section presents a detailed description of the proposed method. Figure 1 shows the overall flow of this model. In Sect. 3.1, we use the sequential three-way decision (S3WD) to fuse the users' expression propensity and text emotional features of comments. In Sect. 3.2, we integrate multiple information and the S3WD feature. And in Sect. 3.3, the multi-features are embedded, where Bi-GRU is used to train for sarcasm detection.

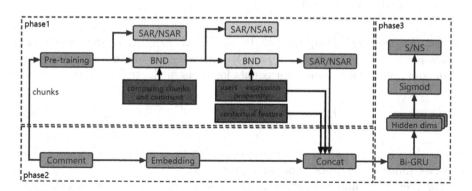

Fig. 1. The overall flow of UEP-S3WD boundary processing for sarcasm detection.

3.1   Phase 1: Fusing Feature by S3WD

The Structure of S3WD. We construct a three-layer sequential structure of S3WD based on the sentiment polarity score and set different thresholds α_i to classify the sequential layers. By initial comparing the sentiment polarity of the chunks in the first sequential layer, the determined results will be sent to the determined SAC regions and NSAR regions, while uncertain results will be sent to the BND region for further processing. In the second sequential layer, comparing the sentiment of the chunks and comment is used to process the BND region, and continuously divides the region. In the third sequential layer, the users' expression propensity is used to process the BND region and complete the final processing of the BND region.

Comment Sentiment Analysis. Sarcasm commonly manifests with a contrastive theme between positive-negative sentiments. In sarcasm comments, negative sentiments are more likely to coexist with neutral or positive ones. In order to capture these spatial patterns, we propose the sentiments of decomposing a comment into separate chunks and analyzing each one separately.

Suppose the comment sequence is $T = T_1, T_2 \cdots T_n$, and the comment is separated two chunks $c = c_{11}, c_{12} \cdots c_{n1}, c_{n2}$. The natural language toolkit (NLTK) is used to separately predict the polarity of sentiment for T and c. Each comment and chunk is predicted with a sentiment score of positive and negative that are used to contrast polarity. T_i^{pos}, c_{ij}^{pos} represents the positive sentiment of comment and chunk, and T_i^{neg}, c_{ij}^{neg} represents the negative sentiment of comment and chunk. In the first sequential layer, we compare the sentiment polarity score of the chunks for dividing the regions. Then, we compare the chunks and the comment in the second sequential layer, the results of contrasting sentiment scores are used as the sarcasm tags.

Users' Expression Propensity. In the third sequential layer, the users' expression propensity serves as a prior feature which is obtained from the sarcastic and non-sarcastic comment counts for users in the training data.

$$U(s_i, ns_i) = \begin{cases} 1 & \Sigma s_i - \Sigma ns_i > 2 \\ 2 & \Sigma s_i - \Sigma ns_i \leq 2 \end{cases} \tag{1}$$

where the sarcastic and non-sarcastic comment is s_i and ns_i separately. And $U(s_i, ns_i)$ means the users' expression propensity.

For previously unseen authors and comments, it is counted as 0. Difference from the Bayesian prior model, more complex factors are considered by the users' expression propensity. For example, even users who tend to be sarcastic sometimes make non-sarcastic comments. Thus, it is necessary to analyze the $U(s_i, ns_i)$ in sarcasm detection.

3.2   Phase 2: Integrating Multi-features

The comments get the label Y_i for training after the treatment of the three-sequential layers. The inputs to the Bi-GRU model are users' comments, which are split into words and punctuation marks and converted to word vectors $w = w_1, w_2 \cdots w_n$, where t is a predefined maximum sequence length. For model learning better, the multi-features are embedded into the word vector and then send as input to Bi-GRU for training. First, considering sentiment context during sarcasm detection, contextual information is further applied.

Each comment $T = T_1, T_2 \cdots T_n$ has a certain number of contextual comments $C = C_1, C_2 \cdots C_n$, they are predicted the combined sentiment score for C_i and T_i by NLTK. The sarcasm context polarity P is obtained as:

$$\overline{C} = \sum_{i=1}^{k} \overline{C_i} \tag{2}$$

$$P_i = \begin{cases} 1 & \overline{C} > \overline{T} \\ 0 & \overline{C} \leq T_i \end{cases} \tag{3}$$

Here, $\overline{C}$ means averaging all contextual sentiment score $\overline{C_i}$, where $\overline{C_i}$ and $\overline{T_i}$ means the combined sentiment score of each comment. Then, we embed all the features together and denoted by F.

$$F = Y_i \oplus U(s_i, ns_i) \oplus P_i \oplus w \tag{4}$$

3.3 Phase 3: Sarcasm Classification

The features F obtained in phase2 are fed into Bi-GRU as input for training after the embedding of multi-features is performed. Since the advantage of the S3WD, the text features, users' information, and contextual context have been fully considered.

4 Experiments

4.1 Datasets

Reddit[1] is a popular social forum and community. We perform our experiments on a large-scale self-annotated corpus for sarcasm, SARC (Khodak et al. [20]). It includes an unprecedented 533M comments. And the corpus is self-annotated in the sense that a comment is considered sarcastic if its author marked it with the "/s" tag. It is vastly larger than past sarcasm datasets, which enables the training of more sophisticated models. Table 1 provides basic statistics on the entire corpus as well as the subreddits that we consider three variants of the SARC[2] dataset in our experiments.

4.2 Baseline Methods

In this subsection, we introduce the comparing algorithms.

- **Bag-of-words:** This model uses an SVM classifier whose input features comprise a comment's word counts. The size of the vector is the vocabulary size of the training dataset.

---

[1] http://reddit.com/reddits.
[2] http://nlp.cs.princeton.edu/SARC.

Table 1. Basic statistics for SARC.

Data	Comments	Sarcastic	Unsarcastic
SARC/main	257082	128541	128541
SARC/politics	13668	6834	6834
SARC/AskReddit	11660	5830	5830

- **Bi-GRU:** It is a variant of the Long Short-Term Memory Network. Combining the forget gate and the input gate into a single update gate, which is simpler than the standard LSTM model.
- **Kolchinski et al. [5]:** They use a Bayesian method that captures a users' raw propensity for sarcasm, and the propensity is used jointly to learn a Bi-GRU model.
- **CASCADE [4]:** It adopts a hybrid approach of both content-based and context-driven modeling for sarcasm detection. The authors used user profiling along with discourse modeling from comments in discussion threads. Then, the information is used jointly to learn a CNN-based model.
- **Du Y et al. [6]:** They propose a dual-channel convolutional neural network that analyzes not only the semantics of the target text but also its sentimental context. In addition, SenticNet is used to add common sense to the LSTM model. The attention mechanism is then applied to take the user's expression habits into account.
- **BERT [26]:** BERT is the Encoder of Bidirectional Transformer. The main innovation of the model is in the pre-train method. Masked LM and Next Sentence Prediction are used to capture the word and sentence level representations respectively.
- **RoBERTa [27]:** RoBERTa is mainly based on BERT with several adjustments: larger batch size and more training data, removal of next predicted loss, longer training sequence, and dynamic adjustment of the Masking mechanism.
- **RCNN-RoBERTa [19]:** This model uses the pre-trained RoBERTa model and a recurrent convolutional neural network(RCNN) to tackle figurative language in sarcasm detection in social media.

4.3 Comparison of Experimental Results

Table 2 shows the means of 5 runs to control for variation deriving from randomness in the optimization process. Our model is highly competitive. It slightly under-performs on the main dataset (only 0.6 worse than CASCADE) but comes out ahead on politics. This is striking because our model does not involve large-scale data pre-processing nor complex analysis and embedding of user characteristics and forum information as CASCADE does. Compared with the simple concatenating of various features, the features obtained by S3WD can fuse multiple information at the sequential level and fully consider various factors. From the results, the advantages of sequential three-branch decision making are also proved.

Table 2. Mean macro-averaged F1 scores based on five runs.

Models	SARC/main	SARC/politics	SARC/AskReddit
Bag-of-words	64.0	60.0	–
Bi-GRU	74.8	74.3	64.3
Kolchinski	75.3	77.6	**69.1**
CASCADE	**77.0**	75.0	–
Du	–	72.0	–
BERT	–	76.0	–
RoBERTa	–	77.0	–
RCNN-RoBERTa	–	78.0	–
UEP-S3WD	76.4	**79.3**	**69.1**

4.4 Ablation Experiments of Each Part

We experiment on different sequential layers of the S3WD as well as concatenating multiple features so as to analyze the importance of the various features present in its architecture. Table 3 and Table 4 provide the results of all the combinations separately.

In Table 3, we test performance for the polarity contrast of chunks only in the first layer (row 1). Next, we include the sentiment of comparing chunks and comments to the second layer (row 2). A major boost in performance is observed when users' expression propensity is introduced in the third layer (row 3). The division of sequential layers is performed by the threshold (α_1, α_2) , and set (α_1, α_2) to (0.8, 0.6). Overall, S3WD consisting of multiple sequential layers with text features and users' expression propensity provides the best performance in all three datasets.

Table 3. Multiple sequential layers ablation analysis.

Sequential layers	SARC/main	SARC/politics	SARC/AskReddit
First layer	74.1	74.2	66.1
Second layer	74.8	74.8	66.1
Third layer	**76.1**	**78.9**	**67.2**

In Table 4, we perform an ablation analysis of the multiple features embedded into Bi-GRU. It can be seen that user sarcasm propensity is an important basis for judgment, and the sequential three-branch feature combining multiple information performs better on the pol and ask datasets and slightly less well on the main dataset.

Table 4. Multiple features ablation analysis.

Multiple features			SARC/main	SARC/politics	SARC/AskReddit
Contextual	User'	S3WD			
✓	–	–	74.0	74.7	65.8
–	✓	–	76.2	78.1	66.7
–	–	✓	76.1	78.9	67.2
✓	✓	–	75.9	78.5	67.9
✓	–	✓	76.0	78.8	67.1
–	✓	✓	75.7	78.9	67.7
✓	✓	✓	**76.4**	**79.3**	**68.0**

4.5 Users' Expression Propensity Analysis

We investigate the users' expression propensity in more detail. The counts of user sarcastic comments and non-sarcastic comments are used to make judgments about users' expression propensity. To avoid arbitrary judgments, we analyze the effect of $(\Sigma s_i - \Sigma ns_i)$ on $U(s_i, ns_i)$ on the final result. As Table 5 shows, it can be seen that the main and political datasets perform best at $(\Sigma s_i - \Sigma ns_i > 2)$, while AskReddit gives the best performance at $(\Sigma s_i - \Sigma ns_i > 1)$. This also proves that the best results cannot be obtained by making direct judgments on S, because even users who tend to be sarcastic sometimes make non-sarcastic comments.

Table 5. Users' expression propensity analysis.

$(\Sigma s_i - \Sigma ns_i)$	SARC/main	SARC/politics	SARC/AskReddit
>0	74.4	77.1	67.7
>1	75.8	77.1	**69.1**
>2	**76.4**	**79.3**	68.0
>3	75.3	77.7	67.3
>4	75.4	77.3	67.7

5 Conclusion

In this paper, we introduced S3WD, which leverages multiple sequential layers for fusing the text feature and users' expression propensity, and contextual information is further applied to consider sentiment context. We obtain state-of-the-art performance on a large-scale Reddit corpus. Our experiments show that users' expression propensity along with the S3WD feature plays a crucial role in the performance of sarcasm detection. Our work expands the application of the S3WD model in the field of sarcasm detection. However, considering the wide application of pre-trained models, we will combine the S3WD model with pre-training for multiple information fusion in the future.

Acknowledgments. This work was supported by the Major Program of the National Social Science Foundation of China (GrantNo. 18ZDA032), the National Natural Science Foundation of China (Grant No. 61876001), China Scholarship Council, and the Natural Science Foundation for the Higher Education Institutions of Anhui Province of China (KJ2021A0039).

References

1. Kreuz, R, Caucci, G.: Lexical influences on the perception of sarcasm. In: Proceedings of the Workshop on Computational Approaches to Figurative Language, pp. 1–4 (2007)
2. Rohanian, O., Taslimipoor, S., Evans, R., et al.: WLV at SemEval-2018 task 3: dissecting tweets in search of irony. In: Proceedings of The 12th International Workshop on Semantic Evaluation, pp. 553–559 (2018)
3. Tay, Y., Tuan, L.A., Hui, S.C., et al.: Reasoning with sarcasm by reading in-between. arXiv preprint arXiv:1805.02856 (2018)
4. Hazarika, D., Poria, S., Gorantla, S., et al.: Cascade: contextual sarcasm detection in online discussion forums. arXiv preprint arXiv:1805.06413 (2018)
5. Kolchinski, Y.A., Potts, C.: Representing social media users for sarcasm detection. arXiv preprint arXiv:1808.08470 (2018)
6. Du, Y., Li, T., Pathan, M.S., et al.: An effective sarcasm detection approach based on sentimental context and individual expression habits. Cogn. Comput. **14**, 1–13 (2021). https://doi.org/10.1007/s12559-021-09832-x
7. Yao, Y.: Three-way decision: an interpretation of rules in rough set theory. In: Wen, P., Li, Y., Polkowski, L., Yao, Y., Tsumoto, S., Wang, G. (eds.) RSKT 2009. LNCS (LNAI), vol. 5589, pp. 642–649. Springer, Heidelberg (2009). https://doi.org/10.1007/978-3-642-02962-2_81
8. Yao, Y.Y., Pedrycz, W., Skowron, A., et al.: A unified framework of granular computing. Wiley, Chichester (2008)
9. Carvalho, P., Sarmento, L., Silva, M.J., et al.: Clues for detecting irony in user-generated contents: oh...!! it's "so easy";-. In: Proceedings of the 1st International CIKM Workshop on Topic-sentiment Analysis for Mass Opinion, pp. 53–56 (2009)
10. Felbo, B., Mislove, A., Søgaard, A., et al.: Using millions of emoji occurrences to learn any-domain representations for detecting sentiment, emotion and sarcasm. arXiv preprint arXiv:1708.00524 (2017)
11. Riloff, E., Qadir, A., Surve, P., et al.: Sarcasm as contrast between a positive sentiment and negative situation. In: Proceedings of the 2013 Conference on Empirical Methods in Natural Language Processing, pp. 704–714 (2013)
12. Joshi, A., Sharma, V., Bhattacharyya, P.: Harnessing context incongruity for sarcasm detection. In: Proceedings of the 53rd Annual Meeting of the Association for Computational Linguistics and the 7th International Joint Conference on Natural Language Processing (Volume 2: Short Papers), pp. 757–762 (2015)
13. Wallace, B.C., Kertz, L., Charniak, E.: Humans require context to infer ironic intent (so computers probably do, too). In: Proceedings of the 52nd Annual Meeting of the Association for Computational Linguistics (Volume 2: Short Papers), pp. 512–516 (2014)

14. Ghosh, D., Fabbri, A.R, Muresan, S.: The role of conversation context for sarcasm detection in online interactions. arXiv preprint arXiv:1707.06226 (2017)
15. Khattri, A., Joshi, A., Bhattacharyya, P., et al.: Your sentiment precedes you: using an author's historical tweets to predict sarcasm. In: Proceedings of The 6th Workshop on Computational Approaches to Subjectivity, Sentiment and Social Media Analysis, pp. 25–30 (2015)
16. Amir, S., Wallace, B.C., Lyu, H., et al.: Modelling context with user embeddings for sarcasm detection in social media. arXiv preprint arXiv:1607.00976 (2016)
17. Yao, Y.: Three-way decisions and cognitive computing. Cogn. Comput. 8(4), 543–554 (2016). https://doi.org/10.1007/s12559-016-9397-5
18. Jia, X., Deng, Z., Min, F., et al.: Three-way decisions based feature fusion for Chinese irony detection. Int. J. Approx. Reason. 113, 324–335 (2019)
19. Potamias, R.A., Siolas, G., Stafylopatis, A.G.: A transformer-based approach to irony and sarcasm detection. Neural Comput. Appl. 32(23), 17309–17320 (2020). https://doi.org/10.1007/s00521-020-05102-3
20. Khodak, M., Saunshi, N., Vodrahalli, K.A.: Large self-annotated corpus for sarcasm. arXiv preprint arXiv:1704.05579 (2017)
21. Yao, Y.: Three-way granular computing, rough sets, and formal concept analysis. Int. J. Approx. Reason. 116, 106–125 (2020)
22. Zhang, Y., Miao, D., Wang, J., et al.: A cost-sensitive three-way combination technique for ensemble learning in sentiment classification. Int. J. Approx. Reason. 105, 85–97 (2019)
23. Yao, Y.: Tri-level thinking: models of three-way decision. Int. J. Mach. Learn. Cybern. 11(5), 947–959 (2019). https://doi.org/10.1007/s13042-019-01040-2
24. Savchenko, A.V.: Fast inference in convolutional neural networks based on sequential three-way decisions. Inf. Sci. 560, 370–385 (2021)
25. Xu, Y., Li, B.: Multiview sequential three-way decisions based on partition order product space. Inf. Sci. 600, 401–430 (2022)
26. Devlin, J., Chang, M.W., Lee, K., et al.: BERT: pre-training of deep bidirectional transformers for language understanding. arXiv preprint arXiv:1810.04805 (2018)
27. Liu, Y., Ott, M., Goyal, N., et al.: RoBERTa: a robustly optimized BERT pre-training approach. arXiv preprint arXiv:1907.11692 (2019)

Concept Reduction of Object-induced Three-way Concept Lattices

Xiuwei Gao[ID], Yehai Xie[ID], and Guilong Liu[(✉)][ID]

School of Information Science, Beijing Language and Culture University,
Beijing 100083, China
liuguilong@blcu.edu.cn

Abstract. Three-way concept lattices are a combination of three-way decision theory and classic concept lattices. Attribute reduction is one of the critical topics in formal concept analysis and has been extensively studied. This paper discusses the concept reduction of object-induced three-way concept lattices. We propose a new type of reduction for coverings and derive its reduction algorithm to identify all reducts. We study concept reduction of object-induced three-way concept lattices and show that such a reduction can be converted into union reduction for coverings.

Keywords: Covering · Formal context · Object-induced three-way concept · Object-induced three-way concept lattice · Union reduction

1 Introduction

Classical rough set theory, proposed by Pawlak [1,2] in 1982, is a useful tool for analyzing vague and uncertain data. The theory has attracted wide attention in both the theory and its applications. However, it has a restricted range of applications since it is based on an equivalence relation. There are many different generalizations for such a theory. For example, by using coverings instead of partitions, Bonikowski [3] proposes the covering rough set model as an expansion of the classical rough set model. Attribute reduction comes from machine learning, and it has received much research in rough set theory as a useful preprocessing method. Pawlak [1] initiated the study of attribute reduction in information systems. Skowron and Rauszer [4] proposed a discernibility matrix-based reduction method. Although this method has high algorithm complexity, it is still an essential attribute reduction method. Zhu and Wang [5] introduced the concept of reduction in coverings. Chen [6] advanced a method to reduce the attributes of covering decision systems by defining the intersection of coverings.

Formal concept analysis, proposed by Wille [7,8], is a useful model for the mathematization of concept and conceptual hierarchy. A formal concept illustrates the relationship between objects and attributes as a model for philosophical conceptions. The family of formal concepts may be interpreted as a concept

Supported by the National Natural Science Foundation of China (No. 61972052) and the Research Funds of Beijing Language and Culture University (No. 21YCX169).

J. Yao et al. (Eds.): IJCRS 2022, LNAI 13633, pp. 265–273, 2022.
https://doi.org/10.1007/978-3-031-21244-4_20

lattice. Attribute reduction is one of the most important issues in formal concept analysis, Zhang [9] researched the attribute reduction in concept lattice based on the discernibility matrix to simplify knowledge representation in the formal context. Li et al. [10] proposed an approach to attribute reduction in formal contexts via a covering rough set theory, and they obtained judgment theorems for determining all attribute reducts in the formal context. After that, Chen et al. [11] studied the relation between the reduction of a covering and the attribute reduction of a concept lattice, and they proved that every reduct of a given formal context could be seen as the reduct of an induced covering. Wu, Leung and Mi [12] considered the attribute reduction problem in decision contexts. As an application, Liu [13] use covering reduction method to identify reducts for object-oriented concept lattices.

The idea of three-way decisions is a common strategy in life, widely used in various decision-making processes. Three-way decisions aim to divide a whole into three parts and adopt different treatments for different parts. Yao [14,15] proposed a unified framework description of the three-way decision theory. Three-way concept analysis (3WCA) is proposed by Qi, Wei and Yao [16,17] based on the three-way decision theory. Qi, Qian and Wei [18] set up the model of three-way concept lattices. An object-induced three-way concept lattice (OE-lattice) combines 3WD and classic concept lattices, which can supply more information. Ren and Wei [19] studied different types of attribute reductions for three-way concept lattices, gave four attribute-induced three-way attribute reductions, and discussed their relationships.

Recently, Wei et al. [20] considered concept reduction for classic concept lattices. With the inspiration of the above research, this paper focuses on the concept reduction of OE-lattices, which can help us understand the similarities and differences between rough set theory and formal concept analysis. Firstly, we present a novel type of reduction for coverings defined as union reduction and propose an algorithm to identify all reducts of a covering. Then we apply this new covering reduction method to the concept reduction of OE-lattices to remove excessive concepts.

The remainder of the paper is organized as follows. In Sect. 2, we briefly review some basic concepts and properties of formal concept analysis and the three-way formal concept analysis. In Sect. 3, we propose a new type of reduction for coverings, and a reduction algorithm to identify all reducts is obtained. In Sect. 4, we study the concept reduction of OE-lattices and provide a reduction algorithm. Section 5 concludes the paper.

2   Preliminaries

This section recalls some essential concepts and properties of formal concept analysis and the three-way concept analysis.

Definition 1 [8]. *Let U be the set of objects and A be the set of attributes. A triple (U, A, I) is called a formal context (for short, context) if I is a relation from U to A.*

In this paper, we assume that $U = \{x_1, x_2, \cdots, x_n\}$ and $A = \{a_1, a_2, \cdots, a_m\}$ be nonempty finite sets, I is a binary relation from U to A, i.e., $I \subseteq U \times A$. $(x_i, a_j) \in I$ expresses that an object x_i has attribute a_j. Given a context (U, A, I), for any $X \subseteq U$ and $B \subseteq A$, a pair of operators can be defined as follows [8]:

$$X^* = \{a \in A | \forall x \in X, (x, a) \in I\}, B^* = \{x \in U | \forall a \in B, (x, a) \in I\} \quad (1)$$

Definition 2 [8]. *Let (U, A, I) be a context, suppose that $X \subseteq U$ and $B \subseteq A$. If $X^* = B$, $B^* = X$, (X, B) is called a formal concept (for short, concept), X is called the extent and B is the intent of the concept (X, B).*

Let $L(U, A, I)$ denote the set of all the concepts of the context (U, A, I). $(L(U, A, I), \leq)$ is a partially ordered set with the following partial order $\leq$. For any $(X_1, B_1), (X_2, B_2) \in L(U, A, I)$,

$$(X_1, B_1) \leq (X_2, B_2) \Leftrightarrow X_1 \subseteq X_2 (\Leftrightarrow B_2 \subseteq B_1) \quad (2)$$

$(L(U, A, I), \wedge, \vee)$ [8] is a lattice and its meet and join operations are as follows. For $(X_1, B_1), (X_2, B_2) \in L(U, A, I)$.

$$(X_1, B_1) \wedge (X_2, B_2) = (X_1 \cap X_2, (B_1 \cup B_2)^{**}) \quad (3)$$

$$(X_1, B_1) \vee (X_2, B_2) = ((X_1 \cup X_2)^{**}, B_1 \cap B_2) \quad (4)$$

Definition 3 [16]. *Let (U, A, I) be a context, $\overline{I}$ is the complementary relation of I, i.e., $\overline{I} = U \times A - I$. For $X \subseteq U$ and $B \subseteq A$, two negative operators are given as follows.*

$$X^{\overline{*}} = \{a | a \in A | \forall x \in X (x\overline{I}a)\} \text{ and } B^{\overline{*}} = \{x \in U | \forall a \in B (x\overline{I}a)\}$$

Combining the operators $*$ and $\overline{*}$, we define

$$X^{\#} = (X^*, X^{\overline{*}}), \text{ for } X \subseteq U \quad (5)$$

$$(B, C)^{\#} = B^* \cap C^{\overline{*}} \text{ for } B, C \subseteq A \quad (6)$$

Definition 4 [16]. *Let (U, A, I) be a context, $X \subseteq U$ and $B, C \subseteq A$, a triple $(X, (B, C))$ is called an object-induced three-way concept(for short, OE-concept) of context (U, A, I), if $X^{\#} = (B, C)$ and $(B, C)^{\#} = X$. X is called the extent and (B, C) is called the intent of the OE-concept $(X, (B, C))$.*

The set of all OE-concepts of (U, A, I) is denoted by $\#(U, A, I)$. If $(X_1, (B_1, C_1))$ and $(X_2, (B_2, C_2))$ are two OE-concepts, the partial order is defined as follow:

$$(X_1, (B_1, C_1)) \leq (X_2, (B_2, C_2)) \Leftrightarrow X_1 \subseteq X_2 \Leftrightarrow (B_2, C_2) \subseteq (B_1, C_1) \quad (7)$$

$\#(U, A, I)$ is a lattice and is called an object-induced three-way concept lattice (for short, OE-lattice). For any $(X_1, (B_1, C_1)), (X_2, (B_2, C_2)) \in \#(U, A, I)$, the infimum and supremum are given by

$$(X_1, (B_1, C_1)) \wedge (X_2, (B_2, C_2)) = (X_1 \cap X_2, ((B_1, C_1) \cup (B_2, C_2))^{\#\#}) \quad (8)$$

$$(X_1, (B_1, C_1)) \vee (X_2, (B_2, C_2)) = ((X_1 \cup X_2)^{\#\#}, (B_1, C_1) \cap (B_2, C_2)) \quad (9)$$

Similarly, attribute-induced three-way concept lattices [16] can be defined. In this paper, we only study the OE-concept lattices, However, our method is also effective for attribute-induced three-way concept lattices. The following is an example of the computation of OE-concepts for a given context.

Example 1. Suppose that the context (U, A, I) is given in Table 1, where $U = \{1, 2, 3, 4\}$ and $A = \{a, b, c, d\}$. All OE-concepts of context (U, A, I) can easily be calculated as Table 2.

Table 1. Context (U, A, I) of Example 1

	a	b	c	d
1	0	0	1	0
2	1	1	0	1
3	0	0	0	1
4	1	0	1	0

Table 2. OE-concepts of Example 1

11 OE-concepts			
α_0	$(\emptyset, (A, A))$	α_1	$(1, (c, abd))$
α_2	$(2, (abd, c))$	α_3	$(3, (d, abc))$
α_4	$(4, (ac, bd))$	α_5	$(24, (a, \emptyset))$
α_6	$(23, (d, c))$	α_7	$(14, (c, bd))$
α_8	$(13, (\emptyset, ab))$	α_9	$(134, (\emptyset, b))$
α_{10}	$(U, (\emptyset, \emptyset))$		

3 Union Reduction of Coverings

This section studies a new type of reduction for coverings and gives its corresponding reduction algorithm to identify all reducts. We show that the type of reduction is different from the covering reduction defined in [5]. In the next section, we will use the algorithm to identify all concept reducts for OE-lattices in a context.

Definition 5. *[3] Let U be a nonempty finite set, $\mathcal{C}$ is a finite family of nonempty subsets of U such that $\cup_{K \in \mathcal{C}} K = U$, then $\mathcal{C}$ is called a covering of U.*

Clearly, a partition of U is a special covering of U.

Definition 6. *Let $\mathcal{C}$ be a covering of U, $\mathcal{C}' \subseteq \mathcal{C}$. $\mathcal{C}'$ is called a union reduct of $\mathcal{C}$ if $\mathcal{C}'$ satisfies the following conditions:*
(1) $\mathcal{C}'$ is a covering.
(2) If $\mathcal{C}'' \subset \mathcal{C}'$, $\cup_{K \in \mathcal{C}''} K \neq U$.

Note that the union reduction is different from the usual covering reduction defined in [5]. Now we derive the reduction algorithm.

Lemma 1. *Let $\mathcal{C}$ be a covering of U with $U = \{x_1, x_2, \cdots, x_n\}$ and $\mathcal{C} = \{K_1, K_2, \cdots, K_r\}$, $m_i = \{K | K \in \mathcal{C}, x_i \in K\}$, $i = 1, 2, \cdots, n$, then $m_i \neq \emptyset$.*

Proof. For each $x_i \in U = \cup_{i=1}^{r} K_i$, thus there exists $x_i \in K_j$ for some $K_j \in \mathcal{C}$ and $K_j \in m_i$, so $m_i \neq \emptyset$. $\qquad\square$

Theorem 1. *Let $\mathcal{C}$ be a covering of U, if $\mathcal{C}' \subseteq \mathcal{C}$, then*

$$\cup_{K \in \mathcal{C}'} K = U \iff m_i \cap \mathcal{C}' \neq \emptyset \tag{10}$$

Proof. If $\cup_{K \in \mathcal{C}'} = U, \forall i, x_i \in U \Rightarrow x_i \in K \in \mathcal{C}' \Rightarrow K \in m_i \cap \mathcal{C}' \Rightarrow m_i \cap \mathcal{C}' \neq \emptyset$. Conversely, it is obvious that $\cup_{K \in \mathcal{C}'} \subseteq U$. Suppose that $x_i \in U$ and $K \in m_i \cap \mathcal{C}' \Rightarrow x_i \in K \in \mathcal{C}' \Rightarrow x_i \in \cup_{K \in \mathcal{C}'} K$. Hence, $U \subseteq \cup_{K \in \mathcal{C}'} K$. $\qquad\square$

Corollary 1. *Let $\mathcal{C}$ be a covering of U and $\mathcal{C}' \subseteq \mathcal{C}$, then $\mathcal{C}'$ is a union reduct of $\mathcal{C}$ if and only if $\mathcal{C}'$ is a minimal subset satisfying $m_i \cap \mathcal{C}' \neq \emptyset$.*

By Corollary 1, we can induce a union reduction algorithm for a covering. Let $U = \{x_1, x_2, \cdots, x_n\}$ and $\mathcal{C} = \{K_1, K_2, \cdots, K_r\}$.

Algorithm 1. A union reduction algorithm for coverings.

Require: A set U and a covering $\mathcal{C}$ of U;
Ensure: All union reducts;
1: Calculate m_i for $i = 1, 2, \cdots, n$;
2: Transform the discernibility function from its conjunctive normal form (CNF) $f = \prod_{m_i \neq \emptyset} (\sum_{a \in m_i} a)$ into the disjunctive normal form (DNF). $f = \sum_{i=1}^{t} (\prod_{a \in D_i} a)$;
3: **return** $Red = \{D_1, D_2, \cdots, D_t\}$;

Next, we use an example to illustrate our algorithm.

Example 2. Let $U = \{1, 2, 3, 4\}$, $K_1 = \{1, 2\}$, $K_2 = \{1\}$, $K_3 = \{2\}$, $K_4 = \{3, 4\}$, $K_5 = \{1, 3, 4\}$, $\mathcal{C} = \{K_1, K_2, K_3, K_4, K_5\}$ is a covering of U. It is easy to see that $m_1 = \{K_1, K_2, K_5\}$, $m_2 = \{K_1, K_3\}$, $m_3 = \{K_4, K_5\}$, $m_4 = \{K_4, K_5\}$. Now we calculate the reducts.

$$f = (K_1 + K_2 + K_5)(K_1 + K_3)(K_4 + K_5)(K_4 + K_5)$$
$$= K_1 K_4 + K_1 K_5 + K_3 K_5 + K_2 K_3 K_4.$$

All union reducts of $\mathcal{C}$ are $\{K_1, K_4\}$, $\{K_1, K_5\}$, $\{K_3, K_5\}$, and $\{K_2, K_3, K_4\}$. However, $\{K_2, K_3, K_4\}$ is a unique covering reduct [5] of $\mathcal{C}$. Thus, the union reduction is different from the covering reduction defined in [5].

4 Concept Reduction of Object-Induced Three-Way Concept Lattices

This section discusses the concept reduction of object-induced three-way concept lattices and uses previous method to identify all concept reducts. This section can be considered as an application of previous section. For a context (U, A, I), we will show that

$$\{X \times (B \cup C)|(X, (B, C)) \in \#(U, A, I)\}$$

is a covering of $U \times A$, thus, there may be some superfluous concepts in $\#(U, A, I)$. Naturally, we try to remove these superfluous concepts in $\#(U, A, I)$. Therefore, we define the concept reduction of OE-lattice and use a concept reduction method to realize the purpose.

Theorem 2. Let $\#(U, A, I)$ be an OE-lattice, then
(1) $(l_I(a), (l_I(a)^*, l_I(a)^{\overline{*}})) \in \#(U, A, I)$ for each $a \in A$.
(2) $(l_{\overline{I}}(a), (l_{\overline{I}}(a)^*, l_{\overline{I}}(a)^{\overline{*}})) \in \#(U, A, I)$ for each $a \in A$.
(3) $\bigcup_{(X,(B,C))\in\#(U,A,I)} X \times (B \cup C) = U \times A$.

Proof. (1) $(l_I(a))^{\#} = (l_I(a)^*, l_I(a)^{\overline{*}})$ is clear. Moreover, $l_I(a)^{**} \cap l_I(a)^{\overline{**}} = a^{***} \cap l_I(a)^{\overline{**}} = a^* \cap l_I(a)^{\overline{**}} = l_I(a) \cap l_I(a)^{\overline{**}} = l_I(a)$ for $a \in A$.
   The proof of part (2) is similar to that of part (1) and we omit it.
   (3) According to Definition 2.4, It is obvious that

$$\bigcup_{(X,(B,C))\in\#(U,A,I)} X \times (B \cup C) \subseteq U \times A \tag{11}$$

   Conversely, for $(x, a) \in U \times A$, we have $(x, a) \in I$ or $(x, a) \notin I$. If $(x, a) \in I$, $(x, a) \in l_I(a) \times (l_I(a)^* \cup l_I(a)^{\overline{*}})$; If $(x, a) \notin I$, $(x, a) \in l_{\overline{I}}(a) \times (l_{\overline{I}}(a)^* \cup l_{\overline{I}}(a)^{\overline{*}})$. Using parts (1) and (2), we have

$$U \times A \subseteq \bigcup_{(X,(B,C))\in\#(U,A,I)} X \times (B \cup C) \tag{12}$$

By (11) and (12), $\bigcup_{(X,(B,C))\in\#(U,A,I)} X \times (B \cup C) = U \times A$. This completes the proof. $\square$

Definition 7. Let $\#(U, A, I)$ be an OE-lattice, $\mathcal{R} \subseteq \#(U, A, I)$, $\mathcal{R}$ is called a concept reduct of $\#(U, A, I)$ if $\mathcal{R}$ satisfies the following conditions:
(1) $\bigcup_{(X,(B,C))\in\mathcal{R}} X \times (B \cup C) = U \times A$,
(2) If $\mathcal{R}' \subset \mathcal{R}$, $\bigcup_{(X,(B,C))\in\mathcal{R}'} X \times (B \cup C) \neq U \times A$.

Theorem 2 (3) tells us that $\{X \times (B \cup C)|(X, (B, C)) \in \#(U, A, I)\}$ is a covering of $U \times A$. The concept reduction of $\#(U, A, I)$ is equivalent to the union reduction of covering $\{X \times (B \cup C)|(X, (B, C)) \in \#(U, A, I)\}$ of $U \times A$. Thus, we can use the union reduction algorithm for coverings to concept reduction of $\#(U, A, I)$.

To obtain a concept reduction algorithm, we suppose that $U = \{x_1, x_2, \cdots, x_n\}$ and $\#(U, A, I) = \{\alpha_1, \alpha_2, \cdots, \alpha_r\}$, where $\alpha_k = (X_k, (B_k, C_k))$. We define the discernibility matrix $M = (c_{ij})_{m \times n}$ as follow:

$$c_{ij} = \{\alpha_k | (x_i, a_j) \in X_k \times (B_k, C_k)\} \tag{13}$$

We derive concept reduction algorithm for OE-lattices as follows.

Algorithm 2. Concept reduction algorithm for OE-lattices.

Require: A context (U, A, I);
Ensure: All concept reducts of $\#(U, A, I)$;
 1: Calculate $\#(U, A, I)$, $\Delta = \#(U, A, I) - (U, (\emptyset, \emptyset)) - (\emptyset, (A, A)) = \{\alpha_1, \alpha_2, \cdots, \alpha_r\}$;
 2: Calculate discernibility matrix $M = (c_{ij})_{n \times m}$;
 3: Transform f from its CNF $f = \prod_{c_{ij}} (\sum_{\alpha \in c_{ij}} \alpha)$ into DNF $f = \sum_{i=1}^{s} (\prod_{\alpha \in B_i} \alpha)$;
 4: **return** $Red = \{B_1, B_2, \cdots, B_s\}$;

Since $(U, (\emptyset, \emptyset))$ and $(\emptyset, (A, A))$ correspond to an empty set, we delete them in our calculation process. We use the following example to illustrate our reduction algorithm.

Example 3. (Continued with Example 1) There are 11 OE-concepts in Example 2.1, we delete $(U, (\emptyset, \emptyset))$ and $(\emptyset, (A, A))$, thus we have $\Delta = \{\alpha_1, \alpha_2, \cdots, \alpha_9\}$. Then we compute the 4×4 discernibility matrix $M = (c_{ij})_{4 \times 4}$ as follow:

$$M = (c_{ij})_{4 \times 4} = \begin{pmatrix} \{\alpha_1, \alpha_8\} & \{\alpha_1, \alpha_7, \alpha_8, \alpha_9\} & \{\alpha_1, \alpha_7\} & \{\alpha_1, \alpha_7\} \\ \{\alpha_2, \alpha_5\} & \{\alpha_2\} & \{\alpha_2, \alpha_5\} & \{\alpha_2, \alpha_6\} \\ \{\alpha_3, \alpha_8\} & \{\alpha_3, \alpha_8, \alpha_9\} & \{\alpha_3, \alpha_6\} & \{\alpha_3, \alpha_6\} \\ \{\alpha_4, \alpha_5\} & \{\alpha_4, \alpha_7, \alpha_9\} & \{\alpha_4, \alpha_7\} & \{\alpha_4, \alpha_7\} \end{pmatrix}$$

The   minimal   subset   is $\{\{\alpha_1, \alpha_8\}, \{\alpha_1, \alpha_7\}, \{\alpha_2\}, \{\alpha_3, \alpha_8\}, \{\alpha_3, \alpha_6\}, \{\alpha_4, \alpha_5\}, \{\alpha_4, \alpha_7\}\}$, then we calculate the discernibility function.

$$f = (\alpha_1 + \alpha_8)(\alpha_1 + \alpha_7)\alpha_2(\alpha_3 + \alpha_8)(\alpha_3 + \alpha_6)(\alpha_4 + \alpha_5)(\alpha_4 + \alpha_7)$$
$$= \alpha_1\alpha_2\alpha_3\alpha_4 + \alpha_1\alpha_2\alpha_4\alpha_6\alpha_8 + \alpha_2\alpha_3\alpha_4\alpha_7\alpha_8 + \alpha_2\alpha_4\alpha_6\alpha_7\alpha_8$$
$$+ \alpha_1\alpha_2\alpha_3\alpha_5\alpha_7 + \alpha_2\alpha_3\alpha_5\alpha_7\alpha_8 + \alpha_2\alpha_5\alpha_6\alpha_7\alpha_8$$

All concept reducts of Example 4.1 are $\{\alpha_1, \alpha_2, \alpha_3, \alpha_4\}$, $\{\alpha_1, \alpha_2, \alpha_4, \alpha_6, \alpha_8\}$, $\{\alpha_2, \alpha_3, \alpha_4, \alpha_7, \alpha_8\}$, $\{\alpha_2, \alpha_4, \alpha_6, \alpha_7, \alpha_8\}$, $\{\alpha_1, \alpha_2, \alpha_3, \alpha_5, \alpha_7\}$, $\{\alpha_2, \alpha_3, \alpha_5, \alpha_7, \alpha_8\}$, and $\{\alpha_2, \alpha_5, \alpha_6, \alpha_7, \alpha_8\}$. Each reduct can induce a covering of $U \times A$.

5   Conclusions

Three-way concept analysis is an extension of formal concept analysis. In this paper, we have presented a novel type of reduction for coverings and proposed a corresponding reduction algorithm. Additionally, we have studied the concept reduction of OE-lattices. We have shown that each OE-lattice of a given context induces a covering so that concept reduction can be transformed into union reduction of the covering. We have given a concept reduction algorithm based on the discernibility matrix. Furthermore, this concept reduction method is equally effective in attribute-induced three-way concept lattices. Our future work will focus on practical applications of the proposed algorithms.

References

1. Pawlak, Z.: Rough sets. Int. J. Comput. Inf. Sci. **11**(5), 341–356 (1982)
2. Pawlak, Z.: Rough Sets: Theoretical Aspects of Reasoning About Data. Kluwer Academic Publishers, Boston (1991)
3. Bonikowski, Z., Bryniarski, E., Wybraniec-Skardowska, U.: Extensions and intentions in the rough set theory. Inf. Sci. **107**(1–4), 149–167 (1998)
4. Skowron, A., Rauszer, C.: The discernibility matrices and functions in information systems. In: Slowinski, R. (ed.) Intelligent Decision Support, Handbook of Applications and Advances of the Rough Sets Theory, pp. 331–362. Kluwer Academic Publishers, Dordrecht (1992)
5. Zhu, W., Wang, F.: Reduction and axiomization of covering generalized rough sets. Inf. Sci. **152**, 217–230 (2003)
6. Chen, D., Wang, C., Hu, Q.: A new approach to attribute reduction of consistent and inconsistent covering decision systems with covering rough sets. Inf. Sci. **177**(17), 3500–3518 (2007)
7. Wille, R.: Restructuring lattice theory: an approach based on hierarchies of concepts. Ordered Sets, 445–470(1982)
8. Ganter, B., Wille, R.: Formal Concept Analysis: Mathematical Foundations. Springer, Heidelberg (1999). https://doi.org/10.1007/978-3-642-59830-2
9. Zhang, W.-X., Wei, L., Qi, J.-J.: Attribute reduction in concept lattice based on discernibility matrix. In: Slezak, D., Yao, J.T., Peters, J.F., Ziarko, W., Hu, X. (eds.) RSFDGrC 2005. LNCS (LNAI), vol. 3642, pp. 157–165. Springer, Heidelberg (2005). https://doi.org/10.1007/11548706_17
10. Li, T., Wu, W.: Attribute reduction in formal contexts: a covering rough set approach. Fundamenta Informaticae **111**(1), 15–32 (2011)
11. Chen, J., Li, J., Lin, Y., Lin, G., Ma, Z.: Relations of reduction between covering generalized rough sets and concept lattices. Inf. Sci. **304**, 16–27 (2015)
12. Wu, W., Leung, Y., Mi, J.: Granular computing and knowledge reduction in formal context. IEEE Trans. Knowl. Data Eng. **21**(10), 1461–1474 (2009)
13. Liu, G.: Using covering reduction to identify reducts for object-oriented concept lattices. Axioms **11**, 1–14 (2022). https://doi.org/10.3390/axioms11080381
14. Yao, Y.: An outline of a theory of three-way decisions. In: Yao, J.T., Yang, Y., Słowiński, R., Greco, S., Li, H., Mitra, S., Polkowski, L. (eds.) RSCTC 2012. LNCS (LNAI), vol. 7413, pp. 1–17. Springer, Heidelberg (2012). https://doi.org/10.1007/978-3-642-32115-3_1
15. Zhang, X., Yao, Y.: Tri-level attribute reduction in rough set theory. Expert Syst. Appl. **190**, 116187 (2022)
16. Qi, J., Wei, L., Yao, Y.: Three-way formal concept analysis. In: Miao, D., Pedrycz, W., Ślęzak, D., Peters, G., Hu, Q., Wang, R. (eds.) RSKT 2014. LNCS (LNAI), vol. 8818, pp. 732–741. Springer, Cham (2014). https://doi.org/10.1007/978-3-319-11740-9_67

17. Yao, Y.: Interval sets and three-way concept analysis in incomplete contexts. Int. J. Mach. Learn. Cybern. **8**(1), 3–20 (2017)
18. Qi, J., Qian, T., Wei, L.: The connections between three-way and classical concept lattices. Knowl.-Based Syst. **91**, 143–151 (2016)
19. Ren, R., Wei, L.: The attribute reductions of three-way concept lattices. Knowl.-Based Syst. **99**, 92–102 (2016)
20. Wei, L., Cao, L., Qi, J., Zhang, W.: Concept reduction and concept characteristics in formal concept analysis. Sci. Sin Inf. **50**, 1817–1833 (2020). (in Chinese)

An Attention-Based Token Pruning Method for Vision Transformers

Kaicheng Luo[1], Huaxiong Li[1(✉)], Xianzhong Zhou[1], and Bing Huang[2]

[1] Department of Control Science and Intelligence Engineering,
Nanjing University, Nanjing, China
huaxiongli@nju.edu.cn

[2] School of Information Sciences, Nanjing Audit University, Nanjing, China

Abstract. Recently, vision transformers have achieved impressive success in computer vision tasks. Nevertheless, these models suffer from heavy computational cost for the quadratic complexity of the self-attention mechanism, especially when dealing with high-resolution images. Previous literature has illustrated the sparsity of attention, which suggests that uninformative tokens could be discarded to accelerate the model with limited influence to precision. As a natural indicator of token importance, attention scores can be intuitively used to extract the discriminative regions in images. Inspired by these facts, we propose an attention-based token pruning framework to address the issue of inefficiency for vision transformers. We divide the transformer blocks in the model into pruning stages, where the integrated weights in multi-attention heads are fused to estimate the importance of token. The computational cost of the model is reduced by dropping redundant patches progressively after each pruning stage. Experiments conducted on ImageNet1k verify the effectiveness of our method, where the models pruned by our module outperform other state-of-the-art models with similar FLOPs. For fine-grained image recognition, our framework also improves both accuracy and efficiency of ViT on CUB200-2011. More significantly, the proposed attention-based pruning module could be simply plugged in to any vision transformer that contains the class token by fine-tuning only 10 epochs or a single epoch, making a reasonable trade-off between accuracy and cost.

Keywords: Efficient transformer · Token pruning · Self-attention mechanism · Computer vision · Fine-grained visual classification

1 Introduction

Transformer, which has been a dominant architecture for natural language processing, brings phenomenal progress in computer vision [4,18]. Despite its impressive performance in various fields, the heavy computational and memory cost brought by the quadratic complexity of the self-attention module has

J. Yao et al. (Eds.): IJCRS 2022, LNAI 13633, pp. 274–288, 2022.
https://doi.org/10.1007/978-3-031-21244-4_21

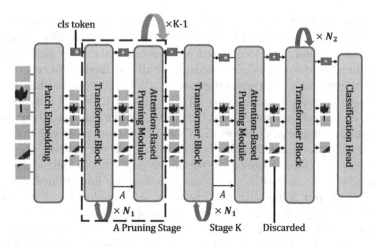

Fig. 1. Illustration of our attention-based token pruning process for Vision Transformer. The pruning modules locate and discard less important patches based on the attention weights from blocks in the same stage, reducing the number of token input to the following transformer layers.

always been a key challenge for ViT. As a consequence, acceleration for transformer is essential to its application on tasks with high-resolution images or long sequences.

Network pruning is an efficient approach to accelerate CNN-type architecture, which is based on the sparsity of channels [13,19]. Similarly, not all tokens are pivotal to the final prediction in vision transformer, which is illustrated by previous literatures about its interpretability [2]. It means abundant computational cost could be saved once those redundant tokens, mainly consisting of background or uninformative patches, are discarded. Apparently, the pruning strategy that determines which tokens to drop is decisive in this framework. Some research propose to use learnable sampler [24] or reinforcement learning [21] to determine discarded tokens. Although a number of these works achieve competitive results, the proposed modules rely on warm up tricks sensitive to hyper parameters and require 30–90 epochs training, which brings considerable excessive cost for tuning.

Raw attention itself shows impressive performance on extracting the informative patches of images. Fine-grained Visual Classification is a challenging task where the in-class difference is huge, raising a higher requirement for extracting discriminative regions. Since ViT is proposed, some researchers use its pretrained model and take advantage of the self-attention mechanism to find vital tokens [11,29]. These transformer-based methods outperform previous CNN-based models on the CUB200-2011 dataset, which indicates the strength of self-attention as a natural indicator of token importance.

Considering that attention weights achieve impressive results on finding informative parts of image, it's reasonable to assume that it has similar ability in locating redundant patches. Based on that assumption, we propose an attention-based token pruning module (ATP), which fuses the information in different attention heads to infer uninformative tokens and drop them. By adding the proposed module between transformer blocks, we divide layers into several stages and execute pruning operations progressively. We apply our module to mainstream vision transformers including ViT, DeiT and LVViT. Experiments are conducted on ImageNet1k and CUB200-2011. Our attention-based token pruning method reduces computational cost of LVViT by 26.0%~31.5% with 0.2%~0.4% loss on accuracy for ImageNet1k. Uninformative patches pruned, the method forces the model to focus more on the remaining important image patches, which brings improvement in both accuracy and efficiency for fine grained visual classification. With our framework, ViT achieves 0.3% higher top-1 accuracy on CUB200-2011 compared to the baseline, cost saved by 37.1%.

In order to verify the effectiveness of our attention-based module, we carry out experiments for different pruning strategies including random and structured sampling (pooling), the result of which shows superiority of the progressive attention-based selector for locating informative areas. In addition, we explore how the length of pruning stages influences the performance of different models. It's noticed that when we use pre-trained weights on ImageNet to fine-tune ViT and DeiT, longer pruning stage leads to solid performance. On the contrary, experiments conducted on LVViT suggest that the model performs better with shorter stages. Considering the differences between these models, we think the main reason is that LVViT uses token-labeling, which applies distillation to all tokens and forces raw attention weights to be better at reflecting the token importance. When it comes to ViT and DeiT, where tokens are not trained with soft labels, longer pruning stages yield integrated attention flow in more layers, which improves the decision quality for pruning. Another point is that both ViT and DeiT contain 12 layers while there are 16 blocks in LVViT-S and 20 blocks in LVViT-M. With the same final pruning rate, more layers in LVViT allow each module to discard a smaller ratio of tokens, which makes the pruning process smoother and lowers the demand for decision quality. Besides, we compare model accuracy under different resolutions, which suggests that the attention-based method loses less accuracy with high resolution input, the pruning ratio for token unchanged. We suppose it's owing to that higher resolution reduces the probability that target objects and uninformative backgrounds are in the same patch. It means some patches including the target, which might have been pruned inevitably under low resolution, could be reserved because areas are split more finely.

2   Related Work

In this section, we present a brief review of efficient transformers and Transformer-based methods for fine-grained image recognition.

2.1   Efficient Transformers

Considering that complexity of self-attention module is the main reason for inefficiency of transformers, there are two ways to improve its efficiency without changing the original framework: designing a new self-attention mechanism with lower complexity and reducing the number of tokens input to the module by pruning strategy.

Improvement of Self-Attention Mechanism. Assume that the input sequence length is T and hidden dimension of transformer block is D. The complexity of the self-attention module is $O(T^2 \cdot D)$ [27]. Researchers improve its efficiency in two ways, applying sparsity and linearization. The inner logic of Sparse Attention is that not every token needs to attend to all other tokens. For instance, Star Transformer [10] uses a combination of local attention and global attention, where only one token is attended to all other tokens. Longformer [1] improves it by applying global attention to more internal nodes. Additional stochastic attention is used in BigBrid [33] to approximate full attention. Theoretically, sparse attention could reduce computational cost by limiting the number of query-key pairs and attain comparable results to baseline. Nevertheless, most of these works are intended for natural language processing while few focus on computer vision [22]. Tt's also challenging to optimize the computation process of sparse matrices on GPU, which makes acceleration less than anticipated in real application. As for linearized attention [15], it usually leads to a larger gap with baseline in performance.

Pruning Strategy for Vision Transformers. Inspired by network slimming methods [13,19] for CNN, VTP [35] applies channel pruning to vision transformer in a simple way that focuses more on feed forward networks rather than self-attention mechanism. There are also some researchers who notice the sparsity of attention, trying to accelerate the model by dropping uninformative patches. DynamicViT [24] train a MLP token sampler via gambel-softmax trick. Pan et al. propose to train a multi-head interpreter by reinforcement learning and emphasize the interpretability of their module [21]. Considering that attention weights are highly correlated in continuous transformer layers, PSViT [32] improves the efficiency of ViT in a different way by reusing the attention calculation.

2.2   Vision Transformer for Fine-Grained Visual Classification

Fine grained visual classification is challenging for the subtle inter-class difference. Previous research, mainly based on CNN backbone, solves the problem by locating discriminative regions [17,31] or feature-encoding method [7,34]. Since vision transformers have achieved impressive performance on image recognition, some works try to replace the CNN backbone with ViT. For example, TransFG [11] integrates raw attention weights of transformer layers to select the most informative image patches. Similar works [29] use different methods to fuse the attention weights but most of them take advantage of the self-attention mechanism, which shows its strength in extracting informative tokens in image.

3   Method

In this section, we first illustrate the architecture of our method and then introduce the attention-based pruning module and the distillation loss for LVViT.

3.1   Overall Framework

Figure 1 illustrates the overall framework of our attention-based token pruning process. Layers in a transformer are split into K pruning stages, each of which contains N_1 transformer blocks and a attention-based token pruning module. Attention weights from the layers in a stage are input to the attention-based token pruning module to determine redundant patches to be discarded. After these pruning stages, there are N_2 transformer blocks conducting inference with low computational cost and memory use. This framework could be plugged into any vision transformer that contains a class token, such as ViT, DeiT and LVViT, simply by fine-tuning for a few epochs.

3.2   Attention-Based Token Pruning Module

The self-attention module in Vision Transformer itself offers a perfect way to evaluate the importance of tokens. With the potential of attention weights fully exploited, it could help discriminate the informative regions in an image, which is vital in the pruning process for Vision Transformer. Precise location of redundant tokens input to the next layer means less loss of accuracy while reducing the cost. Suppose there are N tokens input to the first layer in a pruning stage and H self-attention heads in the model. Tokens input to the last layer of a pruning stage are denoted as follows:

$$\mathbf{x}_{S-1} = \left[x_{S-1}^{cls}, x_{S-1}^1, ..., x_{S-1}^N\right] \in \mathbb{R}^{N \times D_k} \tag{1}$$

where S is the number of blocks in a stage and D_k is the dimension of keys. The attention weights of previous blocks in this stage can be written as:

$$a_l = \left[a_l^0, a_l^1, ..., a_l^{H-1}\right] \in \mathbb{R}^{H \times N \times N} (l \in 0, 1, ..., S-1) \tag{2}$$

where a_l^i refers to attention weights in head i. Given matrix representations of queries Q_i and keys K_i, it could be denoted as:

$$a_l^i = Softmax(\frac{Q_i K_i^T}{\sqrt{D_k}}) \in \mathbb{R}^{N \times N} (i \in 0, 1, ..., H-1) \tag{3}$$

Previous works suggested that the raw attention weights do not necessarily correspond to the relative importance of input tokens especially for higher layers of a model, due to lack of token identifiability of the embeddings. TransFG [11] uses an integrated attention weight of all layers except for the last one, which captures how information propagates from the input layer to the embeddings.

$$a_{final} = \prod_{l=0}^{L-1} a_l \in \mathbb{R}^{H \times N \times N} \tag{4}$$

However, it only chooses the tokens having maximum value in different attention heads as the input of the last layer. Apparently, it couldn't be used as our pruning strategy where tokens should be dropped earlier to reduce computational cost. There are only H tokens input to the next stage, which is not appropriate for losing too much information about target in the early stage. Instead of it, we use the mean attention score of cls token after integrating weights of layers in a stage to evaluate the current importance of tokens:

$$a_{stage}^i = \prod_{l=0}^{S-1} a_l^i \in \mathbb{R}^{N \times N} (i \in 0, 1, ..., H-1) \tag{5}$$

$$o_{stage} = \frac{1}{H} \sum_{i=1}^{H} a_{stage}^{i_{cls}} \in \mathbb{R}^{N \times N} \tag{6}$$

We select a percentage of p tokens that have the highest importance to be input to the next stage, which is to maximize the cumulative importance of tokens to be chosen:

$$\max_{\delta} \sum_{j=1}^{N} \delta_j o_{stage}^j$$

$$s.t. \sum_{j=1}^{N} \delta_j = Np \tag{7}$$

where $\delta_j = \begin{cases} 1 & x_{S-1}^j \text{ is input to the next stage} \\ 0 & x_{S-1}^j \text{ is not input to the next stage} \end{cases}$. It can be realized simply by sorting o_{stage} and choosing p_s percent tokens, concatenated along with the classification token to be input of the next stage. By pruning the tokens with less importance, which are mainly backgrounds or common features, we not only shrink the computational cost and memory use but also force the model to focus on more discriminative regions. Figure 1 illustrates an overall Framework of our attention-based pruning method. Each pruning stage consists of N_1 transformer blocks and one attention-based pruning module. We fix the percentage of each layer to be the same. Consequently, if we suppose the overall sparsity ratio to be P and there are S stages, we have:

$$p_1 = p_2 = ... = p_S = P^{\frac{1}{S}} \tag{8}$$

3.3   Distillation Loss for LVViT

Declining number of token input to transformer would inevitably influences the results model outputs, whether the difference with original output is large or small. To narrow the gap of prediction brought by the pruning process, we apply distillation loss to the LVViT model [14] in our experiments. The model uses soft labels generated from large scale CNN network in the original paper to boost its performance, which provides a perfect channel for knowledge distillation. Since our purpose is to minimize the degradation of accuracy rather than to boost it to another level, we simply use the original model as its teacher model. The training loss for a standard Vision Transformer and other baseline models is:

$$L = CrossEntropy(Softmax(X^{cls}), y^{cls}) \tag{9}$$

where y^{cls} is the ground truth and X^{cls} is the prediction of the model, which is only based on cls token. LVViT [14] utilizes the other tokens by adding an aux-head and generating soft labels for each token using the pretrained model in advance:

$$\begin{aligned} L =& CrossEntropy(Softmax(X^{cls}), y^{cls}) \\ &+\beta\frac{1}{m}\sum_{i=1}^{m}CrossEntropy(Softmax(X^i), y^i_{soft}) \end{aligned} \tag{10}$$

It should be noticed that with our pruning module, a part of the tokens are discarded in the pruning process, in which case m denotes the number of remaining tokens and is equal to NP. For experiments on ImageNet1k, we simply use original LVViT-S and LVViT-M model as Teacher model to generate $y^i_{soft}(i \in 1,...,m)$ and apply knowledge distillation:

$$\begin{aligned} L =& CrossEntropy(Softmax(X^{cls}), y^{cls}) \\ &+\beta\frac{1}{m}\sum_{i=1}^{m}CrossEntropy(Softmax(X^i), Softmax(X^i_{teacher})) \end{aligned} \tag{11}$$

As for experiments on CUB200-2011, we use ViT as baseline, which has achieved success on fine-grained image recognition by fine-tuning, to apply our pruning module. Since it is a transfer learning task where it's uncertain that original model is superior to model with pruning process, we don't add the distillation loss but use the standard cross entropy. Another reason is that ViT does not have a channel for distillation as good as that in LVViT. After experiments, it turns out that the model with our pruning module outperforms the original model, fine-tuned with the same hyper-parameters.

4   Experiments

4.1   Experiments on ImageNet1k

Experiments are mainly conducted on the ImageNet-1k [3]. We first apply our token pruning module on LVViT, a state of the art model that adopts

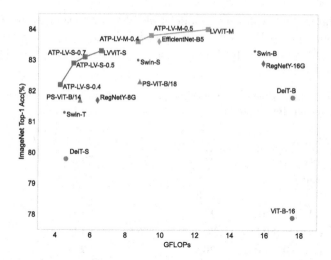

Fig. 2. Model computational cost (FLOPs) and top-1 accuracy on ImageNet1k. We compare our model, which is on the top of LVViT, with state-of-the-art CNN and transformer-like models. With our module, the scaled model achieves higher accuracy than those with similar FLOPs.

the architecture of ViT but becomes more efficient because of modifying and training methods.

Implementation Details. After testing different lengths of stage (result is shown in 4.3), we decide to apply a attention-based pruning module after each transformer block for LVViT. We load weights from the pretrained LVViT model and optimize the architecture for 10 epochs. We also find it work with only one-epoch training, in which case it does not drop too much accuracy(-0.1%). We train the models with 4 GPUs using a batch-size of 512 for ATP-LV-S and 384 for ATP-LV-M. Models are optimized by Adam [16] optimizer with cosine strategy [20], learning rate initialized to $\frac{batchsize}{1024} \times 0.00001$. The distillation parameter β is set to 0.5. We use $K = 15, N_1 = 1, N_2 = 1$ for ATP-LV-S and $K = 19, N_1 = 1, N_2 = 1$ for ATP-LV-M.

Comparison with State-of-the-Art Methods. In Table 1, we compare the performance of our attention-based pruning method on the top of LVViT with other state of the art models. The last number in the model refers to the percentage of patches pruned after entire pruning process. With 50% tokens discarded, ATP-LV-M-0.5 achieves 83.8% accuracy with 9.4 GFLOPs, which outperforms main stream CNN and Transformers including EfficientNet-B5 and Swin-B on accuracy with less FLOPs. To compare with lighter models, ATP-LV-S-0.5 reaches 82.9% accuracy with 5.0 GFLOPs, outperforming Swin-T and PS-ViT-B [32], which is another work to accelerate Vision Transformer by reusing attention calculation. Figure 2 shows model GFLOPs and accuracy on ImageNet1k, which illustrates the advantage of model scaled by our method more clearly. We also apply our approach to DeiT [26], a solid vision transformer baseline, by fine-tuning it for only one epoch and using $K = 3, N_1 = 3, N_2 = 3$. The results are shown in Table 2.

Table 1. Computational cost (FLOPs) and accuracy of our method and existing state of the art models on ImageNet1k.

Models	Params(M)	GFLOPs	Top-1 acc. (%)
DeiT-S [26]	22.1	4.6	79.8
Swin-T [18]	29.0	4.5	81.3
PS-ViT-B/14 [32]	21.3	5.4	81.7
LVViT-S [14]	26.2	6.6	83.3
ATP-LV-S-0.7	26.2	5.5	83.1
ATP-LV-S-0.5	**26.2**	**5.0**	**82.9**
ATP-LV-S-0.4	26.2	4.3	82.2
DeiT-B [26]	86.6	17.6	81.8
PS-ViT-B/18 [32]	21.3	8.8	82.3
RegNetY-16G [23]	84.0	16.0	82.9
Swin-S [18]	50.0	8.7	83.0
EfficientNet-B5 [25]	30.0	9.9	83.6
LVViT-M [14]	55.8	12.7	84.0
ATP-LV-M-0.5	55.8	**9.4**	**83.8**
ATP-LV-M-0.4	55.8	8.7	83.6

Table 2. Computational cost (FLOPs) and accuracy of DeiT using our framework on ImageNet1k.

Models	GFLOPs	Top1 acc. (%)	Top5 acc.
DeiT-B [26]	17.6	81.8	95.6
ATP-De-B-0.6	**13.8**	**81.2**	**95.2**
ATP-De-B-0.5	12.7	80.8	95.0

4.2 Experiments on CUB200-2011

We notice that experiments in fine grained image recognition usually use a high resolution like 448×448, which is a heavy burden for transformer because more tokens are generated, yielding tens of times more complexity for self-attention module than that of a 224×224 image. The results on CUB200-2011 [28] demonstrates that our token pruning method could even improve the accuracy while saving more computational cost because higher resolution leads to more patches where uninformative tokens could be separated more meticulously.

Table 3. Comparison of Our method on the top of ViT with previous state of the art models on CUB-200-2011.

Models	Backbone	Top-1 acc. (%)	GFLOPs
ResNet-50 [12]	ResNet-50	84.5	
GP-256 [30]	VGG-16	85.8	
MaxEnt [6]	DenseNet-161	86.6	
FDL [17]	DenseNet-161	89.1	
PMG [5]	ResNet-50	89.6	
API-Net [36]	DenseNet-161	90.0	
StackedLSTM [8]	GoogleNet	90.4	
ViT-B-16 [4]	ViT-B-16	90.6	79.3
ATP-ViT-B-0.4	**ViT-B-16**	**90.9**	**49.9 (−37.1%)**
ATP-ViT-B-0.3	ViT-B-16	90.8	44.2 (−44.3%)

Implementation Details. In experiments on CUB200-2011, We initialize the model weights by the ViT-B-16 model pre-trained on ImageNet21k. we optimize the model using SGD optimizer [9] with an initial learning rate of 0.03 and a momentum of 0.9. We also use cosine strategy [20] to adjust the learning rate with batch size 16. We use $K = 3, N_1 = 3, N_2 = 3$. The model is fine-tuned for 10,000 steps, where the first 500 steps are warm-up. The input images are resized to 600×600, which is transformed to 448×448 by random crop for training and centering crop for testing.

Comparison with State-of-the-Art Methods. The classification accuracies of CUB-200-2011 are summarized in Table 3. Almost all previous FGIR methods test their performance on this dataset. As is shown in Table 3, ViT itself achieves good performance with pre-trained weights. When we apply our pruning method to ViT, not only does it reduce the computational cost, it also slightly improves model accuracy. With our attention-based token pruning method, ViT achieves similar results in accuracy (+0.2%), reducing the GFLOPs by 25.8% compared to the baseline.

4.3   Ablation Studies and Discussions

Ablation Study for Pruning Strategies. To validate the effectiveness of our attention-based token pruning method, we conduct ablation experiments of different token selecting strategies. To guarantee fairness of comparison, we fix the cost of each strategy to 5.0 GFLOPs. We could learn from Table 4 that our attention-based method does have the advantage of locating informative regions.

Experiments for Length of Pruning Stages. We also discuss how the length of the pruning stage influences the performance of different vision transformers.

Table 4. Comparison of different token sampling strategies on ImageNet1k for LVViT-S.

Sampling strategy	GFLOPs	Top-1 acc. (%)	Top-5 acc. (%)
Original	6.6	83.3	96.4
Random	5.0	81.7	95.6
Structural (pooling)	5.0	82.2	95.8
ATP	**5.0**	**82.9**	**96.1**

As is mentioned in 3.1, the number of pruning stages is denoted as K. N_1 refers to the number of layers in a pruning stage and N_2 is the number of blocks after these pruning stages. N_2 should be at least 1 for two reasons. First, applying a pruning module after the last transformer block could not save computational cost because it won't be the input of any other transformer layer. Second, a typical vision transformer like ViT only uses the class token output by the last transformer block for classification, in which case pruning the other tokens would be meaningless. To avoid extra parameter tuning and focus on the method itself, we just simply keep $N_1 = N_2$ for the experiments. For LVViT-S which contains 16 transformer layers, We use $K = 3, N_1 = 4, N_2 = 4$ and $K = 15, N_1 = 1, N_2 = 1$ respectively to compare models having different lengths of pruning stages. The results indicate that applying the pruning module after each transformer layer except for the last one ($N_2 >= 1$) is better for LVViT. As for DeiT, we conduct experiments using $K = 3, N_1 = 3, N_2 = 3$ and $K = 11, N_1 = 1, N_2 = 1$ because it only has 12 blocks. The situation in experiments for DeiT is quite different, which suggests it benefits from longer pruning stages. In these experiments, the FLOPs of pruned models are fixed to be close for fair comparison. We suppose it's because LVViT has more layers leading to lower pruning ratio in early stages and an aux-head for tokens to apply distillation, which makes raw attention better on reflecting the importance of tokens (Table 5).

Table 5. Comparison of different pruning stages on ImageNet1k for LVViT-S.

Model	K	N_1	N_2	GFLOPs	Top-1 acc. (%)
LVViT-S	–	–	–	6.6	83.3
ATP-LV-S-0.5	3	4	4	5.1	82.7
ATP-LV-S-0.5	15	1	1	5.0	**82.9**
DeiT-B	–	–	–	17.6	81.8
ATP-De-B-0.5	3	3	3	12.7	**80.8**
ATP-De-B-0.5	11	1	1	12.6	80.4

Experiments Under Different Resolutions. During the experiments, We notice that higher resolution could enhance the performance of ATP to another level. Table 6 shows that the results of ViT and our model on CUB-200-2011 using different image resolutions. As can be seen in Table 6, our model reaches similar results than that of ViT when trained with resolutions of 224×224 and 336×336 as well. We also notice that accuracy drops very little (0.2%) when resolution declines from 448 × 448 to 336 × 336. However, it yields only 26.7 GFLOPs when it takes 79.3 GFLOPs for the baseline to get a similar result (90.6%).

Table 6. Comparison of our method on the top of ViT with the baseline on CUB200-2011 using different resolutions.

Resolutions	224×224		336×336		448×448	
Models	Acc.(%)	GFLOPs	Acc.(%)	GFLOPs	Acc.(%)	GFLOPs
ViT-B-16	88.7	17.6	90.0	41.7	90.6	79.3
ATP-ViT-B-0.4	88.6	11.4	**90.7**	**26.7**	**90.9**	**49.9**

The situation is similar in experiments on ImageNet. We compare ATP-De-B with the baseline under different resolutions. The gap between pretrained and pruned models is narrowed to a great extent (Table 7).

Table 7. Comparison of our method on the top of DeiT with the baseline on ImageNet1k using different resolutions.

Resolutions	224×224		384×384	
Models	Top-1 acc. (%)	GFLOPs	Top-1 acc. (%)	GFLOPs
DeiT-B	81.8	17.6	82.8	55.9
ATP-De-B-0.5	80.8	12.7	**82.4**	**39.8**

4.4 Visualization

The attention-based pruning method also has a bonus for its natural interpretability. We visualize our pruning process by recording the decisions ATP makes while testing. Figure 3 demonstrates the process tokens are progressively pruned by modules after transformer blocks, ATP-LV-S-0.5 as an example. Although in deeper layers, information in different tokens has been fused extensively, the visualization still gives a clear sketch of the process redundant patches are gradually discarded.

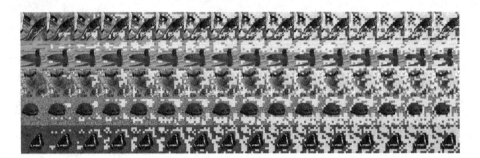

Fig. 3. Visualization of pruned tokens in each stage of our ATP-LV-S. We show original images and visualize the process that less discriminative tokens are progressively pruned by attention-based token pruning module.

5   Conclusion

In this paper we propose an efficient attention-based token pruning (ATP) framework for Vision Transformer. By integrating attention weights in transformer blocks and fusing the information in different attention heads, our ATP module locates the uninformative patches, which are dropped in each pruning stage. Redundant tokens gradually eliminated, our method reduces the computational cost in the deeper transformer layers while sacrificing negligible accuracy, yielding much less extra cost for tuning compared with methods using learnable predictors. It also performs better under higher resolution where the uninformative patches are separated more finely. It could also improve the accuracy of ViT on fine-grained recognition datasets like CUB200-2011, while significantly shrinking the computational cost and memory use.

Acknowledgement. This work was partially supported by the National Natural Science Foundation of China (Nos. 62176116, 62276136, 61876079, 71732003), and the Natural Science Foundation of the Jiangsu Higher Education Institutions of China, No. 20KJA520006.

References

1. Beltagy, I., Peters, M.E., Cohan, A.: Longformer: the long-document transformer. arXiv preprint arXiv:2004.05150 (2020)
2. Chefer, H., Gur, S., Wolf, L.: Transformer interpretability beyond attention visualization. In: Proceedings of the IEEE/CVF Conference on Computer Vision and Pattern Recognition, pp. 782–791 (2021)
3. Deng, J., Dong, W., Socher, R., Li, L.J., Li, K., Fei-Fei, L.: Imagenet: a large-scale hierarchical image database. In: 2009 IEEE Conference on Computer Vision and Pattern Recognition, pp. 248–255. IEEE (2009)
4. Dosovitskiy, A., et al.: An image is worth 16×16 words: transformers for image recognition at scale. arXiv preprint arXiv:2010.11929 (2020)

5. Du, R., et al.: Fine-grained visual classification via progressive multi-granularity training of jigsaw patches. In: Vedaldi, A., Bischof, H., Brox, T., Frahm, J.-M. (eds.) ECCV 2020. LNCS, vol. 12365, pp. 153–168. Springer, Cham (2020). https://doi.org/10.1007/978-3-030-58565-5_10
6. Dubey, A., Gupta, O., Raskar, R., Naik, N.: Maximum-entropy fine grained classification. Adv. Neural Inf. Process. Syst. **31** (2018)
7. Gao, Y., Han, X., Wang, X., Huang, W., Scott, M.: Channel interaction networks for fine-grained image categorization. In: Proceedings of the AAAI Conference on Artificial Intelligence, vol. 34, pp. 10818–10825 (2020)
8. Ge, W., Lin, X., Yu, Y.: Weakly supervised complementary parts models for fine-grained image classification from the bottom up. In: Proceedings of the IEEE/CVF Conference on Computer Vision and Pattern Recognition, pp. 3034–3043 (2019)
9. Goyal, P., et al.: Accurate, large minibatch SGD: training imagenet in 1 hour. arXiv preprint arXiv:1706.02677 (2017)
10. Guo, Q., Qiu, X., Liu, P., Shao, Y., Xue, X., Zhang, Z.: Star-transformer. arXiv preprint arXiv:1902.09113 (2019)
11. He, J., et al.: Transfg: a transformer architecture for fine-grained recognition. arXiv preprint arXiv:2103.07976 (2021)
12. He, K., Zhang, X., Ren, S., Sun, J.: Deep residual learning for image recognition. In: Proceedings of the IEEE Conference on Computer Vision and Pattern Recognition, pp. 770–778 (2016)
13. He, Y., Zhang, X., Sun, J.: Channel pruning for accelerating very deep neural networks. In: Proceedings of the IEEE International Conference on Computer Vision, pp. 1389–1397 (2017)
14. Jiang, Z., et al.: Token labeling: Training a 85.4% top-1 accuracy vision transformer with 56m parameters on imagenet. arXiv e-prints pp. arXiv-2104 (2021)
15. Katharopoulos, A., Vyas, A., Pappas, N., Fleuret, F.: Transformers are rnns: fast autoregressive transformers with linear attention. In: International Conference on Machine Learning, pp. 5156–5165. PMLR (2020)
16. Kingma, D.P., Ba, J.: Adam: a method for stochastic optimization. arXiv preprint arXiv:1412.6980 (2014)
17. Liu, C., Xie, H., Zha, Z.J., Ma, L., Yu, L., Zhang, Y.: Filtration and distillation: enhancing region attention for fine-grained visual categorization. In: Proceedings of the AAAI Conference on Artificial Intelligence, vol. 34, pp. 11555–11562 (2020)
18. Liu, Z., et al.: Swin transformer: hierarchical vision transformer using shifted windows. In: Proceedings of the IEEE/CVF International Conference on Computer Vision, pp. 10012–10022 (2021)
19. Liu, Z., Li, J., Shen, Z., Huang, G., Yan, S., Zhang, C.: Learning efficient convolutional networks through network slimming. In: Proceedings of the IEEE International Conference on Computer Vision, pp. 2736–2744 (2017)
20. Loshchilov, I., Hutter, F.: Sgdr: stochastic gradient descent with warm restarts. arXiv preprint arXiv:1608.03983 (2016)
21. Pan, B., Panda, R., Jiang, Y., Wang, Z., Feris, R., Oliva, A.: Ia-red: interpretability-aware redundancy reduction for vision transformers. Adv. Neural Inf. Process. Syst. **34** (2021)
22. Parmar, N., et al.: Image transformer. In: International Conference on Machine Learning, pp. 4055–4064. PMLR (2018)
23. Radosavovic, I., Kosaraju, R.P., Girshick, R., He, K., Dollár, P.: Designing network design spaces. In: Proceedings of the IEEE/CVF Conference on Computer Vision and Pattern Recognition, pp. 10428–10436 (2020)

24. Rao, Y., Zhao, W., Liu, B., Lu, J., Zhou, J., Hsieh, C.J.: Dynamicvit: efficient vision transformers with dynamic token sparsification. Adv. Neural Inf. Process. Syst. **34** (2021)
25. Tan, M., Le, Q.: Efficientnet: rethinking model scaling for convolutional neural networks. In: International Conference on Machine Learning, pp. 6105–6114. PMLR (2019)
26. Touvron, H., Cord, M., Douze, M., Massa, F., Sablayrolles, A., Jégou, H.: Training data-efficient image transformers & distillation through attention. In: International Conference on Machine Learning, pp. 10347–10357. PMLR (2021)
27. Vaswani, A., Set al.: Attention is all you need. Adv. Neural Inf. Process. Syst. **30** (2017)
28. Wah, C., Branson, S., Welinder, P., Perona, P., Belongie, S.: The caltech-ucsd birds-200-2011 dataset (2011)
29. Wang, J., Yu, X., Gao, Y.: Feature fusion vision transformer for fine-grained visual categorization. arXiv preprint arXiv:2107.02341 (2021)
30. Wei, X., Zhang, Y., Gong, Y., Zhang, J., Zheng, N.: Grassmann pooling as compact homogeneous bilinear pooling for fine-grained visual classification. In: Proceedings of the European Conference on Computer Vision (ECCV), pp. 355–370 (2018)
31. Yang, S., Liu, S., Yang, C., Wang, C.: Re-rank coarse classification with local region enhanced features for fine-grained image recognition. arXiv preprint arXiv:2102.09875 (2021)
32. Yue, X., et al.: Vision transformer with progressive sampling. In: Proceedings of the IEEE/CVF International Conference on Computer Vision, pp. 387–396 (2021)
33. Zaheer, M., et al.: Big bird: transformers for longer sequences. Adv. Neural Inf. Process. Syst. **33**, 17283–17297 (2020)
34. Zheng, H., Fu, J., Zha, Z.J., Luo, J.: Learning deep bilinear transformation for fine-grained image representation. Adv. Neural Inf. Process. Syst. **32** (2019)
35. Zhu, M., Han, K., Tang, Y., Wang, Y.: Visual transformer pruning. arXiv e-prints pp. arXiv-2104 (2021)
36. Zhuang, P., Wang, Y., Qiao, Y.: Learning attentive pairwise interaction for fine-grained classification. In: Proceedings of the AAAI Conference on Artificial Intelligence, vol. 34, pp. 13130–13137 (2020)

Three-Way Approximate Criterion Reduction in Multi-Criteria Decision Analysis

Chengjun Shi[1]([✉])[ID], Zhen Wang[1,2,3][ID], Ling Wei[2,3], and Yiyu Yao[1][ID]

[1] Department of Computer Science, University of Regina,
Regina, SK S4S 0A2, Canada
{csn838,Yiyu.Yao}@uregina.ca
[2] School of Mathematics, Northwest University,
Xi'an 710127, Shaanxi, People's Republic of China
wl@nwu.edu.cn
[3] Institute of Concepts, Cognition and Intelligence, Northwest University,
Xi'an 710127, Shaanxi, People's Republic of China

Abstract. Attribute reduction is one of the major topics in rough set theory. The purpose of attribute reduction is to reduce the dimensionality of data. In multi-criteria decision analysis, criteria are treated as multiple and possibly conflicting points of views on decision alternatives. To reduce the complexity of multi-criteria decision analysis, we raise the problem of criterion reduction. In this paper, we propose the formal definition of criterion reduction and develop a heuristic method to deal with it. At first, we review the definitions of attribute reduction in rough set theory, generalized attribute reduction, and approximate attribute reduction. Then, we discuss the problem of criterion reduction in multi-criteria decision analysis. More specifically, we introduce three-way decision theory and define three-way approximate criterion reducts via a pair of thresholds. Finally, we adopt the point-wise loss function and propose heuristic algorithms to generate three-way approximate criterion reducts. A real-world data set of city rankings is used to validate the proposed method.

Keywords: Rough sets · Attribute reduction · Multi-criteria decision analysis · Criterion reduction · Three-way decision

1 Introduction

In the field of human decision-making problems, a great number of the problems are discussed under an environment of multiple criteria. This significant subset of decision-making problems is called multi-criteria decision-making (MCDM), or multi-criteria decision analysis (MCDA). The investigations of MCDA deal

This work was partially supported by a Discovery Grant from NSERC, Canada. The authors thank reviewers for their constructive comments.

J. Yao et al. (Eds.): IJCRS 2022, LNAI 13633, pp. 289–303, 2022.
https://doi.org/10.1007/978-3-031-21244-4_22

with ranking or sorting a set of decision alternatives by giving a set of possibly conflicting criteria that are mainly related to the descriptions and evaluations of alternatives. The final ranking is generated according to their performances under all criteria [4]. In real-life MCDM problems, we observe that not all criteria contribute equally to the final ranking. Selecting a subset of criteria may produce a different ranking which might be highly close to the original results. In other words, removing certain criteria from the given set may have a marginal effect on the consequence. Furthermore, fewer criteria involved in ranking can significantly reduce decision complexity and cost.

Attribute reduction, as a typical subject in rough set theory and three-way decision theory [3,5,6,8,9,12–14,20,21], is an outstanding technique for feature selection. The objective of attribute reduction in rough sets is to reduce the number of attributes or eliminate certain attributes while the partition over the universe is preserved. One may also view the essential task of attribute reduction as finding out a minimal subset of attributes that is sufficient to have the same classification ability. In a general scenario, an attribute reduct is defined as a subset of attributes that is able to preserve a particular property and must satisfy the sufficiency condition and minimization condition [5,14,23]. With a view to the tolerance of a certain quantity of errors, Ślęzak [10,11] extended rough set theory with information entropy and introduced the Approximate Entropy Reduction Principle (AERP). Fang and Min [2] proposed a framework based on three-way decision to resolve the cost-sensitive approximate attribute reduction problem. Gao et al. [3] presented a method of attribute reduction based on information-theoretic measure by following the core ideas in three-way decision [16,17].

By considering the advances in attribute reduction, we investigate the reduction problem of criteria in MCDA. Greco et al. [4] brought us an insightful observation on the nature of human decision-making, that is, "decision is strongly related to the comparison of different points of view". Their observation reveals another important point: the way, in which humans make a decision, is amalgamating the multidimensional aspects into a single scale of measure. The fundamental thoughts of MCDM have been applied by Zhang and Yao [22] on three-way classifications with game-theoretic rough sets. Most MCDM techniques deal with the decision problems in a closely related way by defining an objective function as a comprehensive view. Then a set of alternatives will be ranked according to the specific function. The motivation of our work is to have a new look at the way to eliminate certain views from multiple points of view or to reduce the set of multiple criteria. It is intuitive and natural for decision-makers and MCDA researchers to take into account fewer points of view, in MCDA, fewer criteria. Despite that, a reduct of criteria will cost fewer computations in deriving the comprehensive function and lead the MCDM models more efficient.

In fact, it might be highly difficult to find a perfect reduct that produces an exactly same ranking as the one produced by full criteria, especially in a large dimensional data set. Three-way decision theory [16–19] gives us a hint to construct three-way approximate criterion reducts that come with two different levels of preservation of ranking. The contributions of this paper consist of three parts. Firstly, we review the concepts and definitions of attribute reduction in rough set theory including the generalized attribute reduction [5,23] and approximate attribute reduction [2,3]. Consequently, we investigate the property of ranking problems and present some formal formulations and definitions of three-way approximate criterion reduction. Lastly, we develop heuristic algorithms to generate three-way approximate criterion reducts.

This paper is organized as follows. In Sect. 2, we review the basic concepts of attribute reduction. In Sect. 3, we discuss the ranking property of MCDM problems and propose the definition of three-way approximate criterion reduct. We also briefly explain the related issues through some relatively small examples. In Sect. 4, we design heuristic algorithms and present the experiment results and comparative analysis. In Sect. 5, we summarize the key points of three-way approximate criterion reduction and point out the future research interests.

2   A Review of Attribute Reduction in Rough Set Theory

In this section, we examine some basic notions of a decision table and concepts of attribute reduction in rough sets. In addition, we discuss the generalized attribute reduction and evaluation-based approximate attribute reduction.

At first, we take a number of concepts from Pawlak's rough sets [8]. Suppose T is an information table, T is made up of a finite nonempty set of objects, a finite nonempty set of attributes, and descriptions of the objects by attributes. The descriptions are obtained by perceiving, observing, and measuring the objects according to the finite set of attributes [23]. The mathematical form of an information table is:

$$T = \{U, AT, \{V_a \mid a \in AT\}, \{I_a \mid a \in AT\}\}. \tag{1}$$

With respect to a subset of attributes $A \subseteq AT$, an indiscernibility relation on the universe U is defined by

$$IND(A) = \{(x, y) \in U \times U \mid \forall a \in A, I_a(x) = I_a(y)\}. \tag{2}$$

The attribute reducts in an information table are defined as follows.

Definition 1 [15]. *Suppose $T = \{U, AT, V, I\}$ is a given information table, a subset of attributes $R \subseteq AT$ is called a reduct, if and only if, the two conditions below are satisfied:*

1. (Sufficiency): $IND(R) = IND(AT)$;
2. (Minimization): $\forall R' \subsetneq R, IND(R') \neq IND(AT)$.

In order to explore a generalized reduct, Zhao et al. [24] interpreted attribute reduction in terms of the properties of a decision table. According to Yao [15], the definition of attribute reduct in rough sets is elucidated in two aspects, namely, a conceptual aspect and a computational aspect. Jia et al. [5] proposed a formal definition of a generalized reduct. In light of their findings, we generally denote the result produced by a set of attributes as $Res(AT)$. For any subset $R \subseteq AT$, $Res(R)$ returns the result which is produced by R. By comparing the results, we can define a general attribute reduct as follows.

Definition 2. *Suppose AT is a given set of attributes, a subset $R \subseteq AT$ is called a reduct, if and only if, R satisfies the following conditions:*

1. (Sufficiency): $Res(R) \triangleq Res(AT)$,
2. (Minimization): $\forall R' \subsetneq R$, $\neg(Res(R') \triangleq Res(AT))$,

where $\triangleq$ means that the two involved results are equivalent to each other.

The sufficiency condition in Definition 2 requires that the two results are quantitatively or qualitatively equivalent. It simply argues that a reduct R must produce the same result as the one produced by using the whole set AT. The second condition ensures that R is the minimum subset satisfying the sufficiency condition. In a general decision-making process, some errors might be allowed. The users may accept a decision result with some tolerance. An approximate reduct is defined by measuring the similarity between $Res(R)$ and $Res(AT)$. In this way, one may view a highly similar result as an acceptable result.

Definition 3. *Suppose AT is a given set of attributes, a subset $R \subseteq AT$ is called an α-approximate reduct, if and only if, R satisfies the following conditions:*

1. (Sufficiency): $Sim(Res(R), Res(AT)) \geqslant \alpha$,
2. (Minimization): $\forall R' \subsetneq R$, $Sim(Res(R'), Res(AT)) < \alpha$,

where Sim is a mapping $\{Res(R) \times Res(AT) \mid R \subseteq AT\} \longrightarrow \mathcal{R}$.

We introduce a threshold α in Definition 3 to determine whether a result is acceptable or not. Basically, the Sim function measures the similarity between $Res(R)$ and $Res(AT)$ and is usually given. A higher value of Sim indicates a higher similarity, that is, $Res(R)$ is closer to $Res(AT)$ and more acceptable from the users' viewpoint.

3   Approximate Criterion Reduction from Three-Way Decision Perspectives

After recalling the concepts of attribute reduction, we now discuss the criterion reduction problem in this section. For more details, we discuss two main difficulties in criterion reduction. Then we propose the definitions of criterion reduction and three-way approximate criterion reduction through a number of assumptions.

3.1 Problem Statement of Criterion Reduction

The analysis of MCDM is established on the basis of a specific table called a multi-criteria decision-making table.

Definition 4. *A multi-criteria decision-making table* (MCDMT) *is a triplet* $\langle A, C, p \rangle$, *where* $A = \{a_1, \ldots, a_n\}$ *is a finite and non-empty set of* n *alternatives,* $C = \{c_1, \ldots, c_m\}$ *is a finite and non-empty set of* m *criteria, and* $p : A \times C \longrightarrow V$ *is function that maps a decision alternative* a_i *and a criterion* c_j *into a value* $p(a_i, c_j) = p_j(a_i) \in V$.

Assumptions:

1. The domain of a criterion $c_j \in C$ is completely pre-ordered.
2. All of the criteria are presented in numerical scales and the ascending or descending orderings are given.

Example 1. Table 1 is part of Movehub City Rankings[1]. It consists of fifteen decision alternatives $A = \{a_1, a_2, \ldots, a_{15}\}$ that represent fifteen different cities. To avoid personal preferences, we implicitly present the cities by symbols a_i instead of their real names. The set of criteria contains six criteria, denoted by

Table 1. An MCDMT

Alternatives	Criteria					
	c_1	c_2	c_3	c_4	c_5	c_6
a_1	70.46	19.07	51.01	86.16	31.87	76.45
a_2	81.89	49.70	82.86	34.31	76.77	24.22
a_3	82.43	54.30	75.00	85.59	60.28	21.35
a_4	65.18	11.25	44.44	83.45	8.61	85.70
a_5	78.12	32.91	67.49	78.07	43.89	33.22
a_6	71.91	22.91	59.55	30.55	40.51	44.53
a_7	80.74	51.24	84.85	18.40	83.76	16.67
a_8	84.52	80.72	36.66	82.08	77.13	30.21
a_9	86.00	63.28	88.43	43.08	90.08	15.34
a_{10}	76.16	33.69	61.67	68.21	57.01	18.18
a_{11}	90.45	50.13	41.12	30.54	65.27	48.31
a_{12}	78.84	44.03	72.69	29.86	52.08	46.59
a_{13}	81.44	52.91	79.63	68.93	80.87	42.45
a_{14}	79.63	49.51	75.28	6.78	78.52	32.08
a_{15}	83.31	68.77	54.17	38.64	65.53	68.58

---

[1] BLITZER posted the data set on Kaggle in 2017 (Version 1: https://www.kaggle.com/datasets/blitzr/movehub-city-rankings), as published by MoveHub (https://www.movehub.com/city-rankings).

$C = \{c_1, c_2, c_3, c_4, c_5, c_6\}$. They are, respectively, "Movehub Rating", "Purchase Power", "Health Care", "Pollution", "Quality of Life", and "Crime Rating". All of the six criteria are in numerical scales and are in the value domain $[0, 100]$. Among them, orderings under c_1, c_2, c_3, and c_5 are ascending, that is, a higher value is better. For example, city a_9 with 90.08 points under c_5 has the best quality of life. While, orderings under c_4 and c_6 are descending, that is, a lower value is better. City a_9 with 15.34 points under c_6 has the best crime rating.

An MCDM method helps us comprehensively rank the alternatives with respect to their performances under the selected criteria. Our purpose of criterion reduction is to preserve the ranking by using as few criteria as possible. In such an MCDMT, we have the following two observations, and they are the difficulties that we have to deal with when constructing criterion reducts.

Difficulty 1. The definition of the property in MCDA is not clear. To describe the ranking property in an MCDMT, we recall the classical attribute reduction in rough sets. An indiscernibility relation forms a partition over the universe and the result is described by the partition. In an MCDMT, our task is to preserve the ability to rank alternatives. The desired result can be represented by the final ranking produced by all criteria. The preservation of the result can be correspondingly explained as producing the same ranking with a subset of criteria.

Example 2. We take alternatives a_1, a_2, a_3 and the method of Weighted Sum Model as an example. For simplicity, we consider that all of the six criteria are equally weighted. Given with the MCDMT, c_1, c_2, c_3, and c_5 are ascending, c_4 and c_6 are descending. The scores for the three alternatives are computed by:

$$Score(a_1) = p_1(a_1) + p_2(a_1) + p_3(a_1) - p_4(a_1) + p_5(a_1) - p_6(a_1) = 9.80,$$
$$Score(a_2) = p_1(a_2) + p_2(a_2) + p_3(a_2) - p_4(a_2) + p_5(a_2) - p_6(a_2) = 232.60,$$
$$Score(a_3) = p_1(a_3) + p_2(a_3) + p_3(a_3) - p_4(a_3) + p_5(a_3) - p_6(a_3) = 165.07.$$

The ranking is $a_2 \succ a_3 \succ a_1$. If we only consider c_1, c_2, and c_3, the new scores are computed by:

$$Score(a_1) = p_1(a_1) + p_2(a_1) + p_3(a_1) = 140.54,$$
$$Score(a_2) = p_1(a_2) + p_2(a_2) + p_3(a_2) = 214.45,$$
$$Score(a_3) = p_1(a_3) + p_2(a_3) + p_3(a_3) = 211.73.$$

The new ranking is $a_2 \succ a_3 \succ a_1$ that is exactly the same as the previous result. When we rank these three alternatives, we can say that the subset $\{c_1, c_2, c_3\}$ produces the same result.

Difficulty 2. The ranking property may not be monotonic. It is difficult to determine whether a single criterion is dispensable or indispensable. We discuss one possible situation to show the non-monotonicity. Suppose function $\mathcal{F}(C)$ returns the ranking produced by using the entire set of criteria C, a possible situation is that $\mathcal{F}(C - \{c_1\}) \neq \mathcal{F}(C)$ or $\mathcal{F}(C - \{c_2\}) \neq \mathcal{F}(C)$, but $\mathcal{F}(C - \{c_1, c_2\}) = \mathcal{F}(C)$. If we only remove a single criterion either c_1 or c_2, we cannot derive the same ranking. However, we may obtain the same ranking by removing c_1 and c_2 simultaneously. This situation disables us to say c_1 is dispensable or c_2 is dispensable, individually. This difficulty makes it insufficient to reduce a set $R \subseteq C$ by only exploring $\forall c \in R, R - \{c\}$. Therefore, it is necessary to explore all of the possible subsets of R if the target is to figure out the optimal/minimal reduct.

Example 3. From Table 1, we select a_{12}, a_{13}, and a_{14} and we still use Weighted Sum Model to rank them. The results are shown as follows.

$Score(a_{12}) = p_1(a_{12}) + p_2(a_{12}) + p_3(a_{12}) - p_4(a_{12}) + p_5(a_{12}) - p_6(a_{12}) = 171.19,$
$Score(a_{13}) = p_1(a_{13}) + p_2(a_{13}) + p_3(a_{13}) - p_4(a_{13}) + p_5(a_{13}) - p_6(a_{13}) = 183.47,$
$Score(a_{14}) = p_1(a_{14}) + p_2(a_{14}) + p_3(a_{14}) - p_4(a_{14}) + p_5(a_{14}) - p_6(a_{14}) = 244.08.$

The final ranking is $\mathcal{F}(C) = a_{14} \succ a_{13} \succ a_{12}$. Then, we test the following three cases:

1. Remove criterion c_4, the ranking is

$$\mathcal{F}(C - \{c_4\}) = a_{13} \succ a_{14} \succ a_{12}.$$

2. Remove criterion c_5, the ranking is

$$\mathcal{F}(C - \{c_5\}) = a_{14} \succ a_{12} \succ a_{13}.$$

3. Remove criteria c_4 and c_5, the ranking is

$$\mathcal{F}(C - \{c_4, c_5\}) = a_{14} \succ a_{13} \succ a_{12}.$$

Removing either c_4 or c_5 individually doesn't produce the same ranking with $\mathcal{F}(C)$, however, removing them together will provide us with a same ranking $\mathcal{F}(C - \{c_4, c_5\}) = \mathcal{F}(C)$. The non-monotonicity implies that we cannot simply examine the individual necessity of each criterion. We have to explore all possible subsets of criteria to find out an optimal reduct.

3.2 Definitions of Criterion Reduct and Three-Way Approximate Criterion Reducts

The criterion reduction can be defined by rewriting Definition 3 in terms of multiple-criteria decision-making. The rankings $\mathcal{F}(R)$ and $\mathcal{F}(C)$ can be viewed as the results $Res(R)$ and $Res(C)$ correspondingly.

Definition 5. *Given an MCDMT, a subset $R \subseteq C$ is called an α-approximate criterion reduct, if and only if, the following conditions are satisfied:*

1. (Sufficiency): $Sim(\mathcal{F}(R), \mathcal{F}(C)) \geqslant \alpha$;
2. (Minimization): $\forall R' \subsetneq R, Sim(\mathcal{F}(R'), \mathcal{F}(C)) < \alpha$.

There have been a lot of well-developed measurements in MCDA. They can be adopted as alternatives for the similarity function in Definition 5. One of the feasible ways is the normalized distance-based performance measure (NDPM) which is widely used in information retrieval. Some other metrics in MCDA include Spearman's rank correlation coefficient, Kendall's τ coefficient, Pearson's correlation coefficient, Mean Average Precision (MAP), and Normalized Discounted Cumulative Gain (NDCG). In learning to rank, point-wise, pair-wise, and list-wise loss functions are involved in comparing two rankings. More specifically in criterion reduction, we concentrate on measuring the similarity between $\mathcal{F}(R)$ and $\mathcal{F}(C)$. For example, the NDPM measures the distance between $\mathcal{F}(R)$ and $\mathcal{F}(C)$, the correlation coefficients can reflect how close $\mathcal{F}(R)$ is to $\mathcal{F}(C)$, the loss functions can be adopted to quantify the essential loss of $\mathcal{F}(R)$ compared to $\mathcal{F}(C)$.

By considering the difficulties mentioned before, $Sim(\mathcal{F}(R), \mathcal{F}(C))$ doesn't necessarily satisfy the monotonicity. The similarity $Sim(\mathcal{F}(R), \mathcal{F}(C))$ may decrease when we add one criterion or some criteria to R, on the other hand, the value of $Sim(\mathcal{F}(R), \mathcal{F}(C))$ may increase when we remove a criterion or some criteria from R.

Example 4. We take Table 1 as an example to illustrate the idea of criterion reduction. Initially, we generate the ranking by using TOPSIS with respect to the whole set of criteria C. For simplicity, we suppose that all of the six criteria are equally weighted. The ranking $\mathcal{F}(C)$ is shown as follows:

$$a_9 \succ a_7 \succ a_2 \succ a_{14} \succ a_{13} \succ a_{12} \succ a_3 \succ a_{11} \succ a_{15} \succ a_8 \succ a_{10} \succ a_5 \succ a_6 \succ a_1 \succ a_4.$$

We simply use Spearman's ranking correlation coefficient as a similarity measurement. The result of Spearman's correlation coefficient is a real number belonging to $[-1, +1]$. The value $+1$ means that the two rankings are exactly the same, while the value -1 means that the two rankings are exactly opposite. In this small example, we set the threshold $\alpha = 0.95$ and explore all possible subsets of C. Finally, we obtain three α-approximate reducts:

$$R_1 = \{c_1, c_2, c_3\}, \quad Sim(\mathcal{F}(R_1), \mathcal{F}(C)) = 0.9571,$$
$$R_2 = \{c_1, c_3, c_5\}, \quad Sim(\mathcal{F}(R_2), \mathcal{F}(C)) = 0.9857,$$
$$R_3 = \{c_2, c_3, c_5\}, \quad Sim(\mathcal{F}(R_3), \mathcal{F}(C)) = 0.9536.$$

The rankings produced by the whole set C and the three approximate reducts are listed in Table 2. The results, plotted in Fig. 1, demonstrate that the criterion

Table 2. The rankings produced by using C, R_1, R_2, and R_3

Rankings	
$\mathcal{F}(C)$	$a_9 \succ a_7 \succ a_2 \succ a_{14} \succ a_{13} \succ a_{12} \succ a_3 \succ a_{11} \succ a_{15} \succ a_8 \succ a_{10} \succ a_5 \succ a_6 \succ a_1 \succ a_4$
$\mathcal{F}(\{c_1, c_2, c_3\})$	$a_9 \succ a_7 \succ a_2 \succ a_{13} \succ a_3 \succ a_{14} \succ a_{15} \succ a_{12} \succ a_8 \succ a_{11} \succ a_5 \succ a_{10} \succ a_6 \succ a_1 \succ a_4$
$\mathcal{F}(\{c_1, c_3, c_5\})$	$a_9 \succ a_7 \succ a_2 \succ a_{13} \succ a_{14} \succ a_3 \succ a_{12} \succ a_{15} \succ a_{11} \succ a_8 \succ a_5 \succ a_{10} \succ a_6 \succ a_1 \succ a_4$
$\mathcal{F}(\{c_2, c_3, c_5\})$	$a_9 \succ a_7 \succ a_{13} \succ a_2 \succ a_{14} \succ a_3 \succ a_{15} \succ a_{12} \succ a_8 \succ a_{10} \succ a_{11} \succ a_5 \succ a_6 \succ a_1 \succ a_4$

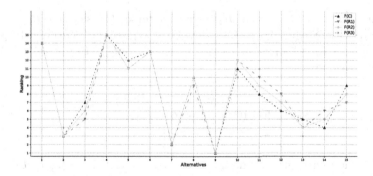

Fig. 1. The ranking results plotting

reduction successfully reduces the number of criteria. Either one of the three reducts uses only half of the criteria and achieves a highly similar ranking.

Three-way decision offers us a novel view to look at the approximate criterion reduction. Instead of using a single threshold α, a pair of thresholds is introduced and results in a pair of approximate criterion reducts. Definition 6 provides a formal description of the three-way approximate criterion reduction.

Definition 6. *Given an* MCDMT *and a pair of thresholds* (α_l, α_h) *with* $\alpha_l < \alpha_h$. *A pair of subsets* (R_l, R_h) *is called three-way approximate criterion reducts, if* $R_l \subseteq R_h$, *where* R_l *is an* α_l-*approximate criterion reduct and* R_h *is an* α_h-*approximate criterion reduct.*

As specified in Definition 6, the pair of reducts satisfies $R_l \subseteq R_h$, where R_l is viewed as a low-level reduct and R_h is a high-level reduct. The low-level reduct R_l is determined by a smaller threshold α_l and R_h is determined by a greater threshold α_h. Recalling the sufficiency condition, the result produced by R_h is better than that one produced by R_l. However, R_l contains fewer criteria resulting in less decision cost, R_h contains more criteria that cost more computations. Three-way approximate criterion reducts naturally form a tripartition over the set C, denoted by $\langle \text{CORE}, \text{ENHC}, \text{SUPF} \rangle$:

1. Core criteria: $\text{CORE} = R_l$,
2. Enhanced criteria: $\text{ENHC} = R_h - R_l$,
3. Superfluous criteria: $\text{SUPF} = R_h^c$.

The three subsets of criteria are pair-wise disjoint and their union is the whole set C. In terms of the tri-partition, we have the following strategies: (i) the set CORE is necessary to be used in an MCDM method that meets decision-maker's minimum requirements; (ii) the ranking or the model performance will be improved if CORE and ENHC are simultaneously selected into ranking; and (iii) the criteria in the set SUPF are not necessary to be considered.

Example 5. In the previous example, we successfully obtain three α-approximate criterion reducts with $\alpha = 0.95$. We set the pair of thresholds as $(\alpha_l = 0.95, \alpha_h =$

0.98) and examine the subset R_1 as an example. Considering $R_l = R_1 = \{c_1, c_2, c_3\}$ with $Sim(\mathcal{F}(R_l), \mathcal{F}(C)) = 0.9571$, a possible higher level approximate reduct is $R_h = \{c_1, c_2, c_3, c_4, c_6\}$, of which, $Sim(\mathcal{F}(R_h), \mathcal{F}(C)) = 0.9964$. The tri-partition over C is:

$$
\begin{aligned}
\text{CORE} &= R_l = \{c_1, c_2, c_3\}, \\
\text{ENHC} &= R_h - R_l = \{c_4, c_6\}, \\
\text{SUPF} &= R_h^c = \{c_5\}.
\end{aligned}
\tag{3}
$$

That is, if the user focuses on decision cost, the set CORE $= \{c_1, c_2, c_3\}$ should be considered to rank the alternatives. If the user aims to improve the result, the ENHC set should be additionally involved in ranking. The selected criteria are CORE $\cup$ ENHC $= \{c_1, c_2, c_3, c_4, c_6\}$ and the decision cost correspondingly increases.

4   Algorithms and Experimental Results

In this section, we introduce a loss function from learning to rank approaches and construct heuristic algorithms based on it. Then, we implement the proposed algorithms on a data set to generate criterion reducts.

4.1   Heuristic Algorithms Based on Loss Function

Although the property of ranking is non-monotonic, we are still able to design a heuristic algorithm to obtain admissible results. According to Chen et al. [1] and Liu [7], "the minimization of loss functions will lead to the maximization of the ranking measures". To illustrate criterion reduct, we adopt the point-wise loss function to construct heuristic algorithms.

Definition 7. *Suppose that rankings $\mathcal{F}(R)$ and $\mathcal{F}(C)$ are given, f is a function that indicates an alternative's position in a ranking. A point-wise loss function is defined by:*

$$
L^{\text{point}}(\mathcal{F}(R), \mathcal{F}(C)) = \sum_{i=1}^{n} (f(\mathcal{F}(R), a_i) - f(\mathcal{F}(C), a_i))^2.
\tag{4}
$$

By using an addition strategy, we develop a heuristic algorithm to generate criterion reducts as shown in Algorithm 1. The function **APPROXREDUCT** requires two parameters, B and α. The parameter B is the initial set which is usually an empty set and α is the threshold given by users. The function repeats adding a single criterion with minimum loss into the set B until the result $\mathcal{F}(B)$ satisfies the **sufficiency** condition. In most rough set attribute reduction algorithms, a second stage is required to delete redundant attributes by examining each of the following proper sets $\forall c \in B, B - \{c\}$. Due to the non-monotonicity in criterion reduction, the final outcome of B may **not** satisfy the **minimization** condition. To minimize B, it is necessary to test every possible proper set of B,

---

Algorithm 1. A heuristic algorithm to compute criterion reduct

---

Input: A multi-criteria decision-making table MCDMT = $\{A, C, p\}$;
       A ranking $\mathcal{F}(C)$;
       A threshold α.
Output: An admissible approximate reduct *Reduct*.
 1: **function ApproxReduct**(B, α)
 2:    **while** True **do**
 3:      **for** c in $C - B$ **do**
 4:        Compute the loss $L^*(\{B \cup \{c\}\})$ (L^* can be either one of the loss functions);
 5:        Select a criterion c_{opt} that comes with minimum loss;
 6:        let $B = B \cup \{c_{\mathrm{opt}}\}$;
 7:      **end for**
 8:      **if** $Sim(\mathcal{F}(B), \mathcal{F}(C)) \geqslant \alpha$ **then**
 9:        **break**;
10:      **end if**
11:    **end while**
12:    **return** B;
13: **end function**
14: let *Reduct* = **ApproxReduct**($\emptyset, \alpha$);
15: **return** an admissible approximate reduct *Reduct*.

Algorithm 2. An addition strategy algorithm to compute three-way approximate criterion reducts

---

Input: A multi-criteria decision-making table MCDMT;
       A ranking $\mathcal{F}(C)$;
       A pair of thresholds (α_l, α_h).
Output: Three-way approximate criterion reducts (R_l, R_h).
 1: Define function **ApproxReduct**;
 2: let $R_l = $ **ApproxReduct**($\emptyset, \alpha_l$);
 3: let $R_h = $ **ApproxReduct**(R_l, α_h);
 4: **return** three-way approximate criterion reducts (R_l, R_h).

---

which is a non-deterministic polynomial problem. In Algorithm 1, we only show the addition strategy to construct a subset satisfying the sufficiency condition. Even though, the set B produced in by **ApproxReduct** is an admissible result that can sufficiently rank the alternatives and effectively reduce the decision cost and complexity.

To generate three-way approximate criterion reducts, we can simply reuse the function **ApproxReduct** from Algorithm 1 and implement it twice with two different settings of parameters. The first call takes an empty set as the initial state and α_l as the threshold and returns a low-level approximate criterion reduct R_l. The second call takes R_l as the initial state and α_h as the threshold and returns a high-level approximate criterion reduct R_h. The two calls of function **ApproxReduct** guarantee that the pair of reducts satisfies $R_l \subseteq R_h$.

4.2 Experimental Results and Analysis

Table 3 is the statistic summary of the entire data set. There are two hundred and sixteen cities and six criteria. The summary contains a number of statistic indices for each criterion including mean, std, and so forth.

Table 3. The statistic summary of city ranking data set

| | Criteria | | | | | |
	Movehub rating	Purchase power	Health care	Pollution	Quality of life	Crime rating
Count	216.0000	216.0000	216.0000	216.0000	216.0000	216.0000
Mean	79.6767	46.4772	66.4428	45.2404	59.9945	41.3386
Std	6.5010	20.6145	14.4164	25.3697	22.0194	16.4164
Min	59.8800	6.3800	20.8300	0.0000	5.2900	9.1100
25%	75.0700	28.8150	59.4200	24.4100	42.7525	29.3750
50%	81.0600	49.2200	67.6850	37.2100	65.1500	41.1400
75%	84.0200	61.6075	77.2075	67.6750	78.6175	51.3275
Max	100.0000	91.8500	95.9600	92.4200	97.9100	85.7000

In this case study, we focus on testing Algorithm 2 to generate three-way approximate criterion reducts because we could simply treat the results R_l and R_h as an α_l-approximate reduct and an α_h-approximate reduct. At first, we apply the settings in the previous examples:

1. Use the whole set of criteria to generate a ranking $\mathcal{F}(C)$, the ranking is shown in Fig. 2, the top three rated cities are Zurich, Dresden, and Dusseldorf.
2. Adopt Spearman's ranking correlation coefficient as the similarity measurement.

The inputs are:

1. An MCDMT;
2. A ranking $\mathcal{F}(C)$;
3. A pair of thresholds $(\alpha_l = 0.9, \alpha_h = 0.98)$.

The outputs are:

$$R_{l1} = \{c_2, c_4, c_5\}, \ Sim(\mathcal{F}(R_{l1}), \mathcal{F}(C)) = 0.9401,$$
$$R_{h1} = \{c_2, c_3, c_4, c_5, c_6\}, \ Sim(\mathcal{F}(R_{h1}), \mathcal{F}(C)) = 0.9926.$$

Based on the outputs, we have the following tri-partition over C:

$$\text{CORE} = \{c_2, c_4, c_5\},$$
$$\text{ENHC} = \{c_3, c_6\},$$
$$\text{SUPF} = \{c_1\}.$$

The rankings $\mathcal{F}(R_{1l})$ and $\mathcal{F}(R_{1h})$ are displayed in Fig. 3.

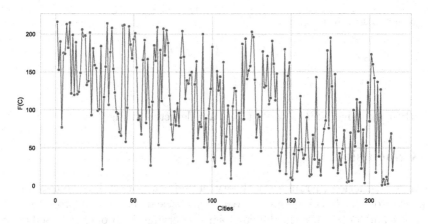

Fig. 2. The ranking $\mathcal{F}(C)$ generated by the whole set C

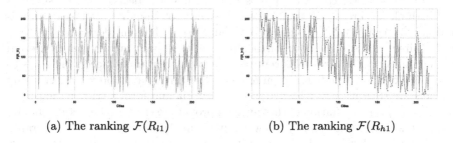

(a) The ranking $\mathcal{F}(R_{l1})$ (b) The ranking $\mathcal{F}(R_{h1})$

Fig. 3. The rankings produced by reducts R_{l1} and R_{h1}

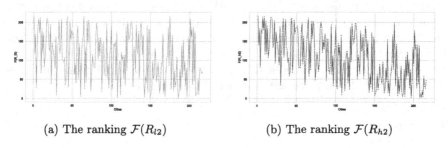

(a) The ranking $\mathcal{F}(R_{l2})$ (b) The ranking $\mathcal{F}(R_{h2})$

Fig. 4. The rankings produced by reducts R_{l2} and R_{h2}

To generate more reducts, we manually set $R_l = $ **ApproxReduct** $(\{c_1\}, 0.9)$ instead of using $\emptyset$ as the initial set. Then, we use the same inputs and obtain another pair of three-way approximate criterion reducts:

$$R_{l2} = \{c_1, c_4, c_5\}, \ Sim(\mathcal{F}(R_{l2}), \mathcal{F}(C)) = 0.9373$$
$$R_{h2} = \{c_1, c_2, c_4, c_5, c_6\}, \ Sim(\mathcal{F}(R_{h2}), \mathcal{F}(C)) = 0.9811.$$

The tri-partition over C is:

$$CORE = \{c_1, c_4, c_5\},$$
$$ENHC = \{c_2, c_6\},$$
$$SUPF = \{c_3\}.$$

And their rankings $\mathcal{F}(R_{l2})$ and $\mathcal{F}(R_{h2})$ are depicted in Fig. 4.

5   Conclusion

Attribute reduction in rough set theory provides a powerful technique for feature selection in MCDA problems. We have checked the desired property in an MCDMT and looked into the difficulties in preserving the property. Based on the concepts of general approximate reduction, we have proposed criterion reduction in a similar way. Moreover, the philosophy of three-way decision theory suggests us to think in threes. By combing the principle of thinking in threes and criterion reduction, we have taken into account users' different decision objectives or requirements and we have further defined three-way approximate criterion reduction. A pair of thresholds represents two different users' needs and leads to two levels of approximate criterion reducts. The three-way approximate criterion reducts are then used to trisect the set of criteria and form a tri-partition.

The proposed definition of three-way approximate criterion reduction enables that a low-level reduct contains fewer criteria and produces a less similar ranking and a high-level one has more criteria and produces a more similar result. In this paper, we have presented the conceptualization and designed heuristic algorithms to demonstrate the validity. We have generated two pairs of three-way approximate criterion reducts. The experimental results have indicated the feasibility of using fewer criteria to produce highly similar rankings. Based on the contributions in this paper, we may explore other potential extensions of attribute reduction in various fields. The underlying idea of thinking in threes may shed new light on three-way approximate reduction.

References

1. Chen, W., Liu, T.Y., Ma, Z.M., Li, H.: Ranking measures and loss functions in learning to rank. Adv. Neural Inf. Process. Syst. **22**, 315–323 (2009)
2. Fang, Y., Min, F.: Cost-sensitive approximate attribute reduction with three-way decisions. Int. J. Approx. Reas. **104**, 148–165 (2019)
3. Gao, C., Wang, Z.C., Zhou, J.: Three-way approximate reduct based on information-theoretic measure. Int. J. Approx. Reas. **142**, 324–337 (2022)
4. Greco, S., Figueira, J., Ehrogtt, M.: Multiple Criteria Decision Analysis. Springer, New York (2016). https://doi.org/10.1007/978-1-4939-3094-4
5. Jia, X.Y., Shang, L., Zhou, B., Yao, Y.Y.: Generalized attribute reduct in rough set theory. Knowl.-Based Syst. **91**, 204–218 (2016)
6. Liu, K.Y., Yang, X.B., Fujita, H., Liu, D., Yang, X., Qian, Y.H.: An efficient selector for multi-granularity attribute reduction. Inf. Sci. **505**, 457–472 (2019)

7. Liu, T.Y.: Learning to rank for information retrieval. Found. Trends® Inf. Retr. **3**, 225–331 (2009)
8. Pawlak, Z.: Rough sets. Int. J. Comput. Inf. Sci. **11**, 341–356 (1982)
9. Qian, Y.H., Liang, J.Y., Pedrycz, W., Dang, C.Y.: Positive approximation: an accelerator for attribute reduction in rough set theory. Artif. Intell. **174**, 597–618 (2010)
10. Ślęzak, D.: Approximate reducts in decision tables. In: Proceedings of the Sixth International Conference on Information Processing and Management of Uncertainty in Knowledge-Based Systems, pp. 1159–1164 (1996)
11. Ślęzak, D.: Approximate entropy reducts. Fundamenta Informaticae **53**, 3–4 (2002)
12. Wang, J., Miao, D.Q.: Analysis on attribute reduction strategies of rough set. J. Comput. Sci. Technol. **13**, 189–192 (1998)
13. Xu, W.H., Zhang, X.Y., Zhong, J.M., Zhang, W.X.: Attribute reduction in ordered information systems based on evidence theory. Knowl. Inf. Syst. **25**, 169–184 (2010)
14. Yao, Y.Y., Zhao, Y.: Attribute reduction in decision-theoretic rough set models. Inf. Sci. **178**, 3356–3373 (2008)
15. Yao, Y.Y.: The two sides of the theory of rough sets. Knowl.-Based Syst. **80**, 67–77 (2015)
16. Yao, Y.: Tri-level thinking: models of three-way decision. Int. J. Mach. Learn. Cybern. **11**(5), 947–959 (2019). https://doi.org/10.1007/s13042-019-01040-2
17. Yao, Y.: The geometry of three-way decision. Appl. Intell. **51**(9), 6298–6325 (2021). https://doi.org/10.1007/s10489-020-02142-z
18. Yao, Y.Y.: Symbols-meaning-value (SMV) space as a basis for a conceptual model of data science. Int. J. Approx. Reas. **144**, 113–128 (2022)
19. Yao, Y.Y.: Human-machine co-intelligent through symbiosis in the (SMV) space. Appl. Intell. (2022). https://doi.org/10.1007/s10489-022-03574-5
20. Zhang, X.Y., Yao, Y.Y.: Class-specific attribute reducts in rough set theory. Inf. Sci. **418–419**, 601–618 (2017)
21. Zhang, X.Y., Yao, Y.Y.: Tri-level attribute reduction in rough set theory. Expert Syst. Appl. **190**, 116187 (2022)
22. Zhang, Y., Yao, J.T.: Multi-criteria based three-way classifications with game-theoretic rough sets. In: Kryszkiewicz, M., Appice, A., Slezak, D., Rybinski, H., Skowron, A., Raś, Z.W. (eds.) ISMIS 2017. LNCS (LNAI), vol. 10352, pp. 550–559. Springer, Cham (2017). https://doi.org/10.1007/978-3-319-60438-1_54
23. Zhao, Y., Luo, F., Wong, S.K.M., Yao, Y.: A general definition of an attribute reduct. In: Yao, J.T., Lingras, P., Wu, W.-Z., Szczuka, M., Cercone, N.J., Slezak, D. (eds.) RSKT 2007. LNCS (LNAI), vol. 4481, pp. 101–108. Springer, Heidelberg (2007). https://doi.org/10.1007/978-3-540-72458-2_12
24. Zhao, Y., Wong, S.K.M., Yao, Y.: A note on attribute reduction in the decision-theoretic rough set model. In: Peters, J.F., Skowron, A., Chan, C.-C., Grzymala-Busse, J.W., Ziarko, W.P. (eds.) Transactions on Rough Sets XIII. LNCS, vol. 6499, pp. 260–275. Springer, Heidelberg (2011). https://doi.org/10.1007/978-3-642-18302-7_14

Close Contact Detection in Social Networks via Possible Attribute Analysis

Huilai Zhi[1]([⊠]), Jinhai Li[2], and Jianjun Qi[3]

[1] Henan Polytechnic University, Jiaozuo 454003, People's Republic of China
zhihuilai@126.com
[2] Kunming University of Science and Technology,
Kunming 650500, People's Republic of China
[3] Xidian University, Xi'an 710071, People's Republic of China
qijj@mail.xidian.edu.cn

Abstract. Discovering close contacts and the key structures in social networks plays a vital role in network analysis. The existing methods for identifying key network structures often suffer from high computational complexity, and lack a clear and reasonable semantic explanation. To tackle this issue, we propose a method for close contact detection by using the technic of formal concept analysis. Specifically, we establish the relationship between social networks and formal contexts, and adopt possible attribute analysis to discover close contacts and identify prime cliques. After that, we discuss the dynamic updating mechanism of close contacts and prime cliques under the evolution of a social network. In addition, we conduct some experiments to show the relationships between the number of prime cliques and the size of social networks, and the feasibility and effectiveness of the proposed updating methods.

Keywords: Close contact detection · Social network · Formal concept analysis · Possible attribute analysis

1 Introduction

Social networks proliferate dramatically in this big data era [2]. Discovering meaningful patterns, unclosing the internal features of sub-structures, and identifying key structures are important tasks in various practical social networks, such as recommendation system modeling [23], information diffusion mechanism detection [34] and private data protection [6], to name just a few.

Generally speaking, the key structures come into being due to close contacts between individuals. In the existing studies, the key structures, such as maximal clique, isolated maximal clique and community, bridge and structure hole spanner have been explored in different applications. Roughly speaking, a maximal clique is a crowd in which no more node can be added to this structure while keeping its structure unchanged

Supported by the Natural Science Foundation of Henan Province under Grant 222300420445, and the Fundamental Research Funds for the Universities of Henan Province under Grant NSFRF210318.

J. Yao et al. (Eds.): IJCRS 2022, LNAI 13633, pp. 304–316, 2022.
https://doi.org/10.1007/978-3-031-21244-4_23

[22]; an isolated clique is a crowd in which each node inside this clique has no connections with the nodes out of this clique [18,24]; a bridge is a specific structure which enables the individuals of isolated communities to interact with each other [5]; bridging nodes are known as structural hole spanners, which can facilitate necessary communications between different communities [1].

It has been proved that discovering key structures from complex graphs is a time-consuming problem, which have attracted the attentions of many researchers. For instance, Hao et al. [13] mined the diversified top-k maximal clique from social internet of things by using formal concepts. Lu et al. [22] explored the complex graph randomly and iteratively to mine maximum cliques.

Community detection is another difficult problem. Jabbour et al. [14] devised an agglomeration method to detect highly overlapping community structure by expanding maximal cliques. Fu et al. [8] used a division strategy to detect communities based on node density and similarity. Yang et al. [29] designed a graph-based label propagation algorithm for community detection. Li et al. [21] described a novel mining strategy to mine useful community structures by integrating center locating and membership optimization.

The technique of deep learning has a powerful ability of managing internal information and logical structures, and has greatly facilitated the study of social network analysis [12]. Cao et al. [3] combined network structures and node contents to realize community detection by deep learning. Tu et al. [26] designed a unified framework for community detection and network representation learning. Cavallari et al. [4] presented a method to discover both finite and infinite communities on graphs. Jin et al. [16] incorporated graph convolutional networks with markov random fields to carry out semi-supervised community detection in attribute networks. Lately, Jin et al. [17] systematically surveyed the existing community detection approaches, including statistical approaches and deep learning based methods.

Nowadays, social life is greatly influenced by the widespread of COVID-19. Early detection and treatment is an effective way to curb the spread of the disease, and close contact detection is frequently used to identify the suspected infected persons. However, there is little studies on this issue. Considering the existing methods, such as statistical approaches and deep learning based methods, which are of high computational complexity and lack of sufficient semantic explanations, we propose a method to detect close contact in social networks based on formal concept analysis [9].

Formal concept analysis is a mathematical theory, which can model a domain and facilitate knowledge discovery by integrating objects with their attributes to build a set of cognitive units, i.e., the so-called formal concepts, and establishing hierarchial structures between them. At present, this theory has been applied in many fields [10,15,19,27], such as open data categorization, logical data analysis, knowledge discovery and decision rule acquisition. Lately, Yan and Li [28] have resort to this theory, and proposed a method for knowledge discovery and updating under the evolution of network formal contexts based on three-way decision. In a word, the existing studies have shown that that formal concept analysis can meet our requirements, and achieving the task of social network analysis.

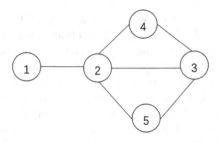

Fig. 1. Graphical representation of a toy social network.

The rest of this paper is structured as follows. Section 2 reviews some basic notions of social networks and formal concept analysis. Section 3 proposes a method for close contacts detection based on necessary attribute analysis. In Sect. 4, a dynamic updating method of close contact detection is presented under the evolution of a social network. A summary of findings and future work come in Sect. 5.

2   Related Theoretical Foundations

In this section, we briefly review some basic notions on social networks and possible attribute analysis.

2.1   The Basic Notions of Social Networks

In this study, a social network is represented by an undirected graph $G = (N, E)$, where N is the set of nodes, and E contains the linkages of each pair of nodes in N. Concretely, if there is a linkage between $x_i \in N$ and $x_j \in N$, we denote this case as $e(x_i, x_j) = 1$; otherwise, it is denoted as $e(x_i, x_j) = 0$. Moreover, we also stipulate that for any $x \in N$, $e(x, x) = 1$.

Example 1. Fig. 1 is an undirected graph, which characterizes a small social network with 5 nodes and 6 edges.

In what follows, close contacts are defined for a single node.

Definition 1. *Let $G = (N, E)$ be the undirected graph of a social network, and $x \in N$. If $e(x, x_i) = 1$, then we call x_i a close contact of x, and denote the set of all close contacts of x by $c(x)$, i.e., $c(x) = \{x_i \mid e(x, x_i) = 1, x_i \in N\}$.*

It is worth noting that for any $x \in N$, we have $x \in c(x)$ by the stipulation raised at the first paragraph of this subsection.

In the rest of the paper, the discussed cliques are always non-empty, i.e., we stipulate that $\emptyset \neq X \subseteq N$.

Definition 2. *Let $G = (N, E)$ be the undirected graph of a social network, and $X \subseteq N$. We call $c(X) = \bigcup\limits_{x \in X} c(x)$ the close contacts of the clique X.*

It should be pointed that although we have reused the function $c(\cdot)$, it makes no confusions, as Definition 2 actually embodies Definition 1.

Definition 3. *Let $G = (N, E)$ be the undirected graph of a social network, and $X \subseteq N$. If $c(X) = A$, and there is no $X' \supset X$ such that $c(X') = A$, we call X a prime clique of G.*

Definition 4. *Let $G = (N, E)$ be the undirected graph of a social network, and $X \subseteq N$. For any $Y \subseteq X$, if $c(Y) = c(X)$, and there is no $Z \supset Y$ such that $c(Z) = c(Y)$, we call Y a skeleton of X. Moreover, if X is also the skeleton of X, we call X a vulnerable clique.*

Definition 5. *Let $G = (N, E)$ be the undirected graph of a social network, and $X \subseteq N$. We call $\delta(X)$ the stability measurement of X, which is defined as*

$$\delta(X) = \frac{|Y \mid c(Y) = c(X), Y \subseteq X|}{2^{|X|}}.$$

Continued with Example 1. It is easy to obtain the close contacts of each node, and the result is listed in Table 1.

Table 1. Close contacts of each node in Example 1.

Node n	Close contact $c(n)$
1	{1,2}
2	{1,2,4,5}
3	{3,4,5}
4	{2,3,4}
5	{2,3,5}

Moreover, by Definition 3, whether a clique is prime can be determined. For instance, $\{3, 4\}$ is not a prime clique, due to the fact that $c(\{3, 4, 5\}) = c(\{3, 4\}) = \{2, 3, 4, 5\}$. In fact, $\{3, 4, 5\}$ is a prime clique, as adding either node 1 or 2 can make the close contacts change to be $\{1, 2, 3, 4, 5\}$.

What is more, by Definition 4, $\{3, 4, 5\}$ has two skeletons, i.e., $\{4, 5\}$ and $\{3\}$, and thus $\{3, 4, 5\}$ is not a vulnerable clique.

In addition, by Definition 5, we can obtain $\delta(\{3, 4, 5\}) = \frac{5}{8}$, as $c(\{3, 4, 5\}) = c(\{4, 5\}) = c(\{3, 5\}) = c(\{3, 4\}) = c(\{3\})$.

2.2 The Basic Notions of Possible Attribute Analysis

The basic setting of formal concept analysis is a formal context $K = (U, V, I)$, which describes a set of objects U by a set of attributes V, and I expresses the relations between U and V. Specifically, $I(x, a) = 1$ shows that the object x possesses an attribute a, while $I(x, a) = 0$ expresses the opposite.

By generalizing the classical model with a possibility theoretic view, many new types of concepts have been proposed. Property oriented concept is one of the most successful generalizations, which has a explicit semantic explanation [32,33].

Definition 6. (*[7,30]*) *Let* $K = (U, V, R)$ *be a formal context. For* $X \in 2^U$ *and* $A \in 2^V$, *possibility and necessary operators* $\Diamond : 2^U \to 2^V$ *and* $\Box : 2^V \to 2^U$ *are respectively defined as:*

$X^\Diamond = \{a \in V \mid X \cap R(a) \neq \emptyset\}$ *and* $A^\Box = \{x \in U \mid R(x) \subseteq A\}$,

where $R(a)$ *is the set of objects that possess the attribute* a, *and* $R(x)$ *is the set of attributes possessed by the object* x.

The pair (X, A) *is called a property oriented concept if* $X^\Diamond = A$ *and* $A^\Box = X$. *Furthermore, all the property oriented concepts make up a lattice structure, which is called property oriented concept lattice, and is denoted by* $PL(K)$.

From a granule description view point, property oriented concepts explicitly exhibit the possible attributes of the extents. Therefore, this analysis method is called possible attribute analysis.

By using possible attribute analysis, the stability measurement of a clique can be computed as follows.

Theorem 1. *Let* $K = (U, V, R)$ *be a formal context of a social network* G, *and* $\emptyset \neq X \subseteq U$. *If the clique* X *has a set of skeletons* $\{Y_i\}_{i \in n}$, *then the stability measurement of* X *is* $\delta(X)$ *where*

$$\sigma(X) = \frac{\sum_{i=1}^{n} 2^{|X-Y_i|} - \sum_{1 \leq i < j \leq n} 2^{|X|-|Y_i \cup Y_j|} + \cdots + (-1)^n 2^{|X-Y_1 \cup Y_2 \cup \cdots \cup Y_n|}}{2^{|X^{\Diamond\Box}|}}.$$

Proof. Since $\{Y_i\}_{i \in n}$ is the set of the skeletons of X , we have $|\{C_i | Y_i \subseteq C_i \subseteq A\}| = 2^{|X-Y_i|}$.

According to inclusion-exclusion principle, we can show that

$$\left| \bigcup_{i=1}^{n} C_i \right| = \sum_{i=1}^{n} 2^{|X-Y_i|} - \sum_{1 \leq i < j \leq n} 2^{|X|-|Y_i \cup Y_j|} + \cdots + (-1)^n 2^{|X-Y_1 \cup Y_2 \cup \cdots \cup Y_n|},$$

By Definition 5, it follows that

$$\sigma(X) = \frac{\sum_{i=1}^{n} 2^{|X-Y_i|} - \sum_{1 \leq i < j \leq n} 2^{|X|-|Y_i \cup Y_j|} + \cdots + (-1)^n 2^{|X-Y_1 \cup Y_2 \cup \cdots \cup Y_n|}}{2^{|X^{\Diamond\Box}|}}.$$

3   Close Contact Detection in Social Networks via Possible Attribute Analysis

The following proposition states that the close contact of a specific clique can be obtained via its possible attributes.

Proposition 1. *Let* $K = (U, V, R)$ *be a formal context of a social network* G, *and* $\emptyset \neq X \subseteq U$. *Then,* $C(X) = X^\Diamond$.

Theorem 2. *Let $K = (U, V, R)$ be a formal context of a social network G, and $\emptyset \neq X \subseteq U$. X is an extent of a property oriented concept of $PL(K)$ if and only if X is a prime clique of G.*

Proof. $\Rightarrow$: Suppose X is an extent of a property oriented concept (X, A). Then, it is followed that X is the biggest set of objects such that $X^\diamond = A$. As $C(X) = X^\diamond$, we can conclude that X is a prime clique of G.

$\Leftarrow$: Suppose X is a prime clique of G, and $c(X) = A$. As X is the biggest set of objects that makes $c(X) = X^\diamond = A$, and A is the set of attributes that are possibly possessed by X, then by the definition of a property oriented concept, we can conclude that (X, A) is a property oriented concept of $PL(K)$.

Theorem 2 actually manifests that by using possible attribute analysis, we can determine the prime cliques of a social network. What is more, if a clique X is prime, we can derive its close contact via a property oriented concept $(X, X^\diamond)$.

Continued with Example 1, the corresponding formal context K of the social network G is shown in Table 2.

Table 2. The formal context K of the social network in Example 1.

	1	2	3	4	5
1	*	*			
2	*	*	*	*	*
3		*	*	*	*
4		*	*	*	
5		*	*		*

The property oriented concept lattice of K is shown in Fig. 2. In this figure, for convenience we omit the brackets and commas in the representations of the sets.

It can be checked that all the extents of $PL(K)$ are prime cliques except $\emptyset$. Concretely, there are 7 prime cliques of G, which are collectively listed with their close contact in Table 3.

4   Close Contact Detection in Dynamic Social Networks

An evolutionary process may occurs from time to time in almost all real social networks, and the relationship between individuals will continuously change as time goes on [25]. In this section, we discuss close contact detection in dynamic social networks.

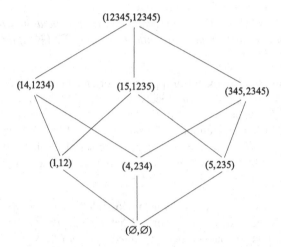

Fig. 2. The property oriented concept lattice $PL(K)$.

Table 3. Prime cliques and their close contacts

Prime clique	Close contact
$\{1, 2, 3, 4, 5\}$	$\{1,2,3,4,5\}$
$\{1, 4\}$	$\{1,2,3,4\}$
$\{1, 5\}$	$\{1,2,3,5\}$
$\{3, 4, 5\}$	$\{2,3,4,5\}$
$\{1\}$	$\{1,2\}$
$\{4\}$	$\{2,3,4\}$
$\{5\}$	$\{2,3,5\}$

4.1   The Dynamic Updating of Close Contacts in Social Networks

Under the settings of possible attribute analysis, the task of dynamic updating of close contacts in social networks is equivalent to dynamic updating of property oriented concept lattice when the formal context changes under the evolution of a social network.

When adding a new objet to a social network, one row and one column will be added to the corresponding formal context. In fact, adding one row is to add one object to the original formal context, and adding one column is to add one attribute to the original formal context. Then, the task of the updating of a property oriented concept lattice can be divided into two steps:

– adding an object to the formal context and updating the lattice;
– adding an attribute to the formal context and updating the lattice.

Inspired by the traditional methods for updating formal concept lattices [11, 20, 31, 35], we can get the ones for updating property oriented concept lattices. Algorithm 1 updates a property oriented concept lattice when adding a new object, and Algorithm 2 updates a property oriented concept lattice when adding a new attribute.

Algorithm 1. The updating of a property oriented concept lattice when adding a new object

Input: A property oriented concept lattice $PL(K)$, and a new object x.
Output: The updated property oriented concept lattice $PL(K')$.
1: Visit the lattice structure of $PL(K)$ from the maximal concept in a descending order.
2: **For** each visited concept (X, A) of $PL(K)$
3:    **If** $A \cup \{x\}^* = A$
4:      **Then** perform updating $(X, A) \rightarrow (X \cup \{x\}, A)$.
5:      **Else**
6:        **If** there is no concept with an intent $A \cup \{x\}^*$
7:          **Then**
8:            (X, A) keeps unchanged.
9:            Create a new concept $(X \cup \{x\}, A \cup \{x\}^*)$.
10:           Insert $(X \cup \{x\}, A \cup \{x\}^*)$ into the current hierarchical structure.
11:        **End If**
12:    **End If**
13: **End For**
14: **Return** the obtained hierarchical structure $PL(K')$.

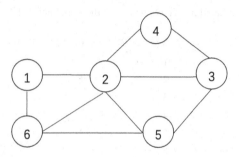

Fig. 3. The updated toy social network when adding a new object 6.

Example 2. Continued with Example 1. We add a new object 6, and correlate it to object 1, 2, and 5. Then, the updated social network can be seen in Fig. 3, and its corresponding formal context is shown in Table 4 .

After updating the property oriented concept lattice, we obtain the result shown in Fig. 4.

By Fig. 4, it can be observed that there are 9 prime cliques of G, including $\{1, 2, 3, 4, 5, 6\}$, $\{2, 3, 5\}$, $\{1, 4\}$, $\{1, 5, 6\}$, $\{3, 4\}$, $\{1, 6\}$, $\{1\}$, $\{4\}$, and $\{5\}$. Beside, it can be seen that there are 5 prime cliques keeping unchanged, i.e., $\{1, 4\}$, $\{2, 3, 5\}$, $\{1\}$, $\{4\}$, and $\{5\}$, which accounts for $\frac{5}{7}$ of the prime cliques of the original network. In other words, only a few prime cliques will change after updating the original network.

Algorithm 2. The updating of a property oriented concept lattice when adding a new attribute

Input: A property oriented concept lattice $PL(K)$, and a new object x.
Output: The updated property oriented concept lattice $PL(K')$.
1: Visit the lattice structure of $PL(K)$ from the minimal concept in an ascending order.
2: **For** each visited concept (X, A) of $PL(K)$
3:    **If** $X \cap \{a\}^* = \emptyset$
4:       **Then** (X, A) keeps unchanged..
5:       **Else**
6:          Perform updating $(X, A) \rightarrow (X, A \cup \{a\})$.
7:          **If** there is no concept with an extent $X - \{a\}^*$
8:          **Then**
9:             Create a new concept $(X - \{a\}^*, A)$.
10:             Insert $(X - \{a\}^*, A)$ into the current hierarchical structure.
11:          **End If**
12:    **End If**
13: **End For**
14: **Return** the obtained hierarchical structure $PL(K')$.

Table 4. The updated formal context K of the social network in Example 2.

	1	2	3	4	5	6
1	*	*				*
2	*	*	*	*	*	*
3		*	*	*	*	
4		*	*	*		
5		*	*		*	*
6	*	*			*	*

Table 5. Experimental results on social networks with varying sizes.

N	10	20	30	40	50
Lattice size	38	510	3438	15790	54974
Run-time(ms)	1	59	1636	38227	107401

4.2 Experimental Analysis

The sizes of social networks (denoted as N), as well as the probability that any two nodes in social networks are related (denoted as P), will greatly affect the running time and the scale of concept lattices. In this subsection, experimental analysis is adopted to show the performance of the proposed method.

In the experiment, we control P and N, and generate the social networks by using random numbers to determine whether any tow nodes are connected.

First, we stipulate $p = 50\%$, and generate social networks with different sizes. The experimental results are shown in Table 5.

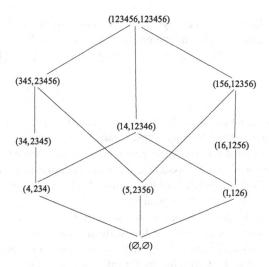

Fig. 4. The updated property oriented concept lattice of Example 2.

And then, we stipulate $N = 30$, and generate social networks with varying probabilities between nodes. The experimental results are shown in Table 6.

Table 6. Experimental results on social networks with varying probabilities between nodes.

P	30%	40%	50%	60%	70%
Lattice size	16520	9717	3283	1058	520
Run-time(ms)	37226	11862	1580	265	76

From the experimental results, it can be observed that a lot of property oriented concepts are extracted with a considerable time consuming even for social networks of relatively small sizes. However, in most cases, it is unnecessary to mine all the patterns in applications, and we only care about a small proportion of them. For example, In epidemiological investigations, we only focus on those who carry pathogens, and those who have close contact with them.

Continued with Example 1, suppose 4 and 5 have been affected. Then, epidemiological investigations can be carried out with these two patients. Concretely, we have that $c(\{4\}) = \{2, 3, 4\}$, $c(\{5\}) = \{2, 3, 5\}$, and $c(\{4, 5\}) = \{2, 3, 4, 5\}$. In other words, individuals 2 and 3 are close contacts.

In fact, with the aids of a property oriented lattice, the same results can also be obtained. However, it is worth noting that with the help of a property oriented lattice, we can derive more different interesting information.

For instance, by a property oriented concept $(\{3, 4, 5\}, \{2, 3, 4, 5\})$, it can be observed that $c(\{3, 4, 5\}) = c(\{4, 5\}) = \{2, 3, 4, 5\}$, which implies that when adding individual 3 into the clique $\{4, 5\}$, the close contacts keeps unchanged. In other words, the arrival of individual 3 can not change the range of close contacts. But if we add other individuals into the clique $\{4, 5\}$, the range of close contacts will change, For instance, if we add individual 2 into the clique $\{4, 5\}$, we have $c(\{2, 4, 5\}) = \{1, 2, 3, 4, 5\}$.

5   Conclusion

Early detection and treatment is an effective way to curb the spread of COVID-19. This paper has proposed a method to detect close contacts by using possible concept analysis. Specifically, we have realized close contact detection based on property oriented concepts. Moreover, we have discussed the dynamic updating of close contacts when the social network changes. Finally, we have pointed out the advantages of exploring a social network by using formal concept analysis.

In the future, the following issues need our further explorations: (1) directed and weighed complex networks are more in line with the needs of practical applications, and further research is strongly needed; (2) incomplete information permeates in almost all application scenarios, how to mine useful information in incomplete environment deserves our efforts; (3) what is more, using parallel computing technic to accelerate knowledge discovery is another important issue.

Acknowledgements. We would like to thank the organization committee of the 2022 International Joint Conference on Rough Sets, which provides us an opportunity to share recent developments in conceptual knowledge discovery and machine learning based on three-way decisions and granular computing.

References

1. Burt, R.S.: Structural holes: the social structure of competition. Harvard University Press, Cambridge, MA, USA (2009)
2. Camacho, D., Panizo-LLedot, À., Bello-Orgaz, G., Gonzalez-Pardo, A., Cambria, E.: The four dimensions of social network analysis: an overview of research methods, applications, and software tools. Inf. Fusion **63**, 88–120 (2020)
3. Cao, J., Jin, D., Yang, L., Dang, J.: Incorporating network structure with node contents for community detection on large networks using deep learning. Neurocomputing **297**, 71–81 (2018)
4. Cavallari, S., Cambria, E., Cai, H., et al.: Embedding both finite and infinite communities on graphs. IEEE Comput. Intell. Mag. **14**(3), 39–50 (2019)
5. Corradini E., Nocera A., Ursino D., Virgili L.: Defining and detecting k-bridges in a social network: the Yelp case, and more. Knowl.-Based Syst. **195**, 105721 (2020)
6. Du, J., Jiang, C., Chen, K.-C., Ren, Y., Poor, H.V.: Community-structured evolutionary game for privacy protection in social networks. IEEE Trans. Inf. Forensics Secur. **13**(3), 574–589 (2018)
7. Duntsch I., Gediga G.: Modal-style operators in qualitative data analysis. In: Proceedings of the 2002 IEEE International Conference on Data Mining, pp. 155–162, Maebashi, Japan (2002). https://doi.org/10.1109/icdm.2002.1183898
8. Fu L., Li F., Li D.: Community division algorithm based on node density and similarity. In: Proceedings of the IEEE International Conference on Artificial Intelligence and Computer Applications, pp. 739–743, Chongqing, China (2020). https://doi.org/10.1109/ICAICA50127.2020.9182596

9. Ganter, B., Wille, R.: Formal Concept Analysis. Springer, Heidelberg (1999). https://doi.org/10.1007/978-3-642-59830-2
10. Gligorijevic, M.F., et al.: Open data categorization based on formal concept analysis. IEEE Trans. Emerg. Top. Comput. **9**(2), 571–581 (2021)
11. Godin, R., Missaoui, R., Alaoui, H.: Incremental concept formation algorithms based on galois (concept) lattices. Comput. Intell. **11**(2), 246–267 (1995)
12. Liu F.Z., et al.: Deep Learning for Community Detection: progress, Challenges and Opportunities. In: Twenty-Ninth International Joint Conference on Artificial Intelligence and Seventeenth Pacific Rim International Conference on Artificial Intelligence, pp. 4981–4987, Yokohama, Japan (2020). https://doi.org/10.24963/ijcai.2020/693
13. Hao, F., Pei, Z., Yang, L.T.: Diversified top-k maximal clique detection in social internet of things. Futur. Gener. Comput. Syst. **107**, 408–417 (2020)
14. Jabbour S., Mhadhbi N., Radaoui B., Sais L.: Detecting highly overlapping community structure by model-based maximal clique expansion. In: Proceedings of IEEE International Conference on Big Data, pp. 1031–1036, Seattle, WA, USA (2018). https://doi.org/10.1109/BigData.2018.8621868
15. Janostik, R., Konecny, J., Krajca, P.: Interface between logical analysis of data and formal concept analysis. Eur. J. Oper. Res. **284**(2), 792–800 (2020)
16. Jin, D., Liu, Z., Li, W., et al.: Graph convolutional networks meet Markov random fields: semi-supervised community detection in attribute networks. Proceed. AAAI Conf. Artif. Intell. **33**(1), 152–159 (2019)
17. Jin D., et al.: A survey of community detection approaches: from statistical modeling to deep learning. IEEE Transactions on Knowledge and Data Engineering. https://doi.org/10.1109/TKDE.2021.3104155
18. Kumar S., Hamilton W.L., Leskovec J., Jurafsky D.: Community interaction and conflict on the web. In: Proceedings of the World Wide Web Conference, pp. 933–943, Lyon, France (2018). https://doi.org/10.1145/3178876.3186141
19. Kumar, C.A.: Knowledge discovery in data using formal concept analysis and random projections. J. Appl. Math. Comput. Sci. **21**(4), 745–756 (2011)
20. Kuznetsov, S.O., Obiedkov, S.A.: Comparing performance of algorithms for generating concept lattices. J. Exp. Theor. Artif. Intell. **14**(2–3), 189–216 (2002)
21. Li, H.-J., Bu, Z., Li, A., Liu, Z., Shi, Y.: Fast and accurate mining the community structure: integrating center locating and membership optimization. IEEE Trans. Knowl. Data Eng. **28**(9), 2349–2362 (2016)
22. Lu, C., Yu, J.X., Wei, H., Zhang, Y.: Finding the maximum clique in massive graphs. Proceed. VLDB Endow. **10**(11), 1538–1549 (2017)
23. Ma J.W., et al.: DBRec: dual-bridging recommendation via discovering latent groups. In: Proceedings 28th ACM International Conference on Information and Knowledge Management, pp. 1513–1522, Beijing, China (2019). https://doi.org/10.1145/3357384.3357892
24. Molter, H., Niedermeier, R., Renken, M.: Enumerating isolated cliques in temporal networks. In: Cherifi, H., Gaito, S., Mendes, J.F., Moro, E., Rocha, L.M. (eds.) COMPLEX NETWORKS 2019. SCI, vol. 882, pp. 519–531. Springer, Cham (2020). https://doi.org/10.1007/978-3-030-36683-4_42
25. Newman, M.E.J.: Networks: An Introduction. Oxford University Press, New York, USA (2010)
26. Tu, C., Zeng, X., Hao, W., et al.: A unified framework for community detection and network representation learning. IEEE Trans. Knowl. Data Eng. **31**(6), 1051–1065 (2019)
27. Wei, L., Liu, L., Qi, J.J., et al.: Rules acquisition of formal decision contexts based on three-way concept lattices. Inf. Sci. **516**, 529–544 (2020)
28. Yan, M.Y., Li, J.H.: Knowledge discovery and updating under the evolution of network formal contexts based on three-way decision. Inf. Sci. **601**, 18–38 (2022)

29. Yang, G., Zheng, W., Che, C., Wang, W.: Graph-based label propagation algorithm for community detection. Int. J. Mach. Learn. Cybern. **11**(6), 1319–1329 (2020)

30. Yao Y.Y.: Concept lattices in rough set theory. In: Proceedings of 23rd International Meeting of the North American Fuzzy Information Processing Society, pp. 796–801, Banff Alberta, Canada (2004). https://doi.org/10.1109/NAFIPS.2004.1337404

31. Zhi, H.L., Li, J.H.: Influence of dynamical changes on concept lattice and implication rules. Int. J. Mach. Learn. Cybern. **9**(5), 705–805 (2018)

32. Zhi, H.L., Qi, J.J., Qian, T., Wei, L.: Three-way dual concept analysis. Int. J. Approximate Reasoning **114**, 151–165 (2019)

33. Zhi, H.L., Li, J.H.: Granule description based knowledge discovery from incomplete formal contexts via necessary attribute analysis. Inf. Sci. **485**, 347–361 (2019)

34. Zhang Y., Lyu T., Zhang Y.: COSINE: Community-preserving social network embedding from information diffusion cascades. In: Proceedings of the Thirty-Second AAAI Conference on Artificial Intelligence, pp. 2620–2627, New Orleans, Louisiana, USA (2018)

35. Zou, L.G., Zhang, Z.P., Long, J.: A fast incremental algorithm for constructing concept lattices. Expert Syst. Appl. **42**(9), 4474–4481 (2015)

Uncertainty in Three-Way Decisions, Granular Computing, and Data Science

A Probabilistic Approach to Analyzing Agent Relations in Three-Way Conflict Analysis Based on Bayesian Confirmation

Mengjun Hu[1]($\boxtimes$)(iD) and Guangming Lang[2]($\boxtimes$)

[1] Department of Computer Science, University of Regina,
Regina, SK S4S 0A2, Canada
mengjun.hu@uregina.ca
[2] School of Mathematics and Statistics, Changsha University of Science
and Technology, Changsha, Hunan 410114, China
langguangming1984@126.com

Abstract. Conflict analysis is commonly based on a conflict situation involving agents and their ratings or attitudes toward a set of issues. Analyzing the relationships between agents is one of the essential topics in conflict analysis. Alliance, conflict, and neutrality are three typical relations. The majority of existing research adopts an auxiliary function that uses $+1$, -1, and 0 to denote these three relations concerning a single issue. An auxiliary function is aggregated for a group of issues, which is primarily limited to taking the average in the existing works. Moreover, computing the values of an auxiliary function is also associated with potential semantics issues. This paper proposes a probabilistic approach to analyzing agent relations, which is very different from the current approaches. Bayesian confirmation is adopted to explore how a rating confirms or disconfirms the alliance/conflict relation between two agents. Accordingly, we construct three regions of confirmatory, disconfirmatory, and neutral ratings. Three types of confirmation rules are induced from these regions and used to devise appropriate strategies in maintaining and developing relations with agents.

Keywords: Three-way conflict analysis · Bayesian confirmation · Agent relation · Three-way decision

1 Introduction

Conflict analysis studies the conflict situations between agents or organizations that arise from their attitudes towards a set of issues. To formulate a conflict situation, Pawlak proposed a situation table that takes a set of agents as rows, a set of issues as columns, and the ratings or attitudes as the table cells [17]. There are commonly three types of ratings, namely, positive, negative, and neutral, which are represented by $+1$, -1, and 0, respectively. The majority of existing research

J. Yao et al. (Eds.): IJCRS 2022, LNAI 13633, pp. 319–333, 2022.
https://doi.org/10.1007/978-3-031-21244-4_24

follows such a three-valued situation table [1, 2, 13–16, 18, 19, 24–26], although an extension to multi-valued situation tables has been presented in [21].

One of the essential topics in conflict analysis is the relationships between agents. There are typically three kinds of agent relationships: alliance, conflict, and neutrality. The use of threes, such as three ratings and three agent relations, links conflict analysis to the theory of three-way decision [20, 22] that exploits a philosophy of thinking in threes, a methodology of working in threes, and a mechanism of processing in threes. In a more concrete and narrow sense, the Trisecting-Acting-Outcome (TAO) model is presented to model the general process of three-way decision [20]. It includes a Trisecting step of dividing a whole into three parts, an Acting step of devising and applying appropriate strategies to process the three parts, and an Outcome step of evaluating and optimizing the overall results. The connections between conflict analysis and three-way decision lead to the topic of three-way conflict analysis, which attracts increasing attention from researchers [4, 12, 14–16, 18, 21, 24–26].

A common way to analyze agent relations is by means of an auxiliary function. Typically, an auxiliary function uses +1, −1, 0 to qualitatively denote the alliance, conflict, and neutrality relations. Regarding a single issue, it is commonly agreed that two agents are allied if they hold the same non-neutral (i.e., +1 or −1) rating, and they are in conflict if one holds a positive and the other holds a negative rating. Different opinions have been argued about the cases when at least one neutral rating of 0 is involved [16, 17, 21]. With respect to a set of issues, an aggregation is often applied to the auxiliary function. A mainly used aggregation is to simply take the average over individual issues, which results in values from the interval $[-1, +1]$. After that, a pair of thresholds, typically one positive and one negative, can be applied to cut off the values into three regions, corresponding to the alliance, conflict, and neutrality relations. Luo et al. [16] point out a semantics issue associated with such an aggregation of an auxiliary function and present a pair of alliance and conflict functions instead to analyze the agent relations. The alliance and conflict functions are aggregated separately by taking the average and are considered together to decide the relation between agents.

While the existing works are confined mainly with the use of auxiliary, alliance, conflict functions and their aggregations, we present a probabilistic perspective of analyzing agent relations based on Bayesian confirmation. Take the alliance relation as an example. For a given agent a, we could formulate a hypothesis that an agent is allied with a. The ratings on a set of issues together form a description or a piece of evidence of an agent. Bayesian confirmation measures are adopted to quantify the degree to which a rating confirms or disconfirms the hypothesis that the corresponding agent is allied with a. Accordingly, we apply two thresholds to construct three regions of ratings that confirm, disconfirm, and are neutral to the hypothesis. Three types of confirmation rules can be correspondingly induced to predict the agent relations based on ratings. This presented approach eliminates the need for any auxiliary function to represent the relations and the associated semantics issues of interpreting the aggregation of a function. Furthermore, the presented probabilistic view is novel and unique in three-way conflict analysis. The ratings over a set of issues are considered together to form a description or a piece of evidence of an agent, rather than aggregating the ratings or an auxiliary function to produce a single value.

The rest of this paper is arranged as follows. Section 2 briefly reviews the basic concepts in Bayesian confirmation and conflict analysis. The presented approach is then discussed in Sect. 3 and illustrated with an example in Sect. 4. Section 5 summarizes the work and discusses possible directions of future work.

2   Overviews of Bayesian Confirmation and Conflict Analysis

This section reviews the basic concepts of Bayesian confirmation and analyzing agent relationships in conflict analysis.

2.1   Bayesian Confirmation

Confirmation theory studies how a piece of evidence e confirms a hypothesis h. In general, the concept of confirmation can be defined through standard logic or probability theory. The logic branch is built on Hempel's work that describes instantial confirmation as a relation between instances and propositions [8]. In a nutshell, an instance confirms a proposition if it satisfies both the condition and the conclusion in the proposition; it disconfirms the proposition if it satisfies the condition but not the conclusion; otherwise, it is neutral/irrelevant to the proposition. For the probability branch, Carnap [3,5] presents that a precise probabilistic explication of confirmation should include the following three aspects:

- Qualitative confirmation: e inductively supports h;
- Comparative confirmation: e supports h more strongly than e' supports h';
- Quantitative confirmation: e inductively supports h to a degree.

In particular, Bayesian confirmation considers the posterior probability $Pr(h|e)$ and the prior probability $Pr(h)$ to interpret the above three aspects. There are two ways to interpret the qualitative confirmation, that is, the absolute and incremental confirmation [6]. The absolute confirmation simply compares the posterior probability with a given threshold. Accordingly, e confirms h if $Pr(h|e)$ is greater than the threshold. The incremental confirmation compares the posterior probability to the prior probability and interprets the qualitative confirmation as follows:

$$\begin{cases} e \text{ confirms } h, & \text{iff } Pr(h|e) > Pr(h); \\ e \text{ is (confirmationally) irrelevant/neutral to } h, & \text{iff } Pr(h|e) = Pr(h); \\ e \text{ disconfirms } h, & \text{iff } Pr(h|e) < Pr(h). \end{cases}$$

For quantitative confirmation, Bayesian confirmation applies various quantitative confirmation measures that involve the posterior probability $Pr(h|e)$ and the prior probability $Pr(h)$. A few equivalent interpretations of the qualitative conditions inspire a series of commonly used quantitative measures. For example, the condition $Pr(h|e) > Pr(h)$ could be equivalently expressed as follows by applying Baye's theorem [6, 9, 10, 23]:

$$Pr(h|e) > Pr(h) \iff \frac{Pr(h|e)}{Pr(h)} > 1 \iff \frac{Pr(e|h)}{Pr(e)} > 1 \iff \frac{Pr(e|h)}{Pr(e|\neg h)} > 1, \quad (1)$$

where $\neg h$ represents the negation of h and all denominators are assumed to be non-zero. These equivalent formulations give rise to the following quantitative confirmation measures [5]:

$$c_d(h, e) = Pr(h|e) - Pr(h),$$
$$c_r(h, e) = \frac{Pr(h|e)}{Pr(h)} = \frac{Pr(e|h)}{Pr(e)},$$
$$c_r^+(h, e) = \frac{Pr(e|h)}{Pr(e|\neg h)}. \quad (2)$$

There is a special number that represents neutrality in a quantitative measure. For example, this number is 0 in $c_d(h, e)$ and 1 in all the other measures in Eq. (2). To be consistent with the qualitative confirmation, the following set of qualitative constraints is usually required in a quantitative measure:

$$c(h, e) \begin{cases} > 0, & \text{iff } e \text{ confirms } h, \\ = 0, & \text{iff } e \text{ is neutral to } h, \\ < 0, & \text{iff } e \text{ disconfirms } h. \end{cases} \quad (3)$$

A quantitative measure is called a relevance measure if it satisfies the above qualitative constraints. Apparently, $c_d(h, e)$ is a relevance measure. The other measures in Eq. (2) could be normalized into the following relevance measures [5–7]:

$$c_{nr}(h, e) = c_r(h, e) - 1,$$
$$c_{nr}^+(h, e) = c_r^+(h, e) - 1,$$
$$c_{lr}(h, e) = \log c_r(h, e),$$
$$c_{lr}^+(h, e) = \log c_r^+(h, e). \quad (4)$$

The comparative confirmation could be easily interpreted through quantitative confirmation measures by comparing $c(h, e)$ and $c(h', e')$.

2.2 Analyzing Agent Relationships in Conflict Analysis

Conflict analysis can be formulated based on a situation table that describes the attitudes of agents toward issues. Formally, a situation table can be represented by the following triplet [17, 21]:

$$S = (A, I, r), \quad (5)$$

where A is a finite nonempty set of agents, I is a finite nonempty set of issues, and $r : A \times I \to \{+1, 0, -1\}$ is a rating function with the ratings interpreted as: for $a \in A, i \in I$,

$$\begin{cases} r(a, i) = +1, & \text{iff } a \text{ is positive on } i; \\ r(a, i) = 0, & \text{iff } a \text{ is neutral on } i; \\ r(a, i) = -1, & \text{iff } a \text{ is negative on } i. \end{cases}$$

Yao [21] discusses an extension of the above three-valued situation table into a many-valued situation table where a rating could take any value between the interval $[-1, 1]$. In this work, we limit our discussion to three-valued situation tables.

One of the essential topics in conflict analysis is to study the relationships between agents. Typically, there are three relationships between two agents, namely, alliance, conflict, and neutrality relations. With respect to a single issue, the followings are generally agreed:

– an agent is self-allied;
– two positive ratings or two negative ratings lead to an alliance relation;
– a positive rating and a negative rating lead to a conflict relation;
– a neutral rating and a non-neutral rating leads to a neutrality relation.

Different opinions have been presented when two neutral ratings are involved. In this case, Pawlak [17] considers a neutrality relation if the two agents are different persons/organizations and an alliance relation if they are actually the same (i.e., an agent is self-allied). Instead, Yao [21] argues that two neutral ratings also express a kind of alliance between agents and, therefore, considers an alliance relation no matter whether the two neutral ratings are from the same agent or different agents. Luo et al. [16] summarize all meaningful possibilities into a general framework, which could cover both Pawlak's and Yao's opinions.

The formal definitions of the relationships between agents are commonly formulated through an auxiliary function. An auxiliary function is usually defined in terms of ratings and uses the three values of $+1$, -1, and 0 to denote the alliance, conflict, and neutrality relations, respectively. For two agents $x, y \in A$ and an issue $i \in I$, the above Pawlak's opinion leads to the following auxiliary function [17]:

$$\Phi_i^P(x, y) = \begin{cases} +1, & r(x, i) \cdot r(y, i) = +1 \text{ or } x = y, \\ -1, & r(x, i) \cdot r(y, i) = -1, \\ 0, & r(x, i) \cdot r(y, i) = 0 \text{ and } x \neq y. \end{cases} \tag{6}$$

In contrast, Yao's auxiliary function is defined as [21]:

$$\Phi_i^Y(x, y) = \begin{cases} +1, & r(x, i) = r(y, i), \\ -1, & r(x, i) \cdot r(y, i) = -1, \\ 0, & r(x, i) \cdot r(y, i) = 0 \text{ and } r(x, i) \neq r(y, i). \end{cases} \tag{7}$$

Using an auxiliary function Φ_i, the three relations with respect to a single issue $i \in I$ can be defined as a trisection of agent pairs in $A \times A$ [21]:

$$R_i^= = \{(x,y) \in A \times A \mid \Phi_i(x,y) = +1\},$$
$$R_i^{\asymp} = \{(x,y) \in A \times A \mid \Phi_i(x,y) = -1\},$$
$$R_i^{\approx} = \{(x,y) \in A \times A \mid \Phi_i(x,y) = 0\}, \tag{8}$$

where $R_i^=$, $R_i^{\asymp}$, and $R_i^{\approx}$ denote the alliance, conflict, and neutrality relations, respectively.

When it comes to a set of issues $J \subseteq I$, the relations between two agents are commonly defined by aggregating their relations on every single issue from J. For example, one may define an aggregated auxiliary function by taking average as [16]:

$$\Phi_J(x,y) = \frac{\sum\limits_{i \in J} \Phi_i(x,y)}{|J|}, \tag{9}$$

where $|\cdot|$ denotes the cardinality. Then the three relations can be defined by applying two thresholds on the aggregated auxiliary function. Formally, for a pair of thresholds (α, β) with $-1 \leq \beta < 0 < \alpha \leq +1$, we have:

$$R_J^= = \{(x,y) \in A \times A \mid \Phi_J(x,y) \geq \alpha\},$$
$$R_J^{\asymp} = \{(x,y) \in A \times A \mid \Phi_J(x,y) \leq \beta\},$$
$$R_J^{\approx} = \{(x,y) \in A \times A \mid \beta < \Phi_J(x,y) < \alpha\}. \tag{10}$$

Then the relations regarding a single issue defined in Equation (8) become special cases of those in Eq. (10) with $J = \{i\}$, $\alpha = +1$, and $\beta = -1$. Luo et al. [16] further consider the aggregations of an auxiliary function with respect to a set of issues, a set of agents as the first parameter, a set of agents as the second parameter, and any combination of the three.

Instead of aggregating an auxiliary function, the relations can also be defined through aggregating the rating function r or a distance function. Pawlak [17] defines a distance function on a single issue through his auxiliary function Φ_i^P. Accordingly, an aggregated distance function over J is defined by taking the average and is used to formulate the alliance, conflict, and neutrality relations. Yao [21] defines a conflict function $c_i : A \times A \rightarrow [0,1]$ that represents the degree of conflict between two agents regarding a single issue i. The conflict function is aggregated over J by taking the average and is used to define three levels of conflict between agents. Lang [12] and Lang and Yao [14] consider a pair of alliance and conflict evaluation functions on J that take values from the intervals $[0,1]$ and $[-1,0]$, respectively. Luo et al. [16] present a pair of alliance and conflict functions on i that take values from the set $\{0,1\}$. The two functions are aggregated separately and considered together in defining the alliance, conflict, and neutrality relations between agents. Zhi et al. [26] explore the relations through consistency and inconsistency measures inspired by formal concept analysis. Sun et al. [18] apply the decision-making approach used in probabilistic rough sets over two universes to determine the attitude of an agent towards a set of issues, which can be further used to formulate the agent relations.

3 Analyzing Agent Relationships Based on Bayesian Confirmation

For a set of issues $J \subseteq I$ of interest, the ratings of an agent toward these issues form a piece of evidence about the agent. We consider a set representation to formulate such a rating, which collects all the pairs of an issue from J and the corresponding rating from $\{+1, 0, -1\}$. For example, for $J = \{i_1, i_2\} \subseteq I$, $\{\langle i_1, +1 \rangle, \langle i_2, 0 \rangle\}$ is a rating on J, which represents a positive rating on the issue i_1 and a neutral rating on i_2. Furthermore, we treat the empty rating, represented by $\emptyset$, as valid, and it is the only rating that could be formulated in a special case of $J = \emptyset$. We define the following concept of rating spaces by considering all the possibilities of a rating on J.

Definition 1. *For a given set of issues $J = \{i_1, i_2, \cdots, i_{|J|}\} \subseteq I$, the rating space RAT_J on J is defined as follows:*

$$\mathrm{RAT}_J = \begin{cases} \{\emptyset\}, & J = \emptyset, \\ \{ \{\langle i_1, v_1 \rangle, \langle i_2, v_2 \rangle, \cdots, \langle i_{|J|}, v_{|J|} \rangle\} \mid v_1, v_2, v_{|J|} \in \{-1, 0, +1\} \}, & J \neq \emptyset, \end{cases} \tag{11}$$

where $| \cdot |$ represents the cardinality of a set.

A rating in RAT_J contains exactly one pair for each issue in J. For simplicity, we will use RAT_i to denote a rating space on a singleton set $\{i\}$. Apparently, the cardinality of the rating space on J is:

$$|\mathrm{RAT}_J| = 3^{|J|}. \tag{12}$$

Consider a simple example of $J = \{i_1, i_2\}$. The corresponding rating space is:

$$\mathrm{RAT}_{\{i_1, i_2\}} = \{\{\langle i_1, +1 \rangle, \langle i_2, +1 \rangle\}, \{\langle i_1, +1 \rangle, \langle i_2, 0 \rangle\}, \{\langle i_1, +1 \rangle, \langle i_2, -1 \rangle\},$$
$$\{\langle i_1, 0 \rangle, \langle i_2, +1 \rangle\}, \{\langle i_1, 0 \rangle, \langle i_2, 0 \rangle\}, \{\langle i_1, 0 \rangle, \langle i_2, -1 \rangle\},$$
$$\{\langle i_1, -1 \rangle, \langle i_2, +1 \rangle\}, \{\langle i_1, -1 \rangle, \langle i_2, 0 \rangle\}, \{\langle i_1, -1 \rangle, \langle i_2, -1 \rangle\}\} \tag{13}$$

We represent the family of ratings on any subset of I as:

$$\mathcal{RAT} = \bigcup_{J \subseteq I} \mathrm{RAT}_J. \tag{14}$$

Based on the above concept of rating spaces, we can generalize the rating function r regarding a single agent and a single issue into a rating function r' that takes a single agent and a set of issues as:

$$r' : A \times 2^I \to \mathcal{RAT}, \tag{15}$$

where 2^I represents the power set of I. For a given set of issues $J \subseteq I$, the rating $r'(a, J) \in \mathrm{RAT}_J$ is a piece of evidence about an agent $a \in A$. For simplicity, we will omit the superscript $'$ and use r to represent both rating functions in our following discussion, where they can be easily distinguished without any confusion.

There are three relations between agents, namely, alliance, conflict, and neutrality. We focus on analyzing the alliance relation in this work. The other two relations can be similarly studied. From the view of confirmation, for a given set of issues $J \subseteq I$, we formulate two complementary hypotheses regarding a given agent $a \in A$: (1) an agent is an ally of a, represented by $R^=_{(a,J)}$; (2) an agent is not an ally of a, represented by $\neg R^=_{(a,J)}$. We follow a few recent works [12,16,21] that use the superscripts $=$, $\times$, and $\approx$ to represent alliance, conflict, and neutrality relations, respectively. It should be noted that an agent does not necessarily have a conflict relation with a in the case of the hypothesis $\neg R^=_{(a,J)}$. They may also have a neutrality relation.

There is a subtle issue that needs to be solved before we could investigate the confirmation relationships between ratings and the above hypotheses. For a rating $rat \in \text{RAT}_J$, the posterior probability $Pr(R^=_{(a,J)}|rat)$ is necessary in calculating a Bayesian confirmation measure $c(R^=_{(a,J)}, rat)$. However, the posterior probability is not available for a rating that does not appear in a given situation table. Therefore, we focus on the set of ratings with respect to J that appear in a given situation table S, denoted as RAT^S_J, and construct the following three confirmation regions regarding the allies of an agent.

Definition 2. *Given a set of issues $J \subseteq I$ and an agent $a \in A$ in a situation table S, we construct the following confirmatory* CON, *disconfirmatory* DIS, *and neutral* NEU *regions regarding the allies of a:*

$$\text{CON}^=(a, J) = \{rat \in \text{RAT}^S_J \mid c(R^=_{(a,J)}, rat) \geq s\},$$

$$\text{DIS}^=(a, J) = \{rat \in \text{RAT}^S_J \mid c(R^=_{(a,J)}, rat) \leq t\},$$

$$\text{NEU}^=(a, J) = \{rat \in \text{RAT}^S_J \mid t < c(R^=_{(a,J)}, rat) < s\}, \tag{16}$$

where c is a Bayesian confirmation measure and s, t are two thresholds satisfying $t < s$.

Extra conditions may be introduced to make the ranges of the two thresholds s and t meaningful. Generally, the value that represents neutrality in a Bayesian confirmation measure should be in between s and t. For example, we would require $t < 0 < s$ if a relevance measure is used and $0 < t < 1 < s$ if c_r in Eq. (2) is used.

We could formulate three types of confirmation rules from the three confirmation regions. Specifically, we have the confirmatory $C^=$, disconfirmatory $D^=$, and neutral $N^=$ rules as follows:

$(C^=)$ $\forall rat \in \text{CON}^=(a, J), rat \rightarrow^C R^=_{(a,J)},$

$(D^=)$ $\forall rat \in \text{DIS}^=(a, J), rat \rightarrow^D R^=_{(a,J)},$ or equivalently, $rat \rightarrow^C \neg R^=_{(a,J)},$

$(N^=)$ $\forall rat \in \text{NEU}^=(a, J), rat \rightarrow^N R^=_{(a,J)},$ \hfill (17) where $\to^C$, $\to^D$, and $\to^N$ represent confirmation, disconfirmation, and neutrality, respectively. For an arbitrary agent b, if $r(b, J) \in \text{CON}^=(a, J)$, or equivalently, we have a confirmatory rule $r(b, J) \to^C R^=_{(a,J)}$, then b is confirmed to be in an alliance relation with a, that is, the rating $r(b, J)$ increases our belief that b is an ally of a. Similarly, if $r(b, J) \in \text{DIS}^=(a, J)$, or equivalently, we have a disconfirmatory rule $r(b, J) \to^D R^=_{(a,J)}$, then b is disconfirmed to be in an alliance relation with a, that is, the rating $r(b, J)$ decreases our belief that b is an ally of a. Otherwise, b is neither confirmed nor disconfirmed to be an ally of a. In other words, its rating $r(b, J)$ does not affect our belief on b being an ally of a.

Our formulation is different from the existing studies on agent relationships in a few aspects. Firstly, we present a rating-oriented approach that trisects the ratings instead of agents or agent pairs. A common idea in existing studies is to trisect all pairs of agents into three regions of the alliance, conflict, and neutrality relations, as introduced in Sect. 2.2. Such an agent-oriented approach may introduce difficulties in analyzing agent relationships. Although the relations between agents are specified, the rules or reasons behind them are not clearly expressed, making it challenging to understand and interpret the relations. In contrast, our approach trisects the ratings and accordingly, the rules can be easily formulated. The whole process is not related to any specific agent, making the rules general and easily applicable to any agent, especially when new agents join the situation. Moreover, the ratings can be considered as representations of agents. Agents with the same rating are reasonably analyzed in the same way, making ratings meaningful and intuitive. Secondly, we adopt a probabilistic view by applying Bayesian confirmation instead of aggregating ratings in existing studies. Furthermore, rather than solely the posterior probability, we compare the posterior and prior probabilities, which reflects the impact of ratings on changing our belief of the relation.

The three types of confirmation rules may help an agent take appropriate strategies in building relationships with others, especially with new agents joining the situation. If an agent b has a rating in the confirmatory region $\text{CON}^=(a, J)$, the rating of b significantly increases the probability of b being an ally of a. Therefore, it is promising for a to invest efforts in building an alliance relation with b. In contrast, if b holds a rating from the disconfirmatory region $\text{DIS}^=(a, J)$, the rating of b significantly decreases the probability of b being an ally of a. In this case, a may expect b to be a potential enemy regarding the set of issues J. Otherwise, if b holds a rating from the neutral region $\text{NEU}^=(a, J)$, the rating of b does not affect the probability of b being an ally of a. Thus, a may look for factors other than ratings in order to pursue an alliance relation with b.

The above idea of taking different strategies to build relationships based on a trisecting relates our framework with the Trisecting-Acting-Outcome (TAO) model of three-way decision proposed by Yao [20]. As the name suggests, the TAO model involves three steps that are shown in Fig. 1: (1) a trisecting step that divides a whole into three parts; (2) an acting step that devises and applies appropriate strategies to deal with the three parts; (3) an outcome-evaluation step that evaluates the overall results from the previous two steps. In contrast to the trisecting step that has been widely studied, the acting and outcome-evaluation steps are only explored in a few recent works [11,22].

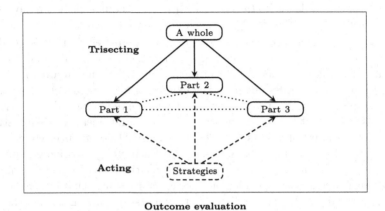

Fig. 1. The TAO model of three-way decision [22]

Following the TAO model in Fig. 1, we summarize our framework of analyzing alliance relation based on Bayesian confirmation as given in Fig. 2. We omit the third step of outcome evaluation as it is not explored in the proposed framework. Furthermore, the evaluation of the analysis of agent relationships is not well studied in the literature. It could be an interesting direction for the future work of our framework.

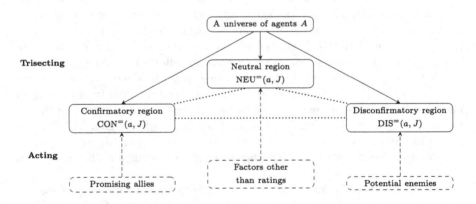

Fig. 2. Analyzing alliance relation based on Bayesian confirmation

4   An Example

We illustrate the presented model with the situation table S given in Table 1, which describes the opinions of fourteen agents on four issues. Specifically, we have $A = \{a_1, a_2, a_3, a_4, a_5, a_6, a_7, a_8, a_9, a_{10}, a_{11}, a_{12}, a_{13}, a_{14}\}$ and $I = \{i_1, i_2, i_3, i_4\}$, denoting the set of agents and the set of issues, respectively.

Table 1. A situation table

	i_1	i_2	i_3	i_4
a_1	+1	−1	0	0
a_2	−1	0	−1	+1
a_3	0	0	−1	0
a_4	−1	−1	−1	−1
a_5	0	+1	0	+1
a_6	−1	0	−1	0
a_7	−1	+1	−1	−1
a_8	0	+1	0	0
a_9	+1	+1	0	+1
a_{10}	−1	−1	−1	0
a_{11}	0	+1	0	−1
a_{12}	−1	0	−1	0
a_{13}	+1	+1	0	+1
a_{14}	−1	0	−1	−1

Let us consider the allies of the agent a_1 with respect to the set of issues $J = \{i_1, i_2, i_3\} \subseteq I$. Suppose the prior probability that an agent is allied with a_1 towards J is $Pr(R^{=}_{(a_1, J)}) = 0.3$. For simplicity, we assume this prior probability is given in this simple illustration. In real applications, this prior probability may be estimated based on the relations between agents in previous conflict situations or factors outside the current conflict situation. Suppose the set of allies of a_1 regarding J turns out to be $R^{=}_J(a_1) = \{a_1, a_5, a_9, a_{11}, a_{12}, a_{13}\}$. Table 2 shows the calculation of the posterior probabilities regarding the ratings in $\text{RAT}^S_J = \{rat_1, rat_2, rat_3, rat_4, rat_5, rat_6, rat_7\}$. For example, for $rat_1 = \{\langle i_1, +1\rangle, \langle i_2, -1\rangle, \langle i_3, 0\rangle\}$, we have:

$$Pr(R^{=}_{(a_1, J)}|rat_1) = \frac{|R^{=}_J(a_1) \cap \{a_1\}|}{|\{a_1\}|} = 1. \tag{18}$$

Suppose we adopt the confirmation measure $c_d(h, e) = Pr(h|e) - Pr(h)$ in Eq. (2). The values of the confirmation measure c_d regarding the ratings in

Table 2. Calculating the posterior probabilities

Rating	Corresponding agents	Posterior probability
$rat_1 = \{\langle i_1, +1\rangle, \langle i_2, -1\rangle, \langle i_3, 0\rangle\}$	$\{a_1\}$	$Pr(R^=_{(a_1,J)}|rat_1) = \frac{1}{1} = 1$
$rat_2 = \{\langle i_1, -1\rangle, \langle i_2, 0\rangle, \langle i_3, -1\rangle\}$	$\{a_2, a_6, a_{12}, a_{14}\}$	$Pr(R^=_{(a_1,J)}|rat_2) = \frac{1}{4} = 0.25$
$rat_3 = \{\langle i_1, 0\rangle, \langle i_2, 0\rangle, \langle i_3, -1\rangle\}$	$\{a_3\}$	$Pr(R^=_{(a_1,J)}|rat_3) = \frac{0}{1} = 0$
$rat_4 = \{\langle i_1, -1\rangle, \langle i_2, -1\rangle, \langle i_3, -1\rangle\}$	$\{a_4, a_{10}\}$	$Pr(R^=_{(a_1,J)}|rat_4) = \frac{0}{2} = 0$
$rat_5 = \{\langle i_1, 0\rangle, \langle i_2, +1\rangle, \langle i_3, 0\rangle\}$	$\{a_5, a_8, a_{11}\}$	$Pr(R^=_{(a_1,J)}|rat_5) = \frac{2}{3} \approx 0.67$
$rat_6 = \{\langle i_1, -1\rangle, \langle i_2, +1\rangle, \langle i_3, -1\rangle\}$	$\{a_7\}$	$Pr(R^=_{(a_1,J)}|rat_6) = \frac{0}{1} = 0$
$rat_7 = \{\langle i_1, +1\rangle, \langle i_2, +1\rangle, \langle i_3, 0\rangle\}$	$\{a_9, a_{13}\}$	$Pr(R^=_{(a_1,J)}|rat_7) = \frac{2}{2} = 1$

Table 1 are computed as:

$$c_d(R^=_{(a_1,J)}, rat_1) = Pr(R^=_{(a_1,J)}|rat_1) - Pr(R^=_{(a_1,J)}) = 1 - 0.3 = 0.7,$$
$$c_d(R^=_{(a_1,J)}, rat_2) = Pr(R^=_{(a_1,J)}|rat_2) - Pr(R^=_{(a_1,J)}) = 0.25 - 0.3 = -0.05,$$
$$c_d(R^=_{(a_1,J)}, rat_3) = Pr(R^=_{(a_1,J)}|rat_3) - Pr(R^=_{(a_1,J)}) = 0 - 0.3 = -0.3,$$
$$c_d(R^=_{(a_1,J)}, rat_4) = Pr(R^=_{(a_1,J)}|rat_4) - Pr(R^=_{(a_1,J)}) = 0 - 0.3 = -0.3,$$
$$c_d(R^=_{(a_1,J)}, rat_5) = Pr(R^=_{(a_1,J)}|rat_5) - Pr(R^=_{(a_1,J)}) = 0.67 - 0.3 = 0.37,$$
$$c_d(R^=_{(a_1,J)}, rat_6) = Pr(R^=_{(a_1,J)}|rat_6) - Pr(R^=_{(a_1,J)}) = 0 - 0.3 = -0.3,$$
$$c_d(R^=_{(a_1,J)}, rat_7) = Pr(R^=_{(a_1,J)}|rat_7) - Pr(R^=_{(a_1,J)}) = 1 - 0.3 = 0.7. \tag{19}$$

By Definition 2, with a pair of thresholds $s = 0.2$ and $t = -0.2$, we construct the three regions as:

$$\text{CON}^=(a_1, J) = \{rat \in \text{RAT}^S_J \mid c(R^=_{(a_1,J)}, rat) \geq 0.2\}$$
$$= \{rat_1, rat_5, rat_7\};$$
$$\text{DIS}^=(a_1, J) = \{rat \in \text{RAT}^S_J \mid c(R^=_{(a_1,J)}, rat) \leq -0.2\}$$
$$= \{rat_3, rat_4, rat_6\};$$
$$\text{NEU}^=(a_1, J) = \{rat \in \text{RAT}^S_J \mid -0.2 < c(R^=_{(a_1,J)}, rat) < 0.2\}$$
$$= \{rat_2\}. \tag{20}$$

We divide the set RAT^S_J into three disjoint parts $\text{CON}^=(a_1, J)$, $\text{DIS}^=(a_1, J)$, and $\text{NEU}^=(a_1, J)$ with respect to the agent a_1. Accordingly, we formulate the confirmation rules as:

$$(\text{C}^=) \quad rat_1 = \{\langle i_1, +1\rangle, \langle i_2, -1\rangle, \langle i_3, 0\rangle\} \rightarrow^C R^=_{(a_1,J)},$$
$$rat_5 = \{\langle i_1, 0\rangle, \langle i_2, +1\rangle, \langle i_3, 0\rangle\} \rightarrow^C R^=_{(a_1,J)},$$
$$rat_7 = \{\langle i_1, +1\rangle, \langle i_2, +1\rangle, \langle i_3, 0\rangle\} \rightarrow^C R^=_{(a_1,J)};$$
$$(\text{D}^=) \quad rat_3 = \{\langle i_1, 0\rangle, \langle i_2, 0\rangle, \langle i_3, -1\rangle\} \rightarrow^D R^=_{(a_1,J)},$$
$$rat_4 = \{\langle i_1, -1\rangle, \langle i_2, -1\rangle, \langle i_3, -1\rangle\} \rightarrow^D R^=_{(a_1,J)},$$
$$rat_6 = \{\langle i_1, -1\rangle, \langle i_2, +1\rangle, \langle i_3, -1\rangle\} \rightarrow^D R^=_{(a_1,J)};$$
$$(\text{N}^=) \quad rat_2 = \{\langle i_1, -1\rangle, \langle i_2, 0\rangle, \langle i_3, -1\rangle\} \rightarrow^N R^=_{(a_1,J)}. \tag{21}$$

Using the rating from an agent on J as a piece of evidence, one can classify the agents in A into three disjoint parts according to the three types of confirmation rules. If an agent $a \in A$ has a rating from $\{rat_1, rat_5, rat_7\}$ on J, it increases the probability of a being allied with a_1; if a has a rating from $\{rat_3, rat_4, rat_6\}$, it decreases the probability of a being allied with a_1; if a has a rating from $\{rat_2\}$, it does not affect the probability of a being allied with a_1. In other words, the agents with the ratings rat_1, rat_5, and rat_7 are more likely to be allied with a_1; the agents with the rating rat_2 are on-average likely to be allied with a_1; the agents with the ratings rat_3, rat_4, and rat_6 are less likely to be allied with a_1. If the agent a_1 intends to win more allies, it must pay more attention to agents with ratings rat_1, rat_2, rat_5, and rat_7.

5 Conclusion and Future Work

This paper applies Bayesian confirmation theory to analyzing agent relationships in three-way conflict analysis. To formulate the approach, we present a formal representation of a rating over a set of issues as a set of issue-rating pairs. From the view of Bayesian confirmation, such a rating is considered as a piece of evidence regarding a corresponding agent. Quantitative Bayesian confirmation measures are adopted to evaluate the degree to which a rating confirms or disconfirms the hypothesis that a corresponding agent is allied with a given agent. By applying a pair of thresholds, we construct three confirmation regions of confirmatory, disconfirmatory, and neutral ratings and induce the corresponding confirmation rules. These rules may help us understand and predict the allies and enemies of a given agent and accordingly, devise appropriate strategies to maintain or develop desired relationships with agents.

The presented approach introduces a probabilistic view of studying agent relations in conflict analysis, which is very different from the existing common idea of aggregating auxiliary functions. There are a few directions for further exploring this probabilistic approach, such as the estimation of prior probability and the computation of the two thresholds. Furthermore, one can analyze the conflict relation in a similar way and synthesize the results with the analysis regarding the alliance relation to arrive at a final decision. This requires a combination of alliance/non-alliance and conflict/non-conflict into alliance/conflict/neutrality, which is a very interesting topic.

Acknowledgement. The authors thank the reviewers for their valuable comments and suggestions. This work is partially supported by the National Natural Science Foundation of China (No. 62076040), Hunan Provincial Natural Science Foundation of China (Nos. 2020JJ3034, 2020JJ4598), Hunan Provincial Key Laboratory of Intelligent Computing and Language Information Processing (No. 2018TP1018), and the Scientific Research Fund of Chongqing Key Laboratory of Computational Intelligence (No. 2020FF04).

References

1. Ali, A., Ali, M.I., Rehman, N.: New types of dominance based multi-granulation rough sets and their applications in conflict analysis problems. J. Intell. Fuzzy Syst. **35**, 3859–3871 (2018)
2. Bashir, Z., Mahnaz, S., Malik, M.G.A.: Conflict resolution using game theory and rough sets. Int. J. Intell. Syst. **36**, 237–259 (2020)
3. Carnap, R.: Logical Foundations of Probability, 1st edn. University of Chicago Press, Chicago (1950)
4. Du, J., Liu, S., Yong, L., Yi, J.: A novel approach to three-way conflict analysis and resolution with pythagorean fuzzy information. Inf. Sci. **584**, 65–88 (2022)
5. Festa, R.: Bayesian confirmation. In: Galavotti, M.C., Pagnini, A. (eds.) Experience, Reality, and Scientific Explanation, WONS, vol. 61, pp. 55–87. Springer, Dordrecht (1999). https://doi.org/10.1007/978-94-015-9191-1_4
6. Fitelson, B.: Studies in Bayesian confirmation theory. Ph.D. thesis, University of Wisconsin (2001). https://fitelson.org/thesis.pdf
7. Greco, S., Słowiński, R., Szczęch, I.: Measures of rule interestingness in various perspectives of confirmation. Inf. Sci. **346–347**, 216–235 (2016)
8. Hempel, C.: Studies in the logic of confirmation. Mind **54**, 97–121 (1945)
9. Hu, M.: Three-way Bayesian confirmation in classifications. Cogn. Comput. (2021). https://doi.org/10.1007/s12559-021-09924-8
10. Hu, M., Deng, X., Yao, Y.: An application of Bayesian confirmation theory for three-way decision. In: Mihálydeák, T., et al. (eds.) IJCRS 2019. LNCS (LNAI), vol. 11499, pp. 3–15. Springer, Cham (2019). https://doi.org/10.1007/978-3-030-22815-6_1
11. Jiang, C., Guo, D., Duan, Y., Liu, Y.: Strategy selection under entropy measures in movement-based three-way decision. Int. J. Approximate Reasoning **119**, 280–291 (2020)
12. Lang, G.: A general conflict analysis model based on three-way decision. Int. J. Mach. Learn. Cybern. **11**(5), 1083–1094 (2020). https://doi.org/10.1007/s13042-020-01100-y
13. Lang, G., Miao, D., Hamido, F.: Three-way group conflict analysis based on pythagorean fuzzy set theory. IEEE Trans. Fuzzy Syst. **28**(3), 447–461 (2020)
14. Lang, G., Yao, Y.: New measures of alliance and conflict for three-way conflict analysis. Int. J. Approximate Reasoning **132**, 49–69 (2021)
15. Li, X., Wang, X., Lang, G., Yi, H.: Conflict analysis based on three-way decision for triangular fuzzy information systems. Int. J. Approximate Reasoning **132**, 88–106 (2021)
16. Luo, J., Hu, M., Lang, G., Yang, X., Qin, K.: Three-way conflict analysis based on alliance and conflict functions. Inf. Sci. **594**, 322–359 (2022)
17. Pawlak, Z.: An inquiry into anatomy of conflicts. Inf. Sci. **109**, 65–78 (1998)
18. Sun, B., Chen, X., Zhang, L., Ma, W.: Three-way decision making approach to conflict analysis and resolution using probabilistic rough set over two universes. Inf. Sci. **507**, 809–822 (2020)
19. Tong, S., Sun, B., Chu, X., Zhang, X., Wang, T., Jiang, C.: Trust recommendation mechanism-based consensus model for Pawlak conflict analysis decision making. Int. J. Approximate Reasoning **135**, 91–109 (2021)
20. Yao, Y.: Three-way decision and granular computing. Int. J. Approximate Reasoning **103**, 107–123 (2018)

21. Yao, Y.: Three-way conflict analysis: Reformulations and extensions of the Pawlak model. Knowl.-Based Syst. **180**, 26–37 (2019)
22. Yao, Y.: Tri-level thinking: models of three-way decision. Int. J. Mach. Learn. Cybern. **11**, 947–959 (2019)
23. Yao, Y., Zhou, B.: Two Bayesian approaches to rough sets. Eur. J. Oper. Res. **251**, 904–917 (2016)
24. Yi, H., Zhang, H., Li, X., Yang, Y.: Three-way conflict analysis based on hesitant fuzzy information systems. Int. J. Approximate Reasoning **139**, 12–27 (2021)
25. Zhang, X., Chen, J.: Three-hierarchical three-way decision models for conflict analysis: a qualitative improvement and a quantitative extension. Inf. Sci. **587**, 485–514 (2022)
26. Zhi, H., Qi, J., Qian, T., Ren, R.: Conflict analysis under one-vote veto based on approximate three-way concept lattice. Inf. Sci. **516**, 316–330 (2020)

Hierarchical Multi-granulation Sequential Three-Way Decisions

Chengxin Hong[1], Jin Qian[1(✉)], Haoying Jiang[1], Zhigang Tong[1], Ying Yu[1], and Caihui Liu[2]

[1] School of Software, East China Jiaotong University,
Nanchang 330013, Jiangxi, China
qjqjlqyf@163.com

[2] Department of Mathematics and Computer Science,
Gannan Normal University, Ganzhou 341000, China

Abstract. In granular computing, a single conditional attribute is usually used as a view to describe the target concept, and each view can choose a specific level of granularity to describe the object in the hierarchical rough set model. However, the existing three-way decision model cannot combine multi-level and multi-view to make decisions, and these models are extremely complicated and difficult to apply. Within the multi-level data, how to obtain a certain decision from different levels and views is the most important issue. To this end, we propose a hierarchical multi-granulation sequential three-way decision model by combining multi-granularity and sequential three-way decisions. Specifically, we construct concept hierarchy tree of conditional attribute, then construct granular view under different levels of granularity, and update the information by multi-step three-way decision-making method. Finally, the experimental results demonstrate that the proposed model can mine the rules of hierarchical decision table. The model will improve the theoretical framework of hierarchical rough set model.

Keywords: Hierarchical rough set · Multi-granulation · Three-way decisions

1 Introduction

Sequential three-way decisions (S3WD) [1] is the closest approach to the way human brain think. The key to sequential three-way decisions is to transform delayed decisions into definite (accept and reject) decisions by adding additional information. Qian et al. [2] combined multi-granularity and three-way decision to implement five multi-granulation sequential three-way decisions models with typical aggregation strategies. Qian et al. [3] combined the hierarchical rough set and three-way decision to propose a hierarchical sequential three-way decisions model.

On the other hand, Granular computing(GrC) is a method to simulate human thinking and solve problems, the cognitive limitation of human beings and often

© The Author(s), under exclusive license to Springer Nature Switzerland AG 2022
J. Yao et al. (Eds.): IJCRS 2022, LNAI 13633, pp. 334–345, 2022.
https://doi.org/10.1007/978-3-031-21244-4_25

divide data into "granule" to observe, analyze and solve problems under different levels of granularity. Thus, how to describe the granule is an important issue [4]. Qian et al. [5] proposed multi-granulation rough set to extend the classical single-granulation rough set, and defined the approximation of the set by using the multiple equivalence relations in the universe. Feng and Miao [6] proposed a hierarchical rough set model to transform one-dimensional data into multi-dimensional data by constructing a concept hierarchy tree. Wu and Leung [7,8] proposed the multi-scale information table using multi-scale granular labeled partition to describe the information granules of the scale. Hao et al. [9] introduced sequential three-way decisions into the multi-scale decision table to study the optimal scale selection problem of dynamic sequential update information.

A view is usually chosen to process data in the classical rough set models, and a specific level is usually selected to describe the target concept for each view in the hierarchical rough set models. Indeed, the existing three-way decision model obtain information that cannot reflect multi-level and multi-view decisions. In other words, the existing hierarchical rough set model cannot solve this type of problem well. It is necessary to consider constructing different levels of granular views in the hierarchical rough set models. The work of this paper provides an in-depth study of this issue.

Three main contributions of this paper. Firstly, concept hierarchy tree of conditional attribute is constructed and hierarchical decision table is defined. Secondly, we construct granular view under different levels of granularity based on indistinguishable relations. Finally, we propose a hierarchical multi-granulation sequential three-way decision model by combining multi-granularity and sequential three-way decisions.

The rest of the paper is organized as follows. Section 2 briefly reviews the Pawlak rough set model, hierarchical decision table, multi-granulation rough set and sequential three-way decisions. In Sect. 3, we construct the hierarchical decision table via granular view. Section 4 proposes a hierarchical multi-granulation sequential three-way decision model by combining multi-granularity and sequential three-way decisions, then designs the corresponding algorithms and explore some properties of the proposed model. Section 5 gives the relevant experiments and conclusions. Finally, the paper ends with conclusions and further work in Sect. 6.

2   Preliminaries

In this section, we will review some basic concepts of Pawlak rough set model, multi-granulation rough set, hierarchical decision table and sequential three-way decisions. For a detailed description, please refer to paper [5,6,10–12].

2.1   Pawlak Rough Set

In general, we use a four-tuple $S = (U, AT = C \cup D, \{V_a | a \in AT\}, \{f_a | a \in A\})$ to represent the information system.

Definition 1. *Consider a partition π_{A_i} induced by a set of conditional attribute A_i, the upper and lower approximations of X with respect to the division A_i are defined as follows:*

$$
\begin{aligned}
\underline{apr}_{\pi_{A_i}}(X) &= \{x \in U | [x]_{A_i} \subseteq X\}, \\
\overline{apr}_{\pi_{A_i}}(X) &= \{x \in U | [x]_{A_i} \cap X \neq \phi\}.
\end{aligned}
\tag{1}
$$

where $|\cdot|$ is the cardinal number of elements in the set.

2.2   Hierarchical Decision Table

Feng and Miao [6] proposed hierarchical rough set model by combining concept hierarchy tree and rough set to describe multi-dimensional data.

Definition 2. *Let $HT = (U, AT = \{a_i^l | i = 1, 2, \ldots, s; l = 1, 2, \ldots, m\} \cup \{d\}, V, f)$ be a hierarchical decision table. The index set $L = (l_1, l_2, \ldots, l_s)$ is called a level combination of conditional attributes, which denotes the combination of the conditional attribute a_i at l_i-th levels, $i = 1, 2, \ldots, s$. Each level combination $L = (l_1, l_2, \ldots, l_s)$ can form a single-level information table $C^L = \{a_1^{l_1}, a_2^{l_2}, \ldots, a_s^{l_s}\}$.*

Definition 3. *Given the $L_1 = (l_1^1, l_2^1, \ldots, l_s^1)$-th decision table and $L_2 = (l_1^2, l_2^2, \ldots, l_s^2)$-th decision table. If $l_i^1 \leq l_i^2$ ($i = 1, 2, \ldots, s$), then L_1 is said to be coarser L_2 or L_2 is the finer L_1, and is denoted as $L_1 \succcurlyeq L_2$. Furthermore, if there exists $i = 1, 2, \ldots, s$ such that $l_i^1 < l_i^2$, then L_1 is said to be strictly coarser L_2 or L_2 is the strictly finer L_1, and is denoted as $L_1 \succ L_2$.*

2.3   QIAN's MGRS

In what follows, the optimistic and pessimistic multi-granulation rough set are briefly reviewed below.

Definition 4. *Given a granular structure $GS = \{A_1, A_2, \cdots, A_q\}$ and $\forall X \subseteq U$, the optimistic multi-granulation lower and upper approximations $\sum_{i=1}^{q} A_i^{O}(X)$ and $\overline{\sum_{i=1}^{q} A_i}^{O}(X)$ are defined as follows:*

$$
\sum_{i=1}^{l} A_i^{O}(X) = \{x \in U : \bigvee_{i=1,2,\ldots q} [x]_{A_i} \subseteq X\},
\tag{2}
$$

$$
\overline{\sum_{i=1}^{q} A_i}^{O}(X) = \sim \sum_{i=1}^{q} A_i^{O}(\sim X).
\tag{3}
$$

where $\sim X$ is the complement of set X.

Definition 5. *Given a granular structure* $GS = \{A_1, A_2, \cdots, A_q\}$ *and* $\forall X \subseteq U$, *the pessimistic multi-granulation lower and upper approximations* $\underline{\sum\limits_{i=1}^{q} A_i}^{P}(X)$

and $\overline{\sum\limits_{i=1}^{q} A_i}^{P}(X)$ *are defined as follows:*

$$\underline{\sum\limits_{i=1}^{l} A_i}^{P}(X) = \{x \in U : \bigwedge_{i=1,2,\ldots q} [x]_{A_i} \subseteq X\}, \tag{4}$$

$$\overline{\sum\limits_{i=1}^{q} A_i}^{P}(X) = \sim \underline{\sum\limits_{i=1}^{q} A_i}^{P}(\sim X). \tag{5}$$

where $\sim X$ *is the complement of set* X.

2.4   Sequential Three-Way Decisions

The sequential three-way decision model is the evolution of the multi-step classical three-way decision [12,13]. In what follows, we briefly review the sequential three-way decisions.

Definition 6. *Given a l-th level of the granular structure* $GS^l = \{A_1^l, A_2^l, \cdots, A_q^l\}$, *and a decision class* D_j. *The lower approximation* $\underline{apr}_{\pi_{A_i^l}}(D_j)$ *and the upper approximation* $\overline{apr}_{\pi_{A_i^l}}(D_j)$ *are defined by*

$$\begin{aligned} \underline{apr}_{\pi_{A_i^l}}(D_j) &= \{x \in U^l | [x]_{\pi_{A_i^l}} \subseteq D_j\}, \\ \overline{apr}_{\pi_{A_i^l}}(D_j) &= \{x \in U^l | [x]_{\pi_{A_i^l}} \cap D_j \neq \emptyset\}. \end{aligned} \tag{6}$$

where $U^1 = U$, $U^{l+1} = BND_{\pi_{A_i^l}}(D_j) = \overline{apr}_{\pi_{A_i^l}}(D_j) - \underline{apr}_{\pi_{A_i^l}}(D_j)$, $[x]_{\pi_{A_i^l}}$ *denotes the equivalence class containing* x *in the partition* U^l/A_i^l.

3   Hierarchical Decision Table via Granular View

The conditional attributes a_i is formed along the $l(a_i) + 1$ hierarchy levels: $0, 1, \cdots, l(a_i)$. Level 0 is the special value of Any (*). The hierarchy is constructed differently due to gaps in a priori knowledge.

We take a conditional attribute as a view, and all conditional attributes are combined as a granular view. However, constructing an ideal granular view is a complex and interesting issue. In Fig. 1, we use aggregation and decomposition operations to construct granular views based on indistinguishable relationships. All hierarchical granular views are represented as a lattice, red arrows indicate finer paths and blue arrows indicate coarser paths. For convenience, $gv_i^{l,t}$ denotes granular view, where l denotes level of granular view and t denotes the number of attributes ascending in the granular view. A multi-level granular structure $GS^{l,t}$ with respect to a sequence of granular view $\{gv_1^{l,t}, gv_2^{l,t}, \ldots, gv_q^{l,t}\}$.

In Fig. 1, the top node $gv_1^{1,0} = \{a_1^0, a_2^0, a_3^0\}$ represents the most generalized granular view, while the bottom node $gv_1^{2,3} = \{a_1^1, a_2^1, a_3^1\}$ denotes the detailed granular view. Note that the node $gv_1^{1,3} = \{a_1^1, a_2^1, a_3^1\}$ is the largest granular view in the level 1, and node $gv_1^{2,0} = \{a_1^1, a_2^1, a_3^1\}$ is the smallest granule in the level 2, the two granular views are the same granular view under different levels of granularity.

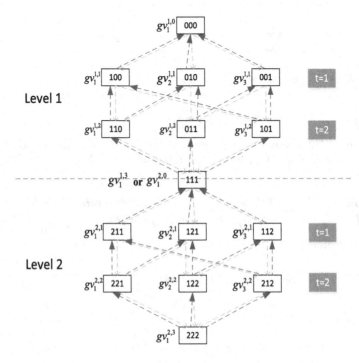

Fig. 1. Granular view under different levels of granularity

4   Hierarchical Multi-granulation Sequential Three-Way Decisions

We combine sequential three-way decisions and multi-granularity to discuss the influence of granular view on decision rule at different levels, then propose a hierarchical multi-granulation sequential three-way decision(HMS3WD) and design algorithms to explore the properties and theorems.

Definition 7. *Given a hierarchical decision table HT, a given decision class D_j, a multilevel granular structure $GS^{l,t} = \{gv_1^{l,t}, gv_2^{l,t}, \ldots, gv_q^{l,t}\}$, $gv_i^{l,t} \in GS^{l,t}$*

be a parallel set of granular view, the lower approximation $\underline{apr}_{\pi_{gv_i^{l,t}}}(D_j)$ *and the*
upper approximation $\overline{apr}_{\pi_{gv_i^{l,t}}}(D_j)$ *are defined by*

$$\underline{apr}_{\pi_{gv_i^{l,t}}}(D_j) = \{x \in U^{l,t} | [x]_{gv_i^{l,t}} \subseteq D_j\},$$

$$\overline{apr}_{\pi_{gv_i^{l,t}}}(D_j) = \{x \in U^{l,t} | [x]_{gv_i^{l,t}} \cap D_j \neq \emptyset\}. \tag{7}$$

where $U^{1,t} = U, U^{l+1,t} = \overline{apr}_{\pi_{gv_i^{l,t}}}(D_j) - \underline{apr}_{\pi_{gv_i^{l,t}}}(D_j)$ *is the gradually reduced*
universe.

Proposition 1. *Given a hierarchical decision table HT, then the three-way decision regions update of the granular view gv with respect to D_j is as follows:*
(1) Positive region

$$POS_{gv_i^{l,t}}(D_j) = \underline{apr}_{\pi_{gv_i^{l,t}}}(D_j)$$

$$= \{x \in U^{l,t} | [x]_{gv_i^{l,t}} \subseteq D_j\}; \tag{8}$$

(2) Boundary region

$$BND_{gv_i^{l,t}}(D_j) = \overline{apr}_{\pi_{gv_i^{l,t}}}(D_j) - \underline{apr}_{\pi_{gv_i^{l,t}}}(D_j)$$

$$= \{x \in U^{l,t} | [x]_{gv_i^{l,t}} \cap D_j \neq \emptyset\}; \tag{9}$$

(3) Negative region

$$NEG_{gv_i^{l,t}}(D_j) = U^{l,t} - \overline{apr}_{\pi_{gv_i^{l,t}}}(D_j)$$

$$= U^{l,t} - POS_{gv_i^{l,t}}(D_j) \cup BND_{gv_i^{l,t}}(D_j). \tag{10}$$

Proof. The equivalence class $[x]_{gv_i^{l,t}}$ of the l-th level will be further divided into the equivalence class $[x]_{gv_i^{l+1,t}}$ of the $(l+1)$-th level. We easily get to know that $[x]_{gv_i^{l,t}} \subseteq [x]_{gv_i^{l-1,t}}$.

(1) For any $x \in POS_{gv_i^{l-1,t}}(D_j)$, we get $[x]_{gv_i^{l-1,t}} \subseteq D_j$. Then, $[x]_{gv_i^{l,t}} \subseteq D_j$ is true when $[x]_{gv_i^{l,t}} \subseteq [x]_{gv_i^{l-1,t}}$. So, we obtain the result that $x \in POS_{gv_i^{l,t}}(D_j)$.

(2) For any $x \in BND_{gv_i^{l,t}}(D_j)$, it is easy to obtain $[x]_{gv_i^{l,t}} \cap D_j \neq \emptyset$ and $[x]_{gv_i^{l,t}} \nsubseteq D_j$, which implies $x \notin NEG_{gv_i^{l,t}}(D_j)$ and $x \notin POS_{gv_i^{l,t}}(D_j)$ due to $[x]_{gv_i^{l,t}} \subseteq [x]_{gv_i^{l-1,t}}$. As a result, $x \in BND_{gv_i^{l-1,t}}(D_j)$ is true.

(3) For any $x \in NEG_{gv_i^{l-1,t}}(D_j)$, we have $[x]_{gv_i^{l,t}} \subseteq D_j$, it is easy to know that $[x]_{gv_i^{l+1,t}} \subseteq D_j$ is true when $[x]_{gv_i^{l,t}} \subseteq [x]_{gv_i^{l-1,t}}$. So, $x \in NEG_{gv_i^{l,t}}(D_j)$. $\square$

In what follows, we construct an algorithm to compute the regions of sequential three-way decisions under a granular structure as shown in Algorithm 1. The main idea of Algorithm 1 is to first deletes the objects belonging to the positive region and negative region under the first level of granularity, and then obtain the update region $U^{2,t} = BND_{gv_i^{1,t}}(D_j)$. For the update region $U^{2,t}$, delete the objects belonging to the positive region and negative region in the next level of granularity, and repeat these steps until the updated universe becomes an empty set or no level of granularity can be computed. It is easy to observe that the time complexity of Algorithm 1 is $O(m|D_j||U|^2)$.

Algorithm 1: Computing the regions of sequential three-way decisions under a granular structure.

---

input : An universal set of object, U; the number of attributes ascending, t; a granular view, $gv_i^{l,t}, l \in \{1, 2, ..., m\}$

output: Three regions, $POS_{gv_i}(D_j)$, $BND_{gv_i}(D_j)$ and $NEG_{gv_i}(D_j)$

1   $POS_{gv_i}(D_j) = \emptyset$, $BND_{gv_i}(D_j) = U$ and $NEG_{gv_i}(D_j) = \emptyset$;

2   $l = 1, U^{1,t} = U$;

3 **for** $l \leftarrow 1$ **to** m **do**

4   if $U^{l,t} = \emptyset$ or $l > m$ turn to 9;

5   Compute $POS_{gv_i^{l,t}}(D_j)$ and $NEG_{gv_i^{l,t}}(D_j)$ according to Definition 7;

6     $POS_{gv_i}(D_j) = POS_{gv_i}(D_j) \cup POS_{gv_i^{l,t}}(D_j)$;

     $NEG_{gv_i}(D_j) = NEG_{gv_i}(D_j) \cup NEG_{gv_i^{l,t}}(D_j)$;

7     $U^{l+1,t} = BND_{gv_i}(D_j) - POS_{gv_i^{l,t}}(D_j) - NEG_{gv_i^{l,t}}(D_j)$;

8     $BND_{gv_i}(D_j) = U^{l+1,t}$;

9 **end**

10 Output $POS_{gv_i}(D_j)$, $BND_{gv_i}(D_j)$ and $NEG_{gv_i}(D_j)$;

Definition 8. *Given a hierarchical decision table HT, a multilevel granular structure $GS^{l,t} = \{gv_1^{l,t}, gv_2^{l,t}, ..., gv_q^{l,t}\}$, $gv_i^{l,t} \in GS^{l,t}$ be a parallel set of granular view. The lower and upper approximations of optimistic hierarchical multi-granulation sequential three-way decision with respect to D_j are defined as*

$$\sum_{i=1}^{l} \overline{gv_i^{l,t}}^{O} (D_j) = \{x \in U : \bigvee_{i=1,2,...q} [x]_{gv_i^{l,t}} \subseteq D_j\}, \tag{11}$$

$$\overline{\sum_{i=1}^{q} gv_i^{l,t}}^{O} (D_j) = \sim \sum_{i=1}^{q} \underline{gv_i^{l,t}}^{O} (\sim D_j). \tag{12}$$

where $U^{1,t} = U, U^{l+1,t} = \overline{\sum_{i=1}^{m} gv_i^{l,t}}^{O} (D_j) - \sum_{i=1}^{m} \underline{gv_i^{l,t}}^{O} (D_j)$ is the gradually reduced universe.

Definition 9. *Given a hierarchical decision table HT, a multilevel granular structure $GS^{l,t} = \{gv_1^{l,t}, gv_2^{l,t}, ..., gv_q^{l,t}\}$, $gv_i^{l,t} \in GS^{l,t}$ be a parallel set of granular view.*

The lower and upper approximations of pessimistic multi-granulation generalized hierarchical decision with respect to D_j are defined as

$$\sum_{i=1}^{l} gv_i^{l,t}\stackrel{P}{} (D_j) = \{x \in U : \mathop{\wedge}_{i=1,2,...l} [x]_{gv_i^{l,t}} \subseteq D_j\}, \tag{13}$$

$$\overline{\sum_{i=1}^{q} gv_i^{l,t}}^P (D_j) = \sim \sum_{i=1}^{q} gv_i^{l,t}\stackrel{P}{} (\sim D_j). \tag{14}$$

where $U^1 = U, U^{l+1} = \overline{\sum_{i=1}^{q} gv_i^{l,t}}^P (D_j) - \sum_{i=1}^{q} gv_i^{l,t}\stackrel{P}{} (D_j)$ *is the gradually reduced universe.*

Definition 10. *Given a multilevel granular structure $GS^{l,t} = \{gv_1^{l,t}, gv_2^{l,t}, \ldots, gv_q^{l,t}\}$, $gv_i^{l,t} \in GS^{l,t}$ be a parallel set of granular view and the decision class partition $\pi_D = \{D_1, D_2, \ldots, D_k\}$, the positive, boundary and negative regions of π_D are defined as follows:*

$$POS^{\Delta}_{GS^{l,t}}(\pi_D) = \mathop{\cup}_{1 \leq j \leq k} \sum_{i=1}^{q} gv_i^{l,t}\stackrel{\Delta}{} (D_j); \tag{15}$$

$$BND^{\Delta}_{GS^{l,t}}(\pi_D) = \mathop{\cup}_{1 \leq j \leq k} (\overline{\sum_{i=1}^{q} gv_i^{l,t}}^{\Delta} (D_j) - \sum_{i=1}^{q} gv_i^{l,t}\stackrel{\Delta}{} (D_j)); \tag{16}$$

$$NEG^{\Delta}_{GS^{l,t}}(\pi_D) = U^{l,t} - POS^{\Delta}_{GS^{l,t}}(\pi_D) \cup BND^{\Delta}_{GS^{l,t}}(\pi_D). \tag{17}$$

where Δ denotes a generalized aggregation strategy.

It should be pointed out that $NEG^{\Delta}_{GS}(\pi_D)$ is empty set. Thus, the negative regions of π_D are not considered in the following.

5   Experiments and Analysis

5.1   Data Sets

In order to evaluate our algorithm, we perform some experiments on a personal computer with windows 10, 1.8 GHz CPU and 8 GB memory. The software is IntelliJ idea 2017.3. The results of the following experiments objective are to compare the size of regions under different levels of granularity. For convenience, we abbreviate the optimistic hierarchical multi-granulation sequential three-way decisions and the pessimistic hierarchical sequential three-way decisions as OHMS3WD, PHMS3WD. The characteristics of the six datasets are described in Table 1.

Table 1. Description of the datasets

No	Dataset	$|U|$	$|C|$	$|V_d|$
1	Abalone	4177	8	28
2	Deal winequality red	1599	11	6
3	Fars	100968	14	8
4	Glass	214	9	6
5	Marketing	6876	6	9
6	Obesity	2111	16	7

It is worth mentioning that we need to reprocess the dataset. We delete the third attribute 'fnlwgt' in Adult because it has noting relevant to the individual wage level judgments, and remove the 1st, 18th to 24th and 26th to 28th attributes in Fars because these are not relevant to the fatal accident results and remove attributes 1st, 2nd, 5th, 11th and 12th of the Marketing since these are not related to income. Then, we use Rosetta software (http://www.lcb.uu.se/tools/rosetta/) to convert the continuous data to discrete values. Finally, we construct the concept hierarchy tree and stratify the experimental data by general social cognition. (some information from Baidu Encyclopedia).

5.2   Comparison of the Positive Regions Under Different Levels of Granularity

In what follows, we compute the number of positive regions and analyze the uncertainty of the boundary regions of the hierarchical multi-granulation sequential three-way decisions. For convenience, we use $OGVl$ and $PGVl$ ($l = 1, 2, 3$) to denote the optimistic and pessimistic strategies to select the granular view at the level l. Figure 2 show the change of the number of positive regions under different levels of granular view.

- The positive regions enlarges as the levels of granular view increases, indicating that detailed granular view are conducive to information judgment.
- The positive regions increases monotonically with ascending number of attributes.

5.3   Comparisons of Uncertainty of the Boundary Regions Under Different Levels of Granularity

We employ deferment rates to evaluate the boundary region quality of sequential three-way decisions as follows:

$$DR^{l,t} = \frac{|BND_{GS^{l,t}}^{\Delta}(\pi_D)|}{|U|}. \tag{18}$$

The experiment results of deferment rate change on six data sets under different levels of granular view are shown in Fig. 3.

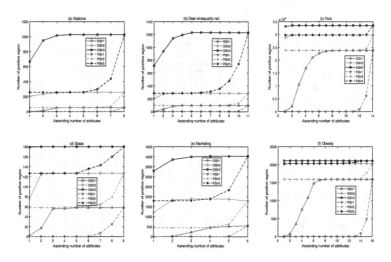

Fig. 2. Positive regions under different granular views

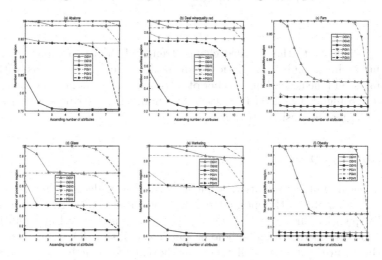

Fig. 3. Uncertainty of the boundary regions under different granular views

- The boundary region reduces monotonically with increasing level of granular views.
- The boundary region reduces monotonically with the ascending number of attribute.

5.4 Comparisons of the Size of the Two Regions Under Different Levels of Granular View

In this subsection, we mainly compare the number of two regions under different strategies and level of granular view.

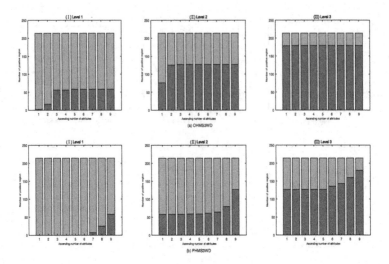

Fig. 4. Optimistic and pessimistic strategies under different levels of granular views on Glass

Figure 4 illustrate the detailed change trends of the two regions under different hierarchical multi-granulation sequential three-way decisions with the increasing level (the ascending number of attributes) of granular view. It is obvious that the higher the level (the ascending number of attributes) of granular view, the larger the positive regions.

6   Conclusions

A hierarchical multi-granulation sequential three-way decision model is proposed, which combines the hierarchical decision table, multi-granulation rough set and sequential three-way decisions. It provides a multi-level and multi-view method for the existing models. The properties of this model are analyzed.

In the future work, we will focus on the extension of hierarchical multi-granulation sequential three-way decision model.

Acknowledgments. The research is supported by the National Natural Science Foundation of China (Nos. 62066014, 62163016, 62166001), Double thousand plan of Jiangxi Province of China, Jiangxi Province Natural Science Foundation of China under Grant Nos. 20202BABL202018, 20212ACB202001, 20202BABy02010.

References

1. Yao, Y.Y.: The superiority of three-way decisions in probabilistic rough set models. Inf. Sci. **181**, 1080–1096 (2011)
2. Qian, J., Liu, C.H., Miao, D.Q., Yue, X.D.: Sequential three-way decisions via multi-granularity. Inf. Sci. **507**, 606–629 (2020)

3. Qian, J., Tang, D.W., Yu, Y., Yang, X.B., Gao, S.: Hierarchical sequential three-way decision model. Int. J. Approx. Reason. **140**, 156–172 (2022)
4. Yao, Y.Y.: Granular computing using neighborhood systems. In: Advances in Soft Computing. Springer-Verlag, London, pp. 539–553 (1999). https://doi.org/10.1007/978-1-4471-0819-1_40
5. Qian, Y.H., Liang, J.Y., Yao, Y.Y., Dang, C.Y.: MGRS: a multi-granulation rough set. Inf. Sci. **180**, 949–970 (2010)
6. Feng, Q.R., Miao, D.Q., Cheng, Y.: Hierarchical decision rules mining. Expert Syst. Appl. **37**, 2081–2091 (2010)
7. Wu, W.Z., Leung, Y.: Theory and applications of granular labelled partitions in multi-scale decision tables. Inf. Sci. **181**, 3878–3897 (2011)
8. Wu, W.Z., Leung, Y.: Optimal scale selection for multi-scale decision tables. Int. J. Approx. Reason. **54**, 1107–1129 (2013)
9. Hao, C., Li, J.H., Fan, M., Liu, W.Q., Tsang, E.C.C.: Optimal scale selection in dynamic multi-scale decision tables based on sequential three-way decisions. Inf. Sci. **415**, 213–232 (2017)
10. Yao, Y.Y., Deng, X.F.: Sequential three-way decisions with probabilistic rough sets. In: Proceedings of the 10th IEEE International Conference on Cognitive Informatics and Cognitive Computing, pp. 120–125 (2011)
11. Pawlak, Z.: Rough sets. Int. J. Comput. Inf. Sci. **11**, 341–356 (1982)
12. Yao, Y.: Granular computing and sequential three-way decisions. In: Lingras, P., Wolski, M., Cornelis, C., Mitra, S., Wasilewski, P. (eds.) RSKT 2013. LNCS (LNAI), vol. 8171, pp. 16–27. Springer, Heidelberg (2013). https://doi.org/10.1007/978-3-642-41299-8_3
13. Yao, Y.Y.: Rough sets and three-way decisions. In: Proceedings of the 10th International Conference on Rough Sets and Knowledge Technology (RSKT2015), Tianjin, China, pp. 62–73 (2015)

KNN Ensemble Learning Integration Algorithm Based on Three-Way Decision

Xinyuan Jia[1(✉)], Yating Li[2], and Pengling Wang[1]

[1] Maths and Information Technology School, Yuncheng University, Yuncheng 044000, China
Jxywyyx369@163.com

[2] Mechanical and Electrical Engineering Department, Yuncheng University, Yuncheng 044000, China

Abstract. The KNN algorithm is affected by overlapping classification, unbalanced data, and K-value selection. It is difficult to apply to some environments with uncertain phenomena. At the same time, the three-way decision is a decision theory that conforms to the human cognitive model with subjective characteristics, so the idea of a three-way decision is introduced into the KNN ensemble learning algorithm and the KNN ensemble learning algorithm based on the three-way decision is proposed. Based on the KNN ensemble learning algorithm, the conditional Probability of each class is calculated and combined with the cost function, which is used to determine the positive domain, negative domain, and boundary domain in the three-way decision theory. This paper performs the three-way decision KNN ensemble learning classification on seven real UCI datasets. The experimental results show that it can effectively improve the classification accuracy and F1-score of the data.

Keywords: KNN · Three-way decision · Ensemble learning algorithm

1 Introduction

In many decision problems, considering the high cost of accepting and rejecting, we often neither accept nor reject, but choose not to commit, which gives us a third decision option. The three-way decision [1–3], by abstracting various decision problems from different disciplines, introduces a third option of non-commitment or delayed decision making, thus avoiding the risks associated with direct acceptance or rejection. The three-way decision can be regarded as an essential intermediate step in -sequential decision-making. The non-commitment option can be studied again, and the two-way decision can be finally obtained through further data collection and analysis.

KNN classification is an effective classification method. However, it may have high misclassification cases in the actual decision-making process. This paper will combine KNN classification, integration algorithm, and three-way decision-making to propose a new multi-classification method. On the one hand, based on the Bayesian decision process, the three-way decision can systematically calculate the threshold value of the

2021 Applied Research Project of Yuncheng University: Specialty Packaging Design and Evaluation based on Three-way Decision and TOPSIS Model (Project No.: CY-2021022).

J. Yao et al. (Eds.): IJCRS 2022, LNAI 13633, pp. 346–360, 2022.
https://doi.org/10.1007/978-3-031-21244-4_26

judgment criterion in KNN ensemble learning classification. On the other hand, the KNN ensemble learning can calculate conditional Probability in the classification.

Many researchers have recently proposed improved methods based on the traditional KNN classification algorithm. Guo [4] et al. proposed a BPSO-Adaboost-KNN method, which has much higher accuracy than the traditional KNN algorithm on unbalanced data sets than than the traditional KNN algorithm. Yu Ying [5] et al. introduced variable precision rough sets into the KNN algorithm, which maintains high classification accuracy while effectively improving the classification efficiency. The KNN model and the multi-representative k-nearest neighbor classification algorithm were proposed by Guo [6] et al. The three-way decision proposed by Professor Yao Yiyu can be regarded as a special decision strategy, which provides a reasonable semantic interpretation of the three domains of the Probability rough set and the decision rough set. For example, Li Meijing et al. combined the three-way decision with the concept lattice, which greatly improved the efficiency of attribute simplification. Zhang Chunying et al. combined the set-to-granule space and the three-way decision idea, which achieved good results in the field of evaluation decision of venture capital. In this paper, we use the idea of the three-way decision combined with the KNN ensemble learning classification model, which can effectively solve the classification problem of uncertain data.

In this paper, we introduce the idea of a three-way decision into the KNN ensemble learning algorithm and propose the KNN ensemble learning algorithm based on the three-way decision, which can avoid the defects of the traditional KNN classification against the high error discrimination and the poor interpretability of the results. Based on adjusting the threshold and KNN parameters, the accuracy of data prediction is improved.

Section 2 describes the KNN classification model and the KNN ensemble learning model. Section 3 introduces the basic model of three-way decision making, especially the calculation of the threshold learning algorithm based on the three-way decision. Section 4 introduces the KNN ensemble learning algorithm based on the three-way decision. The experimental analysis of the algorithm is carried out in Sect. 5.

2 Structural Design of KNN Integration Algorithm

2.1 Introduction of the KNN Classification Model

Algorithms in the field of machine learning initially proposed by Fix and Hodges in the 1950s as a statistical learning method based on sample instances [7]. The K-Nearest Neighbor method was developed and refined by Cover and Hart in subsequent studies and was proposed in 1968 as a theoretically mature machine learning algorithm.

The basic idea of KNN ensemble learning is "one who stays near vermilion gets stained red, and one who stays near ink gets stained black", a sample in the feature space through the distance formula to calculate the nearest K samples, which most of these K samples belong to which category the sample belongs to. As shown in Fig. 1. When K = 4, the nearest 5 samples are 2 squares and 3 triangles, the type of unknown samples is the triangle, when K = 11, the nearest 15 samples are 9 squares and 6 triangles, the type of unknown samples is square. K-nearest neighbor algorithm generally chooses to adjust the model's parameters by the K-fold Cross-validation method.

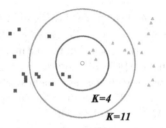

Fig. 1. Interpretation of the O-centered K-nearest neighbor method

The spatial distance formula applied in the KNN algorithm has various formulas for measuring point-to-point distances in space in the mathematical context, such as the Euclidean distance formula and the Manhattan distance formula. Euclidean distance is the most common distance measure, which measures the absolute distance between points in a multidimensional space, and the formula is as follows.

$$d(x, y) = \sqrt{\sum_{i=1}^{n} (x_i - y_i)^2} \tag{1}$$

The Manhattan distance is derived from the city block distance and is the result of summing the distances in multiple dimensions with the following equation.

$$d(x, y) = \sum_{i=1}^{n} |x_i - y_i| \tag{2}$$

KNN algorithms generally have the advantages of simple implementation, easy processing of analysis, high adaptability to information, and easy parallelism, but inevitably suffer from the defects of computationally intensive operation, not very good efficiency, and insensitivity to unbalanced data. The three-way decision proposed in this paper can improve the classification effect to a certain extent by dividing the sample data into three domains, and it also improves the imbalanced data problem.

2.2 Mathematical Model of KNN Algorithm

In this section, the KNN algorithm is introduced in detail. By constructing the mathematical model of the KNN algorithm, the principle of KNN classification is represented through mathematical formulas [8], which are elaborated as follows.

$$T = \{(x_1, y_1), (x_2, y_2), \cdots, (x_N, y_N)\}$$

where $x_i \in X \subseteq R^n$ is the feature vector of the instance, $y_i \in Y = \{c_1, c_2, \cdots, c_k\}$ is the class of the instance, $i = 1, 2, \cdots, N$; the instance feature vector x.

Output: The class y to which the instance x belongs.

1. Find the k points in the training set T that is closest to x according to the given distance metric, and the neighborhood of x covering these k points is denoted as $N(x, k)$.

2. Decide the category y of x in $N(x, k)$ based on the classification decision rule.

$$y_t = \arg \max_{c \in \{c_1, c_2, \cdots, c_k\}} \sum_{x_t \in N(x_T, k)} E(y_i, c) \tag{3}$$

And

$$E(a, b) = \begin{cases} 1 & \text{if}(a = b) \\ 0 & \text{else} \end{cases} \tag{4}$$

Also because

$$p(c_j)_{(x,k)} = \sum_{x_t \in N(x_t,k)} \frac{E(y_i, c_j)}{k} \tag{5}$$

where $p(c_j)_{(x,k)}$ is the probability that the unknown sample belongs to class C_j. The equation can be reduced to.

$$y_t = argmax\{p(c_1)_{(x_t,k)}, p(c_2)_{(x_t,k)}, \cdots, p(c_m)_{(x_t,k)}\} \tag{6}$$

2.3 Introduction to KNN Ensemble Learning

To address the problem of unstable classifiers with low accuracy, scholars propose ensemble learning algorithms, which can obtain a better and more comprehensive, strongly supervised model. Ensemble learning classifiers tend to work better than individual classifiers, even if each classifier is a weak learner. If there are enough weak learners and they are diverse enough, then the final integrated voting classifier can still be a strong learner. The ensemble works best when the predictors are as independent of each other as possible. This paper takes KNN ensemble learning with optimal K values and different effects.

The most intuitive of all ensemble learning methods is majority voting, i.e., "majority rule," which assumes that each base learner is a voter and each category is a competitor. The competitor with the most votes wins, as described below.

Table 1. Predicted classification probability

Classifier	A	B
Classifier 1	60%	40%
Classifier 2	57%	43%
Classifier 3	75%	25%
Classifier 4	45%	55%
Classifier 5	47%	53%

Hard voting - the category with the most votes wins, Classifier 1~ Classifier 3 votes A, Classifier 4 and Classifier 5 votes B, minority obeys majority, and the final predicted result of the voting classifier is A.

Soft voting - the category with high Probability mean wins, let the ratio of the weights of Classifier 1~ Classifier 5 be a:b:c:d:e, then the predicted Probability of A, B is calculated as follows.

$$
\begin{aligned}
A &= \frac{0.6 \times a + 0.57 \times b + 0.75 \times c + 0.45 \times d + 0.47 \times e}{a + b + c + d + e}, \\
B &= \frac{0.4 \times a + 0.43 \times b + 0.25 \times c + 0.55 \times d + 0.53 \times e}{a + b + c + d + e}.
\end{aligned}
\tag{7}
$$

A voting classifier finally predicts the results for the category with a high probability of A, or B. This paper adopts a simple average soft voting method for Ensemble learning.

3   The Basic Model for Three-Way Decision

This section briefly introduces the three-way decision model [9], particularly the derivation of the three-way decision threshold. According to the Bayesian decision process, the three-way decision model comprises two-state sets and three action sets. Assume that state sets denote that an object belongs to X and does not belong to X. The action sets denote the three actions taken to classify x, namely, immediate execution, delayed execution, and rejected execution. Considering the losses (or risks) incurred by taking different actions in different states, a 3×2 loss matrix is thus constructed, as shown in Table 2.

Table 2. Loss matrix for three-way decision

	$X(P)$	$\neg X(N)$
a_P	λ_{PP}	λ_{PN}
a_B	λ_{BP}	λ_{BN}
a_N	λ_{NP}	λ_{NN}

In Table 2, λ_{PP}, λ_{BP}, and λ_{NP} denote the losses when x belongs to X, when action is taken to execute, when the decision is delayed, and when it is not executed, respectively; λ_{PN}, λ_{BN}, and λ_{NN} denote the losses when x belongs to $\neg X$, when action is taken to execute, when the decision is delayed, and when it is not executed, respectively. $P_r(X|[x])$ denotes the conditional probability that the object x belongs to X. Then the expected loss associated with taking a particular individual action can be obtained from the loss function and the conditional probability.

$$
\begin{cases}
R(a_P|[x]) = \lambda_{PP} P_r(X|[x]) + \lambda_{PN} P_r(X|[x]) \\
R(a_B|[x]) = \lambda_{BP} P_r(X|[x]) + \lambda_{BN} P_r(X|[x]) \\
R(a_N|[x]) = \lambda_{NP} P_r(X|[x]) + \lambda_{NN} P_r(X|[x])
\end{cases}
\tag{8}
$$

According to the Bayesian decision principle, the set of actions with the least expected loss is selected as the best action plan. As a result, the following decision rule is obtained.

P. If $R(a_P|[x]) \leq R(a_B|[x]))$ and $R(a_P|[x]) \leq R(a_N|[x]))$ hold simultaneously, then $x \in POS(X)$,

B. If $R(a_B|[x]) \leq R(a_P|[x]))$ and $R(a_B|[x]) \leq R(a_N|[x]))$ hold simultaneously, then $x \in BND(X)$,

N. If $R(a_N|[x]) \leq R(a_P|[x]))$ and $R(a_N|[x]) \leq R(a_B|[x]))$ hold simultaneously, then $x \in NEG(X)$.

Since $P_r(X|[x]) + P_r(\neg X|[x]) = 1$, the rule can be simplified based on the conditional probability $P_r(X|[x])$ and the loss function. Consider a reasonable case of the loss function.

$$\lambda_{PP} \leq \lambda_{BP} < \lambda_{NP}, \lambda_{NN} \leq \lambda_{BN} < \lambda_{PN}$$

Accordingly, the decision rules P, B, and N can be simplified as.

P. If $P_r(X|[x]) \geq \alpha$ and $P_r(X|[x]) \geq \gamma$ hold simultaneously, then $x \in POS(X)$,

B. If $P_r(X|[x]) < \alpha$ and $P_r(X|[x]) < \beta$ hold simultaneously, then $x \in BND(X)$,

N. If $P_r(X|[x]) < \gamma$ and $P_r(X|[x]) \leq \beta$ hold simultaneously, then $x \in NEG(X)$.

where the expressions for the thresholds α, β, and γ are:

$$
\begin{cases}
\alpha = \frac{\lambda_{PN} - \lambda_{BN}}{(\lambda_{PN} - \lambda_{BN}) + (\lambda_{BP} - \lambda_{PP})} \\
\beta = \frac{\lambda_{BN} - \lambda_{NN}}{(\lambda_{BN} - \lambda_{NN}) + (\lambda_{NP} - \lambda_{BP})} \\
\gamma = \frac{\lambda_{PN} - \lambda_{NN}}{(\lambda_{PN} - \lambda_{NN}) + (\lambda_{NP} - \lambda_{PP})}
\end{cases}
\tag{9}
$$

In addition, for the boundary region, the condition in the decision rule B indicates that, $\alpha > \beta$, so

$$\frac{\lambda_{BP} - \lambda_{PP}}{\lambda_{PN} - \lambda_{BN}} < \frac{\lambda_{NP} - \lambda_{BP}}{\lambda_{BN} - \lambda_{NN}}$$

This also implies that $0 \leq \beta < \gamma < \alpha \leq 1$. In this case, the rule is further simplified.

P1. If $P_r(X|[x]) \geq \alpha$ holds, then $x \in POS(X)$,

B1. If $\beta < P_r(X|[x]) < \alpha$ holds, then $x \in BND(X)$,

P1. If $P_r(X|[x]) \leq \beta$ holds, then $x \in NEG(X)$.

4   KNN Emseble Learning Based on Three-Way Decision

4.1   KNN Emseble Learning Classification Model Based on Three-Way Decision

4.1.1   Model Principle

Unbalanced samples, while screening uncertain samples for delayed decisions [10]. Firstly, the best K value is selected for KNN ensemble training, the conditional probability of classification is calculated by the KNN ensemble learning classification model, secondly, the threshold value for each class is calculated according to the threshold calculation formula, and finally, the sample data is divided into three domains according to the calculated conditional probability and threshold value, and the sample data in the edge domain is continuously divided to provide multiple choices for the classification of samples. The three-way decision model especially improves the classification accuracy of uncertain samples, and finally, the samples that cannot be divided in the edge domain are processed for delayed decisions.

4.1.2 Model Design

This section introduces the KNN ensemble learning model based on the three-way decision, which is shown in Fig. 2. The classification rules of the model are as follows.

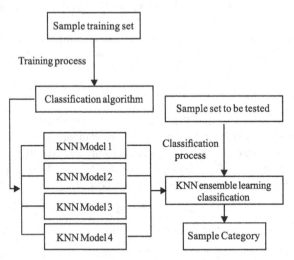

Fig. 2. KNN ensemble learning model based on the three-way decision.

Define the set consisting of the objects we want to study as U. For U containing m classes, the m classes can be expressed as $Class = \{C_1, C_2, \cdots, C_m\}$, where C_i denotes the objects of i classes, so $C_i \cap C_j \neq \emptyset (i \neq j)$, $\bigcup_{i=1}^m C_i = U$ while $\sim C_i = \bigcup_{j=1, j \neq i}^m C_j$, so $x \in U$ will have m states, i.e., the set of states is $\Omega = \{C_1, C_2, \cdots, C_m\}$. Let the set of thresholds corresponding to the category C_i be $(\alpha^{c_i}, \beta^{c_i}, \gamma^{c_i})$ and the set of thresholds corresponding to the category C_j be $(\alpha^{c_j}, \beta^{c_j}, \gamma^{c_j})$ with the following rules.

If $\exists P(C_i|x) \geq \alpha^{C_i}$ and $\forall P(C_j|x) \leq \beta^{C_j}$, then $x \in POS(X)$.

If $P(C_i|x) \leq \beta^{C_i}$ and $P(C_j|x) \leq \beta^{C_j}$, then $x \in POS(X)$.

In the rest of cases, then $x \in BND(X)$.

4.2 Design of KNN Ensemble Learning Algorithm Based on the Three-way Decision

In the traditional KNN algorithm, a clear judgment must be made on the samples' category to be classified for various reasons. Therefore, introducing the idea of the three-way decision and ensemble learning into the KNN classification algorithm will significantly reduce the cost of classification and greatly improve the correctness of classification in massive data classification. Some of the pseudocodes of the KNN ensemble learning algorithm based on the three-way decision are shown in Algorithm 4.1.

Algorithm 4.1. Pseudocode for the three-way decision KNN ensemble learning algorithm

---

Input: Pre-processed raw data set, $POS = []$, $NEG = []$, $BND = []$, k_range = range(1, 31)

Output: The first classification completes the sample prediction categories of positive and negative domains

Step 1 Sample training is carried out to find out the best K value

for k in k_range:

knn = KNeighbors Classifier(n_neighbors=k)

scores = cross_val_score(~)

    k_error.append(1 - scores.mean())

Step 2 Different optimal K values are integrated

clf1 = neighbors. KNeighbors Classifier(~)

  ⋮   ⋮   ⋮

clf4 = neighbors.KNeighborsClassifier(~)

vote_clf=VotingClassifier(estimators=[(clf1),(clf2),(clf3),(clf4)],voting='soft')

for clf in (clf1,···,clf4,vote_clf):   #Iterative classifier

clf.fit()

    predict = clf.predict()

    print(acuracy,precision,recall,F1-score)

Step 3 Calculate the conditional probability of each class of the test set samples and the threshold value of each class $\left(\alpha^{c_i}, \beta^{c_i}\right)$

Step 4 Perform 3 domain divisions

for i in range(number of test set samples):

if Probability of a certain classification$\geq \alpha^{c_i}$ and All other classification probabilities$\leq$ β^{c_j}:

      POS.append(number of test set samples)

If all classification probabilities$\leq \beta_{(c_i, c_j)}$:

      NEG.append(number of test set samples)

    else:

      BND.append(number of test set samples)

Step 5 Output sample prediction category in negative domain and positive domain.

---

5   The Process of Classifying Data

This paper focuses on improving the traditional KNN algorithm, thus proposing a KNN ensemble learning algorithm based on the three-way decision, whose classification process is shown in Fig. 3.

Classification process of KNN ensemble learning algorithm based on the three-way decision.

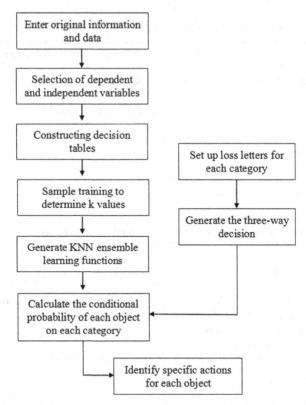

Fig. 3. Flow chart of KNN ensemble learning classification based on the three-way decision

Algorithm 4.2. Classification process of KNN ensemble learning algorithm based on the three-way decision

---

Input: Sample C to be classified
Output: The class of the sample C to be classified

Step 1 Enter the original sample data and perform data processing (fill in missing data, etc.).

Step 2 Perform sample training to select the best 4 to 5 K values.

Step 3 The distance between the unknown samples and the training set data is calculated by Manhattan distance, and the distance values are sorted in ascending order, and the category with the highest frequency of occurrence among the K distance values is taken.

Step 4 Repeat Step 3 for each K value in Step 2 and perform ensemble learning to derive the classification categories.

Step 5 Calculate the conditional probability of each test sample classification

Step 6 According to Equation (9), we calculate each type of threshold and divide the probability set in Step 5 into 3 domains, and record the sample data in the boundary domain as the next discriminator.

Step 7 Repeat Step 2~Step 6 until the data in the boundary domain cannot be divided into the positive and negative domains.

Step 8 Output sample prediction categories for positive and negative domains.

The specific classification process is shown in Algorithm 4.2.

5.1 Experimental analysis of KNN Ensemble Learning Algorithm Based on the Three-way Decision

This section verifies the effectiveness and feasibility of the algorithm in this paper by conducting an experimental analysis of the KNN ensemble learning algorithm based on the three-way decision for the UCI dataset.

Evaluation Indicators

For a binary classification problem, the sample is divided into positive and negative classes, which will generate the following four cases in the real problem.

TP (true positive): The number of samples in the positive class that predict the correct sample.

FN (false negative): The number of samples in the positive class predicting the wrong sample.

FP (false positive): The number of samples in the negative class that predicted the wrong sample.

TN (true negative): The number of samples in the negative class that predict the correct sample. The commonly used classification evaluation metrics are accuracy, precision, recall and F1-score. Details are shown in Table 3

Table 3. Classification evaluation indicators

Evaluation indicators	Definition	Equation
Accuracy	Ratio of the number of correctly classified samples to the total number of samples	$accuracy = \frac{TP+TN}{TP+TN+FP+FN}$
Precision	Ratio of the number of samples correctly classified as positive class samples to the number predicted to be positive class	$precision = \frac{TP}{TP+FP}$
Recall	Ratio of the number of correctly classified positive class samples to the number of positive class samples	$recall = \frac{TP}{TP+FN}$
F1-score	The ratio of 2 times the product of precision rate and recall rate to the sum of precision rate and recall rate	$F1 = \frac{precision*recall*2}{precision+recall}$

5.2 Case Classification Experiment

5.2.1 The Three-way Decision KNN Ensemble Learning-Trial_risk

Take Trial_risk data as an example, after importing the data, 5 missing values are deleted and the remaining missing values are filled with the values of similar data. There are

772 samples in the experiment, 540 samples in the training set account for 70%, and 232 samples in the test set account for 30%. Class 0 means the company is not a fraudulent company, class 1 means the company is fraudulent. The comparison of the four models of UCI trial_risk is carried out. KNN Model 1: KNN classification model. KNN Model 2: KNN ensemble learning classification model. KNN Model 3: KNN classification model with each adjustment of threshold and K value. KNN Model 4: KNN ensemble learning classification model with each adjustment of threshold and K value. The accuracy, precision, recall,and F1-score of the four models are shown in Fig. 4. From the results, it can be seen that the three-way decision idea can improve the classification effect of the model.

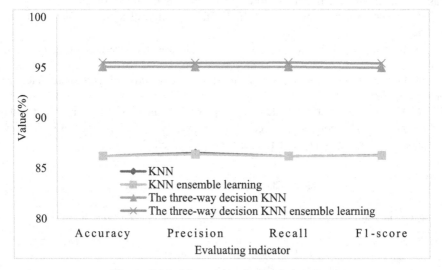

Fig. 4. Trial_risk model evaluation index chart

Table 4. Trial_risk model results evaluation table (%)

Model	Accuracy	Precision	Recall	F1-score
KNN	86.21	86.52	86.21	86.30
KNN ensemble learning	86.21	86.39	86.21	86.27
The three-way decision KNN	95.09	95.07	95.09	95.00
The three-way decision KNN ensemble learning	**95.51**	**95.47**	**95.51**	**95.44**

The accuracy, precision, recall, and F1-score of the four models can be seen in Table 4. Among them, the three-way decision KNN ensemble learning model is the best in terms of accuracy, precision, recall, and F1-score. Overall, the three-way decision KNN ensemble learning model outperforms the three-way decision KNN model. The introduction of the three-way decision and ensemble learning ideas can improve the classification effect of the model.

Combined with Table 4 and Fig. 4, it can be seen that both the K value of KNN and the threshold of the three-way decision have an impact on the experimental results, and the most important of the three-way decision is the selection of the threshold, so this paper further investigates the selection of the K value and the threshold of the three-way decision, and the specific process of dividing the threshold and K value of the three-way decision KNN ensemble learning model of Trial_risk is shown in Table 4, and the threshold 3 decimal places are retained.

From Table 5, it can be seen that the threshold and K values will affect the classification effect of the experiment. If α is larger, the radius of the positive domain is smaller, and the classification of samples within the radius of the positive domain will be more accurate; β is smaller, the number of samples in the boundary domain will be larger, and there will be more nearest neighbors for samples in the boundary domain to determine which category this sample belongs to, and the classification of samples in the boundary domain will be more accurate, but the efficiency of classification will be reduced. By adjusting the size of the threshold (α, β), a balance between accuracy and classification efficiency can be found[9].

Table 5. The three-way decision KNN ensemble learning classification process

$(\alpha_1, \beta_1, \alpha_2, \beta_2, (k))$	POS	BND	NEG
(0.889,0.714,0.700, 0.357,(11,15,21,25,2))	126/131	65/93	8/8
(0.943,0.714,0.760, 0.263,(6,12,14,21))	15/16	52/77	0
(0.962,0.714,0.800,0.211,(6,12,14,15,17))	0/1	51/76	0
(0.962,0.714,0.800,0.211,(6,12,14,15,17))	0	0	0

5.3   UCI Dataset Classification Experiments

To verify the effectiveness of the KNN ensemble learning algorithm based on the three-way decision, this paper conducts the three-way decision KNN ensemble learning classification experiment on the real UCI dataset. Six UCI datasets are selected: Balance, Car, Win, Indian Liver Patient Dataset (ILPD), Ionosphere, Transfusion, and the experimental platform is window7.0, python3.7 in Jupyter-notebook.

The dataset details are shown in Table 6, where the missing value data are processed using the average value of the attribute for padding. KNN, KNN ensemble learning, and the three-way decision KNN are denoted as Algorithm 1, Algorithm 2, and Algorithm 3, respectively. The three algorithms and the algorithm in this paper will be compared in terms of accuracy and F1-score on the six UCI datasets.

The accuracy of the four classification algorithms on the six UCI datasets is shown in Table 7. The observation results show that the algorithm in this paper has the best overall improvement in accuracy on the six UCI datasets.

Table 6. UCI dataset description

Dataset	Sample size	Training set	Test set	Number of decision categories
Balance	625	468	157	3
Car	1663	1247	416	3
Win	178	133	45	3
Indian patient	579	405	174	2
Ionosphere	351	245	106	2
Transfusion	748	523	225	2

Table 7. AUC results of 4 classification algorithms (%)

Dataset	Algorithm 1	Algorithm 2	Algorithm 3	Algorithm of this paper
Balance	91.72	91.72	98.55	**99.30**
Car	84.62	85.58	91.59	**92.57**
Win	73.33	75.56	86.21	**92.31**
Indian patient	63.79	65.52	72.97	**79.17**
Ionosphere	86.79	86.79	90.32	**92.22**
Transfusion	74.22	74.67	82.25	**81.29**

The F1-scores of the four classification algorithms on the six UCI datasets are shown in Table 8. The observation results show that the algorithms in this paper have the best overall F1-score improvement on the six UCI datasets.

Table 8. F1-score results of 4 classification algorithms (%)

Dataset	Algorithm 1	Algorithm 2	Algorithm 3	Algorithm of this paper
Balance	88.70	88.70	98.91	**98.95**
Car	84.67	85.69	91.22	**92.33**
Win	72.88	75.32	83.70	**90.73**
Indian patient	63.18	62.37	69.59	**72.57**
Ionosphere	86.44	86.44	89.66	**91.79**
Transfusion	72.16	72.50	78.08	**76.70**

Combining Table 7 and Table 8, it can be seen that Balance, Car, Win, Indian Patient, Ionosphere dataset this paper algorithm improves the best effect, and Transfusion dataset algorithm 3 improves the best effect. Overall, the three-way decision

KNN ensemble learning algorithm can improve the model's classification accuracy and synthetically improve the accuracy and recall of the model.

To further verify the effectiveness of the KNN ensemble learning algorithm based on the three-way decision. Logistic Regression algorithm, SVM algorithm, Decision Tree algorithm and this paper's algorithm are compared in terms of accuracy on six UCI datasets.

Table 9. AUC results of 4 classification algorithms (%)

Dataset	Logistic regression	SVM	Decision tree	Algorithm of this paper
Balance	90.45	91.72	81.53	**99.30**
Car	70.67	82.45	84.14	**92.57**
Win	68.89	68.89	66.67	**92.31**
Indian patient	71.26	72.41	60.35	**79.17**
Ionosphere	83.02	83.02	82.08	**92.22**
Transfusion	79.11	79.11	74.67	**81.29**

The accuracy of the four classification algorithms on the six UCI datasets is shown in Table 9. The observation results show that the algorithm in this paper has the best overall improvement in accuracy on the six UCI datasets.

6 Conclusions

In order to improve the classification accuracy of the model especially for imbalanced data, this paper proposes the KNN ensemble learning algorithm based on the three-way decision, combining the ideas of ensemble learning and the three-way decision, by constructing a loss function matrix to calculate the threshold α and threshold β, using two exact sets (lower and upper approximation sets) to approximate a probability set, and the upper and lower approximation sets divide the data set into three parts: positive domain, boundary domain and negative domain. It can avoid the defects of the traditional KNN classification against the high error discrimination and the poor interpretability of the results. The experimental results are analyzed by four evaluation metrics: accuracy, precision, recall, and F1-score.

The KNN ensemble learning classifier is used to calculate the conditional Probability of each type of data set. Then the threshold value calculated by the three-way decision idea is used to divide the three domains. The data in the boundary domain is continuously divided until the boundary domain cannot be divided. Observing the experimental results, it can be seen that the KNN ensemble learning algorithm based on the three-way decision generally improves the classification accuracy, F1-score effect, and solves the problem of poor data classification.

Here, the conditional probabilities are calculated by the KNN ensemble learning classifier, which belongs to machine learning. The thresholds generated by the three-way decision rely on the loss function, reflecting human involvement. Therefore, the

KNN ensemble learning algorithm based on the three-way decision proposed in this paper provides a human-computer interaction perspective for solving practical problems. Although the three-way decision theory is widely applied and has very strong universality, the most important aspect of the three-way decision is the threshold value calculation, which often needs to be calculated by experts in the industry to be optimal. The effect of the three-way decision often requires continuous adjustment of parameters for optimization, and there is the problem of long data prediction time. The next step can be to use incremental learning, multi-class neighborhood, and other methods for model optimization to further improve data prediction accuracy and shorten the decision based on the KNN ensemble learning algorithm.

References

1. Yao, Y.: An outline of a theory of three-way decisions. In: Yao, J.T., et al. (eds.) Rough Sets and Current Trends in Computing. LNCS (LNAI), vol. 7413, pp. 1–17. Springer, Heidelberg (2012). https://doi.org/10.1007/978-3-642-32115-3_1
2. Yao, Y.: Rough sets and three-way decisions. In: Ciucci, D., Wang, G., Mitra, S., Wu, W.Z. (eds.) Rough Sets and Knowledge Technology. LNCS (LNAI), vol. 9436, pp. 62–73. Springer, Cham (2015). https://doi.org/10.1007/978-3-319-25754-9_6
3. Yao, Y.Y.: Three-way decision and granular computing. Int. J. Approx. Reason. **103**, 107–123 (2018)
4. Guo, H., et al.: BPSO-Adaboost-KNN ensemble learning algorithm for multi-class imbalanced data classification. Eng. App. Artif. Intell. **49**,176–193 (2015)
5. Yu, Y., et al.: An Improved KNN algorithm based on variable precision rough sets. Pattern Recogn. Artif. Intell. **25**(4),617–623 (2012)
6. Chen, L.F., Guo, G.D.: Multi-representatives learning algorithm for nearest neighbour classification. Pattern Recogn. Artif. Intell. **24**(6), 882–888 (2011)
7. Cover, T.M., Hart, P.E.: Nearest neighbor pattern Classification. In: IEEE Trans. Inf. Theory. **13**(1), 21–27 (1967)
8. Li, H.: Statistical Learning Methods. Tsinghua, Beijing (2012)
9. Pei, Y.P., Shang, A., Liu, M.H., Chen, F.: KNN incremental algorithm based on the three-way decisions. Control Eng. China **27**(04), 656–661 (2020)
10. Zhang, S.C.: Cost-sensitive KNN classification. Neurocomputing. 391p. (prepublish) (2020)

Attribute Reduction of Crisp-Crisp Concept Lattices Based on Three-Way Decisions

Tong-Jun Li[1,2(✉)], Zhen-Zhen Xu[1], Ming-Rui Wu[1], and Wei-Zhi Wu[1,2]

[1] School of Information and Engineering, Zhejiang Ocean University, Zhoushan
316022, Zhejiang, China
`{litj,wuwz}@zjou.edu.cn`
[2] Key Laboratory of Oceanographic Big Data Mining and Application of Zhejiang
Province, Zhejiang Ocean University, Zhoushan 316022, Zhejiang, China

Abstract. knowledge reduction in fuzzy environments is a key and difficult issue in data mining and knowledge discovery, the crisp-crisp variable
threshold concept lattice is one kind of the knowledge structures in formal fuzzy contexts. The focus of this paper is on the attribute reduction
of the crisp-crisp variable threshold concept lattices, the attribute reduction is related to a method of three-way decision based on the crisp-crisp
concepts. Firstly, by an illustrating example, we show the extraction of
three-way decision rules. A lot of the decision rules are redundant, in
order to simplify the nonredundant decision rules and keep their performance invariant, we present a notion of the attribute reduction of formal
fuzzy contexts, then we investigate the properties of the attribute redec-
tion. Subsequently, with reference to the rough sets, the discernibility
matrix and discernibility function of formal fuzzy contexts are presented,
so that all the reducts of a formal fuzzy context can be calculated.

Keywords: Crisp-crisp variable threshold concept lattices · Formal
fuzzy contexts · Attribute reduction · Three-way decisions

1 Introduction

Formal Concept Analysis (FCA), firstly proposed by German professor Wille [1]
in 1982, is a powerful tool for data analysis and knowledge discovery. Now, it is
becoming an important research domain in artificial intelligence, and is widely
used in many fields, for example, machine learning, data mining, and information
retrieval, and so on [2–7].

Concept lattice is a basic mathematics structure in FCA, it is established by
a data set called formal context. A formal context is a binary relation between
two nonempty and finite sets, one is called the object set and another is called
the attribute set. A formal concept is a pair of an object subset and an attribute
subset satisfying two conditions, and all the formal concepts of a formal context
form a complete lattice, which is called concept lattice.

As a main research issue in FCA, the attribute reduction of formal contexts
has been paid much attention, it is very important for the theoretical research and

J. Yao et al. (Eds.): IJCRS 2022, LNAI 13633, pp. 361–375, 2022.
https://doi.org/10.1007/978-3-031-21244-4_27

practical application of concept lattices. Ganter and Wille [8] firstly introduced the idea of reduction, that is (i.e.), deleting some objects or attributes so that the intents or the extents of the formal concepts unchanged, respectively. Zhang et al. [9] presented an approach of deleting some attributes of a formal context, so as to keep the structure of the concept lattice unchanged. Wu et al. [10] examined a type of attribute reduction from the perspective of granular computing. Chen et al. [11] proposed a reduction method in combination with graph theory. Based on the meet-irreducible and join-irreducible elements of a concept lattice, Li et al. [12] presented two kinds of attribute reductions, and investigated the relation between them and other two kinds of attribute reductions.

Formal fuzzy contexts are the extension of classic formal contexts. In other words, when the binary relation of a classic formal context is replaced with a binary fuzzy relation, then a formal fuzzy context is established. Various fuzzy concept lattices were built on formal fuzzy contexts, and some related knowledge reductions have been explored. Burusco and Fuentes-Gonzales [13] firstly introduced fuzzy sets into FCA, so formal fuzzy concepts were defined. Then many researchers put forward various fuzzy concept lattice models [14–16]. Among them, there are four kinds of variable threshold concept lattices firstly proposed by Zhang et al. [17], and Shao et al. [18] studied the knowledge reduction of the variable threshold concept lattices, in which the deletion of objects and attributes are all involved. By means of information entropy, Singh et al. [19] investigated one kind of the attribute reductions of formal fuzzy contexts.

The three-way decision proposed by Yao et al. [20] in 2009 is one kind of approaches of decision making. The main idea is to divide all definable knowledge hidden in the data into three classes, by which two types of decision rules can be extracted, i.e., the certainty decision rules and the possibility decision rules. A lot of attention has been paid on the combination of three-way decision and formal concept analysis. Qi et al. [21,22] firstly proposed three-way concept lattices, and discussed the relationship between them and the classic concept lattices; Qian et al. [23] explored the construction of the three-way concept lattices. Zhi et al. [24] proposed the three-way dual concept lattice by the dual operations in a formal context. Li et al. [25] reviewed the research of concept lattices and three-way decision, and put forward prospect for their combination. Li et al. [26] analyzed and compared concept lattices from the perspective of three-way decisions.

Of course, the three-way concept lattices can be used to make decisions, but the negative attributes will be used inevitably. As we know, there is no studies on the three-way decision based on the classic concept lattices or their fuzzy extensions only using the positive attributes. In real life, people are used to making decisions using positive information. Therefore, this paper explores a method of three-way decisions based on the crisp-crisp variable threshold concept lattices, in which the negative attributes are not used, and studies the corresponding attribute reduction. The rest of the paper is organized as follows. In the next section, we briefly review some basic related notions and knowledge. Section 3 presents an approach of three-way decision based on the crisp-crisp variable threshold concept lattices, and investigates the corresponding attribute reduction. Section 4 concludes the paper with a summary.

2   Preliminaries

For the sake of presentation, let's briefly review some relevant knowledge of the crisp-crisp variable threshold concept lattices.

Let U be a nonempty and finite set, and called the universe of discourse. The set of all crisp subsets of U is denoted by $\mathcal{P}(X)$, and the set of all fuzzy sets on U is denoted by $\mathcal{F}(U)$. Here, a fuzzy set $\tilde{X}$ on U means a mapping from U to $[0,1]$, i.e., $\tilde{X} : U \to [0,1]$, which be also called the membership function of $\tilde{X}$. For $V \subseteq U$ and $\tilde{X} \in \mathcal{F}(U)$, the restriction of $\tilde{X}$ on V, denoted as $\tilde{X}_V$, is a fuzzy sets on V, which satisfies $\tilde{X}_V(x) = \tilde{X}(x)$ for all $x \in V$.

Formal contexts are basic data sets in FCA. A formal context is a triplet (U, A, I), where U is a universe, also called the object set, A a set of features or attributes, also called the attribute set, and I a binary relation between U and A, that is (i.e.), it is a subset of Cartesian product $U \times A$, where $(x, a) \in I$ indicates the object x has the attribute a.

Let (U, A, I) be a formal context, and $X \subseteq U$ and $B \subseteq A$. the pair (X, B) is referred to as a formal concept in (U, A, I) if $X^{*I} = B$ and $B^{*I} = X$, where

$$X^{*I} = \{a \in A | \forall x \in X, (x, a) \in I\},$$
$$B^{*I} = \{x \in U | \forall a \in B, (x, a) \in I\}.$$

Then X and B are called the extent and the intent of (X, B), respectively.

The set of all the formal concepts of (U, A, I) is denoted as $L(U, A, I)$ or simplified as $L(I)$. A specialization-generalization relation on $L(U, A, I)$ can be defined as follows:

$$(X_1, B_1) \leqslant (X_2, B_2) \Longleftrightarrow X_1 \subseteq X_2.$$

Note that $X_1 \subseteq X_2$ is equivalent to $B_2 \subseteq B_1$.

The relation "$\leqslant$" is a partial order on $L(U, A, I)$, and $(L(I), \leqslant)$ is a complete lattice, i.e., the so-called concept lattice. Where the infimum and the supremum are given by

$$(X_1, B_1) \wedge (X_2, B_2) = (X_1 \cap X_2, (B_1 \cup B_2)^{**}),$$
$$(X_1, B_1) \vee (X_2, B_2) = ((X_1 \cup X_2)^{**}, B_1 \cap B_2).$$

It is not difficult to verify that

$$L(U, A, I) = \{(X^{*I*I}, X^{*I}) | X \subseteq U\}$$
$$= \{(B^{*I}, B^{*I*I}) | B \subseteq A\}.$$

Formal fuzzy contexts are extensions of formal contexts in fuzzy environments. A formal fuzzy context is denoted by a triplet $(U, A, \tilde{I})$, and the only difference from a classical formal context is that the binary relation $\tilde{I}$ is a fuzzy relation from the object set U to the attribute set A.

Example 2.1. In order to demonstrate some results obtained in the paper, let us consider an illustrating formal fuzzy context $(U, A, \tilde{I})$ depicted in Table 1, where, $U = \{1, 2, 3, 4, 5, 6\}$, $A = \{a, b, c, d, e\}$, and the fuzzy relation $\tilde{I}$ can be read off in Table 1.

Table 1. A formal fuzzy context $(U, A, \tilde{I})$.

U	a	b	c	d	e
1	0.3	0.2	0.5	0.8	1.0
2	0.2	0.3	0.6	1.0	0.9
3	0.5	0.1	0.6	0.2	0.6
4	1.0	0.7	0.3	0.4	0.2
5	0.8	1.0	0.4	0.3	0.1
6	0.2	0.7	0.6	1.0	0.3

Definition 1. *[17] Let $(U, A, \tilde{I})$ be a formal fuzzy context. For $0 < \delta \leqslant 1$, $X \subseteq U$ and $B \subseteq A$, the operators $*^{\tilde{I}\delta} : \mathcal{P}(U) \longrightarrow \mathcal{P}(A)$ and $*^{\tilde{I}\delta} : \mathcal{P}(A) \longrightarrow \mathcal{P}(U)$ are defined as follows:*

$$X^{*\tilde{I}\delta} = \{a \in A | \forall x \in X, \tilde{I}(x, a) \geqslant \delta\}, B^{*\tilde{I}\delta} = \{x \in U | \forall a \in B, \tilde{I}(x, a) \geqslant \delta\}.$$

Let $(U, A, \tilde{I})$ be a formal fuzzy context and $0 < \delta \leqslant 1$. Denote

$$I^{\delta} = \{(x, a) \in U \times A | \tilde{I}(x, a) \geqslant \delta\}.$$

Then (U, A, I^{δ}) is a classic formal context, and for $X \subseteq U$ and $B \subseteq A$, we have that $X^{*\tilde{I}\delta} = X^{*I\delta}$, $B^{*\tilde{I}\delta} = B^{*I\delta}$. Furthermore, it can be found that

$$X^{*I\delta} = \{a \in A | X \subseteq I^{\delta}a\} = \bigcap\{xI^{\delta} | x \in X\},$$
$$B^{*I\delta} = \{x \in U | B \subseteq xI^{\delta}\} = \bigcap\{I^{\delta}a | a \in B\}.$$

where, $xI^{\delta} = \{a \in A | \tilde{I}(x, a) \geqslant \delta\}, x \in U$; $I^{\delta}a = \{x \in U | \tilde{I}(x, a) \geqslant \delta\}, a \in A$.

Let $(U, A, \tilde{I})$ be a formal fuzzy context and $0 < \delta \leqslant 1$. A formal concept $(X, B) \in L(U, A, I^{\delta})$ is also called a crisp-crisp concept of $(U, A, \tilde{I})$ with the threshold level δ, and sometimes denoted as $(X, B)^{\delta}$. So, crisp-crisp concepts with different thresholds are just formal concepts of classical formal contexts, which are derived from a formal fuzzy context at different thresholds.

Proposition 1. *[8] Let $(U, A, \tilde{I})$ be a formal fuzzy context and $0 < \delta \leqslant 1$. Then for $X, X_1, X_2 \subseteq U$, $B, B_1, B_2 \subseteq A$, we have*

(1) $X_1 \subseteq X_2 \Longrightarrow X_2^{*\tilde{I}\delta} \subseteq X_1^{*\tilde{I}\delta}$, $B_1 \subseteq B_2 \Longrightarrow B_2^{*\tilde{I}\delta} \subseteq B_1^{*\tilde{I}\delta}$;

(2) $(X_1 \cup X_2)^{*\tilde{I}\delta} = X_1^{*\tilde{I}\delta} \cap X_2^{*\tilde{I}\delta}$, $(B_1 \cup B_2)^{*\tilde{I}\delta} = B_1^{*\tilde{I}\delta} \cap B_2^{*\tilde{I}\delta}$;

(3) $X \subseteq X^{*\tilde{I}\delta *\tilde{I}\delta}$ $B \subseteq B^{*\tilde{I}\delta *\tilde{I}\delta}$;

(4) $X^{*\tilde{I}\delta} = X^{*\tilde{I}\delta *\tilde{I}\delta *\tilde{I}\delta}$, $B^{*\tilde{I}\delta} = B^{*\tilde{I}\delta *\tilde{I}\delta *\tilde{I}\delta}$.

For convenience, the concept lattice $L(U, A, I^\delta)$ or $L(I^\delta)$ is also denoted by $L_\delta(U, A, \tilde{I})$ or $L_\delta(\tilde{I})$, and called a crisp-crisp variable threshold concept lattice [17]. The set of the extents of all crisp-crisp concepts with the threshold δ is denoted by $Ext_\delta(U, A, \tilde{I})$ or $Ext_\delta(\tilde{I})$.

Table 2. The formal context $(U, A, I^{0.3})$.

U	a	b	c	d	e
1	1	0	1	1	1
2	0	1	1	1	1
3	1	0	1	0	1
4	1	1	1	1	0
5	1	1	1	1	0
6	0	1	1	1	1

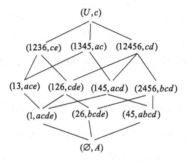

Fig. 1. The crsip-crisp variable threshold concept lattice $L_{0.3}(\tilde{I})$.

Example 2.2. Continuing from Example 2.1, for $\delta_1 = 0.3$ and $\delta_2 = 0.7$, the induced formal contexts $(U, A, I^{0.3})$ and $(U, A, I^{0.7})$ are shown in Tables 2 and 3, and Figs. 1 and 2 are the Hasse diagram of the crisp-crisp variable threshold concept lattices $L_{0.3}(\tilde{I})$ and $L_{0.7}(\tilde{I})$, respectively.

3   Three-Way Decision and Attribute Reduction

In this section, by use of the crisp-crisp variable threshold concept lattices, we present an approach of three-way decision, and illustrate the extraction of decision rules by an example, and according to the obtained decision rules, we study one kind of the attribute reduction of formal fuzzy contexts.

Table 3. The formal context $(U, A, I^{0.7})$.

U	a	b	c	d	e
1	0	0	0	1	1
2	0	0	0	1	1
3	0	0	0	0	0
4	1	1	0	0	0
5	1	1	0	0	0
6	0	1	0	1	0

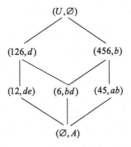

Fig. 2. The crsip-crisp variable threshold concept lattice $L_{0.7}(\tilde{I})$.

3.1   Three-Way Decision Rules Based on Crisp-Crisp Concepts

Let $(U, A, \tilde{I})$ be a formal fuzzy context, and $0 < \delta \leqslant 1$ and $D \subset U$. Denote

$$L_\delta^0(\tilde{I}) = L_\delta(\tilde{I}) - \{(\emptyset, \emptyset^{*I^\delta}), (U, U^{*I^\delta})\},$$
$$CS(D, \tilde{I}(\delta)) = \{(X, B) \in L_\delta^0(\tilde{I}) | X \subseteq D\}.$$

Definition 2. *Let* $(U, A, \tilde{I})$ *be a formal fuzzy context. Selecting* $S = \{\delta_1, \delta_2 \ldots \delta_k\}$, *where* $0 < \delta_i \leqslant 1$ $(1 \leqslant i \leqslant k)$, S *is called a sequence of thresholds. Denote*

$$PS(D, \tilde{I}, S) = \bigcup_{\delta \in S} CS(D, \tilde{I}(\delta)), NS(D, \tilde{I}, S) = \bigcup_{\delta \in S} CS(\sim D, \tilde{I}(\delta)),$$
$$BS(D, \tilde{I}, S) = \bigcup_{\delta \in S} \{(X, B) \in L_\delta^0(\tilde{I}) | X \cap D \neq \emptyset, X \cap (\sim D) \neq \emptyset\},$$

then $PS(D, \tilde{I}, S)$, $NS(D, \tilde{I}, S)$, *and* $BS(D, \tilde{I}, S)$ *are called the positive concept class, the negative concept class, and the boundary concept class of D under the threshold sequence S, respectively. Where, $\sim D$ denotes the complement of D, i.e., $\sim D = \{x \in U | x \notin D\}$.*

It is evident that the set consisting of $PS(D, \tilde{I}, S)$, $NS(D, \tilde{I}, S)$, and $BS(D, \tilde{I}, S)$ is a partition of $\bigcup_{\delta \in S} L_\delta^0(\tilde{I})$.

Example 3.1. Continuing from Example 2.2, let $S = \{0.3, 0.7\}$ and $D = \{1, 2, 3\} \subseteq U$, then from Figs. 1 and 2, it is clear that

$$PS(D, \tilde{I}, S) = \{(13, ace)^{0.3}, (1, acde)^{0.3}, (12, de)^{0.7}\},$$
$$NS(D, \tilde{I}, S) = \{(45, abcd)^{0.3}, (456, b)^{0.7}, (45, ab)^{0.7}, (6, bd)^{0.7}\},$$
$$BS(D, \tilde{I}, S) = \{(1236, ce)^{0.3}, (1345, ac)^{0.3}, (12456, cd)^{0.3}, (126, cde)^{0.3},$$
$$(2456, bcd)^{0.3}, (145, acd)^{0.3}, (26, bcde)^{0.3}, (126, d)^{0.7}\},$$

and it can be seen that the three classes above form a partition of $L^0_{0.3}(\tilde{I}) \cup L^0_{0.7}(\tilde{I})$.

By the positive concepts and the negative concepts of D, some certainty decision rules can be derived, and by the boundary concepts of D, a lot of possibility decision rules can be induced.

Example 3.2. Continuing from Example 3.1, from each of the positive concepts or the negative concepts, a certainty decision rule can be made, and some possibility decision rules are derived from the boundary concepts. So, three groups of decision rules are obtained, which are list as follows:

The first group (induced from $PS(D, \tilde{I}, \{0.3, 0.7\})$):

(r_{11}) $(a, 0.3) \wedge (c, 0.3) \wedge (e, 0.3) \Longrightarrow D,$

(r_{12}) $(a, 0.3) \wedge (c, 0.3) \wedge (d, 0.3) \wedge (e, 0.3) \Longrightarrow D;$

(r_{13}) $(d, 0.7) \wedge (e, 0.7) \Longrightarrow D;$

The second group (induced from $NS(D, \tilde{I}, \{0.3, 0.7\})$):

(r_{21}) $(a, 0.3) \wedge (b, 0.3) \wedge (c, 0.3) \wedge (d, 0.3) \Longrightarrow \sim D,$

(r_{22}) $(b, 0.7) \Longrightarrow \sim D,$

(r_{23}) $(a, 0.7)) \wedge (b, 0.7) \Longrightarrow \sim D,$

(r_{24}) $(b, 0.7) \wedge (d, 0.7) \Longrightarrow \sim D;$

The third group (induced from $BS(D, \tilde{I}, \{0.3, 0.7\})$):

(r_{31}) $(c, 0.3) \wedge (e, 0.3) \longrightarrow D,$

(r_{32}) $(a, 0.3) \wedge (c, 0.3) \longrightarrow D,$

(r_{33}) $(c, 0.3) \wedge (d, 0.3) \longrightarrow D,$

(r_{34}) $(c, 0.3) \wedge (d, 0.3) \wedge (e, 0.3) \longrightarrow D,$

(r_{35}) $(b, 0.3) \wedge (c, 0.3) \wedge (d, 0.3) \longrightarrow D,$

(r_{36}) $(a, 0.3) \wedge (c, 0.3) \wedge (d, 0.3) \longrightarrow D,$

(r_{37}) $(b, 0.3) \wedge (c, 0.3) \wedge (d, 0.3) \wedge (e, 0.3) \longrightarrow D,$

(r_{38}) $(d, 0.7) \longrightarrow D.$

About the meanings of these rules, for example, the certainty decision rule (r_{11}) indicates that if the degrees to which an object x have the attributes, a,

c, and e, are all not less than 0.3, then x must belong to the object set D; and from the possibility decision rule (r_{31}) we know that if the degrees to which an object x have the attributes c and e are all not less than 0.3, then x may belong to the object set D.

It should be point out that some rules may be redundant with respect to the other rules in its group. In the first group, the rule (r_{12}) is redundant with respect to the rule (r_{11}). In the second group, the rules (r_{23}) and (r_{24}) are redundant with respect to the rule (r_{22}). In the third group, the rules (r_{31})-(r_{35}) are redundant with respect to the rules (r_{36}) and (r_{37}).

The unredundant rule are important for theory and application. Moreover, we can see that the concepts with respect to the nonredundant certainty decision rules are maximal in its group, and the concepts with respect to the nonredundant possibility decision rules are minimal.

3.2 Judgement of Attribute Reduction

Definition 3. *Let $(U, A, \tilde{I})$ be a formal fuzzy context and $C \subseteq A$. Denote the restraction of $\tilde{I}$ on $U \times C$ as $\tilde{I}_C$, then the formal fuzzy context $(U, C, \tilde{I}_C)$ is said to be a sub-context of $(U, A, \tilde{I})$.*

Let $(U, A, \tilde{I})$ be a formal fuzzy context, and $0 < \delta \leqslant 1$ and $C \subseteq A$. The two derived operators $*_{\tilde{I}_C^\delta}$ in $(U, C, \tilde{I}_C)$ are rewritten as follows:

$$X^{*\tilde{I}_C^\delta} = \{a \in C | \forall x \in X, \tilde{I}(x,a) \geqslant \delta\}, X \subseteq U;$$
$$B^{*\tilde{I}_C^\delta} = \{x \in U | \forall a \in B, \tilde{I}(x,a) \geqslant \delta\}, B \subseteq C.$$

Of course, $X^{*\tilde{I}_A^\delta}$ and $B^{*\tilde{I}_A^\delta}$ are identical with $X^{*\tilde{I}^\delta}$ and $B^{*\tilde{I}^\delta}$, respectively. In the following, for $a \in A$, we denote $X^{*\tilde{I}_{\{a\}}^\delta}$ and $B^{*\tilde{I}_{\{a\}}^\delta}$ as $X^{*\tilde{I}_a^\delta}$ and $B^{*\tilde{I}_a^\delta}$, respectively.

It can be seen that $X^{*\tilde{I}_C^\delta} = X^{*\tilde{I}^\delta} \cap C$ for all $X \subseteq U$, and $(I^\delta)a = (I_C^\delta)a$ for any $a \in C$, where, I_C^δ is the induced classical relation from $(U, C, \tilde{I}_C)$. Thus, for any $B \subseteq C$, we have that $B^{*\tilde{I}_C^\delta} = \cap\{(I_C^\delta)a | a \in B\} = \cap\{I^\delta a | a \in B\} = B^{*\tilde{I}^\delta}$. Thus, the next proposition can be presented immediately.

Proposition 2. *Let $(U, A, \tilde{I})$ be a formal fuzzy context and $0 < \delta \leqslant 1$. Then for any $C \subseteq A$, we have that $Ext_\delta(\tilde{I}_C) \subseteq Ext_\delta(\tilde{I})$.*

Let $(U, A, \tilde{I})$ be a formal fuzzy context, $0 < \delta \leqslant 1$, $C \subseteq A$, and $D \subset U$. Denote the set of all the maximal elements of $PS(D, \tilde{I}_C, \{\delta\})$ as $PM(D, \tilde{I}_C^\delta)$, the set of all the maximal elements of $NS(D, \tilde{I}_C, \{\delta\}))$ as $NM(D, \tilde{I}_C^\delta)$, and the set of all the minimal elements of $BS(D, \tilde{I}_C, \{\delta\})$ as $BM(D, \tilde{I}_C^\delta)$.

Example 3.3. Continuing from Example 3.1, for $\delta \in S$, we can pick out the elements of $PM(D, \tilde{I}^\delta)$, $NM(D, \tilde{I}^\delta)$ and $BM(D, \tilde{I}^\delta)$ from $PS(D, \tilde{I}^\delta, S)$,

$NS(D, \tilde{I}^\delta, S)$ and $BS(D, \tilde{I}^\delta, S)$, respectively, and list them as follows:

$$PM(D, \tilde{I}^{0.3}) = \{(13, ace)\}, \qquad PM(D, \tilde{I}^{0.7}) = \{(12, de)\};$$
$$NM(D, \tilde{I}^{0.3}) = \{(45, abcd)\}, \qquad NM(D, \tilde{I}^{0.7}) = \{(456, b)\};$$
$$BM(D, \tilde{I}^{0.3}) = \{(145, acd), (26, bcde)\}, \quad BM(D, \tilde{I}^{0.7}) = \{(126, d)\}.$$

Let $(U, A, \tilde{I})$ be a formal fuzzy context, $0 < \delta \leqslant 1$, $C \subseteq A$, and $D \subset U$. Denote

$$PE(D, \tilde{I}_C^\delta) = \{X \subseteq U | (X, X^{*\tilde{I}_C^\delta}) \in PM(D, \tilde{I}_C^\delta)\},$$
$$NE(D, \tilde{I}_C^\delta) = \{X \subseteq U | (X, X^{*\tilde{I}_C^\delta}) \in NM(D, \tilde{I}_C^\delta)\},$$
$$BE(D, \tilde{I}_C^\delta) = \{X \subseteq U | (X, X^{*\tilde{I}_C^\delta}) \in BM(D, \tilde{I}_C^\delta)\},$$
$$PNBE(D, \tilde{I}_C^\delta) = PE(D, \tilde{I}_C^\delta) \cup NE(D, \tilde{I}_C^\delta) \cup BE(D, \tilde{I}_C^\delta).$$

Definition 4. *Let $(U, A, \tilde{I})$ be a formal fuzzy context, and S a sequence of thresholds and $D \subset U$. Then $C \subseteq A$ is referred to as a consistent set for D and S, if for any $\delta \in S$,*

$$PE(D, \tilde{I}_C^\delta) = PE(D, \tilde{I}^\delta), NE(D, \tilde{I}_C^\delta) = NE(D, \tilde{I}^\delta), BE(D, \tilde{I}_C^\delta) = BE(D, \tilde{I}^\delta).$$

And further, if for any $a \in C$, $C - \{a\}$ is not a consistent set for D and S, then C is referred to as a reduct for D and S.

From Example 3.1 we can see that when we construct the nonredundant three-way decision rules in $(U, A, \tilde{I})$ with D and δ, only the concepts of $PM(D, \tilde{I}^\delta)$, $NM(D, \tilde{I}^\delta)$, and $BM(D, \tilde{I}^\delta)$ are used. Let C be a reduct of $(U, A, \tilde{I})$ with D and δ, it can be found that the concepts of $PM(D, \tilde{I}_C^\delta)$ and $PM(D, \tilde{I}^\delta)$ correspond one to one and the corresponding concepts have the same extents, and so do $NM(D, \tilde{I}_C^\delta)$ and $NM(D, \tilde{I}^\delta)$, and $BM(D, \tilde{I}_C^\delta)$ and $BM(D, \tilde{I}^\delta)$. Thus there is also a one-to-one mapping between the nonredundant three-way decision rules in $(U, A, \tilde{I}_C)$ and $(U, A, \tilde{I})$, and it is not difficult to see that the corresponding decision rules have the same support set.

Therefore we can conclude that a reduct C for a target set D and a threshold sequence S is a minimum attributes subset to keep the extents of the maximal positive concepts, the maximal negative concepts, and the minimal boundary concepts unchanged with respect to the original formal fuzzy context and the corresponding decision rules have the same performance.

Theorem 1. *Let $(U, A, \tilde{I})$ be a formal fuzzy context, and S a threshold sequence and $D \subset U$. Then $C \subseteq A$ is a consistent set for D and S, if and only if for any $\delta \in S$ and $X \in PNBE(D, \tilde{I}^\delta)$, we have that $(X, X^{*\tilde{I}_C^\delta}) \in L_\delta(\tilde{I}_C)$.*

Proof. ($\Longrightarrow$) Assume that C be a consistent set for D and S. For any $X \in PE(D, \tilde{I}^\delta)$, by $PE(D, \tilde{I}_C^\delta) = PE(D, \tilde{I}^\delta)$, we have that $X \in PE(D, \tilde{I}_C^\delta)$. Thus, $(X, X^{*\tilde{I}_C^\delta}) \in L_\delta(\tilde{I}_C)$.

By a similar way, we can prove that if $X \in NE(D, \tilde{I}^\delta) \cup BE(D, \tilde{I}^\delta)$, then $(X, X^{*\tilde{I}^\delta_C}) \in L_\delta(\tilde{I}_C)$.

($\Longleftarrow$) Suppose that $(X, X^{*\tilde{I}^\delta_C}) \in L_\delta(\tilde{I}_C)$ for all $X \in PNBE(D, \tilde{I}^\delta)$ and $\delta \in S$. For $\delta \in S$ and $X \in PE(D, \tilde{I}^\delta)$, by the supposition we have that $(X, X^{*\tilde{I}^\delta_C}) \in L_\delta(\tilde{I}_C)$. If there is a $(Y, E) \in L_\delta(\tilde{I}_C)$ such that $X \subset Y \subseteq D$, then by Proposition 2 we have $Y \in Ext_\delta(\tilde{I})$. So, $(Y, Y^{*\tilde{I}^\delta}) \in L_\delta(\tilde{I})$, we can see that $(X, X^{*\tilde{I}^\delta}) \notin PM(D, \tilde{I}^\delta)$. A contradiction occurs! We then conclude that $(X, X^{*\tilde{I}^\delta_C}) \in PM(D, \tilde{I}^\delta_C)$, clearly, $X \in PE(D, \tilde{I}^\delta_C)$. Thus, it is proved that $PE(D, \tilde{I}^\delta) \subseteq PE(D, \tilde{I}^\delta_C)$. Conversely, for any $X \in PE(D, \tilde{I}^\delta_C)$, of course, $X \in Ext_\delta(\tilde{I}_C)$, by Proposition 2 we have that $X \in Ext_\delta(\tilde{I})$. If there is a $(Y, Y^{*\tilde{I}^\delta}) \in PM(D, \tilde{I}^\delta)$ such that $X \subset Y \subseteq D$, by the supposition we have $(Y, Y^{*\tilde{I}^\delta_C}) \in L_\delta(\tilde{I}_C)$, which contradicts $X \in PE(D, \tilde{I}^\delta_C)$. Thus we have that $X \in PM(D, \tilde{I}^\delta)$. Hence, $PM(D, \tilde{I}^\delta_C) \subseteq PM(D, \tilde{I}^\delta)$. Summaring the two contain relations we conclude that $PE(D, \tilde{I}^\delta_C) = PE(D, \tilde{I}^\delta)$.

By the same way, we can prove that for any $\delta \in S$, $NE(D, \tilde{I}^\delta_C) = NE(D, \tilde{I}^\delta)$, $BE(D, \tilde{I}^\delta_C) = BE(D, \tilde{I}^\delta)$. Therefore, by Definition 4 we know that C is a consistent set for D and S.

From Theorem 1 the below conclusion follows immediately.

Corollary 1. *Let $(U, A, \tilde{I})$ be a formal fuzzy context, and S a threshold sequence and $D \subset U$. Then $C \subseteq A$ is a consistent set for D and S, if and only if for any $\delta \in S$, we have that $PNBE(D, \tilde{I}^\delta) \subseteq Ext_\delta(\tilde{I}_C)$.*

Proposition 3. *Let $(U, A, \tilde{I})$ be a formal fuzzy context, $0 < \delta \leqslant 1$, and $B, C \subseteq A$. Then for any $X \subseteq U$, we have that $X^{*\tilde{I}^\delta_{B \cup C} *\tilde{I}^\delta_{B \cup C}} = X^{*\tilde{I}^\delta_B *\tilde{I}^\delta_B} \cap X^{*\tilde{I}^\delta_C *\tilde{I}^\delta_C}$.*

Proof. For any $X \subseteq U$ and $x \in U$, we have

$$x \in X^{*\tilde{I}^\delta_{B \cup C} *\tilde{I}^\delta_{B \cup C}} \Longleftrightarrow X^{*\tilde{I}^\delta_{B \cup C}} \subseteq xI^\delta_{B \cup C} \Longleftrightarrow X^{*\tilde{I}^\delta_B} \subseteq xI^\delta_B, X^{*\tilde{I}^\delta_C} \subseteq xI^\delta_C$$
$$\Longleftrightarrow x \in X^{*\tilde{I}^\delta_B *\tilde{I}^\delta_B}, x \in X^{*\tilde{I}^\delta_C *\tilde{I}^\delta_C} \Longleftrightarrow x \in X^{*\tilde{I}^\delta_B *\tilde{I}^\delta_B} \cap X^{*\tilde{I}^\delta_C *\tilde{I}^\delta_C}.$$

The proof is completed.

By Proposition 3 we have the following corollary.

Corollary 2. *Let $(U, A, \tilde{I})$ be a formal fuzzy context, $0 < \delta \leqslant 1$, and $B, C \subseteq A$. If $B \subseteq C$, then for any $X \subseteq U$, we have that $X^{*\tilde{I}^\delta_C *\tilde{I}^\delta_C} \subseteq X^{*\tilde{I}^\delta_B *\tilde{I}^\delta_B}$.*

Theorem 2. *Let $(U, A, \tilde{I})$ be a formal fuzzy context, and S a threshold sequence and $D \subset U$. Then $C \subseteq A$ is a consistent set for D and S, if and only if for any $\delta \in S$, and $a \in A - C$ and $X \in PNBE(D, \tilde{I}^\delta)$, we have that $X^{*\tilde{I}^\delta_C *\tilde{I}^\delta_C} \subseteq X^{*\tilde{I}^\delta_a *\tilde{I}^\delta_a}$.*

Proof. ($\Longrightarrow$) Assume that C is a consistent set for D and S. For $\delta \in S$, $a \in A - C$, and $X \in PNBE(D, \tilde{I}^\delta)$, by Theorem 1 we have that $X \in L_\delta(\tilde{I}_C)$, which implies that $X^{*\tilde{I}^\delta_C *\tilde{I}^\delta_C} = X$. From $X \in PNBE(D, \tilde{I}^\delta)$, it follows that $X^{*\tilde{I}^\delta *\tilde{I}^\delta} = X$. Thus, $X^{*\tilde{I}^\delta *\tilde{I}^\delta} = X^{*\tilde{I}^\delta_C *\tilde{I}^\delta_C}$. By Proposition 3 we have that $X^{*\tilde{I}^\delta *\tilde{I}^\delta} = X^{*\tilde{I}^\delta_C *\tilde{I}^\delta_C} \cap$

$X^{*\tilde{I}^\delta_{A-C}*\tilde{I}^\delta_{A-c}}$. Thus, we conclude that $X^{*\tilde{I}^\delta_C*\tilde{I}^\delta_C} \subseteq X^{*\tilde{I}^\delta_{A-C}*\tilde{I}^\delta_{A-c}}$. For $a \in A - C$, by Corollary 2 we know that $X^{*^\delta_{A-C}*^\delta_{A-C}} \subseteq X^{*\tilde{I}^\delta_a*\tilde{I}^\delta_a}$. We then conclude that $X^{*\tilde{I}^\delta_C*\tilde{I}^\delta_C} \subseteq X^{*\tilde{I}^\delta_a*\tilde{I}^\delta_a}$ for all $\delta \in S, a \in A - C, X \in PNBE(D, \tilde{I}^\delta)$.

($\Longleftarrow$) Suppose that for any $\delta \in S$, $a \in A - C$, and $X \in PNBE(D, \tilde{I}^\delta)$, we have that $X^{*\tilde{I}^\delta_C*\tilde{I}^\delta_C} \subseteq X^{*\tilde{I}^\delta_a*\tilde{I}^\delta_a}$. By Proposition 3 we know that

$$X^{*^\delta_{A-C}*^\delta_{A-C}} = \bigcap_{a \in A - C} X^{*\tilde{I}^\delta_a*\tilde{I}^\delta_a},$$

then it holds that $X^{*\tilde{I}^\delta_C*\tilde{I}^\delta_C} \subseteq X^{*\tilde{I}^\delta_{A-C}*\tilde{I}^\delta_{A-c}}$. Again according to Proposition 3, we have that $X^{*\tilde{I}^\delta*\tilde{I}^\delta} = X^{*\tilde{I}^\delta_C*\tilde{I}^\delta_C} \cap X^{*\tilde{I}^\delta_{A-C}*\tilde{I}^\delta_{A-c}} = X^{*\tilde{I}^\delta_C*\tilde{I}^\delta_C}$. From $X \in PNBE(D, \tilde{I}^\delta)$, it follows that $X^{*\tilde{I}^\delta*\tilde{I}^\delta} = X$. Hence, we have that $X^{*\tilde{I}^\delta_C*\tilde{I}^\delta_C} = X$, which means $X \in L_\delta(\tilde{I}_C)$. In terms of Theorem 1, we conclude that C is a consistent set for D and S.

From Theorem 2 the following corollary can be obtained.

Corollary 3. *Let $(U, A, \tilde{I})$ be a formal fuzzy context, and S a threshold sequence and $D \subset U$. Then $C \subseteq A$ is a consistent set for D and S, if and only if for any $\delta \in S$ and $X \in PNBE(D, \tilde{I}^\delta)$, we have that $X^{*\tilde{I}^\delta_C*\tilde{I}^\delta_C} \subseteq X^{*\tilde{I}^\delta_{A-C}*\tilde{I}^\delta_{A-C}}$.*

3.3   An Approach of Attribute Reduction

Referring to rough set theory, by introducing the notion of discernibility attribute set into the crisp-crisp variable threshold concept lattices, we give a method for computing the reducts of a formal fuzzy context in the following.

Let $(U, A, \tilde{I})$ be a formal fuzzy context and $0 < \delta \leqslant 1$. For $(X_1, B_1), (X_2, B_2) \in L_\delta(\tilde{I})$, $(X_1, B_1) \prec (X_2, B_2)$ means that (X_1, B_1) is a lower neighbor of (X_2, B_2), or (X_2, B_2) is a upper neighbor of (X_1, B_1), i.e., $(X_1, B_1) \leqslant (X_2, B_2)$, and there is no $(X, B) \in L_\delta(\tilde{I})$ such that $X_1 \subset X \subset X_2$.

Definition 5. *Let $(U, A, \tilde{I})$ be a formal fuzzy context, $0 < \delta \leqslant 1$ and $D \subset U$. For $(X_i, B_i), (X_j, B_j) \in L_\delta(\tilde{I})$, if $X_i \in PNBE(D, \tilde{I}^\delta)$, and $(X_i, B_i) \prec (X_j, B_j)$, then let $DS((X_i, B_i), (X_j, B_j)) = B_i - B_j$, otherwise, let $DS((X_i, B_i), (X_j, B_j)) = \emptyset$. Then $DS((X_i, B_i), (X_j, B_j))$ is referred to as the discernibility attribute set of (X_i, B_i) and (X_j, B_j) for D and δ.*

For $0 < \delta \leqslant 1$, denote

$$DM(D, \tilde{I}^\delta) = \{DS((X_i, B_i), (X_j, B_j)) | (X_i, B_i), (X_j, B_j) \in L_\delta(\tilde{I})\},$$

then $DM(D, \tilde{I}^\delta)$ is called the discernibility matrix of $(U, A, \tilde{I})$ for D and δ.

Example 3.4. Continuing from Example 3.3, we only list the non-empty elements of the discernibility matrace $DM(D, I^{0.3})$ and $DM(D, I^{0.7})$ in Tables 4 and 5, respectively. In each table, the first column on the left consists of the concepts with the extents of $PNBE(D, \tilde{I}^\delta)$, and each concept on the first row above is a upper neighbor of some concepts of the first column on the left.

Table 4. The discernibility matrix $DM(U, \tilde{I}^{0.3})$.

	$(1236, ce)$	$(1345, ac)$	$(145, acd)$	$(2456, bcd)$	$(12456, cd)$	$(126, cde)$
$(13, ace)$	a	e				
$(45, abcd)$			b	a		
$(145, acd)$		d			a	
$(26, bcde)$				e		b

Table 5. The discernibility matrix $DM(U, \tilde{I}^{0.7})$.

	$(126, d)$	$(U, \emptyset)$
$(12, de)$	e	
$(456, b)$		b
$(126, d)$		d

Theorem 3. *Let $(U, A, \tilde{I})$ be a formal fuzzy context, and S a threshold sequence and $D \subset U$. Then $C \subseteq A$ is a consistent attribute set for D and S, if and only if for any $\delta \in S$ and $DS((X_i, B_i)^\delta, (X_j, B_j)^\delta) \neq \emptyset$, we have that $C \cap DS((X_i, B_i), (X_j, B_j)) \neq \emptyset$.*

Proof. ($\Longrightarrow$) Assume that C is a consistent set for D and S. For $\delta \in S$ and $(X_i, B_i), (X_j, B_j) \in L_\delta(\tilde{I})$, if $DS((X_i, B_i), (X_j, B_j)) \neq \emptyset$, then we know that $(X_i, B_i) \prec (X_j, B_j)$, obviously, $X_i \in PNBE(D, \tilde{I}^\delta)$. Because C is a consistent set, by Theorem 1 we have $(X_i, X_i^{*I_C^\delta}) \in L_\delta(\tilde{I}_C)$, from which it follows that $X_i^{*I_C^\delta *I_C^\delta} = X_i$, and since $X_i^{*I_C^\delta} = B_i \cap C$, then $(X_i, X_i^{*I_C^\delta})$ can be rewritten as $(X_i, B_i \cap C)$. Because $(X_j^{*I_C^\delta *I_C^\delta}, X_j^{*I_C^\delta}) \in L_\delta(\tilde{I}_C)$ and $X_j^{*I_C^\delta} = B_j \cap C$, we have $(X_j^{*I_C^\delta *I_C^\delta}, B_j \cap C) \in L_\delta(\tilde{I}_C)$. Noticing $(X_i, B_i) \prec (X_j, B_j)$ and by Proposition 1, it is clear that $X_i \subset X_j \subseteq X_j^{*I_C^\delta *I_C^\delta}$. So, $X_i \subset X_j^{*I_C^\delta *I_C^\delta}$, equivalently, $B_j \cap C \subset B_i \cap C$, from which it follows that $C \cap (B_i - B_j) \neq \emptyset$, i.e., $C \cap DS((X_i, B_i), (X_j, B_j)) \neq \emptyset$.

($\Longleftarrow$) Select $\delta \in S$ and suppose that $C \cap DS((X_i, B_i), (X_j, B_j)) \neq \emptyset$ for all $DS((X_i, B_i)^\delta, (X_j, B_j)^\delta) \neq \emptyset$. For any $X \in PNBE(D, \tilde{I}^\delta)$, by Proposition 1 we have that $X \subseteq X^{*I_C^\delta *I_C^\delta}$. If $X \neq X^{*I_C^\delta *I_C^\delta}$, by $X^{*I_C^\delta *I_C^\delta} \in Ext_\delta(\tilde{I}_C)$ and Proposition 2, we have that $X^{*I_C^\delta *I_C^\delta} \in Ext_\delta(\tilde{I})$. Thus there is a $(Y, E) \in L_\delta(\tilde{I})$ such that $(X, X^{*I^\delta}) \prec (Y, E) \leqslant (X^{*I_C^\delta *I_C^\delta}, X^{*I_C^\delta *I_C^\delta *I^\delta})$. Furthermore, by $X^{*I^\delta} \cap C = X^{*I_C^\delta} \subseteq X^{*I_C^\delta *I^\delta *I^\delta} = X^{*I_C^\delta *I_C^\delta *I^\delta}$, we have $X^{*I^\delta} \supseteq E \supseteq X^{*I^\delta} \cap C$. So, $X^{*I^\delta} - E \subseteq X^{*I^\delta} - C$. By the supposition we have $C \cap (X^{*I^\delta} - E) \neq \emptyset$, which is in conflict with $X^{*I^\delta} - E \subseteq X^{*I^\delta} - C$. Thus, $X = X^{*I_C^\delta *I_C^\delta}$. We then conclude that $PNBE(D, I^\delta) \subseteq Ext_\delta(\tilde{I}_C)$. In terms of Corollary 1 we know that C is a consistent set for D and S.

By the discernibility matrix of a formal fuzzy context, the discernibility function can be defined as follows:

Definition 6. *Let* $(U, A, \tilde{I})$ *be a formal fuzzy context, and* S *a threshold sequence and* $D \subset U$. *For* $\delta \in S$, *denote*

$$DF_\delta(D, \tilde{I}) = \bigwedge\{\vee E | E \in DM(D, I^\delta), E \neq \emptyset\},$$

and

$$DF(D, \tilde{I}, S) = \bigwedge_{\delta \in S} DF_\delta(D, \tilde{I}),$$

then $DF(D, \tilde{I}, S)$ *is referred to as the discernibility function of* $(U, A, \tilde{I})$ *for* D *and* S. *The symbols* $\wedge$ *and* $\vee$ *represent the logical conjunction and disjunction operations, respectively. Here, we set a logical variable for each attribute in* A, *and for simplicity, each logical variable and the corresponding attribute are represented by the same symbol. For example,* $\vee\{a, b, c\}$ *indicates the logical expression* $a \vee b \vee c$. *Here, the symbols,* a, b, *and* c, *represent three logical variables, and it is clear that the corresponding attribute set is* $\{a, b, c\}$.

By Theorem 3, it is not difficult to prove the below theorem.

Theorem 4. *Let* $(U, A, \tilde{I})$ *be a formal fuzzy context, and* S *a threshold sequence and* $D \subset U$. *If the minimal disjunction normal form of the discernibility function* $DF(D, \tilde{I}, S)$ *is* $(\wedge C_1) \vee \cdots \vee (\wedge C_k)$, *then the attribute subsets,* $C_1, \ldots, C_k$, *are all reducts of* $(U, A, \tilde{I})$ *for* D *and* S.

Theorem 4 shows that by means of the discernibility function of a formal fuzzy context, all its attribute reducts can be figured out.

Example 3.5. Continuing from Example 3.4, according to Tables 4 and 5, and by Definition 6, we can gain the discernibility function, and then calculate it by logical operation laws as follows:

$$DF_{0.3}(D, \tilde{I}) = a \wedge e \wedge b \wedge a \wedge d \wedge a \wedge e \wedge b = a \wedge b \wedge d \wedge e,$$

$$DF_{0.7}(D, \tilde{I}) = e \wedge b \wedge d,$$

So,

$$DF(D, \tilde{I}, S) = DF_{0.3}(D, \tilde{I}) \wedge DF_{0.7}(D, \tilde{I})$$
$$= (a \wedge b \wedge d \wedge e) \wedge (e \wedge b \wedge d) = a \wedge b \wedge d \wedge e.$$

According to Theorem 4, we know that there's only one reduct of $(U, A, \tilde{I})$ for D and S, i.e., $\{a, b, d, e\}$.

By the reduct $C = \{a, b, d, e\}$, the nonredundant decision rules in Example 3.2 can be accordingly rewritten as follows:

The first group (induced from $PS(D, \tilde{I}_C, \{0.3, 0.7\})$):

(r'_{11}) (a, 0.3) $\wedge$ (e, 0.3) $\Longrightarrow$ D,   (r'_{13}) (d, 0.7) $\wedge$ (e, 0.7) $\Longrightarrow$ D;

The second group (induced from $NS(D, \tilde{I}_C, \{0.3, 0.7\})$):

(r'_{21}) (a, 0.3) $\wedge$ (b, 0.3) $\wedge$ (d, 0.3) $\Longrightarrow \sim$ D,   (r'_{22}) (b, 0.7) $\Longrightarrow \sim$ D;

The third group (induced from $BS(D, \tilde{I}_C, \{0.3, 0.7\})$):

(r'_{36}) $(a, 0.3) \wedge (d, 0.3) \longrightarrow D,$ (r'_{37}) $(b, 0.3) \wedge (d, 0.3) \wedge (e, 0.3) \longrightarrow D,$
(r'_{38}) $(d, 0.7) \longrightarrow D.$

It can be seen that these rules not only have simpler expressions, but we can also verify that they have the same performance as the original ones.

4   Summaries

In this paper, based on the crisp-crsip variable threshold concept lattices, a method for extracting three-way decision rules was developed. For a given object subset and a sequence of thresholds, according to the three-way decision theory, all the crisp-crisp concepts of a formal fuzzy context were divided into three parts, by which, some certainty and possibility decision rules were made. A notion of attribute reduction of formal fuzzy contexts was presented to simplify the form of the nonredundant decision rules and keep their performance unchanged, and then some judgement therems of the consistent sets were obtained. Subsequently, an approach for computing attribute reducts was proposed by the discarnibility attribute sets and discarnibility function defined in the paper, by which all the reducts of a formal fuzzy context can be figured out. Some notions and conclusions of the paper were illustrated by a lot of examples.

The work of this paper is helpful for expanding the application of three-way decision. In the future, we will introduce the three-way decision into the other fuzzy concept lattices, so as to further combine the three-way decision and the formal concept analysis.

Acknowledgements. This work was supported by grants from the National Natural Science Foundation of China (Nos. 61773349, 61976194).

References

1. Wille, R.: Restructuring lattice theory: an approach based on hierarchies of concepts. Ordered Sets **87**, 445–470 (1982)
2. Osthuizen G.D.: The application of concept lattice to machine learning. Technical report, University of Pretoria, South Africa (1996)
3. Ho, T.B.: Incremental conceptual clustering in the framework of Galois lattice. In: Lu, H., Motoda, H., Liu, H. (eds.) KDD: Techniques and Applications, pp. 49–64. World Scientific, Singapore (1997)
4. Kent R.E., Bowman C.M.: Digital libraries, conceptual knowledge systems and the nebula interface. Technical report, University of Arkansas (1995)
5. Corbett D., Burrow A.L.: Knowledge reuse in SEED exploiting conceptual graphs. In: International Conference on Conceptual Graphs, pp. 56–60. University of New South Wales, Sydney (1996)
6. Siff, M., Reps T.: Identifying modules via concept analysis. In: Harrold M J, Visaggio G, eds. International Conference on Software Maintenance, Bari, Italy, pp. 170–179 (1997)

7. Hu, K.Y., Lu, Y.C., Shi, C.Y.: Advances in concept lattice and its application. J. Tsinghua Univ. Sci. Technol. **40**(9), 77–82 (2000)
8. Ferré, S., Rudolph, S. (eds.): ICFCA 2009. LNCS (LNAI), vol. 5548. Springer, Heidelberg (2009). https://doi.org/10.1007/978-3-642-01815-2
9. Zhang, W.X., Wei, L., Qi, J.J.: Attribute reduction theory and approach to concept lattice. Sci. China Ser. F-Inf. Sci. **48**(6), 713–726 (2005)
10. Wu, W.Z., Leung, Y., Mi, J.S.: Granular computing and knowledge reduction in formal contexts. IEEE Trans. Knowl. Data Eng. **21**(10), 1461–1474 (2009)
11. Chen, J.K., Mi, J.S., Lin, Y.J.: A graph approach for knowledge reduction in formal contexts. Knowl.-Based Syst. **148**(5), 177–188 (2018)
12. Li, T.J., Li, M.Z., Gao, Y.: Attribute reduction of concept lattice based on irreducible elements. Int. J. Wavelets Multiresolut. Inf. Process. **11**(6), 1–24 (2013)
13. Burusco, A., Fuentes-Gonzalez, R.: Concept lattices defined from implication operators. Fuzzy Sets Syst. **114**(3), 431–436 (2000)
14. Li, L.F.: Multi-level interval-valued fuzzy concept lattices and their attribute reduction. Int. J. Mach. Learn. Cybern. **8**, 45–56 (2017)
15. Kumar, S.P., Aswani, K.C.: A method for decomposition of fuzzy formal context. Procedia Eng. **38**, 1852–1857 (2012)
16. He, X.L., Wei, L., She, Y.H.: L-fuzzy concept analysis for three-way decisions: basic definitions and fuzzy inference mechanisms. Int. J. Mach. Learn. Cybern. **9**(11), 1857–1867 (2018)
17. Zhang, W.X., Ma, J.M., Fan, S.Q.: Variable threshold concept lattices. Inf. Sci. **177**(22), 4883–4892 (2007)
18. Shao, M.W., Yang, H.Z., Wu, W.Z.: Knowledge reduction in formal fuzzy contexts. Knowl.-Based Syst. **73**(1), 265–275 (2015)
19. Singh, P.K., Cherukuri, A.K., Li, J.H.: Concepts reduction in formal concept analysis with fuzzy setting using Shannon entropy. Int. J. Mach. Learn. Cybern. **8**(1), 179–189 (2017)
20. Yao, Y.Y.: Three-way decisions with probabilistic rough sets. Inf. Sci. **180**(3), 341–353 (2010)
21. Qi, J., Wei, L., Yao, Y.: Three-way formal concept analysis. In: Miao, D., Pedrycz, W., Ślęzak, D., Peters, G., Hu, Q., Wang, R. (eds.) RSKT 2014. LNCS (LNAI), vol. 8818, pp. 732–741. Springer, Cham (2014). https://doi.org/10.1007/978-3-319-11740-9_67
22. Qi, J.J., Qian, T., Wei, L.: The connections between three-way and classical concept lattices. Knowl.-Based Syst. **91**, 143–151 (2016)
23. Qian, T., Wei, L., Qi, J.: Constructing three-way concept lattices based on apposition and subposition of formal contexts. Knowl.-Based Syst. **116**, 39–48 (2017)
24. Zhi, H., Qi, J., Qian, T., Wei, L.: Three-way dual concept analysis. Int. J. Approx. Reason. **114**, 151–165 (2019)
25. Li, J.H., Deng, S.: Concept lattice, three-way decisions and their research outlooks. Chin. J. Northwest Univ.-Nat. Sci. Ed. **47**(3), 321–329 (2017)
26. Li, L.J., Li, M.Z., Xie, B., Mi, J.S.: Analysis and comparison of concept lattices from the perspective of three-way decisions. Pattern Recogn. Artif. Intell. **29**(10), 951–960 (2016)

Kernelized Fuzzy Rough Sets-Based Three-Way Feature Selection

Xingchen Liu[1], Liuxin Wang[1], Linchao Pan[1], and Can Gao[1,2,3(✉)]

[1] College of Computer Science and Software Engineering, Shenzhen University,
Shenzhen 518060, Guangdong, People's Republic of China
2005gaocan@163.com
[2] Guangdong Key Laboratory of Intelligent Information Processing, Shenzhen
518060, Guangdong, People's Republic of China
[3] SZU Branch, Shenzhen Institute of Artificial Intelligence and Robotics for Society,
Shenzhen 518060, Guangdong, People's Republic of China

Abstract. Feature selection is the process of selecting important features from a dataset. The feature subset formed by important features represents the features of the entire dataset to reduce the complexity of subsequent computations. In recent years, feature selection methods based on rough set theory have been continuously developed, and the approximate quality of kernelized fuzzy rough sets is a better method for evaluating features. However, the heuristic greedy strategy adopted by traditional methods is difficult to guarantee the quality of feature subsets. Based on the idea of three-way decision, this paper proposes fuzzy dependency-based three-way feature selection method. We expand the three potential feature subsets through a differentiated approach and reduce the redundancy among them. Ensemble learning is performed on the three feature subsets to improve the classification performance. The experimental results show that compared with the traditional greedy feature selection method, the proposed feature selection method produces better classification performance, which demonstrates its effectiveness.

Keywords: Kernelized fuzzy rough sets · Three-way decision ·
Feature Selection

1 Introduction

With the continuous growing of the scale of datasets in recent years, a given learning problem and classification task contains a large number of features, and these features are often irrelevant or redundant. Such features will lead to the problems of high computational complexity, weak generalization ability and poor interpretability. Feature selection is an effective technique to alleviate these problems. It reserves highly correlated features and removes redundant and irrelevant features to find the optimal feature subset, and thus improving the performance of models [1]. Therefore, it becomes one of important preprocesses for machine learning, data mining, and pattern recognition etc. [2].

Rough set theories [3] provides an effective method for modeling vague, uncertain, or imprecise data. It uses a pair of exact sets (upper approximation and lower approximation) to describe the uncertainty within the data set.

J. Yao et al. (Eds.): IJCRS 2022, LNAI 13633, pp. 376–389, 2022.
https://doi.org/10.1007/978-3-031-21244-4_28

The attribute reduction methods in this theory remove redundant features in the dataset while maintaining the correlation between feature subsets and decision classes, which coincides with the purpose of feature selection [4]. Based on the rough set theory, the correlation information within data is found without the need for supplementary information, and the number of attributes contained in the data set is reduced, which realizes the feature selection based on rough sets. Presently, there are many extensions in rough sets theory, such as probabilistic rough sets [5], neighborhood rough sets [6], and fuzzy rough sets [7,8], etc.

To deal with the information loss caused by discretizing data, Dubois and Prade defined fuzzy rough sets [7,8] by introducing fuzzy membership functions and extending the membership of elements to [0,1], which provides a high degree of flexibility when dealing with continuous data in fields such as medicine, industry and finance, and can effectively model the ambiguity and uncertainty that exist in the data. On the premise of reserving the advantages of rough sets-based set feature selection for processing high dimensional data, fuzzy rough sets-based feature selection is realized by the fuzzy division of each feature by fuzzy set theory [9]. This method can effectively reduce discrete or continuous noise data, without the cost of adding extra information.

Aiming at the linear inseparability of the data obtained in the real world, that is, there is no dividing hyperplane that can correctly classify the training samples, we use the kernel methods to map the samples from the original space to a higher-dimensional feature space to solve the problem, which makes the samples linearly separable in this feature space. And for a limited-dimensional sample space, there must be a high-dimensional feature space that makes the mapped samples linearly separable. Hu integrated kernel functions with fuzzy rough sets and proposed the model of kernelized fuzzy rough sets, which forms a bridge between kernel machines and rough set-based data analysis [10]. Some generalized feature evaluation functions and attribute reduction algorithms based on the proposed model are shown and the effectiveness of the proposed technique is validated.

The three-way decision [11] theory extends the traditional two-way decision theory and is a decision-making method that conforms to human thinking. In two-way decision, the judgment of objects only stays in two results: acceptance and rejection. However, in practice, people often delay the judgment and decide on objects that they are confident to accept or reject instead of making decisions immediately for uncertain or incomplete information. The three-way decision divides objects into three domains (positive domain, negative domain and boundary domain) according to the decision-making state value by defining the decision function and the threshold of the domain, then constructs the corresponding three-way decision rules [12].

In this paper, we introduce a feature selection method based on kernelized fuzzy rough sets and three-way decision. When constructing feature subsets, how to maintain the maximum relevance for the decision class while minimizing the redundancy between feature subsets is a key issue in feature selection. The three-way strategy we employ is to construct three differentiated subsets of features and to expand the features in the subset from different perspectives. The feature subsets constructed by this strategy tend to be smaller than those constructed by traditional methods. Dependency is an important metric in rough set theory to measure the relevance of features with respect to decision classes. Thus, the dependency gained from new features is used as a reference for our feature subset expansion strategy. Finally, we consider the idea of ensemble learning to construct a multiple feature subsets-based co-classification model.

The rest of the paper is organized as follows. Section 2 presents the notions and properties of the fuzzy rough set model and feature selection. Section 3 shows the three-way attribute reduction algorithm. Experimental analysis is given in Sect. 4. Conclusions come in Sect. 5.

2 Preliminaries

In this section, we will first give some basic definitions, and then review the related work of rough sets, fuzzy rough sets, and kernelized fuzzy rough sets.

2.1 The Notations

Let $I = (\mathbb{U}, \mathbb{A})$ be an information system, where $\mathbb{U} = \{x_1, x_2, ..., x_n\}$ is a nonempty set of finite objects called the universe of discourse and $\mathbb{A}$ is a nonempty finite set of attributes $a : \mathbb{U} \rightarrow V_a$ for every $a \in \mathbb{A}$. For decision systems, $\mathbb{A} = (\mathbb{C}, \mathbb{D})$, where $\mathbb{C}$ is the set of input features and $\mathbb{D}$ is the set of output features. Additionally $a(x), a \in \mathbb{C}, x \in \mathbb{U}$ represents the value of the object x under the attribute a.

2.2 Rough Sets

With any $P \subseteq \mathbb{A}$ there is an associated equivalence relation $IND(P)$:

$$IND(P) = \{(x, y) \in \mathbb{U}^2 \mid \forall a \in P, a(x) = a(y)\} \tag{1}$$

An associated equivalence relation is reflexive, symmetric and transitive. The family of all equivalence classes of $IND(P)$ are denoted by $U/IND(P)$ or U/P for short, which is simply the set of equivalence classes generated by $IND(P)$:

$$\mathbb{U}/IND(P) = \otimes\{\mathbb{U}/IND(\{a\}) \mid a \in P\} \tag{2}$$

where

$$A \otimes B = \{X \cap Y \mid X \in A, Y \in B, X \cap Y \neq \emptyset\} \tag{3}$$

The equivalence classes of the indiscernibility relation with respect to P are denoted $[x]_P, x \in \mathbb{U}$. Let $X \subseteq U$, X can be approximated using only the information contained within P by constructing the P−*lower* and P−*upper* approximations of the classical crisp set X:

$$\underline{P}X = \{x \mid [x]_P \subseteq X\} \tag{4}$$

$$\bar{P}X = \{x \mid [x]_P \cap X \neq \emptyset\} \tag{5}$$

Let P and Q be subsets of condition attributes and decision attributes, respectively, then according to the upper approximation and the lower approximation, then the positive, negative, and boundary regions are defined as:

$$POS_P(Q) = \bigcup_{X \in U/Q} \underline{P}X \tag{6}$$

$$NEG_P(Q) = \mathbb{U} - \bigcup_{X \in U/Q} \bar{P}X \tag{7}$$

$$BND_P(Q) = \bigcup_{X \in U/Q} \bar{P}X - \bigcup_{X \in U/Q} \underline{P}X \tag{8}$$

All objects in the positive region $POS_P(Q)$, must belong to the set X. All objects in the negative region $NEG_P(Q)$, must not belong to the set X. And the objects in the boundary region $BND_P(Q)$, may belong to X. The model of attribute reduction in rough set requires that the positive region of the decision attribute remains unchanged.

If $IND(P) = IND(P - a)$, the attribute $a \in P$ is dispensable in the feature set, otherwise it is indispensable. To achieve attribute reduction, that is, to find the smallest subset P of the conditional attribute set. The minimum subset P needs to satisfy the following two conditions:

(1) $POS_P(Q) = POS_C(Q)$
(2) $\forall a \in P, POS_{P-\{a\}}(Q) = POS_C(Q)$

Then the subset P is a reduct of C.

2.3   Fuzzy Rough Sets

The membership of an object $x \in U$, belonging to the fuzzy positive region can be defined by

$$\mu_{\underline{R_P}X}(x) = \inf_{y \in U} I\left(\mu_{R_P}(x, y), \mu_X(y)\right) \tag{9}$$

Here I is a fuzzy implicator and T is a t-norm. R_P is the fuzzy similarity relation induced by the subset of features P:

$$\mu_{R_P}(x, y) = \bigcap_{a \in P} \{\mu_{R_a}(x, y)\} \tag{10}$$

Many fuzzy similarity relations can be constructed to represent the similarity between objects x and y for feature a, such as

$$\mu_{R_a}(x, y) = 1 - \frac{|a(x) - a(y)|}{|a_{\max} - a_{\min}|} \tag{11}$$

$$\mu_{R_a}(x, y) = \exp\left(-\frac{(a(x) - a(y))^2}{2\sigma_a^2}\right) \tag{12}$$

where σ_a^2 is the variance of feature a. The fuzzy positive region can be defined as

Table 1. Selected t-Norms and their duals (S Conorms)

	Operators T	Operators S
1	$T_M(a,b) = \min(a,b)$	$S_M(a,b) = \max(a,b)$
2	$T_P(a,b) = a \times b$	$S_P(a,b) = a + b - ab$
3	$T_L(a,b) = \max(a+b-1, 0)$	$S_L(a,b) = \min(a+b, 1)$
4	$T_{\cos}(a,b) = \max(ab - \sqrt{1-a^2}\sqrt{1-b^2}, 0)$	$S_{\cos}(a,b) = \min(a+b-ab+\sqrt{2a-a^2}\sqrt{2b-b^2}, 1)$

$$\mu_{POS_{R_P}(Q)}(x) = \sup_{X \in \mathbb{U}/Q} \mu_{R_P}(x) \tag{13}$$

Using the definition of the fuzzy positive region, the new dependency function can be defined as follows:

$$\gamma'_P(Q) = \frac{\sum_{x \in \mathbb{U}} \mu_{POS_{R_P}(Q)}(x)}{|\mathbb{U}|} \tag{14}$$

A fuzzy-rough reduct R can be defined as a subset of features that preserves the dependency degree of the entire dataset, that is, $\gamma'_R(\mathbb{D}) = \gamma'_C(\mathbb{D})$.

2.4 Kernelized Fuzzy Rough Set

Some widely encountered kernel functions satisfying reflexivity, symmetry, and transitivity are:

1. Gaussian kernel: $k_G(x,y) = \exp\left(-\frac{\|x-y\|^2}{\delta_u}\right)$
2. Exponential kernel: $k_E(x,y) = \exp\left(-\frac{\|x-y\|}{\delta}\right)$
3. Rational quadratic kernel: $k_R(x,y) = 1 - \frac{\|x-y\|^2}{\|x-y\|^2+\delta}$

   With the kernel function and the fuzzy operator in Table 1 and Table 2, we can substitute fuzzy relations in fuzzy rough sets. The kernelized fuzzy lower and upper approximation operators are defined as:

1. S-kernel fuzzy lower approximation operator: $\underline{k_S}X(x) = \inf_{y \in U} S(N(k(x,y)), X(y))$;
2. θ-kernel fuzzy lower approximation operator: $\underline{k_\theta}X(x) = \inf_{y \in U} \theta(k(x,y), X(y))$;
3. T-kernel fuzzy upper approximation operator: $\overline{k_T}X(x) = \sup_{y \in U} T(k(x,y), X(y))$
4. σ-kernel fuzzy upper approximation operator: $\overline{k_\sigma}X(x) = \sup_{y \in U} \sigma(N(k(x,y)), X(y))$

   Let the classification be formulated as $<U, A, D>$, where U is the nonempty and finite set of samples, A is the set of features characterizing the classification, D is the class attribute which divides the samples into subset $\{d_1, d_2, ..., d_K\}$. For $\forall x \in U$,

$$d_i(x) = \begin{cases} 0, x \notin d_i \\ 1, x \in d_i \end{cases}$$

We construct the algorithms for computing the fuzzy lower and upper approximations for a given kernel function.

1. $\underline{k_S}d_i(x) = \inf_{y \notin d_i}(1 - k(x, y))$;
2. $\underline{k_\theta}d_i(x) = \inf_{y \notin d_i}\left(\sqrt{1 - k^2(x, y)}\right)$;
3. $\overline{k_T}d_i(x) = \sup_{y \in d_i} k(x, y)$;
4. $\overline{k_\sigma}d_i(x) = \sup_{y \in d_i}\left(1 - \sqrt{1 - k^2(x, y)}\right)$.

The kernelized dependency function is defined as follows:

$$\gamma_B^S(D) = \frac{\left|\cup_{i=1}^I \underline{k_S}d_i\right|}{|U|} \text{ or } \gamma_B^\theta(D) = \frac{\left|\cup_{i=1}^I \underline{k_\theta}d_i\right|}{|U|} \tag{15}$$

The coefficients of classification quality reflect the approximation ability of the approximation space or the ability of the granulated space induced by attribute subset B to characterize the decision.

3 Kernelized Fuzzy Rough Set-Based Three-Way Decision Feature Selection

This section first expounds the problems existing in the heuristic kernelization dependency feature selection strategy, and then describes the feature selection method using the idea of three-way decision.

3.1 Heuristic Feature Selection

Since finding the minimum subset is an NP-hard problem, a heuristic search algorithm is generally used to obtain feature subsets. The maximal dependency(MD) strategy is designed in [13], and its heuristic feature evaluation function is

$$\max_{f \in C-S} \Psi(f, S, D) \tag{16}$$

where $\Psi(f, S, D) = \gamma_B^{S \cup \{f\}}(D) - \gamma_B^S(D)$, C is the initial feature set, S is the selected feature subset, D is the decision feature, and F is a candidate feature.

The purpose of feature selection is to obtain the feature subset with the fewest features under the condition of maintaining the descriptive ability of the feature subset. MD adopts a greedy strategy, that is, adding a candidate feature that maximizes Ψ in each step, so that the dependency of the selected feature subset increases as quickly as possible, and its search can only guarantee a local optimum. The selected feature subset may be too large and redundant, and the quality of the feature subset is difficult to guarantee.

3.2   Feature Selection Based on Three-Way Decision

In order to avoid the problems caused by the greedy strategy and make the feature subset more concise and informative, this paper proposes a three-way decision-based feature selection strategy. In the three-way search, generally each layer maintains three feature subsets, which are used to generate the top three new feature subsets respectively, totaling 9 candidate feature subsets. Then, the top three are selected from the 9 feature subsets, and they are constrained from not originating from the same branch as the 3 feature subsets of the next layer. Three-way feature selection will eventually generate 3 better feature subsets. The method of feature selection and generation of successor is as follows:

$$\bigcup_{i=1}^{w} \max_{f_i \in C - S_i} \Psi\left(f_i, S_i, D\right), \tag{17}$$

$$\Psi\left(f_i, S_i, D\right) = \gamma_B^{S_i \cup \{f_i\}}(D) - \gamma_B^{S_i}(D), \tag{18}$$

C is the conditional feature set, i represents the sequence number of the branch, S_i represents the feature selected by the ith branch, and f_i represents the candidate feature of the ith branch.

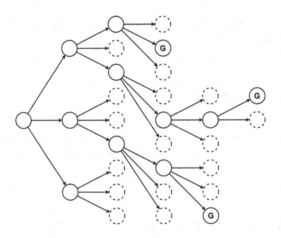

Fig. 1. Three-way feature selection

The idea of three feature selection is shown in Fig. 1. The solid and dashed circle nodes in the figure represent a subset of features. The solid circle indicates that the feature subset will continue to expand, and the dashed circle indicates that the feature subset will not expand. Node G indicates that the feature subset has reached the stopping condition.

The specific descriptions of the three feature selection algorithms are as Algorithm 1.

Algorithm 1. Kernelized fuzzy rough set-based three-way decision feature selection

---

Input: A kernelized fuzzy rough set-based three-way decision system $KFDS =<$ $U, A, D>$, Cutoff threshold θ
Output: Three reduced feature subsets R
1: $subset = \{\{\emptyset\}\}, R = \{\emptyset\}$
2: $k = 3$
3: **while** $flag$ **do**
4:    $flag = FALSE$
5:    **for all** i in $subset$ **do**
6:       **if** $i.dependency > \theta$ **then**
7:          $R.add(i)$
8:          $subset.remove(i)$
9:          $k = k - 1$
10:          $continue$
11:       **end if**
12:       $bestAttrs = getMaxDependencyGainAttrs(A - i)$
13:       **for all** j in $bestAttrs$ **do**
14:          $subset.add(i \cup j)$
15:          $flag = TRUE$
16:       **end for**
17:    **end for**
18:    $subset = subset.getTopK(k)$
19: **end while**

The algorithm first starts with an empty set, and selects the top three features of dependency to form a feature subset of size 1. Next, test whether the current feature subset reaches the threshold. If it reaches the threshold, terminate the expansion of the subset and add it to the output subset set. Otherwise, continue to select the top three features of dependency to expand the subset until the subset There are three feature subsets in the set. In order to maintain the difference of feature subsets, the algorithm constrains that all subsets selected in each round cannot come from the same branch, and existing subsets cannot be selected.

Let the size of the original feature set A in the dataset be N. In the kth round, a feature subset has selected k features, and the time complexity of calculating the dependency gain of the remaining $(N - k)$ features is $O(N - k)$. Then in the worst case, that is, when all features are selected, the total complexity of one feature subset is $O\left(\sum_{k=1}^{N}(N - k)\right) = O\left(N^2\right)$, and the total complexity of three feature subsets is approximately $O(N^2)$.

After obtaining 3 feature subsets, the 3 feature subsets are respectively constructed as homogeneous learners to form three collaborative decision-making models to obtain better learning performance.

3.3   Computational Complexity

The main computational cost of three-branch decision feature selection is from the computation of kernelized dependency with different feature subsets and the selection of features with different branches. Compared with the traditional fuzzy rough set, the usage of kernel functions greatly reduces the storage space and computational cost. With M features, the time complexity of computing the Euclidean distance between a pair of samples is $O(M)$. With N samples, it first spends $O(N)$ to calculate the kernelized lower approximation of each sample, and then merges the lower approximation of all samples by $O(N)$ to obtain the kernelized dependency to measure the quality of feature subsets. In the feature selection process of the three-branch decision, each branch evaluates M features at most and the size of the branch is at most $M - 2$ features. Therefore, the time complexity of computing the kernelized dependency and the feature selection process at different branches are $O(N^2M)$ and $O(M^2)$, respectively. However, the actual computation cost will be much smaller than the theoretical computation cost due to the branch size and the cutoff threshold.

4   Experiment

This part mainly includes the experiment steps and presents an analysis of the model with classification accuracy. We compare the three-way decision model based on kernelized fuzzy rough sets with the traditional greedy algorithm. At the same time, we also make the comparison between soft voting and hard voting for the model in this paper. For each sample, we obtained its three feature subsets obtained, and the closest distance from each feature subset to each class in the data set is calculated and voted, then the closest distance is selected as the feature subset described. Finally, the class to which the majority of feature subsets belong is taken as the class of the sample, this method is called hard voting, while soft voting corresponds to it, the sum of the three feature subsets to the nearest samples of a certain class is taken as the total distance, then the class to which the minimum value belongs can be taken as the class to which it belongs by comparing all distances. In order to facilitate the following representation of the experiment, 'KFRS-FS(S)' is used for soft voting, and 'KFRS-FS(H)' is used for hard voting.

4.1   Datasets and Settings

In this experiment, the specific information of the datasets is shown in Table 3. We summarize the basic information of each dataset as dataset name, number of features and number of samples. At the same time, for the experiment results of each dataset, the average performance of the ten-fold cross-validation method is used as the final performance of our model on the dataset, in order to eliminate the adverse effects of accidental errors in the experiments.

Tests on small-scale datasets show that the kernel-based fuzzy rough set method can extract better feature subsets when the dependency value belongs to $[0.5, 1.0]$. The performance of the algorithm in this paper is compared with that of the greedy algorithm with dependency in $[0.5, 1.0]$. At the same time, for each dataset, the performance difference between soft voting and hard voting is compared. The specific experimental data are shown in Table 5 and Table VI.

Table 2. Residual Implication Induced by the t-Norms and Their Duals

	Residual implication θ	Operator σ
1	$\theta_M(a, b) = \begin{cases} 1, & a \leq b \\ b, & a > b \end{cases}$	$\sigma_M(a, b) = \begin{cases} 0, & a \geq b \\ b, & a < b \end{cases}$
2	$\theta_P(a, b) = \begin{cases} 1, & a = 0 \\ \min(1, b/a), & \text{otherwise} \end{cases}$	$\sigma_P(a, b) = \begin{cases} 1, & a = 0 \\ \max(0, \dfrac{b-a}{1-a}), & \text{otherwise} \end{cases}$
3	$\theta_L(a, b) = \min(b - a + 1, 1)$	$\sigma_L(a, b) = \min(0, b - a)$
4	$\theta_{\cos}(a, b) = \begin{cases} 1, & a \leq b \\ ab + \sqrt{1 - a^2}\sqrt{1 - b^2}, & a > b \end{cases}$	$\sigma_{\cos}(a, b) = \begin{cases} 0, & a > b \\ a + b - ab - \sqrt{2a - a^2}\sqrt{2b - b^2}, & a \leq b \end{cases}$

4.2 Algorithm Performance Comparison

The performance comparison results of the two algorithms on the selected dataset are shown in Table 4. In Table 4, the second column represents the performance of the model in this paper, which is represented by KFRS-FS here, and the third column represents the performance of the classic greedy algorithm, which is represented by GA. It can be seen that the performance of the algorithm in this paper is generally higher than that of the greedy algorithm. Among them, there are more than 5% points of performance improvement in australian, bupa, dnatest, mammographic, spect-train or other datasets, and the improvement is more significant. From overall view, the KFRS-FS algorithm proposed in this

Table 3. Experiment datasets

Dataset	Features	Objects
appendicitis	8	106
australian	15	690
bupa	7	345
dnatest	181	1186
fetal-state	21	2126
german	7	345
haberman	4	306
mammographic	5	748
spectf-train	22	267
vehicle	18	946
wdbc	31	569
weather	5	22

Table 4. Classification accuracy by KFRS-FS and GA

Datasets	KFRS-FS-W1	KFRS-FS-W2	KFRS-FS-W3	KFRS-FS	GA
appendicitis	84.91	86.82	85.73	90.00	87.27
australian	85.22	84.93	85.22	84.78	80.86
bupa	61.71	64.30	66.06	63.42	57.71
dnatest	39.14	38.87	36.76	46.38	38.31
fetal-state	92.05	91.53	91.91	91.97	91.39
german	71.50	71.80	74.00	72.80	71.30
haberman	64.63	71.85	69.57	67.10	64.19
mammographic	73.52	75.00	75.27	75.60	70.66
spectf-train	78.75	78.75	77.75	82.50	76.50
vehicle	70.58	70.58	70.81	73.17	69.27
wdbc	97.19	97.19	97.37	97.54	96.49
weather	85.00	91.67	86.33	90.00	86.66

paper has different degrees of increase in algorithm performance compared with the classical greedy algorithm according to different datasets.

4.3   Analysis of KFRS-FS

This part is mainly aimed at the comparison between soft voting and hard voting inside the KFRS-FS algorithm introduced in this paper, as shown in Table 5 and Table 6, where the second column represents the feature subset distribution obtained by soft voting and hard voting, and the third column represents the performance of soft voting and hard voting. It can be seen that for datasets with fewer features, the performance of soft voting is higher than that of hard voting. On the contrary, the performance of datasets with more features is better than hard voting. It proves that hard voting, which first finds the class to which each feature subset belongs, will have more advantages in the comparison of model performance in the sample space with high dimension while soft voting will ignore the performance of individual feature subsets and try to find an overall performance, this gives soft voting a poor effects in higher dimensions. However, in low-dimensional space, the overall performance will have a better model performance.

In addition, experiments show that the most appropriate cutoff threshold varies in different datasets. When the cutoff threshold is too low, the model can not fully exploit the information in the feature space. When the cutoff threshold is too high, the feature subset may have high redundancy. Both of these will lead to degradation in the performance of the model.

Table 5. Classification accuracy by soft voting

Datasets	Feature subsets by KFRS-FS(S)	Performance of KFRS-FS(S)
appendicitis	[[2, 3], [6, 3], [4, 6]]	85.45
australian	[[6, 2, 4, 7, 1, 5, 8, 10, 3, 0], [13, 6, 2, 4, 7, 1, 5, 8, 10, 3], [9, 6, 1, 4, 7, 2, 5, 8, 10, 3]]	81.15
bupa	[[5, 3, 1, 0, 4], [2, 5, 1, 0, 3], [3, 2, 5, 1, 4]]	63.42
dnatest	[[0, 1], [1, 2], [2, 0]]	39.66
fetal-state	[[6, 1, 7, 0, 12, 3, 9, 13, 14, 20, 4, 16, 10, 11], [1, 12, 7, 0, 3, 13, 9, 20, 6, 4, 16, 10, 11, 17, 8, 15, 2], [16, 1, 12, 7, 3, 14, 13, 9, 6, 20, 0, 10, 17, 4, 8]]	91.97
german	[[9, 5, 4, 0, 8, 7, 6, 2], [1, 3, 9, 4, 5, 0, 8, 7, 6], [3, 9, 4, 2, 5, 0, 8, 7, 6]]	72.29
haberman	[[0, 1], [2, 0], [1, 2]]	67.09
mammographic	[[0, 3, 2], [1, 0, 3], [2, 0, 1]]	68.53
spectf-train	[[40, 26, 1, 5, 4, 2, 7, 21, 33, 24, 3, 22], [42, 40, 3, 8, 1, 33, 21, 4, 7, 28, 2, 14], [41, 40, 28, 1, 4, 21, 8, 3, 7, 33, 2, 24]]	82.50
vehicle	[[8, 17, 14, 15, 9, 0, 12, 2, 16, 7, 3], [11, 17, 14, 15, 9, 0, 2, 12, 16, 3, 7, 1], [6, 17, 14, 15, 9, 0, 2, 12, 16, 3, 7, 1]]	70.23
wdbc	[[22, 27, 21, 11, 24, 20, 8, 7, 18], [27, 0, 21, 11, 24, 20, 8, 6, 18], [20, 27, 21, 11, 24, 18, 22, 8, 9]]	97.01
weather	[[1, 0], [0], [2, 1, 3]]	90.00

Table 6. Classification accuracy by hard voting

Datasets	Feature subsets by KFRS-FS(H)	Performance of KFRS-FS(H)
appendicitis	[[2, 3], [6, 3], [4, 6]]	90.00
australian	[[6, 2, 4, 7, 1, 5, 8], [13, 6, 2, 4, 7, 1, 5], [9, 6, 1, 4, 7, 2, 5]]	84.78
bupa	[[5, 3, 1, 0], [2, 5, 1, 0], [3, 2, 5, 1]]	62.85
dnatest	[[0, 1], [1, 2], [2, 0]]	46.38
fetal-state	[[6, 1, 7, 0, 12, 3, 9, 13, 14, 20, 4, 16, 10, 11, 17, 8], [1, 12, 7, 0, 3, 13, 9, 20, 6, 4, 16, 10, 11, 17, 8, 15, 2], [16, 1, 12, 7, 3, 14, 13, 9, 6, 20, 0, 10, 17, 4, 8, 15]]	91.78
german	[[9, 5, 4, 0, 8, 7, 6, 2, 13, 23, 10, 1], [1, 3, 9, 4, 5, 0, 8, 7, 6, 2, 10, 23], [3, 9, 4, 2, 5, 0, 8, 7, 6, 10, 23, 13]]	72.80
haberman	[[0, 1], [2, 0], [1, 2]]	59.03
mammographic	[[0, 3, 2], [1, 0, 3], [2, 0, 1]]	75.60
spectf-train	[[40, 26, 1, 5, 4], [42, 40, 3, 8, 1], [41, 40, 28, 1, 4]]	73.75
vehicle	[[8, 17, 14, 15, 9, 0, 12, 2, 16, 7, 3, 1, 4], [11, 17, 14, 15, 9, 0, 2, 12, 16, 3, 7, 1, 4], [6, 17, 14, 15, 9, 0, 2, 12, 16, 3, 7, 1, 4]]	73.17
wdbc	[[22, 27, 21, 11, 24], [27, 0, 21, 11], [20, 27, 21, 11]]	97.54
weather	[[1], [0], [2, 1, 3]]	46.66

5   Conclusions

In this paper, the idea of three-way decision is introduced into feature selection based on kernelized fuzzy dependency. From the perspective of multi-branch, multiple feature subsets containing sufficient information and complementarity are obtained, and the classification performance of this method is further improved through ensemble learning. The algorithm proposed in this paper has been performed on benchmark datasets and compared with traditional methods. The experimental results show that the scale of the three feature subsets calculated by the new method is much smaller than the original number of features, which reduces the computational complexity of classification. Moreover, the ensemble learning based on three feature subsets has better classification accuracy on multiple datasets than the traditional kernelized fuzzy rough set feature selection method, indicating that the new method has better classification accuracy. Further research topics include how to extend the three-way decision to the semi-supervised domain, so that the method can be used in more practical situations.

Acknowledgements. The work was supported by the National Natural Science Foundation of China (Nos. 61806127).

References

1. Cervante, L., Xue, B., Shang, L., Zhang, M.: A multi-objective feature selection approach based on binary PSO and rough set theory. In: Middendorf, M., Blum, C. (eds.) EvoCOP 2013. LNCS, vol. 7832, pp. 25–36. Springer, Heidelberg (2013). https://doi.org/10.1007/978-3-642-37198-1_3
2. Li, Y., Li, T., Liu, H.: Recent advances in feature selection and its applications. Knowl. Inf. Syst. **53**(3), 551–577 (2017). https://doi.org/10.1007/s10115-017-1059-8
3. Pawlak, Z.: Rough sets. Int. J. Comput. Inf. Sci. **11**(5), 341–356 (1982)
4. Cervante, L., Xue, B., Shang, L., Zhang, M.: A dimension reduction approach to classification based on particle swarm optimisation and rough set theory. In: Thielscher, M., Zhang, D. (eds.) AI 2012. LNCS (LNAI), vol. 7691, pp. 313–325. Springer, Heidelberg (2012). https://doi.org/10.1007/978-3-642-35101-3_27
5. Qu, Y., Shen, Q., Parthaláin, N.M., Shang, C., Wu, W.: Fuzzy similarity-based nearest-neighbour classification as alternatives to their fuzzy-rough parallels. Int. J. Approx. Reason. **54**(1), 184–195 (2013)
6. Cattaneo, G., et al.: Abstract approximation spaces for rough theories. Rough Sets Knowl. Discov. **1**, 59–98 (1998)
7. Dubois, D., Prade, H.: Rough fuzzy sets and fuzzy rough sets. Int. J. Gen. Syst. **17**(2–3), 191–209 (1990)
8. Dubois, D., Prade, H.: Putting rough sets and fuzzy sets together. In: Słowiński, R. (ed.) Intelligent Decision Support, vol. 11, pp. 203–232. Springer, Dordrecht (1992). https://doi.org/10.1007/978-94-015-7975-9_14
9. Gong, Z., Sun, B., Chen, D.: Rough set theory for the interval-valued fuzzy information systems. Inf. Sci. **178**(8), 1968–1985 (2008)

10. Qinghua, H., Daren, Yu., Pedrycz, W., Chen, D.: Kernelized fuzzy rough sets and their applications. IEEE Trans. Knowl. Data Eng. **23**(11), 1649–1667 (2010)
11. Yao, Y.: Tri-level thinking: models of three-way decision. Int. J. Mach. Learn. Cybern. **11**(5), 947–959 (2020)
12. Gao, C., Zhou, J., Miao, D., Wen, J., Yue, X.: Three-way decision with co-training for partially labeled data. Inf. Sci. **544**, 500–518 (2021)
13. Qinghua, H., Zhang, L., Zhang, D., Pan, W., An, S., Pedrycz, W.: Measuring relevance between discrete and continuous features based on neighborhood mutual information. Expert Syst. Appl. **38**(9), 10737–10750 (2011)

Adaptive K-means Algorithm Based on Three-Way Decision

Yihang Peng[1,2], Qinghua Zhang[1,2(✉)], Zhihua Ai[1,2], and Xuechao Zhi[1,2]

[1] Chongqing Key Laboratory of Computational Intelligence, Chongqing, China
zhangqh@cqupt.edu.cn
[2] Chongqing University of Posts and Telecommunications, Chongqing, China

Abstract. The focus of traditional k-means and its related improved algorithms are to find the initial cluster centers and the appropriate number of clusters, and allocate the samples to the clusters with clear boundaries. These algorithms cannot solve the problems of clusters with imprecise boundaries and inaccurate decisions due to inaccurate information or insufficient data. Three-way clustering can solve this problem to a certain extent. However, most of the existing three-way clustering algorithms divide all clusters into three regions with the same threshold, or divide three regions subjectively. These algorithms are not suitable for clusters with different sizes and densities. To solve the above problems, an adaptive k-means algorithm based on three-way decision is proposed in this paper. First, the traditional clustering results are taken as target set and core region. The distance between each sample in the target set is used as the candidate neighborhood radius threshold. At the same time, neighborhood relationship is introduced to calculate the accuracy of approximation, upper and lower approximation of the target set under the current neighborhood relationship. Second, a boundary control coefficient is defined according to the accuracy of approximation, and as many abnormal data as possible are classified into boundary regions to transform traditional clustering into three-way clustering adapted to different sizes and densities. Finally, five indexes are compared on UCI data set and artificial data set, and the experimental results indicate the effectiveness of the proposed algorithm.

Keywords: Three-way clustering · Three-way decision · Neighborhood · K-means · Accuracy of approximation

1 Introduction

Clustering attempts to classify samples into different clusters according to their similarity. Different clusters should be as far away as possible, and the same cluster should be as close as possible. For decades, many clustering algorithms are proposed. Such as, clustering algorithms based on division: k-means [1] and k-modes [2], hierarchical clustering algorithms: CURE [3] and BIRCH [4], clustering algorithms based on density: DBSCAN [5] and DPC [6].

© The Author(s), under exclusive license to Springer Nature Switzerland AG 2022
J. Yao et al. (Eds.): IJCRS 2022, LNAI 13633, pp. 390–404, 2022.
https://doi.org/10.1007/978-3-031-21244-4_29

Three-way clustering [10,11] is an application field of three-way decision. Compared with traditional clustering, three-way clustering divides the clusters into core region, boundary region and negative region. The samples in the core region clearly belong to a cluster, the samples in the boundary region belong to a cluster with a certain probability, and the samples in the negative region do not belong to a cluster.

Yu et al. [15,17,18,21] introduced the idea of three-way decision into clustering, proposed a three-way clustering method, and also proposed some algorithms to deal with incomplete data and uncertain relation. Wang et al. [13] proposed an algorithm to divide three regions with the stability of each sample. Yao et al. [14] proposed an algorithm with the interval set to represent a cluster. Chen et al. [16,20] proposed a three-way density peak clustering method based on evidence theory to overcome the problem of label propagation errors and an improved DBSCAN based on three-way clustering. Afridi et al. [19] proposed a three-way clustering method based on game theory rough set to solve the uncertainty caused by incomplete and missing data. Wang et al. [22] proposed an effective three-way clustering method, called TWKM, based on disturbance analysis to separate the core region and the boundary region. However, all clusters are divided into three regions with the same threshold in TWKM.

In most of the existing three-way clustering algorithms, all clusters are divided into three regions with the same or subjective threshold, and it is not suitable for clusters with different sizes and densities. Clusters with different sizes and densities should be divided into three regions with different thresholds.

To solve the above problems, in this paper, the idea of three-way decision [23] is introduced and the related concepts of neighborhood [24] are reviewed in brief. An improved adaptive k-means algorithm based on three-way decision is proposed. First, the result of traditional clustering is considered as the core region in this paper. The uncertain data is classified from the core region to the boundary region, and the determined data is still retained in the core region. Second, neighborhood relationship, upper approximation, lower approximation [25] and accuracy of approximation [26,27] are introduced to transform traditional clustering into adaptive three-way clustering. In addition, to identify the appropriate amount of uncertain data, uncertain data can be classified into boundary region, and the determined data can be classified into core region. A boundary control coefficient is defined to find a appropriate radius threshold. Therefore, a three-way k-means algorithm for different cluster sizes and densities is proposed. Finally, the effectiveness of the algorithm is indicated by comparative experiments.

This paper is organized as follows. In Sect. 2, the related concepts and related work are introduced. In Sect. 3, an adaptive k-means algorithm based on three-way decision is proposed. In Sect. 4, the experimental results are analyzed. Finally, the conclusions are drawn in Sect. 5.

2   Related Work

2.1   Neighborhood Rough Set

In neighborhood rough set, for decision information system $IS = (U, A, V, f)$, $U = \{x_1, x_2, \cdots, x_n\}$ represents a non-empty finite set, called the universe. A represents the attribute set. And V is the range. $f : U \times A \rightarrow V$ is an information function, representing the corresponding mapping relationship between samples and attributes.

Definition 1. *Neighborhood Information Granules [24]. Given a decision information system* $IS = (U, A, V, f)$, $U = \{x_1, x_2, \cdots, x_n\}$ *and* $\forall x_i \in U$, *the neighborhood of the sample* x_i *can be defined as:*

$$N(x_i) = \{x_j | x_j \in U, \Delta(x_i, x_j) \leq \mathrm{r}\}, \tag{1}$$

where r is the radius threshold of the sample x_i and $\triangle(\bullet)$ is the distance function. And for $\forall x_1, x_2, x_3 \in U$, $\triangle(\bullet)$ satisfies the following rules:

1) $\triangle(x_1, x_2) \geq 0$,
2) $\triangle(x_1, x_2) = 0$, *only if* $x_1 = x_2$,
3) $\triangle(x_1, x_2) = \triangle(x_2, x_1)$,
4) $\triangle(x_1, x_3) \leq \triangle(x_1, x_2) + \triangle(x_2, x_3)$.

Minkowski distance is a commonly used distance function and defined as follows.

Definition 2. *Minkowski Distance [24]. Given a decision information system* $IS = (U, A, V, f)$, *the Minkowski distance for* $\forall x_1, x_2 \in U$ *under the attribute set A can be defined as:*

$$\Delta_P(x_1, x_2) = \left(\sum_{i=1}^{m} |f(x_1, a_i) - f(x_2, a_i)|^P \right)^{1/P}, \tag{2}$$

where $f(x, a_i)$ represents attribute value of sample x under attribute a_i. When $P = 1$, it is called Manhattan distance. When $P = 2$, it is called Euclidean distance. When $P = \infty$, it is called Chebyshev distance. In this paper, we set $P = 2$.

Definition 3. *Neighborhood Approximation Space [28]. Given a universe U and a neighborhood relation N on U, a tuple $NAS = <U, N>$ is called a neighborhood approximation space.*

Definition 4. *Upper and Lower Approximations of Neighborhood Approximation Space [28]. Given a neighborhood approximation space $NAS = <U, N>$ and set X, where $X \subseteq U$ and U represents a universe, the lower approximation and upper approximation of X in the neighborhood approximation space are respectively defined as:*

$$\underline{NX} = \{x_i | N(x_i) \subseteq X, x_i \in U\}, \tag{3}$$

$$\overline{NX} = \{x_i | N(x_i) \cap X \neq \emptyset, x_i \in U\}. \tag{4}$$

2.2 Three-Way Decision

Yao proposed the concept of three-way decision [29]. It is an extension of the traditional binary-decision model by adding a third region on the basis of the binary-decision model to explain the three regions of the rough set, namely the positive region, negative region and boundary region.

Definition 5. *The Upper and Lower Approximation of Pawlak Rough Set [29]. Suppose U is a universe, and E is an equivalence relation defined on U. Let $apr = (U, E)$ be the approximation space. The division of U under the equivalence relation E is denoted as: $U/E = \{[x]_E \,|\, x \in U\}$, where $[x]$ is an equivalence class containing x. Suppose $\forall X \subseteq U$, its lower approximation and upper approximation are respectively defined as:*

$$\underline{apr}(X) = \{x \in U \mid [x] \subseteq X\}, \tag{5}$$

$$\overline{apr}(X) = \{x \in U \mid [x] \cap X \neq \emptyset\}. \tag{6}$$

The upper and lower approximation divide the universe into three parts, denoted as positive region $POS(X)$, boundary region $BND(X)$ and negative region $NEG(X)$ respectively. It is defined as:

$$POS(X) = \underline{apr}(X) = \{x \in U \mid [x] \subseteq X\}, \tag{7}$$

$$BND(X) = \overline{apr}(X) - \underline{apr}(X), \tag{8}$$

$$NEG(X) = U - \overline{apr}(X). \tag{9}$$

To describe the accuracy of the set, Pawlak et al. [26] proposed the concept of accuracy of approximation. When the accuracy of approximation is equal to 1, the accuracy of the set is the greatest.

Definition 6. *Accuracy of Approximation [27]. Given a universe U and a set X, $X \subseteq U$, the accuracy of approximation of X is defined as:*

$$\delta(X) = \frac{|\underline{apr}(X)|}{|\overline{apr}(X)|}, \tag{10}$$

where $\underline{apr}(X)$ and $\overline{apr}(X)$ denote the lower approximation and upper approximation of set X, respectively.

2.3 Three-Way Clustering

The idea of three-way clustering [9,10,23] is to use three disjoint sets to represent a cluster, called the core region, boundary region and trivial region, respectively. Given a universe U, $C = \{C_1, C_2, \cdots, C_k\}$ is a family clusters of U, where k denotes the number of clusters. The samples in the core region belong to a cluster, the samples in the boundary region belong to a cluster with a certain probability, and the samples in the trivial region certainly do not belong to a cluster. Therefore, a cluster can be represented by core region $POS(C_i)$ and boundary region $BND(C_i)$, where $POS(C_i) \neq \emptyset$, C_i represents a cluster i, $C_i \in C$, and $i = 1, 2, \cdots, k$. $POS(C_i)$ and $BND(C_i)$ satisfy the following condition:

$$\bigcup_{i=1}^{k} (POS(C_i) \cup BND(C_i)) = U. \tag{11}$$

The result of three-way clustering is as follows:

$$TC = \{TC_1, TC_2, \cdots, TC_k\}, \tag{12}$$

where $TC_i = (POS(C_i), BND(C_i))$, $i = 1, 2, \cdots, k$. When $BND(C_i) = \emptyset$, three-way clustering becomes the traditional clustering.

3 Adaptive K-Means Algorithm Based on Three-Way Decision

Suppose U is a universe with n samples. A cluster is represented by a core region and a boundary region in three-way clustering. The samples in the core region clearly belong to a cluster, the samples in the boundary region belong to a cluster with a certain probability. The result of three-way clustering is represented as $TC = \{TC_1, TC_2, \cdots, TC_k\}$, where k denotes the number of clusters, $TC_i = (POS(C_i), BND(C_i))$, $i = 1, 2, \cdots, k$ and C_i represents a cluster i.

In traditional clustering, k-means is a classical algorithm. Related improved algorithms of k-means are to find the initial centers and the appropriate number of clusters. These algorithms simply consider the relationship between samples and clusters. It cannot deal with the problem of clusters with imprecise boundary and inaccurate decision. This problem can be solved by the idea of three-way decision. However, there are several new problems in three-way clustering. It is not appropriate to divide the core and boundary region for all clusters with subjective or same value. Therefore, the main problem is to automatically select the most appropriate threshold for different clusters instead of dividing all clusters into regions with a same threshold. As mentioned before, the samples in boundary region are classified into the core region or the samples in core region are classified into the boundary region. To solve the above problems, an adaptive k-means algorithm based on three-way decision is proposed in this paper. It mainly includes the following two steps. First, the first clustering result is based on traditional clustering method. In this paper, the algorithm of literature [30] is used as the first clustering algorithm, named DCKM. Second, clustering result of the first step is regarded as the target set. There are two cases of the result in the first step. For case 1, all of the clusters are correct samples. For case 2, some clusters include abnormal samples. To divide the boundary region of each cluster, neighborhood and accuracy of approximation are introduced in this paper. The neighborhood radius is selected according to the distance between each cluster sample. The neighborhood of each sample under different neighborhood radius is calculated to obtain the upper approximation, lower approximation and accuracy of approximation. If the accuracy of approximation under each neighborhood radius is all equal to 1, it is determined as case 1. For case 1, the distance between each sample of this cluster will not affect the accuracy of approximation when all these clusters are correct samples. If the accuracy of approximation under some radii is not equal to 1, it is determined as case 2. For case 2, if there are some abnormal samples in this cluster, the distance between some samples of this cluster will affect the accuracy of approximation. For case 1, the core region is the current cluster sample, and the boundary region is an empty set.

For case 2, to make more abnormal samples can be found in this cluster under the condition of high accuracy of approximation. A boundary control coefficient $\lambda_\mu(C_i)$ is defined as follows.

Definition 7. *Boundary Control Coefficient. Given a universe U, the number of clusters k and cluster C_i, $C = \{C_1, C_2, \cdots, C_k\}$ is a family clusters of universe U, where C_i represents a cluster i, $C_i \in C$ and $1 \le i \le k$. The boundary control coefficient is defined as follows.*

$$\lambda_\mu(C_i) = \delta_\mu(C_i) \times \frac{|BND(C_i)|}{|U|}, \tag{13}$$

where μ denotes the neighborhood radius and $\delta_\mu(C_i)$ denotes the accuracy of approximation in cluster C_i under the current neighborhood radius μ. $|BND(C_i)|$ denotes the number of current cluster samples divided into the boundary region, $|U|$ denotes the total number of samples, and $\frac{|BND(C_i)|}{|U|}$ represents the proportion of the number of samples in the boundary region of the current cluster to the total number of samples.

In each cluster, boundary control coefficient can be calculated with different neighborhood radii. To ensure a large accuracy of approximation and find abnormal samples, when $\lambda_\mu(C_i)$ is the largest, the radius is the most appropriate. The number of abnormal samples in the core region is not known in practice. It is necessary to classify the abnormal samples into the boundary regions. Fewer abnormal samples can be divided into boundary region with a larger accuracy of approximation. To find more abnormal samples, the ratio of the number of boundary regions to the total number of samples with different radii can be used as the second variable in Eq. (13). Therefore, the boundary control coefficient $\lambda_\mu(C_i)$ is defined as above.

Some of the data from the R15 dataset [31] are shown in Table 1, where a_1 and a_2 denote attributes. The specific process is as follows.

Neighborhood radius is selected from the distance between each sample in the same cluster, as Table 2. In fact, it is impossible to know case 1 or case 2. Therefore, a judgment is necessary. If the first clustering cluster is $C_1 = \{A, B, C, D\}$, $C_2 = \{E, F, G, H\}$. In cluster C_1, the selection range of radius is the distance between four samples A, B, C and D, such as 0.23, 0.32, 0.33, 0.37, 0.63 and 0.65 from Table 2. For example, when $\mu = 0.23$, $N^{0.23}(A) = \{A, D\}$, $N^{0.23}(B) = \{B\}$,

Table 1. Partial data for R15.

Sample	Attribute		Cluster
	a_1	a_2	
A	9.800	10.132	1
B	10.350	9.768	1
C	10.098	9.988	1
D	9.730	9.910	1
E	12.040	10.028	2
F	12.082	10.044	2
G	12.400	10.156	2
H	11.988	9.926	2

Table 2. The distance between samples.

Sample	Attribute							
	A	B	C	D	E	F	G	H
A	0.00	0.65	0.32	0.23	2.24	2.28	2.59	2.19
B		0.00	0.33	0.63	1.7	1.75	2.08	1.64
C			0.00	0.37	1.94	1.98	2.3	1.89
D				0.00	2.31	2.35	2.68	2.25
E					0.00	0.04	0.38	0.11
F						0.00	0.33	0.15
G							0.00	0.47
H								0.00

Table 3. Description of data sets.

ID	Data set	Number of samples	Dimension
1	Iris	150	4
2	Ionosphere	351	34
3	Seeds	210	7
4	Soybean-small	47	35
5	Wine	178	13
6	Pima	768	8
7	Segmentation	210	19
8	Ecoli	336	6
9	Aggregation	788	2
10	Flame	240	2

Table 4. ACC on ten data sets.

ID	k-means	k-means++	Canopy	DCKM	TWKM	M-RKM	TWD-AKM
1	0.3035	0.3514	0.8533	**0.9266**	0.7400	0.8907	**0.9266**
2	0.1810	0.1906	0.7094	0.7311	0.7122	0.6353	**0.7351**
3	0.1914	0.1419	0.6523	0.8238	0.7476	0.4142	**0.8333**
4	0.3382	0.4170	0.5744	**0.7446**	0.6382	0.7208	**0.7446**
5	0.1835	0.3628	0.6500	0.6516	0.6292	0.4764	**0.6574**
6	0.4986	0.5832	0.6601	0.6966	0.6966	0.6510	**0.6993**
7	0.3266	0.3057	0.3666	**0.5142**	0.4381	0.2529	**0.5142**
8	0.4017	0.4248	0.6250	0.7232	0.5625	0.6529	**0.7352**
9	0.3740	0.3881	0.5051	0.8159	0.7766	0.7941	**0.8249**
10	0.1906	0.2516	0.8333	0.9458	0.8250	0.8000	**0.9542**

$N^{0.23}(C) = \{C\}$, $N^{0.23}(D) = \{A, D\}$, $N^{0.23}(E) = \{E, F, H\}$, $N^{0.23}(F) = \{E, F, H\}$, $N^{0.23}(G) = \{G\}$, $N^{0.23}(H) = \{E, F, H\}$, $\overline{apr^{0.23}}(C_1) = \{A, B, C, D\}$, $apr^{0.23}(C_1) = \{A, B, C, D\}$, $POS_{0.23}(C_1) = \{A, B, C, D\}$, $BND\&_{0.23}(C_1) = \emptyset$, $\overline{\delta_{0.23}(C_1)} = 1$.

After calculation, the accuracy of approximation is always equal to 1 with all radii. Therefore, the value of radius μ has no effect on the final accuracy of approximation. Then, the samples of the cluster are all correct samples. The final core and boundary region of the cluster can be represented as $POS(C_1) = \{A, B, C, D\}$, $BND(C_1) = \emptyset$. Similarly, $POS(C_2) = \{E, F, G, H\}$, $BND(C_2) = \emptyset$. For case 2, the boundary control coefficient is considered to divide different regions in different clusters. If the first clustering is $C_1 = \{A, B, C, D, E, F\}$, $C_2 = \{G, H\}$, the processing is as follows. First, C_1 is taken as an example. According to Table 2, the radius is selected from the distances between the six samples A, B, C, D, E and F. When $\mu = 0.04$, $N^{0.04}(A) = \{A\}$, $N^{0.04}(B) = \{B\}$, $N^{0.04}(C) = \{C\}$, $N^{0.04}(D) = \{D\}$, $N^{0.04}(E) = \{E, F\}$, $N^{0.04}(F) = \{E, F\}$, $N^{0.04}(G) = \{G\}$, $N^{0.04}(H) = \{H\}$, $\overline{apr^{0.04}}(C_1) = \{A, B, C, D, E, F\}$, $apr^{0.04}(C_1) = \{A, B, C, D, E, F\}$, $POS_{0.04}(C_1) = \{A, B, C, D, E, F\}$, $BND\&_{0.04}(C_1) = \emptyset$, $\delta_{0.04}(C_1) = 1$.

Similarly, when μ takes 0.23 and 0.32, $\delta_\mu(C_1) = \frac{4}{7}$, $\lambda_\mu(C_i) = \frac{3}{14}$. When μ takes 0.33, 0.37, 0.63 and 0.65, $\delta_\mu(C_1) = \frac{1}{2}$, $\lambda_\mu(C_i) = \frac{1}{4}$. When μ takes 1.94 and 1.98, $\delta_\mu(C_1) = \frac{2}{8}$, $\lambda_\mu(C_1) = \frac{3}{16}$. When μ takes 1.75 and 1.7, $\delta_\mu(C_1) = \frac{3}{8}$, $\lambda_\mu(C_1) = \frac{15}{64}$. When μ takes 2.24, $\delta_\mu(C_1) = \frac{1}{8}$, $\lambda_\mu(C_1) = \frac{7}{64}$. When μ takes 2.28, 2.31 and 2.35, $\delta_\mu(C_1) = 0$, $\lambda_\mu(C_1) = 0$. If the corresponding accuracy of approximation is not equal to 1 under some neighborhood radii, it can be determined as case 2. It is necessary to classify the abnormal samples into boundary region. When the boundary control coefficient is the largest, the corresponding radius is the best choice. If there are abnormal samples in a cluster, the accuracy of approximation is not equal to 1 in the actual situation. However, the accuracy of approximation is equal to 1 under some neighborhood radii. Therefore,

---

Algorithm 1: Adaptive k-means algorithm based on three-way decision

---

Input: universe U

Output: Clustering results of data sets

1 The first clustering result of data sets U:
2 $cls_1 = DCKM(U) = \{C_1, C_2, \cdots, C_k\}$;
3 **for** $i=1$ to k **do**
4 $\quad$ $C_{current} = C_i, flag = 1$;
5 $\quad$ calculate the distance between each sample of the current cluster and
6 $\quad$ put it into the distance matrix $Dist_i$;
7 $\quad$ **for** j in $C_{current}$ **do**
8 $\quad\quad$ **for** m in $Dist_i$ **do**
9 $\quad\quad\quad$ $\mu_{current} = m$;
10 $\quad\quad\quad$ calculate $N^{\mu_{current}}$ of each sample;
11 $\quad\quad\quad$ calculate $\overline{apr^{\mu_{current}}}(C_{current})$, $\underline{apr^{\mu_{current}}}(C_{current})$,
$\quad\quad\quad\quad$ $\delta_{\mu_{current}}(C_{current})$;
12 $\quad\quad\quad$ **if** each $\delta_{\mu_{current}}(C_{current}) = 1$ **then**
13 $\quad\quad\quad\quad$ flag = 1;
14 $\quad\quad\quad\quad$ $POS_{\mu_{current}}(C_{current}) = \{\underline{apr^{\mu_{current}}}(C_{current})\}$;
15 $\quad\quad\quad\quad$ $BND_{\mu_{current}}(C_{current}) = \emptyset$;
16 $\quad\quad\quad\quad$ remove $C_{current}$ from D ;
17 $\quad\quad\quad$ **else**
18 $\quad\quad\quad\quad$ flag = -1;
19 $\quad\quad\quad\quad$ $max_\delta \leftarrow max(\delta_{\mu_{current}}(C_{current}))$;
20 $\quad\quad\quad\quad$ $\mu \leftarrow \mu_{current}$;
21 $\quad\quad\quad\quad$ $POS_{\mu_{current}}(C_{current}) \leftarrow \{\underline{apr^{\mu_{current}}}(C_{current})\}$;
22 $\quad\quad\quad\quad$ $BND_{\mu_{current}}(C_{current}) \leftarrow$
$\quad\quad\quad\quad$ $\{\overline{apr^{\mu_{current}}}(C_{current}) - \underline{apr^{\mu_{current}}}(C_{current})\}$;
23 $\quad\quad\quad$ **end**
24 $\quad\quad$ **end**
25 $\quad$ **end**
26 **end**
27 **return** $TC = \{(POS(C_1), BND(C_1)), \cdots, (POS(C_k), BND(C_k))\}$

---

the radius is not appropriate. To classify the abnormal samples into the boundary region, in the case of large approximation accuracy, the boundary control coefficient is introduced. The calculation process of other clusters is the same as above. Thus, as the above process, each cluster is divided into core and boundary regions adapted to respective densities and sizes.

4   Experiments

To evaluate the effectiveness of the proposed algorithm, eight UCI datasets [31] and two artificial datasets [32] are chosen in this section to compare with the traditional methods and improved methods.

Table 5. ARI on ten data sets.

ID	k-means	k-means++	Canopy	DCKM	TWKM	M-RKM	TWD-AKM
1	0.4302	0.4692	0.6615	0.7686	0.3711	0.7420	**0.8016**
2	0.1679	0.1679	0.1727	0.1791	0.1286	0.1662	**0.1853**
3	0.4805	0.1575	0.4647	0.5511	0.3964	0.2953	**0.5626**
4	0.2965	0.5006	0.2966	**0.5742**	0.4089	0.5123	**0.5742**
5	0.3054	**0.3711**	0.3694	0.3498	0.2453	0.3605	0.3584
6	0.1152	0.1743	0.1743	0.2140	0.1491	0.1875	**0.2164**
7	0.1729	0.2712	0.1749	0.3039	0.1804	0.1368	**0.3061**
8	0.3224	0.4273	0.3839	**0.5340**	0.1862	0.4368	**0.5340**
9	0.3740	0.3714	0.3451	**0.7054**	0.5333	0.6543	0.6837
10	0.4544	0.4296	0.4422	0.7937	0.4200	0.4007	**0.8088**

Table 6. Jaccard on ten data sets.

ID	k-means	k-means++	Canopy	DCKM	TWKM	M-RKM	TWD-AKM
1	0.1214	0.1500	0.7287	**0.8633**	0.4371	0.7599	**0.8633**
2	0.1016	0.1074	0.1280	0.5775	0.1687	0.5519	**0.5811**
3	0.1107	0.1801	0.5309	0.7073	0.4534	0.3967	**0.7143**
4	0.1773	0.1654	0.5344	**0.5932**	0.5033	0.5279	**0.5932**
5	0.1065	0.2338	0.4220	0.4833	0.3459	0.3537	**0.4896**
6	0.3505	0.4235	0.2284	0.5344	0.2899	0.4925	**0.5376**
7	0.1699	0.1582	0.3124	**0.3461**	0.2902	0.2514	**0.3461**
8	0.1548	0.1702	0.4481	0.5664	0.3691	0.5514	**0.5812**
9	0.1940	0.4760	0.4913	0.6891	0.6195	0.6925	**0.7019**
10	0.1136	0.1586	0.4101	0.8972	0.4676	0.2867	**0.9048**

Table 7. AMI on ten data sets.

ID	k-means	k-means++	Canopy	DCKM	TWKM	M-RKM	TWD-AKM
1	0.5887	0.5869	0.7441	0.7717	0.5873	0.6261	**0.7874**
2	0.1231	0.1231	**0.5496**	0.1463	0.5530	0.4655	0.1516
3	0.5511	0.2989	0.4840	0.5401	0.5669	0.2612	**0.5758**
4	0.5344	0.6579	0.4029	**0.6988**	0.4687	0.6116	**0.6988**
5	0.3973	0.4226	**0.4895**	0.4350	0.4590	0.4318	0.4525
6	0.0631	0.0284	0.4927	0.1279	**0.5344**	0.4826	0.1299
7	0.3089	0.4687	0.2244	**0.4801**	0.2804	0.1448	0.4738
8	0.5862	**0.5939**	0.4545	0.5286	0.3913	0.5448	0.5286
9	0.7381	0.7388	0.3378	0.7407	0.6348	0.7075	**0.7532**
10	0.3989	0.5372	0.7142	0.7176	0.7021	0.6667	**0.7319**

Table 8. RI on ten data sets.

ID	k-means	k-means++	Canopy	DCKM	TWKM	M-RKM	TWD-AKM
1	0.7762	0.7728	0.8464	0.9114	0.7093	0.8307	**0.9124**
2	0.5841	0.5841	0.5865	0.6067	0.5889	0.5368	**0.6904**
3	0.7343	0.6494	0.7258	0.8004	0.7177	0.6745	**0.8057**
4	0.5929	0.7502	0.5929	**0.8094**	0.7354	0.7607	**0.8094**
5	0.6177	**0.7100**	0.6072	0.6561	0.6457	0.6693	0.6596
6	0.5610	0.5507	0.5507	0.5767	0.5767	0.5450	**0.5789**
7	0.6636	0.6942	0.6622	0.7530	0.7053	0.4616	**0.7669**
8	0.7856	**0.8066**	0.6724	0.7903	0.6788	0.4816	0.7903
9	0.8439	0.8435	0.6531	**0.8972**	0.8479	0.8939	0.8749
10	0.7278	0.7051	0.7210	0.8971	0.7100	0.7018	**0.9047**

4.1   The Evaluation Index

1) Accuracy (ACC)

$$ACC = \sum_{c=1}^{k} \frac{n_c^j}{n},\tag{14}$$

where n_c^j is the number of common objects in the cluster c and n represents the number of samples. After one-to-one matching is obtained, its matching category is j. A higher value indicates the better clustering.

2) Rand Index (RI)

$$RI = \frac{a+b}{C_2^{n_{samples}}},\tag{15}$$

where $a + b$ denotes the number of samples belonging to the same cluster. a is the number of samples in the same class of real data and predicted data. b is the number of samples in the different classes of real data and predicted data. $C_2^{n_{samples}}$ denotes the number of any two samples combined into one class.

3) Adjusted Rand Index (ARI)

$$ARI = \frac{RI - E\{RI\}}{\max(RI) - E\{RI\}},\tag{16}$$

similar to RI, a larger value means a better match with the real situation. $E\{RI\}$ represents the expectation of RI and $\max(RI)$ represents the maximum value of RI.

4) Jaccard

$$Jaccard(A, B) = \frac{A \cap B}{A \cup B},\tag{17}$$

where A and B denote two sets for comparing the similarity between two samples. The higher values indicate the higher similarity between the two samples.

5) Adjusted Mutual Information (AMI)

$$AMI(U,V) = \frac{MI(U,V) - E\{MI(U,V)\}}{F(H(U),H(V)) - E\{MI(U,V)\}}, \tag{18}$$

where $E\{MI(U,V)\}$ is the expectation of the mutual information $MI(U,V)$, $F(H(U),H(V)) = \frac{H(U)+H(V)}{2}$. $H(U)$ and $H(V)$ represent the information entropy of U and V, respectively. U and V represent real label and predicted label, respectively.

4.2 Experimental Illustration

To indicate the effectiveness of the proposed algorithm in this paper, the experimental results of k-means, k-means++, Canopy, DCKM [30], M-RKM [35] and TWKM [22] are analyzed with the algorithm in this paper on five evaluation metrics. The proposed algorithm in this paper is called TWD-AKM. The relevant descriptions of the data sets are shown as Table 3. In this experiment, all core and boundary regions are regarded as a clustering result. The experimental results of k-means, k-means++, Canopy, DCKM, TWKM, M-RKM and TWD-AKM are based on 10 datasets, as shown in Tables 4, 5, 6, 7 and 8, with the optimal results of the experiments in bold. The initial cluster centers of k-means, k-means++, Canopy and TWKM are selected randomly, and the randomness of the results is large. Therefore, in this experiment, when calculating the relevant index, 200 times are randomly selected and then the average value is taken as the final result.

The problem of neighborhood radius selection is to be involved in the proposed algorithm in this paper. Meanwhile, the selection range is the distance between each sample in the cluster of the first clustering. From Tables 4, 5, 6, 7 and 8, the proposed algorithm in this paper is superior to other algorithms in most data sets in terms of ACC, ARI, Jaccard, AMI and RI. Each cluster has a clear boundary, the cluster is marked as the core region and boundary region, and the abnormal samples are classified into boundary region. At the same time, the core region and boundary region of all clusters are not divided by a same threshold. According to the experimental results, the proposed algorithm in this paper has better performance on most of data sets. The experimental results of ACC and Jaccard have better performance on most data sets, because only the samples of the core region are considered in the calculation. The proposed algorithm in this paper performs poorly on the data sets, such as Wine, Aggregation and Ecoli. Although initial centers are randomly selected with many times, the number of selected clusters does not match the actual number of clusters and the number of selected clusters is less than the actual number of clusters. The value of ACC is small when the algorithm in this paper selects the real clusters. In Eq. (14), a represents the number of samples of real data and prediction data in the same class. b represents the number of real data and predicted data in the different classes. However, the real data is not in the same class and the predicted data is in the same class. It causes the value of $a + b$ to be small.

Therefore, when the number of selected clusters is closer to the real situation, the values of ARI and RI are small. In conclusion, the proposed algorithm has better performance.

5   Conclusions

As an extension of traditional clustering, three-way clustering has better performance in dealing with data with imprecise boundaries. However, in three-way clustering, it is inappropriate for most algorithms to divide the core region and the boundary region for each cluster with the same threshold, because the same threshold will cause the samples in the core region to be divided into the boundary region or the samples in the boundary region to be divided into the core region. To divide different clusters into appropriate core regions and boundary regions, in this paper, the concept of neighborhood relationship is introduced and the boundary control coefficient is defined to classify abnormal samples into boundary region. And an appropriate threshold can be obtained by this coefficient. Thus, traditional clustering is transformed into a three-way clustering adapted to different sizes and densities. Compared with most of the three-way clustering algorithms, the proposed algorithm in this paper divides the core and boundary regions with different thresholds. To a certain extent, the problem of sample misclassification is avoided. Therefore, the proposed algorithm is more suitable for clusters with different sizes and densities. Finally, the effectiveness of the algorithm is indicated by experiments.

Acknowledgments. This work was supported in part by the National Key Research and Development Program of China (No. 2020YFC2003502), the National Natural Science Foundation of China (No. 61876201), and Chongqing Talents Program (No. CQYC202210202215).

References

1. Wang, Y.L., Luo, X., Zhang, J.: An improved algorithm of k-means based on evolutionary computation. Intell. Autom. Soft Comput. **26**, 961–971 (2020)
2. Yuan, F., Yang, Y.L., Yuan, T.T.: A dissimilarity measure for mixed nominal and ordinal attribute data in k-modes algorithm. Appl. Intell. **50**, 1–12 (2020)
3. Lorbeer, B., Kosareva, A., Deva, B.: Variations on the clustering algorithm BIRCH. Big Data Res. **11**, 44–53 (2018)
4. Li, Y.F., Jiang, H.T., Lu, J.Y.: MR-BIRCH: a scalable mapreduce-based BIRCH clustering algorithm. J. Intell. Fuzzy Syst. **40**, 5295–5305 (2021)
5. Scitovski, R., Sabo, K.: DBSCAN-like clustering method for various data densities. Pattern Anal. Appl. **23**, 541–554 (2020)
6. Lin, J.L., Kuo, J., Chuang, H.: Improving density peak clustering by automatic peak selection and single linkage clustering. Symmetry **12**, 1168 (2020)
7. Zhang, H.R., Fan, M., Shi, B.: Regression-based three-way recommendation. Info. Sci. **37**, 444–461 (2017). https://cn.overleaf.com/project8
8. Li, L., Jin, Q., Yao, B., Wu, J.: A rough set model based on fuzzifying neighborhood systems. Soft. Comput. **52**, 2381–2410 (2020)

9. Chu, X.L., Sun, B.Z., Li, X.: Neighborhood rough set-based three-way clustering considering attribute correlations: an approach to classification of potential gout groups. Inf. Sci. **535**, 28–41 (2020)
10. Wang, P.X., Yao, Y.Y.: CE3: a three-way clustering method based on mathematical morphology. Knowl.-Based Syst. **155**, 54–65 (2018)
11. Yu, H., Zhang, C., Wang, G.Y.: A tree-based incremental overlapping clustering method using the three-way decision theory. Knowl.-Based Syst. **91**, 189–203 (2016)
12. Yu, H., Chang, Z.H., Wang, G.Y.: An efficient three-way clustering algorithm based on gravitational search. Int. J. Mach. Learn. Cybern. **11**, 1003–1016 (2020)
13. Wang, P.X., Yang, X.B.: Three-way clustering method based on stability. IEEE Access. **9**, 33944–33953 (2021)
14. Yao, Y.Y.: Three-way decisions with probabilistic rough sets. Inf. Sci. **180**, 341–353 (2010)
15. Yu, H.: A framework of three-way cluster analysis. In: Rough Sets, IJCRS 2017, vol. 10314, pp. 300–312 (2017). https://doi.org/10.1007/978-3-319-60840-2-22
16. Yu, H., Chen, L.Y., Yao, J.T.: A three-way density peak clustering method based on evidence theory. Knowl.-Based Syst. **211**, 106532 (2021)
17. Yu, H., Su, T., Zeng, X.H.: A three-way decisions clustering algorithm for incomplete Data. In: Rough Sets and Knowledge Technology, RSKT 2014, vol. 8818, pp. 765–776 (2014). https://doi.org/10.1007/978-3-319-11740-9
18. Yu, H., Yun, C., Lingras, P.: A three-way cluster ensemble approach for large-scale data. Int. J. Approx. Reason. **115**, 32–49 (2019)
19. Afridi, M.K., Azam, N., Yao, J.T.: A three-way clustering approach for handling missing data using GTRS. Int. J. Approx. Reason. **98**, 11–24 (2018)
20. Yu, H, Chen, L.Y., Yao J.T.: A three-way clustering method based on an improved DBSCAN Algorithm. Physica A: Stat. Mech. Appl. **535**, 122289 (2019)
21. Yu, H., Wang, X.C., Wang, G.Y.: An active three-way clustering method via low-rank matrices for multi-view data. Inf. Sci. **507**, 823–839 (2020)
22. Wang, P., Shi, H., Yang, X., Mi, J.: Three-way k-means: integrating k-means and three-way decision. Int. J. Mach. Learn. Cybern. **10**(10), 2767–2777 (2019). https://doi.org/10.1007/s13042-018-0901-y
23. Yu, H.: Three-way cluster analysis. Peak Data Sci. **5**, 31–35 (2016)
24. Wang, C.Z., Shi, Y.P., Fan, X.D.: Attribute reduction based on k-nearest neighborhood rough sets. Acoustic Bull. **106**, 18–31 (2019)
25. Pawlak, Z., Wong, S.K.M., Ziarko, W.: Rough sets: probabilistic versus deterministic approach. Int. J. Man Mach. Stud. **29**, 81–95 (1988)
26. Singh, P.K., Tiwari, S.: Topological structures in rough set theory: a survey. Hacettepe J. Math. Stat. **49**, 1270–1294 (2020)
27. Xia, J.R.: The granular accuracy of approximation for the rough sets. Appl. Math. A J. Chin. **27**, 248–252 (2012)
28. Hu, Q.H., Yu, D.R., Xie, Z.X.: Numerical attribute based on neighborhood granulation and rough approximation. J. Softw. **19**, 640–649 (2018)
29. Xu, Y., Tang, J.X., Wang, X.S.: Three sequential multi-class three-way decision models. Inf. Sci. **539**, 62–90 (2020)
30. Zhang, G., Zhang, C.C., Zhang, H.Y.: Improved k-means algorithm based on density canopy. Knowl.-Based Syst. **145**, 289–297 (2018)
31. Dua, D., Graff, C.: UCI Machine Learning Repository (2019). http://archive.ics.uci.edu/ml

32. Fränti, P., Sieranoja, S.: K-means properties on six clustering benchmark datasets. Appl. Intell. **48**(12), 4743–4759 (2018). https://doi.org/10.1007/s10489-018-1238-7

33. Wang, H., Chai, X.H.: Sequential three-way decision of tolerance-based multi-granularity fuzzy-rough sets. IEEE Access **7**, 180336–180348 (2019)

34. Zhang, S.Y., Li, S.G., Yang, H.L.: Three-way convex systems and three-way fuzzy convex systems. Inf. Sci. **510**, 89–98 (2020)

35. Sivaguru, M., Punniyamoorthy, M.: Performance-enhanced rough k-means clustering algorithm. Soft. Comput. **25**, 1595–1616 (2021)

3WS-ITSC: Three-Way Sampling on Imbalanced Text Data for Sentiment Classification

Yu Fang[1,2], Zhao-Chen Li[1], Xin Yang[3,4], and Fan Min[1,5(✉)]

[1] School of Computer Science, Southwest Petroleum University, Chengdu 610500,
China
minfan@swpu.edu.cn
[2] School of Computer, Faculty of Engineering, Universiti Teknologi Malaysia,
81310 UTM Johor Bahru, Johor, Malaysia
[3] Department of Artificial Intelligence, School of Economic Information Engineering,
Southwestern University of Finance and Economics, Chengdu 611130, China
[4] Financial Intelligence and Financial Engineering Key Laboratory of Sichuan
Province, Southwestern University of Finance and Economics, Chengdu 611130,
China
[5] Institute for Artificial Intelligence, Southwest Petroleum University,
Chengdu 610500, China

Abstract. Sentiment analysis is an important research direction of natural language processing. The data imbalance is a critical issue in text sentiment classification task. That arises the problem of high misclassification cost. This paper proposes a three-way sampling sentiment classification model for imbalanced text data to reduce the misclassification cost. Specifically, the model extracts boundary points through three-way sampling and collaborates with cost-sensitive learning for action on sampled results. Firstly, in order to reduce sampling time, the text data is converted into a one-dimensional vector by bag mapping. Secondly, three-way sampling is used to obtain boundary points that can characterize the majority class. Finally, a sequential three-way sentiment classification algorithm is used to predict sentiment polarity. The experimental results show that the proposed model outperforms state-of-the-art sentiment classification methods in the scenario of extremely imbalanced test data.

Keywords: Imbalanced text data · Sentiment classification · Bag mapping · Three-way sampling · Sequential three-way sentiment classification

1 Introduction

The purpose of text sentiment analysis is to reveal people's emotional tendencies. It has become the focus of attention of major merchants and consumers. Senti-

This work was supported by the National Natural Science Foundation of China (62006200); The Southwest Petroleum University Postgraduate English Course Construction Project (No. 2020QY04); Central Government Funds of Guiding Local Scientific and Technological Development (No. 2021ZYD0003).

J. Yao et al. (Eds.): IJCRS 2022, LNAI 13633, pp. 405–419, 2022.
https://doi.org/10.1007/978-3-031-21244-4_30

ment analysis is an important approach for businesses and industry. It helps to attract customers and improve the quality of products or services. At present, it has been applied to many fields such as recommendation system [1] and public opinion analysis [2,7].

Based on deep learning, sentiment classification has been a research hot spot in recent years. Deep learning methods such as convolutional networks [3] and recurrent neural networks [16] have achieved in text modeling. However, when the sample is imbalanced, the training model parameters would be shifted, and the classification quality for majority sample is unsatisfied. Fan et al. [8] provide a cost-sensitive text sentiment analysis method based on sequential three-way decision, which introduces dynamic characteristics of the decision-making process. Kübler [13] investigates feature selection methods in imbalanced data sentiment analysis. Sayyed [18] proposes an application of sampling methods for sentiment analysis on two extreme imbalanced datasets. Ghosh [10] finds that minority over-sampling on live tweets of Twitter can overcome imbalanced classification problem.

Classification methods for imbalanced data can generally be divided into two categories: training data and model parameter oriented ways [5]. The training data oriented method includes random over-sampling (RAMO) and random under-sampling (RAMU).

In this paper, we propose a three-way under-sampling method on imbalanced text data for sentiment classification, namely, 3WS-ITSC. The motivations are: 1) to solve the problems of information loss and 2) to improve the efficiency of sampling.

To conquer the shortcomings of information loss due to the sampling, firstly, three-way sampling method trisects the entire data area into: positive region, negative region, and boundary region. Then, the data points drop into the boundary region are sampled since they can precisely describe the distribution of data.

The process of sampling text data can be represented by word vectors directly. However, it brings inefficiency and high computational cost. In recent years, multi-instance learning (MIL) has already been widely applied in many domains, such as sentiment classification [14,17], image retrieval [4,6], and image classification [15,21]. Inspired by MIL, we design an efficient sampling model by converting the text data into a multi-instance bag structure and mapping each sampled data into a new vector. That can dramatically reduce computational cost.

The experiments are undertaken on 5 review datasets to verify the performance of 3WS-ITSC. In general, the results show that the 3WS-ITSC model is superior to imbalance text data for sentiment classification problem by comparing with the state-of-the-art methods. It has higher stability and feasibility for extremely imbalanced text data.

The remainder of this paper is organized as follows. Section 2 introduces the preliminaries of our model, including three-way sampling (3WS), bag mapping method and the sequential three-way sentiment classification model applied in this paper. Section 3 presents our model framework and algorithm. Section 4 describes the experimental process, the datasets used, the experimental results, and discussions. Section 5 concludes and points out the future studies.

2 Preliminaries

2.1 Three-Way Sampling (3WS)

The idea of TAO (trisecting-acting-outcome) model of three-way decision (3WD) [24, 26] is to divide the result into three decision regions and adopts the strategy of 'divide and conquer'. By introducing a pair of thresholds (α, β), $\beta < \alpha$, on the evaluation function v, the constructed three regions are as follows:

$$
\begin{aligned}
\text{POS}_{(\alpha,\beta)}(v) &= \{x \in U \mid v(x) \geq \alpha\}, \\
\text{NEG}_{(\alpha,\beta)}(v) &= \{x \in U \mid v(x) \leq \beta\}, \\
\text{BND}_{(\alpha,\beta)}(v) &= \{x \in U \mid \beta < v(x) < \alpha\},
\end{aligned}
\tag{1}
$$

where POS, NEG, BNG represents positive region, negative region, boundary region respectively. The value $v(x)$ is called the decision status value of x and may be interpreted as the probability or possibility.

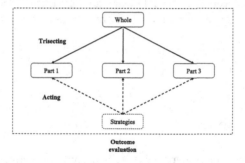

Fig. 1. TAO model of three-way decision [26]

The frame of TAO model is shown in Fig. 1. First, TAO divides a universe into three parts, then takes different actions for different parts, and finally optimizes trisecting and acting for a desirable outcome. Trisecting involves a division of a whole into three parts, acting applies a set of strategies to process the three parts, and outcome evaluation measures the effectiveness of three-way decision [26].

3WS [9] is a sampling method which is proposed according to the characteristics of the boundary regions. The main problem is to determine how to create representative boundary regions which accurately describe the distribution of the data. Support vector data description (SVDD) is a way to describe the distribution of data [19]. A hypersphere is constructed in a high-dimensional space by a Gaussian function. All support vectors on the boundary of the hypersphere are the same distance from the center.

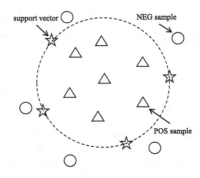

Fig. 2. Schematic of three-way sampling

The idea of 3WS is inspired by SVDD. Less than the radius of the hyper-sphere is the positive region, larger than the radius of the hyper-sphere is the negative region, and the sample equal to the radius of the hyper-sphere is the support vector, that is, the boundary region. The sampling by 3WS is the support vector selected by SVDD. As shown in Fig. 2, the five-pointed star represents the support vector and the sampled point. To speed up the sampling calculation, 3WS adopts a fast incremental SVDD learning algorithm (FISVDD) [11], which is more efficient than existing SVDD algorithms. The definition of the three-way sampling is [11]:

$$POS\,(U) = \left\{ x \in U \mid \|x - \theta\|^2 < R^2 \right\},$$
$$NEG\,(U) = \left\{ x \in U \mid \|x - \theta\|^2 > R^2 \right\}, \tag{2}$$
$$BND\,(U) = \left\{ x \in U \mid \|x - \theta\|^2 = R^2 \right\},$$

where R, θ represents the hyper-sphere radius, hyper-sphere center respectively.

2.2   Bag Mapping

By considering the time and effect of the algorithm, the vector of locally aggregated descriptors representation (miVLAD) algorithm [20,22] could map the original MIL bags to the new vector representations efficiently. Therefore, each of text data can be regarded as a MIL bag, and each word in the text can be regarded as an instance. The purpose of applying the multi-instance bag mapping method is to reduce the computational time caused by 3WS.

The size of the dimension after data mapping is determined by the number of clusters in miVLAD. If the number of clusters selected is n_c, then the dimension of the vector after mapping is $n_c * 50$. Let X_i represent the i-th bag, x_{ij} represent the j-th instance in the bag, and c_k represent the i-th cluster center For each bag X_i, to calculate the difference v_{ik} between the instances x_{ij} and c_k in the bag, the attribute value of v_{ik} is calculated as follows:

$$vec_{ikl} = \sum_{x_{ij} \in \Omega} (x_{ijl} - c_{kl}), \tag{3}$$

where Ω represents all instances in the same cluster as the cluster center c_k. If there are n_c cluster centers, there are n_c vectors, which are then combined to form a new vector. The rules of combination are as follows: First, expand v_i into a vector of $50 * n_c$, and then process the vector as in Eq. 4.

$$vec_{i\cdot l} = sign\,(vec_{i\cdot l})\,\sqrt{|vec_{i\cdot l}|},$$
$$vec_i = \frac{vec_i}{\|vec_i\|_2}. \tag{4}$$

2.3   Three-Way Sentiment Classification (Three-Way SC)

In sentiment analysis, the concept of a three-way decision has been applied in [8,27]. For imbalanced data, the cost-sensitive learning approach is a model parameter-oriented strategy. The essential of cost-sensitive learning is to assign less weight to positive class samples and more weight to negative class samples. Three-way SC can be categorized as a cost-sensitive method. This strategy increases the cost of the minority class and enables the model pay more attention to the minority class.

Definition 1. *Let $Gr = \{Gr_1, Gr_2 \ldots Gr_s\}$ be the s levels of granular structure, where $Gr_j = \{U_j, G_j, \alpha_j, \beta_j, v_j(x)\}$, $j = 1, 2 \ldots s$. At j-th level, the U_j denotes the processing objects, the G_j denotes the size of granules, the (α_j, β_j) is the pair of thresholds, $v_j(x)$ is the evaluation function.*

Since $v(x)$ denotes the likelihood or possibility of x, and according to Definition 1, the three-way sentiment classification (three-way SC) rules are defined as follows:

$$\mathrm{POS}^j_{(\alpha_j)}(v) = \{x \in U_j \mid v(x) > \alpha_j\},$$
$$\mathrm{BND}^j_{(\alpha_j, \beta_j)}(v) = \{x \in U_j \mid \alpha_j \geq v(x) \geq \beta_j\}, \tag{5}$$
$$\mathrm{NEG}^j_{(\beta_j)}(v) = \{x \in U_j \mid v(x) < \beta_j\}.$$

Table 1 shows the six cost types for the j-th granularity level of the cost-sensitive sentiment classification task: the true acceptance λ^j_{PP} (correctly classify a positive sample), the true rejection λ^j_{NN} (correctly classify a negative sample), the boundary acceptance λ^j_{BP} (classifying a positive sample after delay), the boundary rejection λ^j_{BN} (classifying a negative sample after delay), the false rejection λ^j_{NP} (positive samples are misclassified as negative samples), the false acceptance λ^j_{PN} (negative samples are misclassified as positive samples).

The decision cost is regarded as inequality. Evaluating negative samples (minority class) as positive samples in imbalanced text data classification is

Table 1. Three-way sentiment classification cost matrix

Decision	Real sentimental polarity	
	Positive (P)	Negative (N)
Positive decision (D_P)	λ_{PP}^j	λ_{PN}^j
Delayed decision (D_B)	λ_{BP}^j	λ_{BN}^j
Negative decision (D_N)	λ_{NP}^j	λ_{NN}^j

more expensive than evaluating positive samples (majority class) as negative samples. The cost of making a bad decision is higher than the cost of delaying a decision, and the cost of delaying a decision is higher than the cost of making a correct decision, which is expressed as [23]:

$$0 \leq \lambda_{PP}^j \leq \lambda_{BP}^j < \lambda_{NP}^j,$$
$$0 \leq \lambda_{NN}^j \leq \lambda_{BN}^j < \lambda_{PN}^j, \tag{6}$$
$$\lambda_{NP}^j < \lambda_{PN}^j.$$

Definition 2. *Let n_{PN}^j represents the number of negative samples misclassified as positive samples, n_{NP}^j represents the number of positive samples misclassified as negative samples, n_{PP}^j represents the number of correctly classified positive samples, n_{NN}^j represents the number of correctly classified negative samples at the j level. The following formulas can be used to compute the total cost (TC) and the average cost (AC):*

$$TC = \sum_{j=1}^{S} \left(\lambda_{PN}^j n_{PN}^j + \lambda_{NP}^j n_{NP}^j \right),$$
$$AC = \frac{TC}{\sum_{j=1}^{S} \left(n_{PP}^j + n_{NP}^j + n_{PN}^j + n_{NN}^j \right)}. \tag{7}$$

2.4   Sequential Three-Way Sentiment Classification (S3WSC)

Existing sentiment classification models only make static decisions, ignoring the dynamic features of the decision-making process. Sequential three-way decision is a dynamic decision-making model based on 3WD [25]. It is used to simplify complex problems from the perspective of granular computing, granulate a complex problem, process each refined problem one by one, create a multi-layered sequential procedure, and finally achieve the goal from basic to complex. The created s granular layers are represented by $Gr_j (j = 1, 2, 3, ..., s)$. Each particular layer has two (α_j, β_j) and satisfies the following criteria [23]:

$$0 \leq \beta_1 \leq \beta_2 \leq \cdots \leq \beta_s \leq \alpha_s \leq \alpha_{s-1} \leq \cdots \leq \alpha_1 \leq 1. \tag{8}$$

According to Definition 1, S3WSC is divided into a granular structure with s layers. Except for the last one, which is a two-way decision (2WD) layer, the rest are 3WD layers. In the 3WD layer the samples in the boundary region must be further examined. The samples in the positive and negative regions will be used as the classification result. The positive and negative regions eventually approach the boundary region with finer granularity (Fig. 3).

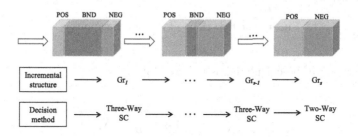

Fig. 3. S3WSC framework

3   Three-Way Sampling on Imbalanced Text Data for Sentiment Classification

3.1   Three-Way Sampling on Imbalanced Text Data

Before adapting 3WS on imbalanced text data, a series of text processing tasks, such as word segmentation, stop word removes, data format conversion, and data are converting into MIL bag structure, are required.

A complex matrix can be used to represent each sample data when a 50 dimensional Glove word vector is used to represent each word. However, the three-way sampling computational time will be excessively long as a result of this. This study introduces a fast bag mapping algorithm that maps data to new vector representations, accelerates 3WS on the bulk of the training set's classes.

The modification of text data is depicted in Fig. 4. After the text data is converted into a matrix, the matrix is mapped to a simple vector by the bag mapping algorithm. The number of clusters n_c is fixed to 1 in this paper, and text data is transformed into a 50-dimensional vector according to Eqs. 3 and 4. This paper maps all text data to a simple vector.

3.2   Sentiment Classification Through Three-Way Sampling

The three-way sampling model is adapted to sentiment classification in this subsection. Based on the cost matrix in Subsect. 2.3, the threshold pair for the

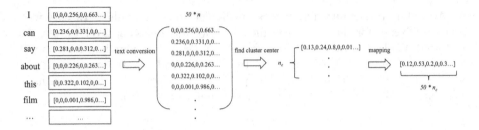

Fig. 4. The flow chart of text data mapping

granular layer can be calculated as follows [23]:

$$\alpha_j = \frac{\lambda_{PN}^j - \lambda_{BN}^j}{\left(\lambda_{PN}^j - \lambda_{BN}^j\right) + \left(\lambda_{BP}^j - \lambda_{PP}^j\right)},$$

$$\beta_j = \frac{\lambda_{BN}^j - \lambda_{NN}^j}{\left(\lambda_{BN}^j - \lambda_{NN}^j\right) + \left(\lambda_{NP}^j - \lambda_{BP}^j\right)}.$$

(9)

The three-way SC is defined in Definition 2.3 except for the last granular layer. In the last granular layer, the pair of (α_s, β_s) meets $\alpha_s = \beta_s = 0.5$. The two-way sentiment classification (two-way SC) is defined as follows:

$$\text{POS}_{(\alpha_s)}^s(v) = \{x \in U_s \mid v(x) \geq \alpha_s\},$$
$$\text{NEG}_{(\alpha_s)}^s(v) = \{x \in U_s \mid v(x) < \alpha_s\}.$$

(10)

Algorithm 1 demonstrates the 3WS-ITSC construction process, and the Table 2 shows the complexity of Algorithm 1. Because the proposed model introduces deep neural network models, the time complexity of 3WS-ITSC is $O(s * (m * n + m * v^2 + l + train))$, s stands for the number of stages, m for the number of training set samples at each step, and n for the number of bags, v for the number of support vectors computed by FISVDD, and l for the number of testing set samples at each stage. There are five key steps in 3WS-ITSC:

Step 1: Initialization. After receiving the first training set and all test sets, the model creates three empty sets (positive, negative, and boundary regions).
Step 2: Bag mapping. Bag mapping method converts matrix data to new vector
Step 3: Sampling. 3WS on the majority class, and use the sampled data to update the model.
Step 4: Three-Way sentiment classification. The model decides on the test set and divides it into positive, boundary, and negative regions. The data in the boundary region are the test set for the next epoch.
Step 5: Two-Way sentiment classification. Repeat Step 1 to 4 until the object data in the boundary region is binary classified at the end.

Algorithm 1. The algorithm of 3WS-ITSC.

Require:

Training samples $A = A_1 \bigcup \cdots \bigcup A_i \bigcup \cdots \bigcup A_s (1 \leqslant i \leqslant s)$;

Testing samples T;

Threshold pair (α_i, β_i) at i-th level;

Ensure: Classification results POS and NEG;

1: // Step 1. Initialize parameters.
2: $D = T$, $B = \varnothing$, $V = \varnothing$, $POS = \varnothing$, $NEG = \varnothing$, $BND = \varnothing$, initialize model M;
3: **for** $(i \in [1..s])$ **do**
4:    // Step 2. Bag mapping.
5:    **for** $(j \in [1..m])$ **do**
6:       Map the $A_{ij} \in A_i$ to the new vector V_{ij} according to Equation (3) and Equation (4);
7:    **end for**
8:    // Step 3. Three-Way Sampling for majority class.
9:    3WS on V_i to obtain B;
10:    Update model M with B;
11:    Calculate $v_i(x)$ by model M;
12:    **if** $i \neq s$ **then**
13:       // Step 4. Three-Way sentiment classification.
14:       $POS_i = \{x \in D \mid v_i(x) \geq \alpha_i\}$;
15:       $BND_i = \{x \in D \mid \beta_i < v_i(x) < \alpha_i\}$;
16:       $NEG_i = \{x \in D \mid v_i(x) \leq \beta_i\}$;
17:       $C = BND_i$;
18:    **else**
19:       // Step 5. Two-Way sentiment classification.
20:       $POS_i = \{x \in D \mid v_i(x) \geq 0.5\}$;
21:       $NEG_i = \{x \in D \mid v_i(x) < 0.5\}$;
22:    **end if**
23:    $POS = POS \bigcup POS_i$;
24:    $NEG = NEG \bigcup NEG_i$;
25: **end for**
26: **return** POS and NEG.

Table 2. Computational complexity of Algorithm 1

Lines	Complexity	Description
Lines 5–7	$O(m * n)$	Bag mapping
Line 9	$O(m * v^2)$	Three-Way Sampling for majority class
Line 10	$O(train)$	Update model
Lines 12–22	$O(l)$	Decision-making process
Total	$O(s * (m * n + m * v^2 + l + train))$	

4   Experiments

In this section, we test the 3WS-ITSC model on IMDB, Yelp, and Amazon review datasets, using three common sentiment classification methods as baseline (fastText [12], TextCNN [3], TextRNN [16]). To verify the effectiveness of our proposed model, we compare it with RAMU-ITSC and the unsampled imbalanced text sentiment classification model (UNS-ITSC). All algorithms are preformed with the same software and hardware configuration (CPU:AMD EPYC 7543 32-Core Processor; RAM:30GB; GPU:RTX A5000; Memory:24GB; Linux 5.4.0-91-generic; python3.8). All the code for the experiments are available at https://github.com/Z-C-Lee/3WS-ITSC.

4.1   Datasets and Parameters Settings

Table 3 shows the dataset briefings of IMDB, Yelp, and Amazon, where N represents the total number of samples, L represents the average length of the sample text, $dist(+, -)$ represents the positive and negative class distributions, pos represents the number of samples in the few-sample category, neg represents the number of samples in the majority class, $test$ represents the number of our test set. It also lists the dataset's imbalance ratio.

Table 3. The description of datasets

Name	N	$dist(+, -)$	Imbalanced ratio	L	pos	neg	$test$
IMDB	$37,500$	$(0.66, 0.34)$	$2:1$	242	$25,000$	$12,500$	$12,500$
Yelp	$486,998$	$(0.74, 0.26)$	$2.8:1$	130	$358,672$	$128,326$	$81,166$
Amazon/Arts	$25,761$	$(0.80, 0.20)$	$5:1$	73	$21,496$	$4,265$	$8,587$
Amazon/Ele	$1,051,775$	$(0.84, 0.16)$	$5.5:1$	79	$887,698$	$164,077$	$175,296$
Amazon/Jew	$53,574$	$(0.86, 0.14)$	$6:1$	87	$46,060$	$7,514$	$17,858$

IMDB[1]: It is a dataset of movie reviews for binary sentiment classification. The dataset contains $50,000$ samples, with $25,000$ positive and $25,000$ negative samples. $12,500$ negative samples are removed in order to achieve an imbalanced state.

Yelp Review Polarity[2]: The restaurant review on Yelp. The sample data with a score of 5 is used as positive sample, and the sample with a score of 1 is used as our negative sample. A total of $486,998$ samples were collected, with $358,672$ being positive and $128,326$ being negative.

Amazon Review Polarity[3]: We select 3 datasets from the Amazon Reviews dataset: 1) Customer feedback on Amazon electronics product. The dataset contains $1,051,775$ samples, with $887,698$ positive and $164,077$ negative samples.

---

[1] http://ai.stanford.edu/~amaas/data/sentiment/.

[2] https://www.yelp.com/dataset.

[3] http://snap.stanford.edu/data/web-Amazon-links.html.

2) Arts product reviews contain 25, 761 samples, with 21, 496 positive and 4, 265 negative samples. 3) Jewelry reviews contain 53, 574 samples, with 46, 060 positive and 7, 514 negative samples.

In our model, several parameters must be set: the Gaussian kernel bandwidth σ, the number of granular layers s, and the number of clusters C in the bag mapping. The value of σ is more smaller, and the more samples are sampled. The dimension of the mapped samples increases as the number of clusters C increases. The parameter values for each dataset in this model are shown in Table 4. The table also indicates how many samples will be sampled at each granularity level (n_{sample}).

Table 4. The parameter values set for each dataset

Name	σ	s	C	n_{sample}
IMDB	0.28	4	1	3, 000
Yelp	0.27	20	1	9, 000
Amazon/Arts	0.3	4	1	2, 000
Amazon/Ele	0.31	20	1	13, 000
Amazon/Jew	0.3	4	1	5, 000

Table 5. The cost values for each dataset

Name	λ_{PP}	λ_{PN}	λ_{BP}	λ_{BN}	λ_{NP}	λ_{NN}
IMDB	0	300	40	40	150	0
Yelp	0	420	40	40	150	0
Amazon/Arts	0	750	40	40	150	0
Amazon/Ele	0	825	40	40	150	0
Amazon/Jew	0	900	40	40	150	0

Refer to the settings of cost value in [8], these settings are for a balanced dataset. Set the weight of the cost based on the proportion of positive and negative samples in the training set, as shown in Table 5.

4.2   Results and Analysis

The F_1-score and AC are evaluation metrics in this paper. The lower the AC, the lower the misclassification cost of the representative model. So a lower AC value is recommended. The addition of the AC value makes our research more practical.

Table 6. Comparison of 3WS, RAMU and unsampled F_1

Classifier	Dataset	3WS-ITSC	RAMU-ITSC	UNS-ITSC
TextRNN	IMDB	$0.8543_{\pm 0.0023}$●	$0.8523_{\pm 0.0067}$	$0.8513_{\pm 0.0170}$
	Yelp	$0.9442_{\pm 0.0011}$●	$0.9402_{\pm 0.0038}$	$0.9359_{\pm 0.0067}$
	Amazon/Arts	$0.9087_{\pm 0.0007}$●	$0.9078_{\pm 0.0016}$	$0.9085_{\pm 0.0003}$
	Amazon/Ele	$0.9497_{\pm 0.0007}$●	$0.9475_{\pm 0.0017}$	$0.9473_{\pm 0.0020}$
	Amazon/Jew	$0.9247_{\pm 0.0009}$	$0.9249_{\pm 0.0006}$	$0.9253_{\pm 0.0001}$●
TextCNN	IMDB	$0.8929_{\pm 0.0016}$●	$0.8923_{\pm 0.0034}$	$0.8925_{\pm 0.0024}$
	Yelp	$0.9615_{\pm 0.0007}$●	$0.9612_{\pm 0.0009}$	$0.9529_{\pm 0.0006}$
	Amazon/Arts	$0.8932_{\pm 0.0034}$●	$0.8785_{\pm 0.0132}$	$0.8922_{\pm 0.0048}$
	Amazon/Ele	$0.9553_{\pm 0.0001}$●	$0.9552_{\pm 0.0002}$	$0.9507_{\pm 0.0005}$
	Amazon/Jew	$0.9122_{\pm 0.0040}$	$0.9186_{\pm 0.0046}$	$0.9233_{\pm 0.0005}$●
FastText	IMDB	$0.9106_{\pm 0.0007}$●	$0.9064_{\pm 0.0032}$	$0.9106_{\pm 0.0014}$
	Yelp	$0.9680_{\pm 0.0007}$●	$0.9679_{\pm 0.0008}$	$0.9663_{\pm 0.0011}$
	Amazon/Arts	$0.9054_{\pm 0.0015}$	$0.9038_{\pm 0.0022}$	$0.9088_{\pm 0.0006}$●
	Amazon/Ele	$0.9511_{\pm 0.0004}$	$0.9526_{\pm 0.0017}$●	$0.9484_{\pm 0.0026}$
	Amazon/Jew	$0.9234_{\pm 0.0008}$	$0.9240_{\pm 0.0007}$●	$0.9233_{\pm 0.0003}$

Table 7. Comparison of 3WS, RAMU and unsampled AC

Classifier	Dataset	3WS-ITSC	RAMU-ITSC	UNS-ITSC
TextRNN	IMDB	$52.4304_{\pm 1.9415}$●	$54.5304_{\pm 2.1962}$	$57.3600_{\pm 2.7171}$
	Yelp	$35.6266_{\pm 2.1883}$●	$39.3009_{\pm 3.0079}$	$43.2820_{\pm 5.6914}$
	Amazon/Arts	$73.7615_{\pm 0.1784}$●	$73.9187_{\pm 0.1565}$	$74.1598_{\pm 0.3590}$
	Amazon/Ele	$57.2253_{\pm 0.6456}$●	$64.2841_{\pm 5.0772}$	$66.2499_{\pm 3.2894}$
	Amazon/Jew	$124.5055_{\pm 0.2821}$●	$124.7558_{\pm 0.1460}$	$124.9104_{\pm 0.0349}$
TextCNN	IMDB	$33.8808_{\pm 0.7831}$●	$34.8720_{\pm 1.2424}$	$38.1408_{\pm 1.1145}$
	Yelp	$22.2724_{\pm 0.4777}$●	$22.5235_{\pm 0.9008}$	$28.6482_{\pm 0.6937}$
	Amazon/Arts	$72.2697_{\pm 0.7930}$●	$73.4191_{\pm 0.3732}$	$72.5737_{\pm 0.5572}$
	Amazon/Ele	$43.6877_{\pm 0.7285}$●	$45.1405_{\pm 0.3887}$	$58.8385_{\pm 1.1550}$
	Amazon/Jew	$122.5925_{\pm 0.6101}$●	$123.2932_{\pm 0.6855}$	$123.7553_{\pm 0.3431}$
FastText	IMDB	$29.5800_{\pm 0.2540}$●	$30.6648_{\pm 2.5412}$	$29.8704_{\pm 1.3362}$
	Yelp	$15.7599_{\pm 0.4187}$●	$16.0076_{\pm 0.9766}$	$19.4546_{\pm 1.1543}$
	Amazon/Arts	$72.0950_{\pm 0.5140}$●	$72.4199_{\pm 0.6604}$	$73.5204_{\pm 0.2225}$
	Amazon/Ele	$38.4875_{\pm 0.2373}$●	$40.2155_{\pm 1.5779}$	$62.0771_{\pm 4.6134}$
	Amazon/Jew	$123.7631_{\pm 0.0739}$●	$123.7815_{\pm 0.3635}$	$124.2619_{\pm 0.3251}$

Table 6 illustrates the F_1 value of FastText, TextRNN and TextCNN on the dataset in Subsect. 4.1. Except for the FastText performance on the Amazon dataset, the results show that 3WS-ITSC outperforms RAMU-ITSC and

UNS-ITSC. When compared with RAMU-ITSC, 3WS-ITSC has no discernible improvement in F_1, but it is accompanied by a high degree of stability.

The AC of FastText, TextRNN, and TextCNN on the dataset are shown in Table 7. The results show that the average cost of 3WS-ITSC on all datasets is lower and more stable than RAMU-ITSC and UNS-ITSC on all classifiers. As the imbalanced ratio of data increases, the improvement of the value of F_1 is less obvious or even not improved, but the AC value is always better than RAMU-ITSC and UNS-ITSC. That means 3WS-ITSC model is more practical than RAMU-ITSC model.

The effectiveness of our model can be observed in Fig. 5, the smaller the plane area, the more efficient the model. Among all the classifiers, the effectiveness of 3WS-ITSC improves more significantly as the data imbalance is increasing. According to Fig. 5(a), the performance of 3WS-ITSC on TextRNN is significantly better than RAMU-ITSC and UNS-ITSC. 3WS-ITSC has slight improvement over RAMU-ITSC on TextCNN and FastText, as shown in Fig. 5(b) and Fig. 5(c), evidently a significant improvement over UNS-ITSC.

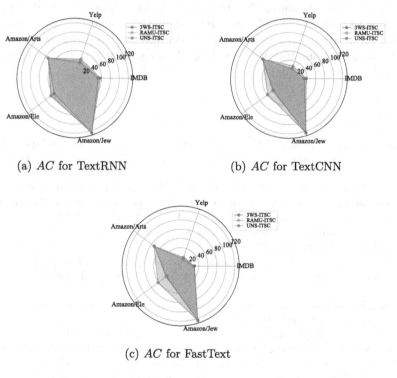

(a) AC for TextRNN

(b) AC for TextCNN

(c) AC for FastText

Fig. 5. The comparison chart for average cost (AC)

In the Amazon dataset, more negative samples are predicted as positive samples due to the extreme imbalance ratio. Thus, while this type of data will have a higher F_1 value, it will also have a higher AC value. On extremely imbalanced data, our method obtains AC value as small as possible while keeping F_1 stable. However, the AC value of our method is not obvious on extremely unbalanced and small sample data.

5   Conclusion and Further Work

In this paper, we propose a 3WS-ITSC model. It provides a solution to the sentiment classification problem with regard to imbalance text data. This paper collaborates the MIL bag mapping method, convert text data to a simpler vector representation. The three-way sampling method divides the data into three regions, of which the boundary region serves as sampling points. 3WS extracts boundary points effectively and describes the spatial structure of the data accurately. The obtained boundary points of the majority class are used to train the classifier.

Although the bag mapping operation, to a large extend, can reduce the sampling computational cost, there is still a significant time-consuming problem when the amount of samples is large. Since the text data are granulated using a random assignment method, the sample points generated may not adequately explain the spatial distribution of the entire dataset. In future work, we will investigate how to improve the sampling efficiency of massive text data and the spatial structure of the data.

References

1. Abbasi-Moud, Z., Vahdat-Nejad, H., Sadri, J.: Tourism recommendation system based on semantic clustering and sentiment analysis. Expert Syst. Appl. **167**, 114324 (2021)
2. Chen, X., Zhang, W., Xu, X., Cao, W.: A public and large-scale expert information fusion method and its application: mining public opinion via sentiment analysis and measuring public dynamic reliability. Inf. Fusion **78**, 71–85 (2022)
3. Chen, Y.: Convolutional neural network for sentence classification. Master's thesis, University of Waterloo (2015)
4. Chen, Y., Bi, J., Wang, J.Z.: Miles: multiple-instance learning via embedded instance selection. IEEE Trans. Pattern Anal. Mach. Intell. **28**(12), 1931–1947 (2006)
5. Chen, Z., Guo, W.: Text classification based on depth learning on unbalanced data. J. Chin. Comput. Syst. **41**(1), 1–5 (2020)
6. Conjeti, S., Paschali, M., Katouzian, A., Navab, N.: Deep multiple instance hashing for scalable medical image retrieval. In: Descoteaux, M., Maier-Hein, L., Franz, A., Jannin, P., Collins, D.L., Duchesne, S. (eds.) MICCAI 2017. LNCS, vol. 10435, pp. 550–558. Springer, Cham (2017). https://doi.org/10.1007/978-3-319-66179-7_63
7. El Barachi, M., AlKhatib, M., Mathew, S., Oroumchian, F.: A novel sentiment analysis framework for monitoring the evolving public opinion in real-time: case study on climate change. J. Clean. Prod. **312**, 127820 (2021)
8. Fan, Q., Liu, D., Ye, X.Q.: Cost-sensitive text sentiment analysis based on sequential three-way decision. Pattern Recogn. Artif. Intell. **33**(8), 732–742 (2020)

9. Fang, Y., Cao, X.M., Wang, X., Min, F.: Three-way sampling for rapid attribute reduction. Inf. Sci. **609**, 26–45 (2022)

10. Ghosh, K., Banerjee, A., Chatterjee, S., Sen, S.: Imbalanced Twitter sentiment analysis using minority oversampling. In: 2019 IEEE 10th International Conference on Awareness Science and Technology (iCAST), pp. 1–5. IEEE (2019)

11. Jiang, H., Wang, H., Hu, W., Kakde, D., Chaudhuri, A.: Fast incremental SVDD learning algorithm with the gaussian kernel. In: Proceedings of the AAAI Conference on Artificial Intelligence, vol. 33, pp. 3991–3998 (2019)

12. Joulin, A., Grave, E., Bojanowski, P., Mikolov, T.: Bag of tricks for efficient text classification. arXiv preprint arXiv:1607.01759 (2016)

13. Kübler, S., Liu, C., Sayyed, Z.A.: To use or not to use: feature selection for sentiment analysis of highly imbalanced data. Nat. Lang. Eng. **24**(1), 3–37 (2018)

14. Li, Y., Yin, C., Zhong, S.h., Pan, X.: Multi-instance multi-label learning networks for aspect-category sentiment analysis. arXiv preprint arXiv:2010.02656 (2020)

15. Liu, G.H., Yang, J.Y., Li, Z.: Content-based image retrieval using computational visual attention model. Pattern Recogn. **48**(8), 2554–2566 (2015)

16. Liu, P., Qiu, X., Huang, X.: Recurrent neural network for text classification with multi-task learning. arXiv preprint arXiv:1605.05101 (2016)

17. Lutz, B., Pröllochs, N., Neumann, D.: Sentence-level sentiment analysis of financial news using distributed text representations and multi-instance learning. arXiv preprint arXiv:1901.00400 (2018)

18. Sayyed, Z.A.: Study of sampling methods in sentiment analysis of imbalanced data. arXiv preprint arXiv:2106.06673 (2021)

19. Tax, D.M., Duin, R.P.: Support vector data description. Mach. Learn. **54**(1), 45–66 (2004)

20. Wei, X.S., Wu, J., Zhou, Z.H.: Scalable algorithms for multi-instance learning. IEEE Trans. Neural Netw. Learn. Syst. **28**(4), 975–987 (2016)

21. Wei, X.S., Ye, H.J., Mu, X., Wu, J., Shen, C., Zhou, Z.H.: Multi-instance learning with emerging novel class. IEEE Trans. Knowl. Data Eng. **33**(5), 2109–2120 (2019)

22. Yang, M., Zhang, Y.X., Wang, X., Min, F.: Multi-instance ensemble learning with discriminative bags. IEEE Trans. Syst. Man Cybern. Syst. **52**(9), 5456–5467 (2021)

23. Yang, X., Li, Y., Li, Q., Liu, D., Li, T.: Temporal-spatial three-way granular computing for dynamic text sentiment classification. Inf. Sci. **596**, 551–566 (2022)

24. Yao, Y.: An outline of a theory of three-way decisions. In: Yao, J., et al. (eds.) RSCTC 2012. LNCS (LNAI), vol. 7413, pp. 1–17. Springer, Heidelberg (2012). https://doi.org/10.1007/978-3-642-32115-3_1

25. Yao, Y.: Granular computing and sequential three-way decisions. In: Lingras, P., Wolski, M., Cornelis, C., Mitra, S., Wasilewski, P. (eds.) RSKT 2013. LNCS (LNAI), vol. 8171, pp. 16–27. Springer, Heidelberg (2013). https://doi.org/10.1007/978-3-642-41299-8_3

26. Yao, Y.: Three-way granular computing, rough sets, and formal concept analysis. Int. J. Approximate Reason. **116**, 106–125 (2020)

27. Zhang, Y., Zhang, Z., Miao, D., Wang, J.: Three-way enhanced convolutional neural networks for sentence-level sentiment classification. Inf. Sci. **477**, 55–64 (2019)

Rough-Fuzzy Clustering Based on Adaptive Weighted Values and Three-Way Decisions

Ge Yuan[1,2], Jie Zhou[1,3(✉)], and Qiongbin Chen[1,2]

[1] National Engineering Laboratory for Big Data System Computing Technology, Shenzhen University, Guangdong 518060, China
jie_jpu@163.com
[2] College of Computer Science and Software Engineering, Shenzhen University, Guangdong 518060, China
[3] SZU Branch, Shenzhen Institute of Artificial Intelligence and Robotics for Society, Shenzhen 518060, Guangdong, China

Abstract. In the rough set-based clustering models, prototypes are iteratively updated by weighting the importance of core and boundary regions. The weighted value w_l is often pre-defined and fixed for all prototype calculations. In this way, the characteristics of data structures are not considered when assigning the weighted values. Some uncertainties may arise in the clustering processes, especially when the densities and sizes of different clusters are discrepant. In this study, an automatic mechanism for adaptively adjusting the weighted value w_l is introduced which adheres to the distributions of approximation region partitions, and the uncertainties caused by the user-defined weighted values can be reduced. Based on the generated approximation region partitions of each cluster, an absolute boundary region is formed in which the samples are classified guided by the notion of three-way decisions. The validity of the proposed method is demonstrated by some benchmark data sets from UCI repository.

Keywords: Rough-fuzzy clustering · Approximation regions · Adaptive weighted values · Three-way decisions

1 Introduction

Rough set theory [1] often divides a universe into three approximation regions related to a given concept based on a predefined relationship among attributes (features). Since upper and lower approximate operators are used to describe uncertain concepts, an uncertain concept can be depicted with two crisp sets. Based on the idea of rough set approximation regions, Lingras and West proposed a rough C-means (RCM) clustering method [2], in which the contributions of lower approximation region (core region) and boundary region of each cluster are weighted to calculate prototypes in iteration processes. However, this method does not well describe the memberships of samples belonging to different clusters, and the structural characteristics of the sample space, especially the case of overlapping, are not detected. On this basis, Mitra extended RCM to a rough-fuzzy C-means (RFCM) clustering method [3]. In this method, the approximation regions of each

© The Author(s), under exclusive license to Springer Nature Switzerland AG 2022
J. Yao et al. (Eds.): IJCRS 2022, LNAI 13633, pp. 420–429, 2022.
https://doi.org/10.1007/978-3-031-21244-4_31

cluster are partitioned and measured by fuzzy membership functions. Maji [4] further considered that the core region of each cluster should have the same contribution in the iterative calculation process of prototypes, and then a variation of RFCM is presented.

The clustering methods based on rough sets involve two key parameters: the approximation region partition threshold Δ and the weighted coefficient w_l ($w_b = 1 - w_l$). Different values of threshold Δ will result in different approximation regions of each class, thus the accuracy and efficiency of iteration processes may be affected. The studies in [5, 6] introduced the idea of approximate region partition based on shadowed sets [7, 8], which transforms the selection of partition thresholds in the RFCM clustering process into an uncertainty balance optimization problem. Li et al. [9] introduced a rough C-means method based on decision rough set models [10], in which the expected loss function from samples to clusters is formed as the approximation region partition principle by fully considering the neighborhood information of each sample. Sarkar et al. [11] used the middle value between the largest and second largest fuzzy membership degrees of all sample points as the approximation region partition threshold, and achieved competitive experimental results.

Although some researches have focused on the determination of approximation region separation threshold Δ, there are few literatures concentrated on the weighted coefficient w_l that measures the contribution of core regions. Intuitively, the core region of each cluster has the most important role for updating the prototype of this cluster. Therefore, the value w_l is generally expected to be larger, such as $w_l = 0.95$. However, different values of Δ will result in different core and boundary regions, especially for the data sets with different density distributions. If the prototype calculations overemphasize the contribution of core regions, the contribution of the boundary regions may be weakened excessively, thus the obtained prototypes may be deviated.

According to the distribution characteristics of approximation regions of each class in the clustering process, this paper introduces an adaptive mechanism for determining the weights of each approximate region when updating the prototypes. In this way, the weights that measure the contribution of different approximation regions are dynamically adjusted rather than user predefined. Furthermore, different clusters will correspond to different weighted values, which is different from the traditional rough set -based clustering methods that have a fixed value for all clusters over all iteration steps.

Moreover, based on the approximation region partition of each cluster, an absolute boundary region over all clusters is formed in which the samples are classified by using three-way decisions as the post process rather than being grouped based on the obtained memberships directly. The theory of three-way decisions [12–14], introduced by Prof. Yao, aims at a unified, discipline-independent framework to study fundamental notions, concepts, and applications of decision-making with three options. Three-way decisions have been successfully exploited in extensive research areas [15–17]. The three decision options, i.e., accepting, rejecting and noncommitment, can be used here to guide the classification of samples in the formed absolute boundary region.

The rest of this paper is organized as follows: Sect. 2 briefly reviews the rough-fuzzy clustering algorithms. Section 3 presents a dynamic adjustment mechanism for determining weights based on the approximation region distribution of each cluster, and an absolute boundary region is formed in which the samples are classified by using three-way decisions as the post process. Section 4 exhibits experimental results. Section 5 gives the conclusions.

2 Preliminaries

Given a data set $x_1, x_2, \cdots, x_N, x_j \in \Re^M$ $(j = 1, 2, \cdots, N)$ is divided into C ($1 < C < N$) groups $G_1, G_2, \cdots, G_C$, the prototype of each class is represented as $v_1, v_2, \cdots, v_C$, and the degree of data point x_j belonging to the class G_i is represented as u_{ij}.

The rough C-means (RCM) clustering method uses the following principles to update the calculation of prototypes [2]:

$$v_i = \begin{cases} w_l A_1 + w_b B_1 & \text{if } \underline{R}G_i \neq \emptyset \wedge R_b G_i \neq \emptyset \\ B_1 & \text{if } \underline{R}G_i = \emptyset \wedge R_b G_i \neq \emptyset \\ A_1 & \text{if } \underline{R}G_i \neq \emptyset \wedge R_b G_i = \emptyset \end{cases}, \tag{1}$$

where $A_1 = \frac{\sum_{x_j \in \underline{R}G_i} x_j}{card(\underline{R}G_i)}$, $B_1 = \frac{\sum_{x_j \in R_b G_i} x_j}{card(R_b G_i)}$. $card(X)$ Represents the cardinality of X. $\underline{R}G_i$ And $\overline{R}G_i$ represent the lower approximation region (core region) and the upper approximation region of cluster G_i with respect to the feature set R, respectively. $R_b G_i = \overline{R}G_i - \underline{R}G_i$ Represents the boundary region of cluster G_i with respect to R. w_l ($0.5 < w_l \leq 1$) and $w_b = 1 - w_l$ respectively measure the contribution of the core region and the boundary region in the iterative calculation of prototypes. In order to determine the core and boundary regions of each cluster, RCM adopts the following principles:

Denote the minimum distance and the second minimum distance of sample x_j over all prototypes as $\|x_j - v_p\|$ and $\|x_j - v_q\|$, respectively. If $\|x_j - v_q\| - \|x_j - v_p\| \leq \Delta$, then $x_j \in \overline{R}G_p$ and $x_j \in \overline{R}G_q$, otherwise $x_j \in \underline{R}G_p$. Δ is a predefined threshold.

Mitra [3] further extended RCM to rough-fuzzy C-means (RFCM), and the prototypes are updated as follows:

$$v_i = \begin{cases} w_l A_2 + w_b B_2 & \text{if } \underline{R}G_i \neq \emptyset \wedge R_b G_i \neq \emptyset \\ B_2 & \text{if } \underline{R}G_i = \emptyset \wedge R_b G_i \neq \emptyset \\ A_2 & \text{if } \underline{R}G_i \neq \emptyset \wedge R_b G_i = \emptyset \end{cases}, \tag{2}$$

where $A_2 = \frac{\sum_{x_j \in \underline{R}G_i} u_{ij}^m x_j}{\sum_{x_j \in \underline{R}G_i} u_{ij}^m}$, $B_2 = \frac{\sum_{x_j \in R_b G_i} u_{ij}^m x_j}{\sum_{x_j \in R_b G_i} u_{ij}^m}$, $\sum_{i=1}^C u_{ij} = 1, 0 < \sum_{j=1}^N u_{ij} < N$. The computation of u_{ij} is the same as in the classical fuzzy C-means (FCM) [18]. In order to determine the core and boundary regions of each cluster, RFCM involves the following principles:

Denote the maximum membership degree and the second maximum membership degree of point x_j to all prototypes as u_{pj} and u_{qj}, respectively. If $u_{pj} - u_{qj} \leq \Delta$, then $x_j \in \overline{R}G_p$ and $x_j \in \overline{R}G_q$, otherwise $x_j \in \underline{R}G_p$. In RCM and RFCM, different values of Δ will result in different approximation region partitions, and then affect the prototype calculations. Therefore, it is necessary to select a reasonable partition threshold Δ based on the characteristics of the data set itself.

A dynamic adjustment mechanism for determining Δ based on shadowed sets was provided in [5], which alleviated the uncertainty caused by the predefined value Δ. The partition threshold Δ of each cluster G_i is optimized as follows:

$$
\alpha_i = \min_{\alpha}(F_i) = \min_{\alpha} \left| \sum_{j:u_{ij} \leq \alpha} u_{ij} + \sum_{j:u_{ij} \geq \max_j(u_{ij}) - \alpha} (1 - u_{ij}) \right.
$$

$$
\left. - card \left\{ x_j | \alpha < u_{ij} < \max_j(u_{ij}) - \alpha \right\} \right|. \tag{3}
$$

Then the approximation regions of each cluster are formed as:

$$
\underline{R}G_i = \left\{ x_j | u_{ij} \geq \max_j(u_{ij}) - \alpha_i \right\},
$$

$$
R_b G_i = \left\{ x_j | \alpha_i < u_{ij} < \max_j(u_{ij}) - \alpha_i \right\}. \tag{4}
$$

According to formula (3), the partition threshold determination is transformed to an uncertainty balance optimization problem, and each cluster can independently optimize the partition threshold that adheres to its own structures. Formula (3) adopts Pedrycz's uncertainty balance criterion [7]. Essentially, shadowed sets are a special form of the three-way approximations of fuzzy sets [8, 19]. The three-way approximation of fuzzy sets can also be constructed based on other criteria, such as minimum distance, minimum cost, entropy balance, etc., the detailed description can refer to [20, 21]. It is worth stressing here that the determination of partition thresholds based on shadowed sets can be integrated into Maji's and Mitra's RFCM directly [5], which are denoted as SRFCM$_1$ and SRFCM$_2$, respectively.

3   Adaptive Weighted Value and Absolute Boundary Region

The existing rough set-based clustering methods have not focused on the impacts caused by the weighted value w_l. Generally, the weighted value w_l is fixed to a larger value in the interval (0.5, 1], then the density and the size of each cluster are not considered. To this end, this section presents an adaptive mechanism to adjust the weighted value w_l in the iteration processes which adheres to the properties of approximation regions of each cluster.

Let the ratio of the number of samples in the core region to the upper approximate region of the class G_i in the clustering process be η_i, which is formulated as follows:

$$
\eta_i = \frac{card\left(\underline{R}G_i\right)}{card\left(\underline{R}G_i\right) + card(R_b G_i)}. \tag{5}
$$

$card\left(\underline{R}G_i\right)$ measures the number of samples that belong to G_i definitely, $card\left(\underline{R}G_i\right) + card(R_b G_i)$ measures the number of samples that belong to class G_i possibly. If $\underline{R}G_i = \varnothing$, then $\eta_i = 0$; if $R_b G_i = \varnothing$, then $\eta_i = 1$. The smaller the value of $card(R_b G_i)$, the fewer samples that represent the overlapping area between clusters. Based on the Gaussian function, an adaptive adjustment function of w_{l_i} of the cluster G_i is defined as follows:

$$w_{l_i} = e^{-\frac{(1-\eta_i)^2}{\delta}}. \tag{6}$$

δ is the adjustment bandwidth. When $\eta_i = 1$, then $w_{l_i} = 1$. In this case, all contributions come from the core region of G_i, which is consistent with RCM and RFCM methods. When $\eta_i \to 0$, the core region of G_i becomes $\varnothing$. Since w_{l_i} satisfies $w_{l_i} > 0.5$, so it has $e^{-\frac{1}{\delta}} > 0.5$. Therefore, it can be deduced that $\delta > 1.4427$.

The relationship between w_l and η when varying the value of δ is shown in Fig. 1. When the value of η is fixed, the lager the value δ is, the larger the value of w_l will be. In the extreme case, when $\delta \to \infty$, no matter what the value of η is, there is $w_l \to 1$. Thus, the information in the boundary region will be completely neglected. The smaller the value δ, the smaller the value of w_l will be. In this case, the contribution of the boundary region will be enhanced. Consequently, it is necessary to reasonably consider the contribution of different approximation regions, and $\delta = 2$ can be regarded as a trade-off choice, that is, enhancing the contribution of the boundary region while fully preserving the contribution coming from the core region.

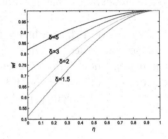

Fig. 1. Relationship between w_l and η

Based on the dynamic adjustment mechanism of w_l, a rough-fuzzy clustering algorithm based on adaptive weights can be constructed, which is described as follows:

Algorithm 1 : Rough-fuzzy clustering algorithm based on adaptive weights

---

Step1: Initialize cluster center points v_i $(i = 1,2,\cdots,C)$;

Step2: **While** not converge

2.1: Calculate items u_{ij} $(i = 1,2,\cdots,C, j = 1,2,\cdots,N)$;

2.2: Based on the shadowed sets, calculate the partition threshold α_i for each cluster G_i;

2.3: Based on α_i, obtain the core and boundary regions of each cluster G_i;

2.4: Calculate the weights w_{l_i} of each cluster G_i based on (6);

2.5: Using formula (2), update prototypes v_i;

End While

Step3: Return v_i and u_{ij} $(i = 1,2,\cdots,C, j = 1,2,\cdots,N)$.

Steps 2.2 and 2.3 establish an optimal method for determining approximation region partition thresholds Δ based on the shadowed sets. Step 2.4 involve an adaptive adjustment mechanism for determining the weights w_l according to the approximation region distributions. By introducing the dynamical adjustment of the thresholds Δ and weights w_l, the obtained prototypes in the clustering process approach to their intrinsic positions.

The samples in the core region of each cluster belong to this cluster definitely, but not necessarily for the samples in the boundary region. For a given data set, there may be some samples that not belong to the core regions of any clusters. The partition of these samples only based on the obtained prototypes or membership degrees may be unreasonable. The set composed of these samples is called as absolute boundary region, which is formulated as follows:

Definition 1: For a given data set X, its absolute boundary region is defined as:

$$R_{ab}X = \{x | if \forall G_i (i = 1, 2, \cdots, C), x \notin \underline{R}G_i\}.$$

After performing Algorithm 1, the samples that not belong to the absolute boundary region can be partitioned according to their memberships directly. However, the samples belonging to the absolute boundary region can not be classified immediately. In other words, according to the three-way decision theory, the belongness of these samples is not recommended since we have not enough information at this time. Fortunately, the clustering results of the samples that not belong to the absolute boundary region provide new evidences for grouping absolute boundary samples. This post process is described in Algorithm 2:

Algorithm 2: The classification of samples in the absolute boundary region

---

Step1: Form the absolute boundary region $R_{ab}X$ according to the results obtained

by Algorithm 1;

Step2: **While** $R_{ab}X \neq \emptyset$ **do**

2.1: $\{p, q\} = \min_{\{i,j\}} \big(dist(x_i, x_j) | x_i \in R_{ab}X, x_j \in \{X - R_{ab}X\} \big)$;

2.2: $label(x_p) = label(x_q)$;

2.3: $R_{ab}X = R_{ab}X - \{x_p\}$;

End While.

In Step 2.2, the cluster labels of samples in the absolute boundary region are obtained according to their nearest neighbor that is not in the absolute boundary region. In this way, the labels of samples that are not in the absolute boundary region provide additional information for classifying the absolute boundary samples, which detects the data distributions and reduces the uncertainty caused by the absolute boundary samples.

4 Experiment Analysis

The proposed algorithm is verified by a synthetic data set and some data sets from UCI repository. The synthetic data set involves two-dimensional Gaussian distribution data points within three clusters, the mean values are$\mu_1 = [-5, -5]$, $\mu_2 = [20, 20]$, $\mu_3 = [20, -10]$, respectively, and the standard deviation values are 8, 7, and 3, respectively. Three groups contain 200, 400, and 200 data points, respectively. The data set is visualized in Fig. 2. The proposed Algorithm 1 is denoted as ASRFCM. Furthermore, when Algorithm 2 is involved, the method is denoted as ASRFCM +. The fuzzy coefficient $m = 2$ and it is fixed for all methods and data sets. The validity indices ACC and NMI are exploited to evaluate the clustering methods. The larger the ACC and NMI [20, 22], the better the clustering results.

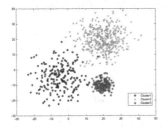

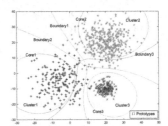

Fig. 2. Synthetic data set   **Fig. 3.** The approximation regions of each cluster obtained by ASRFCM

After performing ASRFCM, the generated approximation regions of each cluster are shown in Fig. 3. It can be found that the core regions can capture the important parts of each cluster. The samples in the boundary regions of Cluster1 and Cluster2 have contribution for calculating the prototypes to some extent. It is interesting that some samples coming from Cluster1 and Cluster 2 belong to the absolute boundary region. According to the adaptive mechanism for adjusting weighted values, the values of w_l of Cluster1, Cluster2, and Cluster 3 are 0.9796, 0.9909, and 0.9728, respectively. The values of w_l are determined according to the approximation region distribution rather than the predefined values. In this way, the contribution of different approximation regions can be measured precisely.

The ACC obtained by ASRFCM for the synthetic data set is 0.964. However, after performing the post process under the three-way decisions, the ACC value obtained by ASRFCM+ archives 0.983. It indicates that the validity of the post process, i.e., the classification results of samples in the core region of each cluster provide useful evidence guiding the classification of samples in the absolute boundary region.

To further verify the proposed notions, the data sets Iris, Wine, Banknote, Thyroid and Flowmeters in the UCI repository are selected for experiments. The fuzzy C-means(FCM) [18], rough-fuzzy C-means(RFCM)[3], shadowed C-means(SCM) [23], and two types of shadowed set-based rough-fuzzy C-means (SRFCM) [5] are involved for comparison. The validity values obtained by different methods are shown in Tables 1 and 2.

The average validity values of each method over all selected data sets are listed in the last row of Tables 1 and 2. It can be found that ASRFCM + achieves the best performance over all methods. Although the performance of ASRFCM is not always better than the compared methods, such as the results obtained by ASRFCM is not better than the one obtained by RFCM on Iris data set, the ASRFCM achieves better average validity values. It means that the optimal selection of the partition thresholds Δ and the adaptive adjustment of the weights w_l are beneficial for calculating the prototypes precisely, and data structures can be detected well in the iteration processes.

Table 1. ACC values obtained by different methods

Data Sets	FCM	RFCM	SCM	SRFCM$_1$	SRFCM$_2$	ASRFCM	ASRFCM +
Iris	0.8933	0.8933	0.8933	0.8933	0.8933	0.8867	**0.9067**
Wine	0.6850	0.7020	0.7020	0.7020	0.7020	0.7020	**0.7070**
Banknote	0.6093	0.6042	0.6086	0.5918	0.6028	0.6108	**0.6130**
Thyroid	0.7953	0.8326	0.8233	0.8279	0.8233	0.8744	**0.8744**
Flowmeters-C	0.3923	0.4917	0.3591	0.5193	0.4530	0.5028	**0.5414**
Avg	0.6751	0.7048	0.6773	0.7069	0.6949	0.7153	**0.7285**

Table 2. NMI values obtained by different methods

Data Sets	FCM	RFCM	SCM	SRFCM$_1$	SRFCM$_2$	ASRFCM	ASRFCM +
Iris	0.7496	0.7582	0.7582	0.7582	0.7582	0.7507	**0.8057**
Wine	0.4160	0.4280	0.4230	0.4230	0.4230	0.4230	**0.4310**
Banknote	0.0292	0.0252	0.0281	0.0180	0.0243	0.0294	**0.0308**
Thyroid	0.3491	0.3759	0.3658	0.3730	0.3653	0.5121	**0.5590**
Flowmeters-C	0.1706	0.2724	0.1329	0.2903	0.2315	0.2512	**0.3106**
Avg	0.3429	0.3719	0.3416	0.3725	0.3605	0.3933	**0.4274**

After exploiting the post process under the three-way decisions, ASRFCM + outperforms other methods over all data sets, which indicates that the classification of samples in the absolute boundary region plays an important role for improving the final clustering accuracy. The uncertainty caused by these samples is difficult to overcome if no additional information is involved. In this case, if we partition the samples in the absolute boundary region directly based on the obtained prototypes, the misclassification rate may increase. Fortunately, the samples in the core region of each cluster provide determinated label information which guide the further grouping operation for the samples in the absolute boundary region.

5 Conclusion

This study presents an adaptive mechanism for adjusting the weighted values w_l in the rough set-based clustering methods. The proposed dynamic mechanism adheres to the approximation region distributions in the iteration processes. Furthermore, guided by the notion of three-way decisions, the samples in the absolute boundary region are partitioned based on the clustering results of samples in the core region of each cluster rather than based on the obtained prototypes or memberships directly. In this way, the clustering results of samples in the core regions provide useful evidence to reduce the uncertainty caused by the samples in the absolute boundary region. The experimental results with a synthetic data set and some data sets from UCI repository demonstrate the effectiveness of the proposed notions. How to extend the proposed methods to deal with high-dimensional scenarios is our next works.

Acknowledgements. This work was supported by the Shenzhen Science and Technology Program (No. JCYJ20210324094601005), the National Natural Science Foundation of China (No. 62076164), and Guangdong Basic and Applied Basic Research Foundation (No. 2021A1515011861).

References

1. Pawlak, Z.: Rough sets. Int. J. Comput. Inform. Sci. **11**(5), 341–356 (1982)
2. Lingras, P., West, C.: Interval set clustering of web users with rough k-means, Technical Report 2002-002, Department of Mathematics and Computer Science, St. Mary's University, Halifax, Canada (2002)
3. Mitra, S., Banka, H., Pedrycz, W.: Rough–fuzzy collaborative clustering. IEEE Trans. Syst. Man Cybern. Part B **36**(4), 795–805 (2006)
4. Maji, P., Pal, S.K.: Rough set based generalized fuzzy c-means algorithm and quantitative indices. IEEE Trans. Syst. Man Cybern. Part B **37**, 1529–1540 (2007)
5. Zhou, J., Pedrycz, W., Miao, D.Q.: Shadow sets in the characterization of rough-fuzzy clustering. Pattern Recogn. **44**(8), 1738–1749 (2011)
6. Zhou, J., Lai, Z.H., Miao, D.Q., et al.: Multigranulation rough-fuzzy clustering based on shadowed sets. Inf. Sci. **507**, 553–573 (2020)
7. Pedrycz, W.: Shadow sets: representing and processing fuzzy sets. IEEE Trans. Syst. Man Cybern. Part B **28**, 103–109 (1998)
8. Yao, Y.Y., Yang, J.L.: Granular rough sets and granular shadowed sets: three-way approximations in Pawlak approximation spaces. Int. J. Approximate Reasoning **142**, 231–247 (2022)
9. Li, F., Ye, M., Chen, X.D.: An extension to rough C-means clustering based on decision-theoretic rough sets model. Int. J. Approximate Reasoning **55**, 116–129 (2014)
10. Yao, Y.Y., Zhao, Y.: Attribute reduction in decision-theoretic rough set model. Inf. Sci. **178**, 3356–3373 (2008)
11. Sarkar, J.P., Saha, I., Maulik, U.: Rough possibilistic type-2 fuzzy C-means clustering for MR brain image segmentation. Appl. Soft Comput. **46**, 527–536 (2016)
12. Yao, Y.Y., Wang, S., Deng, X.F.: Constructing shadowed sets and three-way approximations of fuzzy sets. Inf. Sci. **412–413**, 132–153 (2017)
13. Yao, Y.: The geometry of three-way decision. Appl. Intell. **51**(9), 6298–6325 (2021). https://doi.org/10.1007/s10489-020-02142-z
14. Yao, Y.Y.: Three-way granular computing, rough sets, and formal concept analysis. Int. J. Approximate Reasoning **116**, 106–125 (2020)
15. Yu, H., Wang, X., Wang, G., et al.: An active three-way clustering method via low-rank matrices for multi-view data. Inf. Sci. **507**, 823–839 (2020)
16. Min, F., Zhang, Z.H., Zhai, W.J., et al.: Frequent pattern discovery with tri-partition alphabets. Inf. Sci. **507**, 715–732 (2020)

17. Yue, X.D., Zhou, J., Yao, Y.Y., et al.: Shadowed neighborhoods based on fuzzy rough transformation for three-way classification. IEEE Trans. Fuzzy Syst. **28**(5), 978–991 (2020)
18. Bezdek, J.C.: Pattern Recognition with Fuzzy Objective Function Algorithms. Kluwer Academic Publishers, Norwell (1981)
19. Deng, X.F., Yao, Y.Y.: Decision-theoretic three-way approximations of fuzzy sets. Inf. Sci. **279**, 702–715 (2014)
20. Zhou, J., Pedrycz, W., Gao, C., et al.: Principles for constructing shadowed sets: a comparative evaluation based on unsupervised learning. Fuzzy Sets Syst. **413**, 74–98 (2021)
21. Pedrycz, W.: From fuzzy sets to shadowed sets: interpretation and computing. Int. J. Intell. Syst. **24**, 48–61 (2009)
22. Zhou, J., Pedrycz, W., Yue, X., et al.: Projected fuzzy C-means clustering with locality preservation. Pattern Recogn. **113**, 107748 (2021)
23. Mitra, S., Pedrycz, W., Barman, B.: Shadowed c-means: integrating fuzzy and rough clustering. Pattern Recogn. **43**(4), 1282–1291 (2010)

Author Index

Printed in the United States
by Baker & Taylor Publisher Services